Electrolyte Solutions

SECOND REVISED EDITION

R. A. Robinson
Late Emeritus Professor of Chemistry
University of Malaya

R. H. Stokes
Emeritus Professor of Chemistry
University of New England
Armidale, New South Wales, Australia

DOVER PUBLICATIONS, INC.
Mineola, New York

Copyright

Published in the United Kingdom by David & Charles, Brunel House, Forde Close, Newton Abbot, Devon, TQ12 4PU.

Bibliographical Note

This Dover edition, first published in 2002, is an unabridged republication of the fifth impression of the Second Revised Edition of the work originally published in 1970 by Butterworth & Co. (Publishers) Ltd., London. A new Preface has been specially prepared by Dr. Stokes for this edition.

Library of Congress Cataloging-in-Publication Data

Robinson, R. A. (Robert Anthony), 1904–
Electrolyte solutions / R.A. Robinson, R.H. Stokes.
p. cm.
"This Dover edition, first published in 2002, is an unabridged republication of of the fifth impression of the second revised edition of the work originally published in 1970 by Butterworth & Co. (Publishers) Ltd., London. A new preface has been specially prepared by Dr. Stokes for this edition."
ISBN 0-486-42225-9 (pbk.)
1. Electrolyte solutions. I. Stokes, R. H. (Robert Harold), 1918– II. Title.

QD565 .R63 2002
541.3'72—dc21 2002023777

Manufactured in the United States of America
Dover Publications, Inc., 31 East 2nd Street, Mineola, N.Y. 11501

PREFACE TO THE DOVER EDITION

Robert Anthony Robinson (1904–1979) was my physical chemistry teacher in Auckland (New Zealand), and supervised my first research work for the M.Sc. degree. We collaborated intermittently there until 1946, when I left for Australia and he shortly afterwards to the (then) University of Malaya in Singapore. The collaboration continued by correspondence for many years, and he spent a year in the Armidale department shortly after his retirement from Singapore in 1959. He later held several research appointments in England and the United States.

The first edition (1955) of *Electrolyte Solutions* was written while I was in Perth (Western Australia), which had an excellent airmail service to Singapore. We decided to confine ourselves mainly to topics on which we had first-hand experience, and to include all the numerical data we were continually looking up in the course of our research work, as well as tabulations of reliable fundamental data on electrolytes. This seems to have met a need, and the book in its subsequent editions continues to be widely cited. It was the subject of a Current Contents Citation Classic in 1988, when long out of print.

Why not produce a completely new version rather than the present reprint of the 1970 revision? It would require several volumes to do justice to the huge expansion of the field in recent decades, particularly in the area of solutions in non-aqueous solvents. Aqueous solutions at high temperatures and pressures have also received much attention, and there have been major theoretical advances. We are content to have played a part in laying foundations and drawing attention to gaps in knowledge.

It should be noted that all values quoted are in pre-SI units; see the "Preface to Reprinted Edition" (p. v) and the "Table of Important Constants" (p. xv) for details.

I thank Dover Publications for making our work available again to students and research workers. There is still no substitute for measured fact.

R. H. Stokes
Armidale
January 2002

PREFACE TO REVISED EDITION

In preparing this revised edition we have, with one exception, made only minor changes in the body of the text. The exception is the section on 'hydrogen ion activity' in Chapter 12 which now appears in a new form. There is also an addendum which consists partly of references to recent work which we think might be drawn to the reader's attention for more extensive reading, and partly to short notes on recent data determinations which we hope will be of use to workers in this field. To these we have added a few, more extended, synopses of recent work which we hope will be a stimulus to further research.

R. A. Robinson
R. H. Stokes

PREFACE TO REPRINTED EDITION

Since the text was written before the SI system was in use, a change to SI would entail changes on almost every page and would greatly increase the cost of this reprint. Readers needing to make this conversion should note the following:

(1) our *litre* is the pre SI litre, 1000·028 cm^3
(2) our *calorie* is 4·184 J
(3) our *ohm* (in conductance values) is the old International Ohm, for reasons stated on pp. 96–97
(4) our *volt* is the absolute volt.

R. A. Robinson
R. H. Stokes

PREFACE TO THE SECOND EDITION

In preparing this second edition we have made only a small increase in the length of the text. Some new experimental and theoretical developments are dealt with; increased discussion of some frontier subjects is now possible; and some topics of less importance have been omitted or given less space.

The treatment of conductivity in Chapter 7 is considerably changed, concentrated solutions being deferred to the expanded Chapter 11, where a more detailed account of viscosity also appears. Though the book is still primarily about aqueous solutions, some recent work on conductance in non-aqueous solvents has been added. The appendices have been revised and expanded by nearly fifty per cent; in particular the tables of ionization constants of weak electrolytes are now much more extensive. The index is more detailed and, we hope, more useful.

Professor Fuoss' important new contribution to the theory of ion association appeared just too late to be fully incorporated into Chapter 14. This development is therefore briefly treated in Appendix 14.3.

Certain deficiencies in the notation have been corrected; we acknowledge with thanks the helpful criticism of Professor E. A. Guggenheim in this regard. Other errors in the first edition, kindly pointed out by various correspondents, have also been eliminated. To the acknowledgements in the preface to the first edition, we would add the name of Dr. W. J. Hamer, particularly for his discussions with us of the pH problem.

R. A. Robinson
R. H. Stokes

February, 1959.

PREFACE TO THE FIRST EDITION

In writing a book on electrolytes, any attempt to describe all the properties of interest would lead either to excessive length or else to a sacrifice of detail and depth, and we have therefore restricted ourselves in the main to a discussion of certain properties which seem to us to be fundamental. The worker in the field of polarography, for example, will not find specific mention of his subject but he will find a substantial body of fact and theory which is essential to the interpretation of his data. As the sub-title indicates, the main topics treated are conductance, chemical potential and diffusion. The first of these constitutes the main characteristic whereby electrolytes are distinguished from other solutions, and its fundamental importance needs no elaboration.

Of all the thermodynamic properties, the Gibbs free energy is the most useful for the treatment of equilibrium conditions and we have laid emphasis on quantities such as activities and ionization constants, which are related in a simple way to the Gibbs free energy. Moreover, the interionic attraction theory gives directly an expression for the chemical potential of the electrolyte, so that one naturally prefers to make tests of the theory by means of activity data rather than by derived quantities such as the heat content or heat capacity.

Considerable space is given to the treatment of diffusion in electrolyte solutions for two reasons: partly because this important and rapidly growing section of the subject is not fully discussed in other textbooks, most of the precise experimental results having been published only recently, and partly because the theory of diffusion is of great intrinsic interest as one of the simplest irreversible processes, providing a link between conductance and free energy studies.

The theoretical treatment of these three properties is based on the interionic attraction theory of Debye and Hückel, and especially on its later developments by Onsager and Fuoss and by Falkenhagen. We have illustrated the limiting equations of this theory by reference to data for very dilute solutions, but our main concern as practical electrochemists has been with solutions of laboratory concentrations and one of our aims has therefore been to demonstrate the surprising adequacy of the theory when allowance is made for the dimensions of the ions. As a result of the recent work of Falkenhagen and others, we are now in the unexpected position that, at least for uni-univalent electrolytes, the theory of conductivity is more exact and successful than the theory of chemical potential. Indeed, with reasonable assumptions about the effect of viscosity, the equations now available account for the conductivity of simple aqueous strong electrolytes up to astonishingly high concentrations. With the chemical potential, however, whilst a large part is known to be played in concentrated solutions by ion-solvent interaction, there are many complicating factors which are not yet fully understood.

Since water is the commonest, cheapest and most easily purified of solvents, and because of its great biological importance, it is natural that most of the experimental work on electrolytes has utilized this solvent; we have recognized this by dealing mainly with aqueous solutions although we have introduced recent data of good accuracy for non-aqueous solutions whenever possible.

Extensive appendices and tables are given, including functions and constants useful in computations, as well as compilations of accurate experimental data, especially for concentrated solutions. A reasonably full account is also given of the experimental methods whereby these data were obtained, in order to illustrate the accuracy and the limitations of modern technique.

We acknowledge with gratitude the help we have received from a number of friends; one of us is greatly indebted to the inspiration of Professor H. S. Harned over a period of many years; the other owes a comparable debt to Dr. J. N. Agar, particularly for access to his unpublished work on the theory of irreversible processes. In common with all electrochemists, we have been profoundly influenced by the works of Professors Harned and Owen, Dr. MacInnes, and Professor Guggenheim; and amongst others whose assistance is gladly acknowledged are Dr. R. G. Bates, Professor C. W. Davies, Professor A. R. Gordon, Professor H. N. Parton and Professor T. F. Young. Mrs. Jean M. Stokes has given valuable help in the preparation of the manuscript and the revision of proofs.

The difficulties imposed upon us by geographical separation from each other and from the country of publication have been greatly alleviated by the courtesy and efficiency of the editorial staff of Butterworths Scientific Publications, to whom we express our sincere thanks.

R. A. Robinson
R. H. Stokes

August, 1954.

CONTENTS

LIST OF APPENDICES

LIST OF PRINCIPAL SYMBOLS

(With references to page or equation where first introduced.)

A	Constant of Debye–Hückel equation for the activity coefficient. (Eq. 9.7, App. 7.1)
A_n	Function in theory of the electrophoretic effect. (Eq. 7.8)
A_1, A_2, A_3	Constants of viscosity. (Eq. 11.30 and 11.31 and 11.34)
B	Coefficient of ion-size term in the Debye–Hückel theory. (Eq. 7.37, 9.7, App. 7.1; p. 171)
B_1	Coefficient of the relaxation term in the theory of conductivity. (pp. 143, 171, App. 7.1)
B_2	Coefficient of the electrophoretic term in the theory of conductivity. (pp. 143, 171, App. 7.1)
C_A, C_B	Concentrations of substances A and B in moles/unit volume (Ch. 11 only).
$\bar{C}_{P(A)}, \bar{C}_{P(B)}$	Partial molal heat capacities at constant pressure of solvent and solute respectively. (Eq. 2.31, 2.35)
D	Diffusion coefficient (Eq. 2.53, 2.54); optical density (Ch. 12)
D^*	Self-diffusion or tracer-diffusion coefficient. (pp. 12, 315 seq.)
E	Electromotive force, usually of cell without liquid junction. (p. 40)
E_t	Electromotive force of concentration-cell with transport. (pp. 111, 202)
F	Faraday.
G	Gibbs free energy.
$\bar{G}_i$	Chemical potential or partial molal free energy of substance i. (Eq. 2.1)
$\bar{H}_i$	Partial molal heat content of substance i. (Eq. 2.29, 2.34)
I	Ionic strength, $\frac{1}{2}\Sigma c_i z_i^2$ usually with c_i expressed as mole/l. (p. 143)
J	Flux of matter, in diffusion theory. (p. 46, Eq. 2.53)
$\bar{J}_A, \bar{J}_B$	Relative partial molal heat capacities of solvent and solute. (Eq. 2.32)

K_N, K_m, K_c Equilibrium constants on the mole fraction, molality and molarity scales of concentration respectively. (p. 39)

K_a, K_b Acidic and basic ionization constants. (pp. 338, 342)

K_w Ionization constant of water. (p. 363, App. 12.2)

K_{sp} Specific conductance. (p. 41)

$\bar{L}_A$, $\bar{L}_B$ Relative partial molal heat contents. (p. 35)

M Abbreviation for 'molal'.

$\boldsymbol{N}$ Avogadro number.

N Abbreviation for 'normal'.

$\mathcal{N}_A$, $\mathcal{N}_B$ Mole fractions of A and B. (p. 31)

Q Valency-type factor in definition of activity. (Eq. 2.13, App. 2.1); Interaction coefficient in viscosity. (Eq. 11.33)

$\boldsymbol{R}$ Gas constant.

R Isopiestic ratio. (Eq. 8.1)

S Entropy.

$S_n(\kappa a)$ Integral in theory of electrophoretic effect. (Eq. 7.5, p. 170)

T Absolute temperature.

W_A, W_B Molecular weights of substances A and B.

X Electric field intensity. (p. 136)

Z Impedance. (p. 88)

a_A, a_B Activity of substances A and B. (Eq. 2.2, 2.4)

a Mean diameter of ions. (p. 79)

c Concentration in moles per litre (molarity); occasionally this symbol is used for volume concentration in general or for equivalents per litre (normality) but in such cases specific mention is made of the usage.

$\boldsymbol{e}$ Protonic charge, *i.e.* charge equal and opposite in sign to that of an electron.

f Activity coefficient on mole fraction scale. (Eq. 2.9)

g Rational osmotic coefficient. (Eq. 2.15)

h Hydration number.

$\boldsymbol{k}$ Boltzmann's constant.

k_1, k_2, k_A Forces in theory of electrophoretic effect. (p. 134)

k The quantity $2{\cdot}303\boldsymbol{R}T/\boldsymbol{F}$. (p. 190)

ln, log	Logarithms to bases e and 10 respectively.
m	Moles of solute per kg. of solvent (molality).
n_i	Number of particles of species i per unit volume. (Ch. 4 and 7)
n_A, n_B	Numbers of moles of substances A and B in system. (Eq. 2.1)
q	Critical distance for ion-pair formation (Ch. 14 only). (Eq. 14.1)
q	Mobility function in theory of relaxation effect. (Eq. 7.10)
t_1, t_2	Transport numbers of cation and anion respectively. (Eq. 2.52)
u	Absolute mobility of particle. (p. 42)
u'	Mobility of ion under unit electrical potential gradient. (p. 43)
y	Activity coefficient on molarity scale. (Eq. 2.9)
z_1, z_2	*Algebraic* valencies of cations and anions respectively. (p. 27)
Δ_n	nth order electrophoretic correction to diffusion coefficient. (Eq. 11.17)
Λ	Equivalent conductivity of electrolyte. (p. 41)
Π	Osmotic pressure. (Eq. 2.17)
α	Degree of dissociation.
γ	Activity coefficient on molality scale. (Eq. 2.9)
ε	Dielectric constant; extinction coefficient (Ch. 12)
η	Viscosity.
η_{rel}	Relative viscosity.
κ	Quantity proportional to square root of ionic strength and having dimensions of reciprocal length. (Eq. 4.12)
λ_1, λ_2	Ionic equivalent conductivities. (Eq. 2.43)
μ	Dipole moment.
ν	Number of moles of ions formed from 1 mole of electrolyte.
ϕ	Molal osmotic coefficient. (Eq. 2.16)
ω	Angular frequency.

TABLE OF IMPORTANT CONSTANTS

F	Faraday	96486·8 coulomb (g. equivalent)$^{-1}$
N	Avogadro number	$6{\cdot}02252 \times 10^{23}$ mol^{-1}
c	Velocity of light	$2{\cdot}997925 \times 10^{10}$ cm sec^{-1}
e	Protonic charge	$1{\cdot}60210 \times 10^{-19}$ coulomb $= 4{\cdot}80298 \times 10^{-10}$ c.g.s. e.s.u. of charge
k	Boltzmann's constant	$1{\cdot}38054 \times 10^{-16}$ erg degree^{-1} molecule^{-1}
R	Gas constant	8·3143 absolute joule degree^{-1} mol^{-1} 1·98717 calorie degree^{-1} mol^{-1}
V_0	Ideal gas molar volume	22413·6 cm^3 mol^{-1} (0°C and 1 standard atmosphere)

Ref. Cohen, E. R. and DuMond, J. W. M., *Rev. mod. Phys.* 37 (1965) 538.

---

1 absolute ohm	=	0·999505 international ohm
1 absolute volt	=	0·999670 international volt
1 absolute ampere	=	1·000165 international ampere
1 calorie	=	4·1840 absolute joule
Ice-point	=	273·150 K
1 standard atmosphere	=	760 Torr = 1·01325 bar
	=	101325 N m^{-2}

The values quoted above differ slightly from those given in our 1965 reprinting, but the changes are not great enough to justify recalculation of quantities appearing in the text and equations.

1

PROPERTIES OF IONIZING SOLVENTS

CLASSICAL theories of solutions were built upon the analogy between the solute particles and the molecules of an imperfect gas, the solvent being regarded as a mere provider of the volume in which the solute particles moved. The striking progress made by the modern theory of liquids is based on a very different model: the liquid is seen as a disordered solid in which short-range order persists, though the long-range order characteristic of the solid state has been lost in thermal agitation. Solute and solvent appear on an equal footing and it is only in the limit of extreme dilution, when the solvent molecules so outnumber those of the solute that we can regard the solvent as virtually unchanged, that the classical view-point remains acceptable. The most noteworthy successes of the modern theory have, however, been in the theory of non-polar and uncharged molecules; in the case of electrolyte solutions, the more sensational properties are still too readily attributed only to the nature of the solute. One should not forget that it is the solvent which enables the electrolyte to display its peculiarities; that it plays an active part in producing, from the electrically inert crystal, liquid or gas, the mobile charged particles which force themselves on our attention.

Since water is by far the most important of the ionizing solvents, and all but a very small part of the immense body of factual knowledge about electrolytes refers to aqueous solutions, we shall begin with an account of the structure of water and such of its properties as are relevant to the behaviour of electrolyte solutions.

THE WATER MOLECULE

Spectroscopic studies of the isolated water molecule[1] in the gaseous state have established that the H—O—H bond angle is 105°, and the O—H internuclear distance 0·97 Å (*Figure 1.1*). The isolated molecule has a dipole moment of $1{\cdot}87 \times 10^{-18}$ e.s.u., acting along the bisector of the H—O—H angle with the negative end towards the oxygen nucleus. This dipole moment was treated by BERNAL and FOWLER[2], in their pioneer work on the structure of water and ice, as due to an effective charge of $-e$ (e = protonic charge) situated 0·15 Å from the oxygen nucleus, with $+\,0{\cdot}5e$ at each hydrogen nucleus. A more elaborate model due to VERWEY[3] replaces the tripolar charge distribution of Bernal and Fowler (*Figure 1.2a*) by the quadrupole arrangement shown in *Figure 1.2b*;

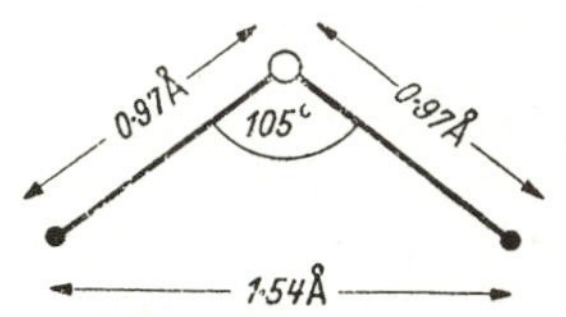

Figure 1.1. Internuclear distances and bond angle of the water molecule

this has led to a very satisfactory prediction of the lattice energy of the ice crystal.

LIQUID WATER

In the liquid state water exhibits properties characteristic of an associated liquid to an extent more marked than do the hydrides of elements close to oxygen in the Periodic Table. To illustrate this

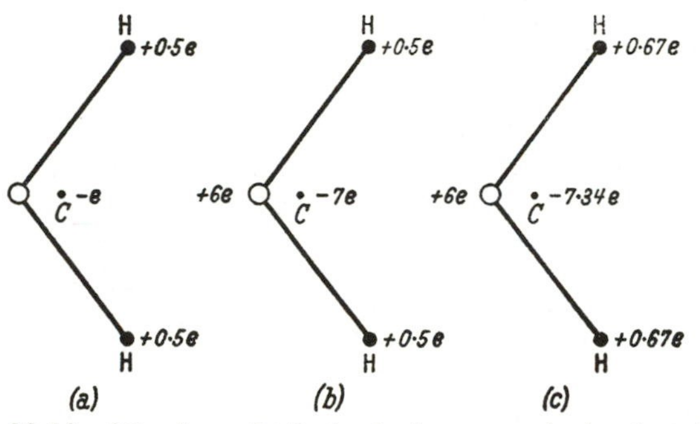

Figure 1.2. Models of the charge distribution in the water molecule. In each model C is taken as the centre of the molecule. The distances OC are not drawn to scale

(*a*) Bernal and Fowler OC = 0·15 Å

(*b*) Verwey Model I OC = 0·022 Å $\widehat{HCH}$ = 107° 10′

(*c*) Verwey Model II OC = 0·049 Å $\widehat{HCH}$ = 109° 44′

we can quote some physical properties of ammonia, water, hydrogen fluoride and hydrogen sulphide.

	NH_3	H_2O	HF	H_2S
Melting point	− 78°	0°	− 84°	− 86°
Boiling point	− 33°	100°	20°	− 60°
Entropy of vaporization (cal deg^{-1} mole^{-1})	23·2	26·1	24·9	21·2

Thus liquid water has a comparatively high boiling point, suggesting the presence of strong intermolecular forces in the liquid state, which make it difficult for the molecule to escape into the vapour phase. The high melting point suggests that there is a kind of quasi-crystalline structure in the liquid so that the solid state can be formed with ease in spite of the comparatively high thermal energy.

The densities in the solid and liquid state at 0° are 0·9168 and 0·99987 g/ml. respectively, so that water contracts by 8·3 per cent on fusion. It contracts by a further 0·012 per cent on heating to 4°, at which temperature the density has a maximum value. The specific heat of ice at 0° is 0·5026 cal/g[3a] compared with 1·0081 cal/g for liquid water at the same temperature; the specific heat has a minimum value of 0·9986 cal/g[3b] at 34·5°.

Its dielectric constant (78·30 at 25°) is high compared with most liquids; hydrogen cyanide has a dielectric constant of 106·8, formamide 109·5 and sulphuric acid 101 at 25°; the value for hydrogen fluoride is 83·6 at 0°. Apart from these four liquids, however, even the more polar of the common liquid solvents are characterized by much lower dielectric constants; 59 for acetamide at 83°, 52 for hydrazine at 25° and 22 for ammonia at its boiling point are examples. Non-polar liquids have dielectric constants of the order of 2.

Even after the most rigorous purification water has a small electrical conductance; so-called 'equilibrium' water has a specific conductivity of $0{\cdot}75 \times 10^{-6}\ \Omega^{-1}\ \mathrm{cm}^{-1}$ at 18° due mainly to dissolved carbon dioxide in equilibrium with the carbon dioxide of the atmosphere. Kohlrausch and Heydweiller[(4)] reported a specific conductivity of $\sim 0{\cdot}04 \times 10^{-6}\ \Omega^{-1}\ \mathrm{cm}^{-1}$ at 18° for highly purified water. This conductance is attributable to a slight dissociation of the water molecules: $H_2O \rightarrow H^+ + OH^-$ or $2H_2O \rightarrow H_3O^+ + OH^-$, and can be explained by assuming that the concentration of hydrogen and hydroxyl ions is $0{\cdot}8 \times 10^{-7}$ equivalents per litre at 18° and 1×10^{-7} equivalents per litre at 25°.

In liquid water, the volume per molecule at room temperature is very nearly 30 Å^3. If water consisted of close-packed spherical molecules, the diameter needed to give this volume would be 3·48 Å. In fact, however, x-ray analysis of liquid water indicates[(5)] that the nearest-neighbour distance (expressed as the O–O internuclear distance) is 2·90 to 3·05 Å in the temperature range 0–80° (*Figure 1.3*). It follows that the molecules are far from close packed, or the volume per molecule would be much smaller. Instead of the twelve nearest neighbours characteristic of close packing, the x-ray data show that the average number of nearest neighbours ranges from 4·4 to 4·9 over this temperature interval. Morgan and Warren have also found x-ray evidence for a set of second nearest neighbours at the expected distance of about 4·5 Å from the central molecule considered, but this set becomes less clearly defined as the temperature rises, and fades out above 30°, indicating that the range of the ordering effects is being reduced by thermal agitation.

Liquid water retains, in fact, over short ranges and for short periods, the tetrahedrally coordinated structure of ice. This view,

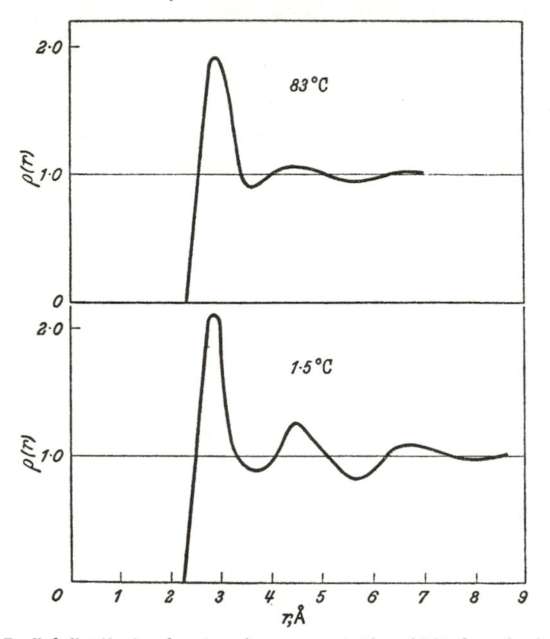

Figure 1.3. Radial distribution functions for water at 1·5° *and* 83° *from the data of Morgan and Warren. The function* ρ(*r*) *gives the probability of finding the centre of a water molecule in a volume-element distant r from a chosen central molecule; its absolute value is adjusted so as to become unity at large values of r, which is equivalent to taking the volume element as the average molecular volume in the liquid*

first put forward by Bernal and Fowler, is the modern and more satisfactory alternative to older views in which the associated nature of water was explained by assuming the presence of various polymerized forms such as 'dihydrol' $(H_2O)_2$ and 'trihydrol' $(H_2O)_3$.

This ice-like structure is believed to be maintained by 'hydrogen bonds', which are essentially electrostatic in nature and result from the especially favourable charge distribution and geometry of the water molecule. As *Figures 1.1* and *1.2* show, the bond angle of water is very close to the tetrahedral angle (109° 28′) and Verwey's model II in particular is ideally adapted to a 4-coordinated structure. The fact that Verwey's calculations lead to values within 1 kcal/mole of the experimental one (10·8 kcal/mole) for the energy of vaporization of ice is strong evidence for the adequacy of a purely electrostatic picture of the intermolecular forces.

Other evidence in favour of the tetrahedrally coordinated structure of water is found in the Raman and infra-red spectra[1]. The main intermolecular Raman band occurs at a frequency-displacement $\Delta\nu = 152$–$225\ cm^{-1}$; this has been shown to arise from the 'breathing' mode of vibration (*i.e.*, contraction and expansion of the tetrahedron). A band in the infra-red at 160–175 cm^{-1} which disappears in dilute solutions of water in dioxane[6] also appears to be due to intermolecular vibrations. Other Raman bands at 60, 500 and 700 cm^{-1} are attributed to rotational oscillations (librations) of the molecule which are not vigorous enough to break the electrostatic bonds with its neighbours. It has been suggested that only one mode of free rotation can occur to any substantial extent in water at ordinary temperatures; this is about the axis which lies in the plane of the three nuclei and bisects the H—O—H angle. The variation of the Raman intensities with temperature suggests that this free rotation becomes important rather suddenly in the vicinity of 40°C.

Liquid water, then, must be pictured as a rather loosely 4-coordinated structure, held together by electrostatic forces arising from the special charge distribution and shape of the water molecule. The association between a molecule and its neighbours can be only temporary, as the structure is continually being broken by thermal agitation, but it must have sufficient permanence to persist over small regions, for times long in comparison with the period of x-rays or even of infra-red radiation. Such times, however, need be only of the order of 10^{-12} sec so that we need feel no surprise that the viscosity of water, for example, is only moderately higher than that of simpler liquids with small molecules. Many of the anomalous properties of water find a natural explanation in terms of its structure. The maximum density at 4° can be attributed to the competition between two opposing effects, the gradual breaking down of the rather open ice-like structure to a somewhat closer-packed structure (as indicated by the increase of the average number of nearest neighbours with temperature) and a simultaneous increase with temperature of the average centre-to-centre distance. The abnormally high dielectric constant is due to the mutual interaction of the electrostatic fields of the molecules which, because of the favourable orientation of the molecular dipoles, leads to a considerable increase in the effective polarization in the liquid state as compared with the vapour.

THE DIELECTRIC CONSTANT AND DIPOLE MOMENT OF POLAR LIQUIDS

If two parallel conducting plates have on their surfaces electric charges of density $+\sigma$, $-\sigma$ respectively, the field intensity between them has *in vacuo* the value:

$$E_v = 4\pi\sigma$$

With an insulating medium between the plates the field strength drops to a value:

$$E = 4\pi\sigma/\varepsilon_s$$

where ε_s, a constant for all reasonably low field intensities, is called the 'static dielectric constant' of the medium, and is always greater than unity. The molecular process responsible for this reduction in the effective field strength is the displacement of electric charges within the molecules; a negative charge appears at the surface of the dielectric in contact with the positive plate, and a positive charge of the same magnitude at the surface in contact with the negative plate. These induced charges in the dielectric reduce the effective charge-density on the plates from σ to σ/ε_s; they may therefore be represented as a polarization, P, of the dielectric, given by:

$$P = \sigma - \sigma/\varepsilon_s = \sigma\left(1 - \frac{1}{\varepsilon_s}\right)$$

The total field, E, inside the dielectric may be represented as the sum of the original field ($4\pi\sigma$) existing *in vacuo* and the polarization field $-4\pi P$. The former is called the electric displacement, D, given by:

$$D = 4\pi\sigma$$

so that

$$D = E + 4\pi P = E\varepsilon_s \qquad \ldots.(1.1)$$

and

$$\varepsilon_s = 1 + \frac{4\pi P}{E}$$

If the plates are separated by a distance d, two opposite elements of the surface of the dielectric, of area δA, will carry charges $+P\delta A$ and $-P\delta A$, and will therefore constitute a dipole of moment $P\delta A \,.\, d$ and of volume $\delta A \,.\, d$; P is therefore the dipole moment per unit volume of the dielectric. The problem of calculating the dielectric constant, ε_s, from the molecular properties of the medium is thus that of calculating the polarization P.

This polarization is the sum of two types: (*a*) that due to the distortion of electronic distributions within atoms and of atomic configurations within molecules, which is called distortion polarization, and (*b*) that due to the partial lining up, under the field, of already existing permanent molecular dipoles, which is called orientation-polarization. The distortion-polarization occurs with extreme rapidity, even in fields alternating with the frequency of light-waves, and is independent of temperature; the orientation-polarization, on the other hand, involves the rotation of molecules, a slower process and one which is furthermore opposed by the thermal agitation and is therefore temperature-dependent. The polarization due to orientation of permanent dipoles is relatively easily dealt with for the case where the dipoles are so far apart that their mutual interaction can be neglected.

For dealing with the distortion-polarization, it is convenient to introduce a quantity, α, the molecular polarizability, which is defined as the time-average dipole moment induced in the molecule by a field of unit intensity. If there are N_0 molecules/cm^3, the contribution, P_d, of the distortion polarization to the total polarization P will be:

$$P_d = N_0 \alpha F,$$

where F is the actual field acting on the molecule. This field F, the internal field, is unfortunately not easily evaluated for liquids or solids except in the case where permanent dipoles are absent ($P = P_d$) and the molecular interactions can be neglected. It is not identical with either of the field-quantities E or D. For gases or such ideal liquids, however, a simple electrostatic calculation shows that it is given by:

$$F = E + \frac{4\pi P}{3}$$

whence:

$$P_d = N_0 \alpha \left(E + \frac{4\pi P}{3} \right) \qquad \ldots.(1.2)$$

But since also

$$D = E + 4\pi P$$

by definition, P can be eliminated from (1.1) and (1.2) giving

$$\frac{D - E}{D + 2E} = \frac{4\pi N_0}{3} \alpha$$

Now the dielectric constant is defined by $D = \varepsilon_s E$, so that this result becomes:

$$\frac{\varepsilon_s - 1}{\varepsilon_s + 2} = \frac{4\pi N_0}{3} \alpha$$

This relation, which it must be emphasized holds only for non-polar molecules in the absence of molecular interactions, is known as the Clausius-Mosotti formula. In electromagnetic theory the dielectric constant is found to be related to the refractive index by Maxwell's relation $\varepsilon_s = n^2$; this formal identification is legitimate as long as it is made clear that ε_s must have the value which would be found by measuring it in a field of optical frequency; that is, as long as only electronic polarization is involved. In general, however, it is convenient to define ε_0 as that part of the static dielectric constant which arises from the distortion polarization, and write $\varepsilon_0 = n^2$ for optical frequencies. The polarizability, α, is then given by:

$$\frac{n^2 - 1}{n^2 + 2} = \frac{4\pi N_0}{3} \alpha$$

The first treatment of orientation-polarization, due to Debye, was modelled on Langevin's theory of paramagnetism. DEBYE[7] assumed that the internal field $F = E + 4\pi P/3$ (the Clausius-Mosotti internal field) would also be adequate for this case, though realizing its limitations due to the neglect of molecular interactions. His formula is therefore applicable only to the polar gases or to dilute solutions of polar substances. The average orientation polarization is calculated on the assumption that the energy of the oriented dipoles is distributed according to the Boltzmann distribution expression, neglecting molecular interactions. The average moment per molecule, $\overline{m}$, is shown to be related to the actual permanent dipole moment μ_0 by the Langevin formula:

$$\overline{m} = \mu_0 \left(\coth \frac{\mu_0 F}{kT} - \frac{kT}{\mu_0 F} \right)$$

which for all ordinary field-strengths approximates closely to:

$$\overline{m} \doteq \frac{\mu_0^2 F}{3kT}$$

The total polarization per unit volume is therefore:

$$P = N_0 F \left(\alpha + \frac{\mu_0^2}{3kT} \right)$$

Hence by an argument closely similar to that given above for the case of distortion polarization alone, Debye's equation:

$$\frac{\varepsilon_s - 1}{\varepsilon_s + 2} = \frac{4\pi N_0}{3} \alpha + \frac{4\pi N_0}{3} \cdot \frac{\mu_0^2}{3kT}$$

is obtained. This equation is experimentally verified for polar gases and dilute solutions, and is indeed the basis for the evaluation of dipole moments from dielectric constant measurements on such systems. Its acknowledged failure for polar liquids was shown by ONSAGER[8] to be due to the inadequate nature of the Clausius-Mosotti expression for the internal field. Onsager proposed that only a part of this field should be active in orienting the dipoles; this part he called the 'cavity field'. The remaining part, the 'reaction field', should remain parallel to the dipole moment, and thus enhance both the permanent and induced dipole moments. On this basis, he arrived at the equation:

$$\frac{(\varepsilon_s - n^2)(2\varepsilon_s + n^2)}{\varepsilon_s(n^2 + 2)^2} = \frac{4\pi N_o}{3} \cdot \frac{\mu_o^2}{3kT}$$

which, for easier comparison with Debye's equation, may be re-written as:

$$\frac{\varepsilon_s - 1}{\varepsilon_s + 2} = \frac{4\pi N_o}{3}\alpha + \frac{4\pi N_o}{3} \cdot \frac{\mu_o^2}{3kT} \cdot \frac{3\varepsilon_s(n^2 + 2)}{(2\varepsilon_s + n^2)(\varepsilon_s + 2)}$$

This clearly reduces to Debye's formula if $\varepsilon_s \approx n^2$ which is the case for dilute solutions or for gases, but gives very different, and better, results for polar liquids, where ε_s differs considerably from n^2. It is still inadequate, however, for the so-called 'associated liquids' such as water and alcohols, and these are precisely the liquids of most interest in connection with electrolytes.

KIRKWOOD[9] has extended Onsager's theory to deal with these liquids by taking detailed account of the short-range interactions which hinder the rotation of the molecular dipoles. His result is:

$$\frac{(\varepsilon_s - 1)(2\varepsilon_s + 1)}{9\varepsilon_s} = \frac{4\pi N_o}{3}\alpha + \frac{4\pi N_o}{3}\frac{\mu^2}{3kT}g \qquad \text{....(1.3)}$$

In this formula the factor g must be calculated from a suitable model for the liquid in question. It is given by:

$$g = 1 + z\,\overline{\cos\gamma}$$

where z is the average number of nearest neighbours and $\overline{\cos\gamma}$ is the averaged cosine of the angle between adjacent dipoles. (Using x-ray data for water, a value of $g \approx 2{\cdot}5$ is obtained at ordinary temperatures.) It must also be recognized that the value of μ in Kirkwood's formula is not quite the same as the dipole moment μ_o of the isolated molecule, owing to further polarization by its neighbours. One approximation[10] used to allow for this effect with spherical molecules is:

$$\mu = \frac{n^2 + 2}{3} \mu_o$$

and with this approximation the only essential difference between Kirkwood's formula and Onsager's lies in the factor g. OSTER and KIRKWOOD[11] were able to calculate the dielectric constants of water and alcohols within about 10 per cent by using equation (1.3).

Some of the difficulty of establishing a satisfactory theory of the dielectric constant of polar liquids arises from uncertainty about the correct value for the distortion-polarizability α, or in other words the part ε_o of the static dielectric constant. The 'optical' value $\varepsilon_o = n^2$ is usually taken, but measurements at high radio-frequencies[14] suggest a value of about $\varepsilon_o = 5$ for water as against $n^2 = 1{\cdot}79$. HARRIS and ALDER[12] explain the high radio-frequency estimate as due to the fact that nuclear vibrations are still present at these frequencies. They also make some criticisms of the relevance of Onsager's cavity field to the calculation of the distortion polarization. They use a different field for this purpose and increase the rigour of Kirkwood's evaluation of g, obtaining the result:

$$\frac{\varepsilon_s - 1}{\varepsilon_s + 2} = \frac{4\pi N_o \alpha}{3} + \frac{4\pi N_o}{3} \cdot \frac{\mu^2}{3kT} \cdot \frac{9\varepsilon_s}{(2\varepsilon_s + 1)(\varepsilon_s + 2)} g \quad \ldots .(1.4)$$

which may also be written as:

$$\varepsilon_s - 1 = 4\pi N_o \left[\frac{3\varepsilon_s}{2\varepsilon_s + 1} \cdot \frac{g\mu^2}{3kT} + \frac{\varepsilon_s + 2}{3} \alpha\right]$$

or as $$\varepsilon_s - 1 = 4\pi N_o \frac{3\varepsilon_s}{2\varepsilon_s + 1} \cdot \frac{g\mu^2}{3kT} + (\varepsilon_s + 2) \frac{\varepsilon_o - 1}{\varepsilon_o + 2}$$

(with $\varepsilon_o = n^2$)

For water, they calculate g from a model due to POPLE[13], in which each water molecule is bonded to four others, but bending of the O—H—O bonds is permitted. The factor g ranges from 2·60 at 0° to 2·46 at 83°. The agreement with experimental values of ε_s is within about 2 per cent over the temperature range 0°–80°, which is extremely satisfactory for such a complicated liquid as water. Good results are also obtained with alcohols, for which Oster and Kirkwood's model of chain-wise association through 'hydrogen bonds', giving $g = 2{\cdot}57$, is used; the calculated values are a few per cent higher than the observed. Harris and Alder have also made the reverse calculation by computing g from the observed dielectric constants, refractive indices and dipole moments of a number of liquids, using both Kirkwood's formula and their own. The two sets of values do not differ greatly, but those derived from equation (1.4) are perhaps slightly more reasonable: acetone and chloroform, believed to be unassociated liquids, give $g = 1{\cdot}0$; the strongly associated liquid hydrogen cyanide gives $g = 3{\cdot}6$; nitrobenzene and pyridine give $g = 0{\cdot}8$ and $0{\cdot}7$ respectively, indicating 'contra-association' of the dipoles as opposed to the head-to-tail or 'co-association' in hydrogen cyanide.

The static dielectric constant ε_s includes a large contribution due to orientation of the permanent dipoles in the applied field. The orientation process requires a finite time, and consequently the dielectric constant decreases as the frequency of the applied field increases. The orientation of the molecules against the viscous forces leads to an energy dissipation in alternating fields, which may be formally dealt with by the use of a complex dielectric constant. Provided that only one orientable dipolar species is involved, the complex dielectric constant, ε, for an angular frequency ω obeys the equation:

$$\varepsilon = \varepsilon_o + \frac{\varepsilon_s - \varepsilon_o}{1 + i\omega\tau} \qquad \ldots.(1.5)$$

where τ is the relaxation time for the orientation process, *i.e.*, the time for the orientation polarization to fall to e^{-1} of its value after the removal of the applied field. Measurements of ε at various frequencies (in the region of substantial change) can therefore determine both ε_o and τ. Such measurements have been made on water and heavy water by COLLIE, HASTED and RITSON[14] using radar techniques at wavelengths of 1·25, 3, and 10 cm over a temperature range of 0° to 75°. Their work leads to conclusions of great value in interpreting the nature of water.

First, equation (1.5) was found to hold with a single value of the relaxation time τ for each temperature. This implies that only one orientable molecular species is involved; the presence of polymeric forms such as 'dihydrol' would necessitate a range of relaxation times at each temperature. Secondly, the relaxation times vary with temperature very nearly in accordance with a theoretical result due to Debye:

$$\tau = \frac{4\pi\eta r^3}{kT} \qquad \ldots.(1.6)$$

where r is the radius of the orientable particle. *Table 1.1* illustrates the striking and, indeed, unexpectedly good agreement of the experimental results with equation (1.6). The 'molecular radius' compares

Table 1.1

Dielectric Relaxation Time and Viscosity of Water—A Test of the Debye Relation

t°C	$\tau \times 10^{12}$ sec	$\eta \times 10^{2}$ poise	$\frac{\tau T}{\eta} \times 10^{7}$ sec deg poise^{-1}	r(Å) by Eq. 1.6
0	17·7	1·787	2·71	1·44
10	12·6	1·306	2·73	1·44
20	9·5	1·002	2·78	1·45
30	7·4	0·798	2·81	1·45
40	5·9	0·653	2·83	1·46
50	4·8	0·547	2·84	1·46
60	4·0	0·467	2·85	1·46
75	3·2	0·379	2·84	1·48

Data from Collie, C. H., Hasted, J. B. and Ritson, D. M., *Proc. phys. Soc.*, 60 (1948) 145.

well with that obtained from x-ray measurements, *viz.*, 1·38 Å. These results too must be interpreted as showing that the only entity undergoing dipole orientation in water is the simple H_2O molecule. If polymeric forms such as $(H_2O)_2$ or $(H_2O)_3$ were present to any significant extent, their proportions would presumably change considerably with temperature and lead to a marked temperature-dependence of the molecular radius as calculated from Debye's relation. Further weight is given to this work by the comparison of the dielectric relaxation times for water and heavy water: the workers already quoted have shown that the ratio τ_{D_2O}/τ_{H_2O} is equal to the viscosity ratio η_{D_2O}/η_{H_2O} (within the experimental error of 2 per cent) at 10°, 20°, 30° and 40°.

Another type of measurement which gives some insight into the nature of the kinetic entities in water is the study of the self-diffusion coefficient (see Chapter 10). We quote in *Table 1.2* some results due to Wang[15] from which a 'molecular radius' of the diffusing particle can be computed by means of the Einstein-Stokes relation:

$$D^* = \boldsymbol{k}T/(6\pi\eta r) \qquad \ldots.(1.7)$$

where D^* is the (self) diffusion coefficient and the other quantities have the same meanings as before.

It will be seen that the values of D^* obtained by using heavy water as tracer are appreciably different from those using H_2O^{18} tracer; but each series shows the constancy of $D^*\eta/T$. The question of which series best represents the actual self-diffusion coefficient is not yet settled; experimental errors in work of this kind are larger than in other diffusion measurements.

Although the molecular radius so calculated (0·8–1·1 Å) is too small, its constancy leaves little doubt that we are dealing with the

Table 1.2

Self-diffusion Coefficient of Water Using Heavy Water as Tracer

t°C	$D^* \times 10^5$ cm² sec⁻¹	$\eta \times 10^2$ poise	$\frac{D^*\eta}{T} \times 10^{10}$ dyn deg⁻¹	r (Å) (Eq. 1.7)
0	1·00	1·787	6·54	1·12
5	1·20	1·516	6·55	1·13
15	1·61	1·138	6·36	1·15
25	2·13	0·890	6·36	1·15
35	2·76	0·719	6·44	1·13
45	3·45	0·596	6·46	1·13
55	4·16	0·504	6·39	1·14

Self-diffusion Coefficient of Water Using H_2O^{18} *as Tracer*

t°C	$D^* \times 10^5$ cm² sec⁻¹	$\frac{D^*\eta}{T} \times 10^{10}$ dyn deg⁻¹	r (Å) (Eq. 1.7)
0	1·33	8·70	0·84
5	1·58	8·61	0·85
15	2·14	8·45	0·87
25	2·83	8·45	0·86
35	3·55	8·28	0·88
45	4·41	8·26	0·88
55	5·41	8·31	0·88

Data from WANG, J. H., *J. Amer. chem. Soc.*, 73 (1951) 510; WANG, J. H., ROBINSON, C. V. and EDELMAN, I. S., *ibid.*, 75 (1953) 466.

motion of the same molecular species at each temperature. The low value as compared with the known radius of 1·38 Å is probably due to the inadequacy of Stokes' law for the motion of particles of molecular dimensions.

We conclude this section by drawing attention to Appendix 1.1 in which we have tabulated those properties of water which a considerable experience of calculations on electrolyte solutions has shown us to be most often needed. These are the density, dielectric constant, vapour pressure and viscosity, at intervals between 0° and 100°C.

Appendix 1.2 gives the densities, dielectric constants and viscosities of a number of non-aqueous solvents, most of them at 25°.

THE EFFECT OF IONS ON THE STRUCTURE AND PROPERTIES OF WATER

It has been shown in the preceding section that the distinctive properties of water can be largely explained in terms of electrostatic forces arising from the charge distribution of the water molecule, combined with the fact that its bond angle is close to the tetrahedral angle. Since the simple ions have dimensions, and bear charges, comparable to those of the water molecule, it is only to be expected that the structure of water will be considerably modified in ionic solutions. It is perhaps less obvious that the presence of *any* solute should alter the character of water, yet there is good reason to suppose that this is the case. The most important evidence in this direction comes from the study of the solubility and temperature coefficient of solubility of simple non-polar gas molecules, and has been ably summarized by FRANK and EVANS[16]. Their paper should be studied very carefully by those who want a full account of the argument; we present here only some of the more important features. From the solubility data, it is possible to compute the entropy lost in the process of solution of a gas in a liquid. For simple non-polar gases in non-polar solvents, this entropy loss is in the range 10–15 cal degree^{-1} mole^{-1} (adjusted to refer to standard states of one atmosphere for the gas and a hypothetical mole fraction of unity for the solution). For solutions of such gases in water, however, the entropy loss is much larger, being in the range 25–40 cal degree^{-1} mole^{-1}. Furthermore, while the entropy lost on solution of these gases in non-polar solvents varies but little with temperature, that for their aqueous solutions decreases rapidly as the temperature is raised. Now the entropy of a system may be regarded as a measure of the degree of disorder prevailing; the extra entropy lost in the formation of aqueous solutions of non-polar gases, as compared with simpler solutions, means that the water structure becomes more ordered through the influence of the dissolved molecules. In the picturesque words of Frank and Evans, 'the water builds a microscopic iceberg round the non-polar molecule'. At the higher temperatures, this effect is naturally less marked, for then the forces responsible for the regular structure can no longer compete with the thermal agitation.

It must be admitted that this conclusion is unexpected, but the thermodynamic evidence is too strong to dismiss. Comparison with the case of an impurity atom introduced into a perfect crystal lattice is a reminder that we should beware of interpreting the quasi-crystalline picture of water too literally, for in this case the foreign atom, by producing lattice dislocations, tends to destroy the existing long-range regularity. In water, as the x-ray evidence quoted in an earlier section shows, regularity extends only over a few molecular diameters, and it is at least not difficult to imagine an increase of regularity. The effect might indeed be pictured as an increase in the average life of the short-range tetrahedral configurations, due to the reluctance of the solute particle to get out of the way. This speculation (which is the present authors') is consistent with the fact that at room temperatures the extra entropy loss is greatest for the heaviest and largest solute molecules, such as radon and chloroform.

In view of the existence of this 'iceberg effect' even for non-polar solutes in water, it is clear that we must be prepared for considerable complications when we turn our attention to aqueous ionic solutions. Here an intense electrical field due to the ionic charge is superimposed on the normal interaction between solvent and solute. At the small distances involved, the field intensity is of the order of a million volts per centimetre; Coulomb's law, even if we insert the bulk dielectric constant of water (~ 80), gives a field of $0{\cdot}5 \times 10^6$ V/cm at a distance 6 Å from the centre of a univalent ion. Furthermore, under the conditions of dielectric saturation obtaining among water molecules in contact with the ion, the bulk dielectric constant is certainly too large, so the field intensity acting on the first layer of water molecules is probably an order of magnitude greater than that given by the above expression.

In very dilute solutions it is permissible to think of the effects produced by a single ion on successive layers of water molecules, but in more concentrated solutions one meets the difficulty that 'the further off from England the nearer is to France'. It is instructive to estimate the average separation of the ions in a solution, assuming as a rough guide that the ions are arranged on a cubic lattice, at least as a time-average. One finds that for a 1 : 1 electrolyte at a concentration c moles per litre, the average interionic distance is

Table 1.3

Average Separation of Ions in a Solution of a 1 : 1 *Electrolyte*

c (mole/l.):	0·001	0·01	0·1	1·0	10·0
Separation (Å):	94	44	20	9·4	4·4

$9{\cdot}4\,c^{-1/3}$ Å, giving the results shown in *Table 1.3* for various concentrations. These figures show that in a one molar solution there can be few water molecules distant by more than two or three molecular diameters from some ion; it is reasonable to talk of successive layers of water molecules round one particular ion only below about 0·1 molar.

Bearing this in mind, we may examine in more detail the effect of ions on the structure of water, again making reference to the admirable analysis given by FRANK and EVANS[16]. They present entropies of solution for a number of ions; part of their data is reproduced in *Table 1.4.*

Table 1.4

Entropy of Solution of Monatomic Ions in Water (after FRANK *and* EVANS[16]*)*

Ion	ΔS cal deg^{-1} mole^{-1}	ΔS^{st} cal deg^{-1} mole^{-1}
F^-	− 40·9	− 3·5
Cl^-	− 26·6	+ 10·2
Br^-	− 22·7	+ 13·9
I^-	− 18·5	+ 17·9
H^+	− 38·6	
Li^+	− 39·6	− 1·1
Na^+	− 33·9	+ 4·0
K^+	− 25·3	+ 12·0
Rb^+	− 23·1	+ 14·1
Cs^+	− 21·3	+ 15·7
Mg^{++}	− 84·2	
Ca^{++}	− 65·5	
Sr^{++}	− 63·7	
Ba^{++}	− 55·6	
Al^{+++}	− 133	
Fe^{+++}	− 120	

ΔS = entropy increase in passing from the hypothetical gas state at 1 atm. to a hypothetical mole fraction of unity in solution. This is a different standard state from that employed in Chapter 3.

ΔS^{st} = calculated contribution to ΔS due to the effect of ions on the structure of water.

The significance of these entropy data may be illustrated by taking the case of potassium chloride as typical. The standard entropy loss per mole is $25{\cdot}3 + 26{\cdot}6 = 51{\cdot}9$ cal deg^{-1}, whereas the corresponding figure for two gram-atoms of argon (the fairest comparison, since both ions have the argon structure) is $2 \times 30{\cdot}2 = 60{\cdot}4$ cal deg^{-1}. The net effect of the ionic charges is evidently to reduce the entropy loss; that is, to promote increased disorder in the water. This effect appears in spite of the fact that in the immediate vicinity of the ion there must surely be a layer of rather firmly oriented water molecules, probably four in number for most of the monatomic and monovalent ions. This firmly-held layer can be regarded as being in a 'frozen' condition, and Frank and Evans estimate that its formation would result in an entropy loss of about 12 cal deg^{-1} $mole^{-1}$. To this figure must be added two other sources of entropy loss: first, an amount due to the reduction of free volume when the (gas) ion enters the solution, conservatively estimated at 20 cal deg^{-1} $mole^{-1}$; and secondly, a contribution due to the partial orientation of water molecules in layers beyond the first, which may be computed by an equation due to LATIMER[17]:

Entropy loss per mole (due to dielectric orientation)

$$-\Delta S_D/(\text{cal deg}^{-1}) = \frac{22z^2}{r_i/(\text{Å}) + 2{\cdot}8}$$

where r_i is the radius of the bare ion, to which 2·8 Å is added to allow for the first rigidly-held layer of water molecules. In other words, a boundary is drawn round the ion outside the first layer of water molecules; within this boundary, the entropy loss is computed by assuming that the water molecules are rigidly held as in ice, while outside it the medium is treated as a classical dielectric continuum with the ordinary dielectric constant. This picture is substantially consistent with recent treatments of the dielectric constant near an ion (see p. 20). These three approximately calculable entropy losses can be combined and subtracted from the experimental values in the second column of *Table 1.4*, giving a remainder (column 3) called by Frank and Evans the 'structure-breaking entropy', ΔS^{st}. It is seen that for all the alkali and halide ions except the smallest (Li^+ and F^-) this structural entropy term corresponds to a considerable increase of disorder, which is greatest for the largest ions. It appears therefore that beyond the first layer of water molecules there is a region where the water structure is broken down; it is pointed out that this could arise from the manner in which the first layer of water molecules is arranged. Round a positive ion, the water molecules would be oriented with all the hydrogens outwards; they could not, therefore, all participate in the normal tetrahedral water arrangement (even if the dimensions of the central ion were close to those of a water molecule) for this arrangement would require two of the water molecules to be oriented with the hydrogens inwards. Frank and Evans support their argument for this structure-breaking effect by a number of other considerations, notably of viscosity and heat capacity data. For polyvalent monatomic ions such as Al^{+++}, the entropy loss is much greater; part of this increase is ascribed to an extension of the 'frozen' region to layers beyond the first.

THE EFFECT OF IONS ON THE DIELECTRIC CONSTANT OF WATER

Quantitative knowledge of the effect of ions on the dielectric constant of water has for many years been recognized as vitally important to the understanding of the forces operating in electrolyte solutions. This information was, however, until recently extremely difficult to obtain; it was not, in fact, known for certain whether the dielectric constant was increased or decreased. This is due to the experimental difficulty of measuring the dielectric constant of a conducting medium. Fortunately, the development of wave guide techniques for measurements at frequencies of the order of 10^{10} c/sec has at last made it possible to determine the dielectric constant of even such highly conducting liquids as electrolytes two molar in concentration with an accuracy of a few per cent. HASTED, RITSON and COLLIE[18], whose important work on the dielectric properties of water and heavy water has been discussed in a previous section, have also made very valuable studies of the dielectric properties of aqueous electrolyte solutions. They find that for all the electrolytes studied (fourteen in number, and including 1 : 1, 2 : 1, 1 : 2 and 3 : 1 valency types) the dielectric constant falls linearly as the electrolyte concentration is increased. This linear drop holds in most cases up to about 2 N, after which, in the case of sodium chloride (the only case studied above 2 N), the drop is less than that demanded by the linear relation. The dielectric relaxation time is also decreased in an approximately linear manner with increasing concentration. The latter effect appears to be consistent with the views of Frank and Evans on the structure-breaking effects of ions, which would have the result that re-orientation of the water molecules could take place more readily. The dielectric relaxation time of water is, however, increased by the addition of polar organic molecules[19]. The explanation of this observation may be connected with the 'iceberg effect' proposed by Frank and Evans to account for the entropy of aqueous solutions of non-polar gases; it appears to be difficult to measure the change in dielectric relaxation time for the latter solutions owing to the low solubility and the consequent smallness of the change. *Table 1.5* summarizes the results of Hasted, Ritson and Collie in the form of a constant $\bar{\delta}$ for each solute at 25°: this quantity is half the molar depression of the dielectric constant and is defined by:

$$\varepsilon = \varepsilon_w + 2\bar{\delta}c$$

where ε_w is the static dielectric constant of water (78·30 at 25°), ε is that of the solution and c the concentration in moles per litre.

Table 1.5

Molar Depression of Dielectric Constant of Water by Electrolytes at 25°

$$\varepsilon = \varepsilon_w + 2\bar{\delta}c$$

	$\bar{\delta}$ (l/mole)		$\bar{\delta}$ (l/mole)
HCl	− 10	NaI	− 7·5
LiCl	− 7	KI	− 8
NaCl	− 5·5	$MgCl_2$	− 15
KCl	− 5	$BaCl_2$	− 14
RbCl	− 5	$LaCl_3$	− 22
NaF	− 6	NaOH	− 10·5
KF	− 6·5	Na_2SO_4	− 11

Data from HASTED, J. B., RITSON, D. M. and COLLIE, C. H., *J. chem. Phys.*, 16 (1948) 1.

The quantity $\bar{\delta}$ is approximately additive for the separate ions and may therefore be represented by:

$$2\bar{\delta} = \bar{\delta}_1 + \bar{\delta}_2 \quad \text{for a 1 : 1 electrolyte}$$
$$2\bar{\delta} = \bar{\delta}_1 + 2\bar{\delta}_2 \quad \text{for a 2 : 1 electrolyte}$$
$$2\bar{\delta} = \bar{\delta}_1 + 3\bar{\delta}_2 \quad \text{for a 3 : 1 electrolyte}$$

etc., but any such subdivision of the observed $\bar{\delta}$ values is of course subject to the arbitrary fixing of $\bar{\delta}_1$ and $\bar{\delta}_2$ for one solute. Hasted, Ritson and Collie suggested:

$$\bar{\delta}_{Na^+} = -8 \text{ l/mole}, \quad \bar{\delta}_{Cl^-} = -3 \text{ l/mole}$$

on the reasonable ground that a positive ion would bind the water molecules in such a way as to leave them less free to rotate than would a negative ion.

Dielectric saturation

In the derivation of Debye's relation between the dipole moment and the dielectric constant of polar liquids, the Langevin formula:

$$\overline{m} = \mu_o \left(\coth \frac{\mu_o F}{\boldsymbol{k}T} - \frac{\boldsymbol{k}T}{\mu_o F} \right)$$

is involved. At ordinary field strengths $\mu_o F \ll \boldsymbol{k}T$ and the approximate expansion of this function to the first power of $\frac{\mu_o F}{\boldsymbol{k}T}$ is adequate, giving:

$$\overline{m} = \frac{\mu_o^2 F}{3\boldsymbol{k}T}$$

At very high field strengths, however, the Langevin function approaches unity asymptotically, giving ultimately $\overline{m} = \mu_o$ when all the dipoles are completely oriented by the field. The Langevin function also enters into the more elaborate calculations by Onsager and others discussed on pp. 9–11; in all models, therefore, this dielectric saturation effect is to be expected. Now the electrical field near an ion is quite intense enough to cause a marked dielectric saturation in surrounding water molecules, with a consequent reduction in the dielectric constant as measured by an external applied field. As a result, the dielectric constant of an electrolyte solution falls as the concentration is increased.

The theoretical importance of a knowledge of the variation of dielectric constant with electrolyte concentration lies in its relevance to the calculation of interionic forces which are discussed in detail in Chapter 9. We shall, however, deal here with the question of how the microscopic dielectric constant varies with the distance from an ion. This question was discussed by Sack[20] and by Debye[7], but as they employed the Clausius-Mosotti expression for the 'cavity field', which is now discredited for polar liquids, it is better to consider only a more modern treatment due to Ritson and Hasted[21]. They have calculated the dielectric constant of water as a function of distance from a point electronic charge, using two different models, one based on Onsager's expression for the dielectric constant, and the other on an empirical modification of Kirkwood's expression. Both models lead to very similar values for the local dielectric constant. There is a region of complete dielectric saturation up to about 2 Å from the point charge, where the dielectric constant has the value of four or five arising from electron and atom polarization only. This is followed by a region of rapid rise ending at about 4 Å from the point charge, and thereafter the dielectric constant is practically stationary at its ordinary bulk value. Since most simple ions have radii in the range 0·5–2 Å, and the water molecule has a diameter of 2·8 Å, it is clear that for monovalent ions the region of appreciable dielectric saturation is confined to the first layer of water molecules round the ion. Ritson and Hasted, however, treat this first-layer saturation as complete only round positive ions; the first shell round negative ions is given the bulk dielectric constant. This admittedly somewhat exaggerated distinction is based on their contention that the molecules in the first layer round a negative ion have greater freedom of rotation than those round a positive ion. It seems equally likely that the difference in dielectric saturation round positive and negative ions is merely a matter of ionic size, for the monovalent negative ions they consider are in fact the halide ions which have radii 1·3–2·2 Å, whereas the positive ions are those of the alkali-metals of radii 0·6–1·6 Å. The important conclusion to which they come is, however, that for monovalent ions the observed lowering of the bulk dielectric constant arises almost entirely in the first layer of water molecules. Polyvalent cations are frequently small and monatomic, and for these this saturation effect certainly extends beyond the first layer of water molecules; but polyvalent anions of any stability are polyatomic and hence large, and for these the position is far from clear.

Schellman[22] has calculated the dielectric saturation effect near an ion in water, using a detailed molecular model of the region near the ion combined with a classical dielectric model at greater distances. He finds that the dielectric saturation effect should be very much smaller than the classical model alone would give, *e.g.*, at 5 Å from a monovalent ion the dielectric constant is only 0·4 per cent less than its ordinary value, and even at 2 Å it is only reduced by about 17 per cent. These conclusions give strong support to the practice of using the ordinary dielectric constant of water in calculating ionic interactions, even in comparatively concentrated solutions.

When considering the significance of the values of the dielectric constant depression given in *Table 1.5*, it should be noted that in most cases the concentration range studied was 0·5–2 N; a few solutions (hydrochloric acid, sodium hydroxide, potassium iodide, potassium fluoride) were examined at 0·2 or 0·25 N. At these concentrations most of the water molecules would be no more than three molecular diameters away from an ion. One need therefore feel no surprise that the linear relation begins to fail about 2 N.

Haggis, Hasted and Buchanan[(19)] consider that the dielectric decrement owes its origin primarily to the prevention of the rotation of water molecules. From a more detailed theoretical consideration they estimate the average number ($n_{irr.}$) of water molecules thus 'irrotationally bound' by the solute particles; this number is close to zero for uncharged solute molecules, and ranges from four to six for the alkali-metal halides, *e.g.*, $n_{irr.} \approx$ four for rubidium chloride and ammonium chloride, five for potassium chloride and six for sodium chloride and lithium chloride. These numbers need not necessarily be the same as the number of molecules moving with the ions as a single kinetic entity, which we regard as 'true' hydration numbers.

Further information about the effect of ions on the solvent has resulted from measurements of nuclear magnetic resonance in electrolyte solutions. The protons of the water molecule are shielded magnetically by the electron cloud and any influence that displaces this electron cloud will change the nuclear magnetic resonance. Shoolery and Alder[23] express this shift as:

$$\delta = 10^7 \,(H_{H_2O} - H_{sample})/H_{H_2O}$$

where H is the applied magnetic field necessary to produce resonance in a constant radio-frequency field. These shifts are proportional to the concentration (except at high concentrations) and δ can be expressed as the sum of cationic and anionic effects in the form:

$$\delta = (\nu_1\delta_1 + \nu_2\delta_2)m$$

Some of these shifts, based on $\delta_2(ClO_4^-) = -0{\cdot}85$ kg/mole and expressed as $\delta_1/|z_1|$ or $\delta_2/|z_2|$, are shown in *Figure 1.4.*

An ion may affect the electron density in a water molecule in

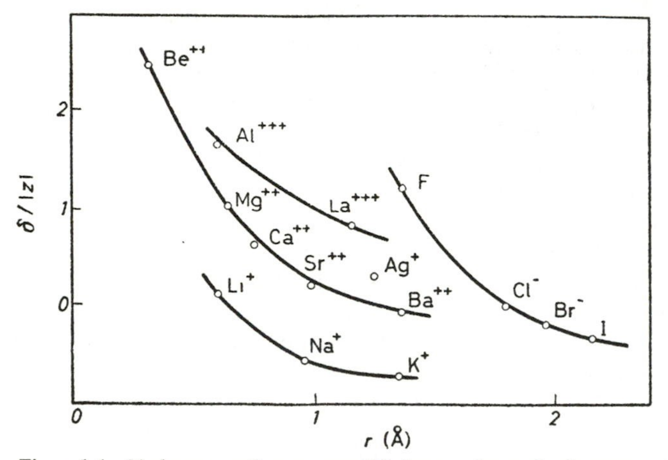

Figure 1.4. Nuclear magnetic resonance shift due to cations and anions versus ionic radius

two ways. A water molecule solvated to a positive ion should on the average be oriented with the oxygen atom towards the ion because the water dipole acts towards the oxygen atom. The positive ion increases the electron drift towards the oxygen atom leaving the protons less shielded. A water molecule solvated to a negative ion should be oriented in the opposite direction with the oxygen away from the ion; the charge on the ion will however increase the electron density in the vicinity of the oxygen atom, again leaving the protons less shielded. In both cases, therefore, the polarization due to solvation should result in a positive shift of the nuclear magnetic resonance. The other effect enters because on solvation a water molecule must break at least one hydrogen bond to another water molecule; in this process the mutually induced dipoles disappear and the electron density around the protons increases. Structure breaking of the solvent therefore results in a negative shift. These ideas are in accord with the experimental findings. Positive shifts are found with the smaller ions which can approach the solvating molecule sufficiently close for the polarization effect to be marked; as would be expected, the positive shift is emphasized for divalent ions and even more so for trivalent ions whether they be positively or negatively charged. A negative shift occurs when the structure breaking effect takes charge; this is found with the larger ions. The silver ion, however, has a higher value than might be expected and the halide ions are in marked contrast to the alkali metals; thus K^+ shows a shift of $-$ 0·71 kg/mole whilst F^-, with almost the same radius, gives 1·20 kg/mole. These experiments bring to light a remarkable power of halide ions in modifying the structure of water.

REFERENCES

[1] MAGAT, M., *Trans. Faraday Soc.*, 33 (1937) 114
[2] BERNAL, J. D. and FOWLER, R. H., *J. chem. Phys.*, 1 (1933) 515
[3] VERWEY, E. J. W., *Rec. Trav. chim. Pays-Bas*, 60 (1941) 887
[3a] GIAUQUE, W. F. and STOUT, J. W., *J. Amer. chem. Soc.*, 58 (1936) 1144
[3b] OSBORNE, N. S., STIMSON, H. F. and GINNINGS, D. C., *J. Res. nat. Bur. Stand.*, 23 (1939) 197
[4] KOHLRAUSCH, F. and HEYDWEILLER, A., *Z. phys. Chem.*, 14 (1894) 317
[5] MORGAN, J. and WARREN, B. E., *J. chem. Phys.*, 6 (1938) 666
[6] CARTWRIGHT, C. H. and ERRERA, J., *Proc. roy. Soc.*, 154 A (1936) 138
[7] DEBYE, P., *Phys. Z.*, 13 (1912) 97; "Polar Molecules," Chemical Catalog Co. Inc. (1929)
[8] ONSAGER, L., *J. Amer. chem. Soc.*, 58 (1936) 1486
[9] KIRKWOOD, J. G., *J. chem. Phys.*, 7 (1939) 911
[10] FRÖHLICH, H., "Theory of Dielectrics," Oxford University Press (1949)
[11] OSTER, G. and KIRKWOOD, J. G., *J. chem. Phys.*, 11 (1943) 175; KIRKWOOD, J. G., *Trans. Faraday Soc.*, 42 A (1946) 7
[12] HARRIS, F. E. and ALDER, B. J., *J. chem. Phys.*, 21 (1953) 1031
[13] POPLE, J. A., *Proc. roy. Soc.*, 205 A (1951) 163
[14] COLLIE, C. H., HASTED, J. B. and RITSON, D. M., *Proc. phys. Soc.*, 60 (1948) 145
[15] WANG, J. H., *J. Amer. chem. Soc.*, 73 (1951) 510; WANG, J. H., ROBINSON, C. V. and EDELMAN, I. S., *ibid.*, 75 (1953) 466
[16] FRANK, H. S. and EVANS, M. W., *J. chem. Phys.*, 13 (1945) 507
[17] LATIMER, W. M., *Chem. Rev.*, 18 (1936) 349
[18] HASTED, J. B., RITSON, D. M. and COLLIE, C. H., *J. chem. Phys.*, 16 (1948) 1
[19] HAGGIS, G. H., HASTED, J. B. and BUCHANAN, T. J., *ibid.*, 20 (1952) 1452
[20] SACK, H., *Phys. Z.*, 27 (1926) 206; 28 (1927) 299
[21] RITSON, D. M. and HASTED, J. B., *J. chem. Phys.*, 16 (1948) 11
[22] SCHELLMAN, J. A., *J. chem. Phys.*, 26 (1957) 1225
[23] SHOOLERY, J. N. and ALDER, B. J., *ibid.*, 23 (1955) 805

2

BASIC CONCEPTS AND DEFINITIONS

ACTIVITY COEFFICIENTS, STANDARD STATES AND CONCENTRATION SCALES FOR ELECTROLYTE SOLUTIONS

THE total Gibbs free energy, G, of a fixed quantity of electrolyte solution of given composition is dependent only on the temperature and pressure no matter what conventions we adopt for expressing partial molal quantities such as the activities of the components. If this fact is kept in mind, much of the confusion which frequently besets the worker in this field with regard to the various types of activity coefficient can be avoided.

We shall denote the solvent and solute by subscripts A and B respectively. The 'solute' will be taken to mean the anhydrous solute, following the usual convention. The partial molal Gibbs free energies, or chemical potentials, of the solvent and solute are then:

$$\left.\begin{aligned}\bar{G}_A &= \left(\frac{\partial G}{\partial n_A}\right)_{n_B, T, P} \\ \bar{G}_B &= \left(\frac{\partial G}{\partial n_B}\right)_{n_A, T, P}\end{aligned}\right\} \quad \ldots\ldots (2.1)$$

where n_A, n_B denote the number of moles of solvent and solute in the system. It will sometimes be convenient to use the subscript w to refer to the solvent if we are dealing with an aqueous solution. Since we are more interested in the variation of chemical potential with composition than in its absolute value, it is usual to express these quantities as a difference between the absolute value and that which holds in some specified standard state. The standard state is indicated by a superscript zero, $\bar{G}_A^0$, $\bar{G}_B^0$. The choice of standard state is entirely at our discretion: it may be a pure component, a saturated solution or some entirely hypothetical solution. In the case of mixed liquids which are non-electrolytes, for example, the standard state for each component is usually taken to be that component in the pure state; this choice preserves symmetry between the two components, which is useful in the study of these systems.

For electrolyte solutions, the standard state to which the free energy of the solvent is referred is invariably the pure solvent at the same temperature and pressure. The activity of the solvent, a_A, is then defined by:

$$\bar{G}_A - \bar{G}_A^0 = \boldsymbol{R}T \ln a_A \qquad \ldots.(2.2)$$

Since the pure solvent can exist in equilibrium with its vapour at pressure p_A^0, and the solution with the solvent vapour at partial pressure p_A, we have also, assuming the vapour to be ideal:

$$\left.\begin{aligned} \bar{G}_A^0 &= \bar{G}_A^0(v) + \boldsymbol{R}T \ln p_A^0 \\ \text{and} \qquad \bar{G}_A &= \bar{G}_A^0(v) + \boldsymbol{R}T \ln p_A \end{aligned}\right\} \qquad \ldots.(2.3)$$

where $\bar{G}_A^0(v)$ is the molal free energy of the vapour in the standard state of one atmosphere pressure at the temperature T.

It follows from (2.2) and (2.3) that:

$$a_A = p_A/p_A^0$$

Strictly, the ratio p_A/p_A^0 should be replaced by the ratio of fugacities, p_A^*/p_A^{*0}. However, the vapour pressures of most commonly used electrolyte solutions are small enough for this to make a negligible difference. (The position is not that the vapour is so ideal that $p = p^*$ but rather that the vapour pressures of solution and solvent are always of similar magnitude so that the correction factor is practically the same for the solution and the pure solvent.)

For electrolytes, however, the pure solute is not a very practical choice as a standard state, since it is frequently a solid or liquid with properties very different from those of solutions. Instead, it is the practice to use as standard states certain hypothetical solutions. The position resembles that which we adopt in discussing the free energy of gases, where the standard state is taken as that of an 'ideal' gas (which is of course a hypothetical concept) at one atmosphere pressure or other unit pressure. For electrolytes, the standard state is, in much the same way, a hypothetical solution (having certain properties which we shall state later) at unit concentration on some chosen scale and at the temperature and pressure of the solution. The chemical potential ascribed to this standard state naturally depends on which concentration-scale we adopt. The scales in common use are:

(a) The molal scale (m = moles of solute per kilogram of solvent).
(b) The molar scale (c = moles of solute per litre of solution).
(c) The mole fraction scale (N_B = moles of solute divided by the total number of moles in the system).

On each of these scales we can define an activity for the solute, using parenthesized letters to emphasize that the activity of the solute, and the free energy in the standard state, depend on the scale chosen:

$$\left.\begin{aligned}\bar{G}_B &= \bar{G}^0_B(m) + \boldsymbol{R}T \ln a_B(m) \\ &= \bar{G}^0_B(c) + \boldsymbol{R}T \ln a_B(c) \\ &= \bar{G}^0_B(N) + \boldsymbol{R}T \ln a_B(N)\end{aligned}\right\} \quad \ldots.(2.4)$$

It should be noted that (provided we always calculate the number of moles of solute on the basis of the anhydrous substance) the quantity $\bar{G}_B$ is unique for a given solution, pressure and temperature. Equations (2.4) are so far no more than definitions; we have not yet stated the properties by which the various standard states are characterized. Before doing this it is convenient to separate the activities a_B into factors referring to the separate ions and the concentrations on the appropriate scales. It is clearly desirable that the chemical potential of the solute treated as a whole should be equal to the sum of the values for the separate ionic constituents. It should be remembered, however, that the concept of the chemical potential of one species of ion is something of a mathematical fiction. This quantity can be defined by the equation:

$$\bar{G}_i = \left(\frac{\partial G}{\partial n_i}\right)_{n_A,\, n_j,\, T,\, P} \quad \ldots.(2.5)$$

where i refers to one kind of ion, and A and j to the solvent and the other ions respectively. Physically, however, the operation implied by equation (2.5) cannot be performed, for it means adding to the solution a quantity of one kind of ion only. Even if this could be done, it would result in an enormous increase in energy of the solution due to the self-energy of the electric charge involved[(1)], an effect which we do not wish to be concerned with, since it depends on the shape of the portion of solution considered. This self-energy change could of course be exactly cancelled by the subsequent addition of the equivalent amount of the oppositely charged ion, when the resultant total free energy change would be that due to the addition of a quantity of the electrically neutral electrolyte as implied by equation (2.1). We may thus agree to discuss the free energy change due to the addition of one species of ion only, neglecting the self-energy effect, provided that we end with formulae involving only electrically equivalent amounts of cations and anions. We can then write for each ionic species i,

$$\bar{G}_i = \bar{G}^0_i + \boldsymbol{R}T \ln a_i \quad \ldots.(2.6)$$

Let one mole of electrolyte give in the ionized state ν_1 moles of cations of valency z_1 and ν_2 moles of anions of valency z_2. From the condition of electrical neutrality:

$$\left.\begin{aligned} \nu_1|z_1| &= \nu_2|z_2| = -\nu_2 z_2 \\ \text{also} \qquad \bar{G}_B &= \nu_1\bar{G}_1 + \nu_2\bar{G}_2 \\ \bar{G}_B^0 &= \nu_1\bar{G}_1^0 + \nu_2\bar{G}_2^0 \end{aligned}\right\} \qquad \ldots(2.7)$$

so that by using (2.4), (2.6) and (2.7) we obtain:

$$a_B = a_1^{\nu_1}a_2^{\nu_2} \qquad \ldots.(2.8)$$

Formulae of the type of (2.8) will hold, with different values of the activities, for each different concentration scale.

Now for each ionic species we define an 'activity coefficient' obtained by dividing the ionic activity a_1 or a_2 by the concentration of the ion on the appropriate scale, *e.g.*,

$$\left.\begin{aligned} &\text{molal scale:} & a_1(m) &= \gamma_1 m_1 \\ &\text{molar scale:} & a_1(c) &= y_1 c_1 \\ &\text{mole fraction scale:} & a_1(N) &= f_1 N_1 \end{aligned}\right\} \qquad \ldots.(2.9)$$

where γ, y and f are called respectively the molal, molar and rational activity coefficients. The activity on the mole fraction scale, and consequently f, are dimensionless quantities. It is usual to regard γ and y also as dimensionless quantities which implies that $a(m)$ and $a(c)$ have the dimensions of molality and molarity respectively. It follows that the arbitrary constants $\bar{G}^0(m)$ and $\bar{G}^0(c)$ contain concealed terms in $\boldsymbol{R}T\ln$ (mole kg^{-1}) and $\boldsymbol{R}T\ln$ (mole litre^{-1}); these do not in practice cause any difficulty. The ionic concentrations are simply related to those of the electrolyte as a whole by the equations:

$$\left.\begin{aligned} m_1 &= \nu_1 m \\ c_1 &= \nu_1 c \\ N_1 &= \nu_1 N_B{}^* \end{aligned}\right\} \qquad \ldots.(2.10)$$

with similar relations for the anion.

Equation (2.8) therefore gives, using (2.9) and (2.10):

$$a_B(m) = (\nu_1^{\nu_1}\nu_2^{\nu_2})m^{\nu}\gamma_1^{\nu_1}\gamma_2^{\nu_2} \qquad \ldots.(2.11)$$

with exactly similar results for other concentration scales. The symbol $\nu(= \nu_1 + \nu_2)$ denotes the total number of moles of ions given by one mole of electrolyte.

The individual ionic activity coefficients in (2.11) occur as a product, each raised to powers which satisfv the condition of

* But see p. 31.

electrical equivalence. In order to simplify the appearance of (2.11) we now introduce a 'mean ionic activity coefficient' defined by:

$$\gamma_{\pm}^{\nu} = \gamma_1^{\nu_1}\gamma_2^{\nu_2} \qquad \ldots.(2.12)$$

so that (2.11) becomes:

$$a_B(m) = (\nu_1^{\nu_1}\nu_2^{\nu_2})(m\gamma_{\pm})^{\nu} = (Qm\gamma_{\pm})^{\nu} \qquad \ldots.(2.13)$$

where Q is a convenient abbreviation for $(\nu_1^{\nu_1}\nu_2^{\nu_2})^{1/\nu}$. A mean activity $a_{\pm}$ can also be defined by $a_{\pm}^{\nu} = a_B$ and a mean ionic molality $m_{\pm}$ by:

$$m_{\pm}^{\nu} = (\nu_1^{\nu_1}\nu_2^{\nu_2})m^{\nu} \qquad \ldots.(2.14)$$

These formulae look rather clumsy, but take quite simple forms when the numerical values corresponding to various valency types are substituted as shown in Appendix 2.1. Though the molality scale is chosen here, similar formulae with the same numerical quantities apply to the other scales.

The various mean ionic activity coefficients $\gamma_{\pm}$, $y_{\pm}$ and $f_{\pm}$, are of great use in the study of solutions, and occur so often that they are frequently abbreviated to γ, y and f without subscripts when there is no danger of confusion.

We are now ready to assign to the hypothetical standard states for electrolyte solutions the properties which will make them most useful.

The standard state for each concentration scale is so chosen that the mean ionic activity coefficient on that scale approaches unity when the concentration is reduced to zero. This applies to every temperature and pressure.

In the standard state, $\bar{G}_B = \bar{G}_B^0$ by definition; hence from equation (2.4), $a_B = 1$, *i.e.*, the standard state is a state in which the solute is at unit activity. However, *it is not the state in which, in the actual solution, the solute has unit activity.* For example, at 25° a 1·734 M solution of potassium chloride has a mean molal activity coefficient of 0·577 so that its activity is:

$$a_B = (1{\cdot}734 \times 0{\cdot}577)^2 = 1{\cdot}000$$

This is *not* the standard state for potassium chloride on the molality scale, but is merely a numerical accident. At another temperature, a solution of this composition would have an activity different from unity. We may compare with the case of a gas which at some temperature and pressure happens to conform exactly with the equation $PV = \boldsymbol{R}T$; this does not make it an ideal gas. The state used as the standard state for gases is that of a hypothetical (ideal) gas at a pressure of one atmosphere. Similarly the standard state for electrolyte solutions is that of a hypothetical solution at a 'mean molality' (or molarity or mole fraction if these scales are used) of unity. This hypothetical solution also has the 'ideal' quality that the mean activity coefficient of the ions is unity at all temperatures and pressures. It is often referred to as the 'hypothetical molal solution', though as just indicated, a more correct term would be 'hypothetical *mean* molal solution'. The expressions 'hypothetical mean molar solution' and 'solution with hypothetical mean mole fraction of unity' are likewise applicable to the standard states for the molar and mole fraction scales respectively.

Another misconception which must be avoided is that of regarding the standard state as one of infinite dilution. Certainly, at infinite dilution the activity coefficient is unity, as it is in the standard state; but the partial molal free energy, which involves a term in the logarithm of the concentration, is negatively infinite at infinite dilution. It will be shown later that the partial molal heat content, heat capacity and volume of the solute are the same in the hypothetical standard state as they are at infinite dilution of the actual solution.

OSMOTIC COEFFICIENTS

At 25° a 2 M solution of potassium chloride has a water activity of 0·9364; since the mole fraction of water is 0·9328, the rational activity coefficient of the water is $f_A = 1{\cdot}004$, a figure which fails to emphasize the departure from ideality indicated by the activity coefficient of the solute, $f_\pm = 0{\cdot}614$. Splitting the activity of the solvent into concentration and activity coefficient factors seldom proves informative. Instead we define the osmotic coefficient which may be:

(a) the 'rational' coefficient, g, defined by:

$$\ln a_A = g \ln N_A = -g \ln \left(1 + \frac{\nu m W_A}{1000}\right) \qquad \ldots .(2.15)$$

where W_A is the molecular weight of the solvent. Expanding in a series, we get:

$$\ln a_A = -g \left[\frac{\nu m W_A}{1000} - \frac{1}{2}\left(\frac{\nu m W_A}{1000}\right)^2 + \ldots\right]$$

(b) the molal osmotic coefficient, ϕ, defined by:

$$\ln a_A = -\frac{\nu m W_A}{1000}\phi \qquad \ldots .(2.16)$$

Thus, for 2 M potassium chloride $g = 0{\cdot}944$ and $\phi = 0{\cdot}912$. Should the solution contain more than one solute species, equation

(2.16) is still valid provided the term νm in this definition of ϕ is replaced by a summation over all the solutes present.

The osmotic pressure, Π, of a solution can be expressed to a good approximation by the formula:

$$-\ln a_A = \frac{\Pi \bar{V}_A}{\mathbf{R}T} = -g \ln N_A = \frac{\nu m W_A}{1000} \phi \quad \ldots .(2.17)$$

where $\bar{V}_A$ is the partial molal volume of the solvent. For a dilute solution (2.17) approximates to:

$$\Pi \approx g\mathbf{R}T \frac{\nu m W_A}{1000 \bar{V}_A} \approx \nu g \mathbf{R}Tc$$

The osmotic coefficient is, therefore, related to the Van't Hoff factor, i, of classical solution theory, by $\nu g \approx i$. The molal osmotic coefficient is more exactly related to the osmotic pressure, by

$$\Pi = \frac{\nu \mathbf{R}TW_A}{1000\bar{V}_A} \phi m$$

RELATION BETWEEN ACTIVITY COEFFICIENTS ON DIFFERENT SCALES

It is often necessary to convert activity coefficients from one scale to another. The required relations are readily established from the various definitions, remembering that the quantity $\bar{G}_B$ is the same whatever scale is used. As an example, the relation between activity coefficients on the molal and molar scales is derived below. The other relations, which may be obtained in a similar way, are then given without proof. We have:

$$\left.\begin{aligned} \bar{G}_B &= \bar{G}^0_B(m) + \mathbf{R}T \ln a_B(m) \\ &= \bar{G}^0_B(c) + \mathbf{R}T \ln a_B(c) \end{aligned}\right\} \quad \ldots .(2.4)$$

Also:

$$a_B(m) = Q^\nu (m\gamma_\pm)^\nu$$

$$a_B(c) = Q^\nu (cy_\pm)^\nu \quad \ldots .(2.13)$$

where Q is the numerical factor given for the various valency types in Appendix 2.1. Hence:

$$G^0_B(m) + \nu \mathbf{R}T \ln m + \nu \mathbf{R}T \ln \gamma_\pm = \bar{G}^0_B(c) + \nu \mathbf{R}T \ln c + \nu \mathbf{R}T \ln y_\pm$$

and

$$\ln \gamma_\pm = \ln \frac{c}{m} + \ln y_\pm + \frac{\bar{G}^0_B(c) - \bar{G}^0_B(m)}{\nu \mathbf{R}T} \quad \ldots .(2.18)$$

The last term on the right is a constant for a given temperature and pressure and may be evaluated by making use of the property defining the standard states, viz., $c \to 0$, $m \to 0$, $\gamma_{\pm} \to 1$, $y_{\pm} \to 1$. In addition it follows from the definitions of molality and molarity that as $c \to 0$, $\frac{c}{m} \to d_0$, d_0 being the density of the pure solvent.

Hence as $c \to 0$, equation (2.18) gives:

$$\ln d_0 + \frac{\bar{G}_B^0(c) - \bar{G}_B^0(m)}{RT} = 0$$

so that (2.18) may be written:

$$\ln \gamma_{\pm} = \ln \frac{c}{m} + \ln y_{\pm} - \ln d_0$$

or

$$\gamma_{\pm} = \frac{cy_{\pm}}{md_0} \qquad \ldots .(2.19)$$

which is the required relation. In case the quantity $\frac{c}{m}$ is not directly available, it may be computed from the density, d, of the solution by either of the relations:

$$c = \frac{md}{1 + 0{\cdot}001\, mW_B} \quad \text{or} \quad m = \frac{c}{d - 0{\cdot}001\, cW_B} \qquad \ldots .(2.20)$$

where W_B is the molecular weight of the solute.

When we are considering the electrolyte solute as a whole, though the concepts of molality and molarity are quite unambiguous, there is a logical difficulty in defining the idea of the mole fraction of the solute as a whole. Either we take it as the ratio of the total number of solute particles (ions) to the total number of ions plus molecules of solvent, or else we take it as the ratio of formula weights of solute to the total number of formula weights of solute ions plus solvent, *i.e.*, if m is the molality of the solute, and W_A the molecular weight of solvent, are we to say that the 'mole fraction of solute' is:

$$N_B = \frac{\nu m}{\nu m + 1000/W_A}$$

or

$$N_B = \frac{m}{\nu m + 1000/W_A} \qquad \ldots .(2.21)$$

Fortunately, it does not matter much, as the relation between the rational activity coefficient and the others is unaffected by the choice. Here we shall choose the definition (2.21), which has the advantage of preserving the close similarity of form of equations (2.10) to (2.14) for the different scales, and ensures that the numerical factors of Appendix 2.1 will apply to the mole fraction scale as well as to the molar and molal scales. However, it is $(N_A + \nu N_B)$ and not $(N_A + N_B)$ which equals unity if equation (2.21) is chosen. The concept of the mole fraction of each ionic species, and hence of the ionic mean mole fraction, is unambiguous and consistent with this choice.

The relations between the three kinds of activity coefficients are summarized below:

$$\left.\begin{aligned} f_\pm &= \gamma_\pm(1 + 0{\cdot}001\,\nu W_A m) \\ f_\pm &= y_\pm \frac{d + 0{\cdot}001\,c(\nu W_A - W_B)}{d_0} \\ \gamma_\pm &= \frac{d - 0{\cdot}001\,cW_B}{d_0} y_\pm = \frac{c}{md_0} y_\pm \\ y_\pm &= (1 + 0{\cdot}001\,mW_B)\frac{d_0}{d}\gamma_\pm = \frac{md_0}{c}\gamma_\pm \end{aligned}\right\} \quad \ldots (2.22)$$

(ν = number of moles of ions formed by the ionization of one mole of solute;

W_A = molecular weight of solvent; W_B that of solute;

d = density of solution; d_0 that of pure solvent;

m = moles of solute per kilogram of solvent; c = moles of solute per litre of solution;

$f_\pm, \gamma_\pm, y_\pm$ = mean rational, molal and molar activity coefficients respectively.)

For a solution containing more than one electrolyte (or other solute), it can be shown that for each solute:

$$\left.\begin{aligned} f_\pm &= \gamma_\pm(1 + 0{\cdot}001\,W_A\Sigma\nu m) \\ f_\pm &= y_\pm \frac{d + 0{\cdot}001(W_A\Sigma\nu c - \Sigma cW_B)}{d_0} \\ \gamma_\pm &= \frac{d - 0{\cdot}001\Sigma cW_B}{d_0} y_\pm = \frac{c}{md_0} y_\pm \\ y_\pm &= (1 + 0{\cdot}001\Sigma mW_B)\frac{d_0}{d}\gamma_\pm = \frac{md_0}{c}\gamma_\pm \end{aligned}\right\} \quad \ldots (2.23)$$

The summations are to be made over all the solute species. For a mixed solvent containing weight fractions, x and $(1 - x)$, of solvents of molecular weight, $W_{A'}$ and $W_{A''}$ respectively, the quantity W_A in these equations must be replaced by

$$\left(\frac{x}{W_{A'}} + \frac{1-x}{W_{A''}}\right)^{-1}$$

The relation between the two osmotic coefficients is:

$$g \ln N_A = -\frac{\nu m W_A}{1000}\phi$$

which for most purposes can be written as the approximation:

$$\phi \approx g\left(1 - \frac{1}{2}\frac{\nu m W_A}{1000}\right)$$

At 2 M potassium chloride ($\phi = 0{\cdot}912$ at 25°) this approximation gives $g = 0{\cdot}946_0$ instead of the correct value of $0{\cdot}944_4$ and 1·081 instead of 1·071 at 4·8 M($\phi = 0{\cdot}988$).

THE GIBBS-DUHEM EQUATION

Since the chemical potential is the partial molal derivative of the Gibbs free energy, the Gibbs-Duhem equation applies:

$$S\mathrm{d}T - V\mathrm{d}P + \Sigma n_i \mathrm{d}\bar{G}_i = 0$$

where n_i indicates the number of moles of the ith species, the summation covering both solvent and solute species. For the restricted case of a system maintained at constant temperature and pressure:

$$n_A \mathrm{d}\bar{G}_A + n_B \mathrm{d}\bar{G}_B + n_C \mathrm{d}\bar{G}_C + \ldots = 0$$

and for a solution containing only one solute,

$$n_A \mathrm{d}\bar{G}_A = -n_B \mathrm{d}\bar{G}_B \qquad \ldots.(2.24)$$

Multiplying each side of equation (2.24) by $(1000/W_A n_A)$ we get:

$$(1000/W_A)\mathrm{d}\bar{G}_A = -m\mathrm{d}\bar{G}_B \qquad \ldots.(2.25)$$

In addition,

$$N_A \mathrm{d}\bar{G}_A = -N_B \mathrm{d}\bar{G}_B$$

These very important results are true not only for the chemical potential but for all partial molal quantities like partial molal volumes, entropies, heat contents, *etc.*

If the partial molal free energy of the solvent has been measured over a concentration range (and we shall demonstrate later that many methods are available) equation (2.25) enables us to obtain information about the free energy of the solute, although there may be difficulties of computation in the integration of the equation, to which we shall refer in subsequent chapters. The converse operation is also possible. For an aqueous electrolyte solution equations (2.2), (2.4), (2.13) and (2.25) give:

$$\begin{aligned} -55{\cdot}51 \mathrm{d} \ln a_W &= m\mathrm{d}\bar{G}_B/(\boldsymbol{RT}) \\ &= \nu m \mathrm{d} \ln (\gamma_{\pm} m) \qquad \ldots.(2.26) \end{aligned}$$

a form of the Gibbs-Duhem equation of which much use will be made in later chapters. As, by definition,

$$\nu m\phi = -55{\cdot}51 \ln a_W, \qquad \ldots.(2.16)$$

equation (2.26) may easily be transformed into:

$$(\phi - 1)\frac{\mathrm{d}m}{m} + \mathrm{d}\phi = \mathrm{d} \ln \gamma$$

which on integration gives:

$$\ln \gamma = (\phi - 1) + \int_0^m (\phi - 1)\mathrm{d} \ln m \qquad \ldots.(2.27)$$

if it is remembered that the limiting value of ϕ at infinite dilution is unity. Alternatively,

$$\nu m\phi = -55{\cdot}51 \ln a_W = \int_0^m \nu m \mathrm{d} \ln (\gamma m)$$

whence
$$\phi = 1 + \frac{1}{m}\int_0^m m\mathrm{d} \ln \gamma \qquad \ldots.(2.28)$$

If the activity coefficient can be expressed in the form (see Chapter 9):

$$-\ln \gamma = \frac{\alpha\sqrt{m}}{1 + \beta\sqrt{m}}$$

then
$$1 - \phi = \frac{\alpha\sqrt{m}}{3}\sigma(\beta\sqrt{m})$$

where
$$\sigma(x) = \frac{3}{x^3}[(1 + x) - 2 \ln (1 + x) - 1/(1 + x)]$$

The function $\sigma(x)$ is tabulated in Appendix 2.2.

THE RELATION OF THE PARTIAL MOLAL HEAT CONTENT, HEAT CAPACITY AND VOLUME TO THE ACTIVITY COEFFICIENT

The partial molal heat content of a solute in solution is given by the equation:

$$\bar{H}_B = -T^2\frac{\partial}{\partial T}\left(\frac{\bar{G}_B}{T}\right) \qquad \ldots.(2.29)$$

the differentiation being carried out at constant pressure and composition. From the definition of the activity coefficient of an electrolyte on the molality scale:

$$\bar{G}_B = \bar{G}_B^0 + \boldsymbol{R}T \ln a_B$$
$$= \bar{G}_B^0 + \nu\boldsymbol{R}T \ln Q + \nu\boldsymbol{R}T \ln m + \nu\boldsymbol{R}T \ln \gamma_\pm$$

where Q is the numerical coefficient of Appendix 2.1.

Hence $$\frac{\partial}{\partial T}\left(\frac{\bar{G}_B}{T}\right)_{m,\,P} = \frac{\partial}{\partial T}\left(\frac{\bar{G}_B^0}{T}\right)_P + \nu\boldsymbol{R}\left(\frac{\partial \ln \gamma_\pm}{\partial T}\right)_{m,\,P}$$

so that $$\bar{H}_B = \bar{H}_B^0 - \nu\boldsymbol{R}T^2\left(\frac{\partial \ln \gamma_\pm}{\partial T}\right)_{m,\,P} \qquad \ldots.(2.30)$$

where $\bar{H}_B^0$ is the partial molal heat content in the standard state. Now at infinite dilution, $\gamma_\pm = 1$ at all temperatures, so that

$$\bar{H}_B^\infty = \bar{H}_B^0$$

that is to say, the partial molal heat contents have the same value in the standard state and at infinite dilution. Partial molal heat contents are usually expressed relative to infinite dilution, when they are called relative partial molal heat contents and are denoted by the symbol, $\bar{L}_B$:

$$\bar{L}_B = \bar{H}_B - \bar{H}_B^0$$

or: $$\bar{L}_B = -\nu\boldsymbol{R}T^2\left(\frac{\partial \ln \gamma_\pm}{\partial T}\right)_{m,\,P}$$

The mole fraction scale may also be used to express the activity coefficient in this formula; but the molar scale is unsuitable since the composition of a solution of fixed molarity varies with the temperature.

It will be noted that if we were to use, as the standard state, an 'actual' state of unit activity having a mean molality $m_\pm$ and mean activity coefficient $\gamma_\pm$, such that $m_\pm\gamma_\pm = 1$, the differentiation with respect to a temperature at constant composition could not be carried out in a meaningful way, as the composition of this actual standard solution would have to change with temperature, inversely to the change in $\gamma_\pm$.

The partial molal heat capacity, $\bar{C}_{(P)B}$ is naturally defined by:

$$\bar{C}_{(P)B} = \left(\frac{\partial \bar{H}_B}{\partial T}\right)_{m,\,P} = \bar{C}_{(P)B}^0 - \nu\boldsymbol{R}\left(T^2\frac{\partial^2 \ln \gamma_\pm}{\partial T^2} + 2T\frac{\partial \ln \gamma_\pm}{\partial T}\right)_{m,\,P} \qquad \ldots.(2.31)$$

Again $$\bar{C}_{(P)B} = \bar{C}_{(P)B}^0$$

and a relative partial molal heat capacity, $\bar{J}_B$, can be used in a similar way to $\bar{L}_B$:

$$\bar{J}_B = \bar{C}_{P(B)} - \bar{C}_{P(B)}^0 \qquad \ldots(2.32)$$

The relation between the partial molal volume of the solute and the activity coefficient is of minor importance, since pressures are seldom different from atmospheric. It is:

$$\bar{V}_B = \left(\frac{\partial \bar{G}_B}{\partial P}\right)_{m,\,T}$$

$$= \left(\frac{\partial \bar{G}_B^0}{\partial P}\right)_T + \nu \boldsymbol{R}T \left(\frac{\partial \ln \gamma_\pm}{\partial P}\right)_{m,}$$

$$= \bar{V}_B^0 + \nu \boldsymbol{R}T \left(\frac{\partial \ln \gamma_\pm}{\partial P}\right)_{m,\,T} \qquad \ldots.(2.33)$$

where $\bar{V}_B^0$ is the partial molal volume in the standard state, which once more can be shown to equal that at infinite dilution $\bar{V}_B^\infty$, because we have imposed on the standard state the requirement that it results in $\gamma_\pm \to 1$ as $m \to 0$ for all pressures as well as all temperatures.

The result (2.33) is not used for the determination of $\bar{V}_B$, which can be much more simply obtained from density measurements, but (2.30) and (2.31) are often used in the estimation of heat contents and heat capacities from electromotive force measurements, or conversely, in converting activity coefficients from one temperature to another.

The corresponding relations for the solvent are usually expressed in terms of its activity a_A:

$$\bar{H}_A = \bar{H}_A^0 - \boldsymbol{R}T^2 \left(\frac{\partial \ln a_A}{\partial T}\right)_{m,\,P} \qquad \ldots.(2.34)$$

$$\bar{C}_{(P)A} = \bar{C}_{(P)A}^0 - \boldsymbol{R} \left[T^2 \frac{\partial^2 \ln a_A}{\partial T^2} + 2T \frac{\partial \ln a_A}{\partial T}\right]_{m,\,P} \qquad \ldots.(2.35)$$

$$\bar{V}_A = \bar{V}_A^0 + \boldsymbol{R}T \left(\frac{\partial \ln a_A}{\partial P}\right)_{m,\,T} \qquad \ldots.(2.36^*)$$

and here, in virtue of the choice of the pure solvent as the standard state, the meanings of $\bar{H}_A^0$, $\bar{C}_{(P)A}^0$ and $\bar{V}_A^0$ are obvious.

Since $\bar{G}_B = \bar{G}_B^0 + \boldsymbol{R}T \ln (Qm)^\nu + \nu \boldsymbol{R}T \ln \gamma_\pm$,

$$\left(\frac{\partial \bar{G}_B}{\partial T}\right)_{m,\,P} = \left(\frac{\partial \bar{G}_B^0}{\partial T}\right)_P + \nu \boldsymbol{R} \ln (Qm\gamma_\pm) + \nu \boldsymbol{R}T \left(\frac{\partial \ln \gamma_\pm}{\partial T}\right)_{m,\,P}$$

the partial molal entropy is given by:

$$\bar{S}_B = \bar{S}_B^0 - \nu \boldsymbol{R} \ln (Qm\gamma_\pm) - \nu \boldsymbol{R}T \left(\frac{\partial \ln \gamma_\pm}{\partial T}\right)_{m,\,P} \qquad \ldots.(2.37)$$

* The formula: $\bar{V}_A = W_A \Big/ \left(d - c\frac{\partial d}{\partial c}\right)$ is useful for calculating the partial molal volume of the solvent from density data.

This is important in showing that the partial molal entropy of the solute approaches infinity as the concentration decreases, so that $\bar{S}_B^{\infty}$ is not equal to $\bar{S}_B^0$. Thus neither the chemical potential nor the partial molal entropy at infinite dilution is equal to that in the standard state; the former becomes negatively infinite like the chemical potential of a perfect gas, ($\bar{G} = \bar{G}^0 + \boldsymbol{R}T \ln P$) whilst the entropy becomes positively infinite. But from each can be split a term which has equal values at infinite dilution and in the standard state: thus the term $\boldsymbol{R}T \ln \gamma_{\pm}$ is in a way the non-ideal contribution to the chemical potential and is zero both at infinite dilution and in the hypothetical standard state of unit molality.

A question that may raise some difficulty concerns the activity coefficient of an electrolyte which can be treated as completely dissociated by one school of thought and as only partially dissociated from another point of view. How are the activity coefficients calculated on these different assumptions related to one another? Consider a system consisting of a kilogram of solvent and m moles of solute completely dissociated into $m_1 = \nu_1 m$ cations and $m_2 = \nu_2 m$ anions. The total free energy of the system is:

$$G = \frac{1000}{W_A}\bar{G}_A + \nu_1 m\bar{G}_1 + \nu_2 m\bar{G}_2$$

We might, however, regard the solute as forming an 'aggregate', (intermediate ion, neutral molecule or complex ion) from its simple ions. If the solute has the formula, $M_{\nu_1}X_{\nu_2}$, let the formula of the aggregate be $M_{n_1}X_{n_2}$; furthermore, let a fraction $(1-\alpha)$ of the cations form the aggregate. Then we have the following concentrations:

cations: $\quad m_1' = \alpha\nu_1 m$

anions: $\quad m_2' = \left[\nu_2 - (1-\alpha)\dfrac{n_2\nu_1}{n_1}\right] m$

aggregate: $\quad m_{12}' = (1-\alpha)\dfrac{\nu_1}{n_1}m$

the primes drawing attention to the fact that some quantities have different meanings according as we adopt the idea of complete as against partial dissociation.

Then:

$$G = \frac{1000}{W_A}\bar{G}_A + \alpha\nu_1 m\bar{G}_1' + \left[\nu_2 - (1-\alpha)\frac{n_2\nu_1}{n_1}\right] m\bar{G}_2' + (1-\alpha)\frac{\nu_1}{n_1}m\bar{G}_{12}'$$

The total free energy is independent of any views of the nature of the dissociation and so is the chemical potential of the solvent, so that:

$$\nu_1\bar{G}_1 + \nu_2\bar{G}_2 = \alpha\nu_1\bar{G}_1' + \left[\nu_2 - (1-\alpha)\frac{n_2\nu_1}{n_1}\right]\bar{G}_2' + (1-\alpha)\frac{\nu_1}{n_1}\bar{G}_{12}' \quad \ldots.(2.38)$$

But since the ions are in equilibrium with the aggregate:

$$\bar{G}_{12}' = n_1\bar{G}_1' + n_2\bar{G}_2' \quad \ldots.(2.39)$$

and it follows that:

$$\nu_1\bar{G}_1 + \nu_2\bar{G}_2 = \nu_1\bar{G}_1' + \nu_2\bar{G}_2'$$

Now expressing the chemical potentials in terms of the molal activity coefficients:

$$\nu_1\bar{G}_1^0 + \nu_2\bar{G}_2^0 + \boldsymbol{R}T(\nu_1 \ln \gamma_1 m_1 + \nu_2 \ln \gamma_2 m_2)$$
$$= \nu_1\bar{G}_1'^0 + \nu_2\bar{G}_2'^0 + \boldsymbol{R}T(\nu_1 \ln \gamma_1' m_1' + \nu_2 \ln \gamma_2' m_2')$$

But $\bar{G}_1^0$ and $\bar{G}_1'^0$ are identical since both refer to the same hypothetical molal solution of the ions, so that:

$$(\gamma_1 m_1)^{\nu_1}(\gamma_2 m_2)^{\nu_2} = (\gamma_1' m_1')^{\nu_1}(\gamma_2' m_2')^{\nu_2}$$

or $$(\gamma_1\nu_1 m)^{\nu_1}(\gamma_2\nu_2 m)^{\nu_2} = (\gamma_1'\alpha\nu_1 m)^{\nu_1}\left[\gamma_2'\left\{\nu_2 - (1-\alpha)\frac{n_2\nu_1}{n_1}\right\}m\right]^{\nu_2}$$

or $$\gamma_\pm^\nu = \alpha^{\nu_1}\left[1 - (1-\alpha)\frac{n_2\nu_1}{n_1\nu_2}\right]^{\nu_2}\gamma_\pm'^\nu$$

In this general case the problem would require very careful handling, but it becomes simpler if the aggregate is electrically neutral, *i.e.*, if $n_2\nu_1 = n_1\nu_2$. (An even simpler case occurs when $n_1 = \nu_1$, $n_2 = \nu_2$, *i.e.*, the aggregate $M_{n_1}X_{n_2}$ is identical with the molecule $M_{\nu_1}X_{\nu_2}$.) Under these conditions:

$$\gamma_\pm = \alpha\gamma_\pm' \quad \ldots.(2.40)$$

This relation will find considerable use in later sections; we may have measured the stoichiometric activity coefficient $\gamma_\pm$ of a binary electrolyte, the calculation from the experimental data having been made on the assumption that the electrolyte is fully dissociated. If we have reason to believe that the electrolyte is actually dissociated only to the extent α, then the mean ionic activity coefficient, $\gamma_\pm'$, is a better measure and it is related to the determined quantity, $\gamma_\pm$, by the simple relation (2.40). Furthermore, if $n_1 = \nu_1$ and $n_2 = \nu_2$, equations (2.38) and (2.39) give:

$$\nu_1\bar{G}_1 + \nu_2\bar{G}_2 = \nu_1\bar{G}_1' + \nu_2\bar{G}_2' = \bar{G}_{12}'$$

Introducing the molal activity coefficients and the definition of K_m, the dissociation constant on the molal scale:

$$\bar{G}'^0_{12} - n_1\bar{G}'^0_1 - n_2\bar{G}'^0_2 = \boldsymbol{R}T \ln K_m$$

we can derive the equation:

$$(\gamma_\pm m_\pm)^\nu = (\gamma'_\pm m'_\pm)^\nu = K_m \gamma'_{12} m'_{12}$$

Similar relations are valid for other concentration scales:

$$(f_\pm N_\pm)^\nu = (f'_\pm N'_\pm)^\nu = K_N f'_{12} N'$$

$$(y_\pm c_\pm)^\nu = (y'_\pm c'_\pm)^\nu = K_c y'_{12} c'_{12}$$

but it should be noted that the three dissociation constants are not identical; instead, we have:

$$K_N = K_m(0{\cdot}001\, W_A)^{\nu-1} \qquad \ldots . (2.41)$$

and

$$K_c = K_m d_0^{\nu-1} \qquad \ldots . (2.42)$$

The distinction is important when discussing the dissociation constant of a weak acid: for example, the dissociation constant of acetic acid at 25° is $1{\cdot}749 \times 10^{-5}$ on the molarity scale, but $1{\cdot}754 \times 10^{-5}$ on the molality scale. In solvents other than water, K_c may differ substantially from K_m.

THE RELATION BETWEEN THE FREE ENERGY CHANGE AND THE POTENTIAL OF A GALVANIC CELL

Consider a galvanic cell operating under reversible conditions at constant temperature and pressure. An example is the cell:

$$H_2 \text{ (1 atm.)} \mid HCl \mid AgCl, Ag,$$

by which we mean a cell with hydrochloric acid at a given concentration as electrolyte and two electrodes, one of which is a hydrogen electrode (*i.e.*, hydrogen bubbling round platinized platinum) and the other is a layer of silver chloride deposited on silver (an effective substitute for the less manageable chlorine electrode). The natural or spontaneous cell reaction is:

$$\tfrac{1}{2}H_2 + AgCl \rightarrow Ag + HCl$$

and if this cell is used as a battery to generate current, hydrogen gas dissolves as hydrogen ions at the left-hand electrode and silver chloride decomposes to give chloride ions at the right-hand electrode. Hydrogen ions travel from left to right through the cell and chloride ions in the opposite direction, 'positive current' passes through the external circuit from the right-hand to the left-hand electrode and the right-hand electrode has a higher potential than the left-hand electrode. With aqueous 0·1 M hydrochloric acid as electrolyte, the potential of this cell at 25° is 0·3524 V.

For the reversible operation of such a cell, the electromotive force of the cell must be balanced by an opposing electromotive force from an external source such as a potentiometer; now consider an infinitesimal departure from balance, such that the spontaneous cell reaction proceeds to an infinitesimal extent and an infinitesimal quantity, δq, of electricity (measured in coulombs) passes. Let ΔG be the *increase* in free energy of the cell constituents on the passage of $\boldsymbol{n}$ faradays of electricity. In the example cited above,

$$\Delta G = \bar{G}^0_{\mathrm{Ag}} + \bar{G}_{\mathrm{HCl}} - \tfrac{1}{2}\bar{G}^0_{\mathrm{H_2}} - \bar{G}^0_{\mathrm{AgCl}}$$

if $\boldsymbol{n} = 1$. The reaction being a spontaneous one, ΔG must be numerically negative, *i.e.*, there must be a decrease in free energy. It is the quantity $(-\Delta G)\dfrac{\delta q}{\boldsymbol{n}F}$ which appears as electrical work, $E\delta q$, when an infinitesimal quantity of electricity passes through the circuit under reversible conditions. Hence:

$$-\Delta G = \boldsymbol{n}EF,$$

which is the fundamental equation for the potential of a reversible cell.

It is desirable, whenever possible, to write the cell in such a way that, when the spontaneous cell reaction proceeds, positive current goes from left to right through the cell and in the opposite direction in the external circuit. The numerical value of the potential will then be positive. In some cases it may not be known in which direction the spontaneous cell reaction does go: we can write the cell in either one of two ways and there may be little to decide in favour of either, but *whichever way the cell is written, E, the potential of the cell, must be used to mean the excess potential of the right-hand electrode over that of the left-hand electrode.* If experiment reveals that the right-hand electrode is at a lower potential than the left-hand one, then we assign a numerically negative value to E. For example, we would say $E = -0{\cdot}3524$ V for the cell:

$$\mathrm{Ag, AgCl} \mid 0{\cdot}1\ \mathrm{M\ HCl} \mid \mathrm{H_2}\ (\text{at } 25^\circ).$$

But a great deal of trouble can be avoided if E is always taken as the excess potential of the right-hand with respect to the left-hand electrode, recognizing that an excess potential may be numerically a negative quantity

Standard Cell Potentials

The standard cell potential is the potential of a cell in which all substances involved in the cell reaction are in the standard states. The standard cell potential is a function of the temperature and pressure, and it depends on the nature of the solvent and on the scale we use to define the standard state, *i.e.*, the molal, molar or mole fraction scale. The standard potential of the H_2 | HCl | AgCl, Ag cell on the molal scale is 0·2224 V at 25° and one atmosphere pressure with water as the solvent.

It is conventional to regard the standard potential of the hydrogen electrode as zero. The standard potential of the silver–silver chloride electrode is then 0·2224 V. It should be written Cl^- | AgCl, Ag, $E^\circ = 0{\cdot}2224$ V.

UNITS AND DIMENSIONS FOR CONDUCTANCE

The accepted unit of electrical resistance is the absolute ohm, which is 10^9 electromagnetic units of resistance. (The volt is 10^8 e.m.u. of potential and the ampere 10^{-1} e.m.u. of current.) In the electromagnetic system the dimensions of resistance are $[LT^{-1}]$. The resistance R of a uniform conductor is directly proportional to its length, l, and inversely to its cross-sectional area A. This may be expressed by:

$$R = \rho l/A$$

where the constant of proportionality ρ is called the specific resistance of the material for the temperature in question. Clearly ρ has the dimensions of resistance multiplied by length, *i.e.*, its units are ohm-centimetres. Its basic dimensions in the electromagnetic system are $[L^2T^{-1}]$. The reciprocal of the specific resistance is the specific conductance, denoted by K_{sp}:

$$K_{sp} = \frac{1}{\rho} = \frac{l}{AR}$$

and has dimensions $[L^{-2}T]$ in the electromagnetic system.

In electrolyte solutions, another variable, the concentration, has a dominating effect on the conductivity. It is convenient therefore to divide the specific conductivity by the concentration thus arriving at a quantity:

$$\Lambda = K_{sp}/c$$

If c is a concentration in moles per unit volume (usually per c.c.) Λ is a molar conductivity but if c is expressed in equivalents per unit volume (and again the c.c. is usually adopted as the unit of volume) a more useful quantity, the equivalent conductivity, ensues.

Λ will be taken to mean equivalent conductivity unless it is stated be a molar conductivity. The resulting dimensions for Λ in the electromagnetic system are [equiv.$^{-1}$ LT], the conventional units being cm^2 Ω^{-1} equiv.$^{-1}$.

Since the current results from the motion in opposite directions of oppositely charged ions, the equivalent conductivity can be considered as the sum of two ionic conductivities:

$$\Lambda = \lambda_1 + \lambda_2 \qquad \ldots.(2.43)$$

At infinite dilution:

$$\Lambda^0 = \lambda_1^0 + \lambda_2^0 \qquad \ldots.(2.44)$$

and Kohlrausch's law of the Independent Migration of Ions states that when each ion is moving in a medium where the ions are so far apart that they are without influence on one another, then λ_1^0 depends only on the nature of the cation and properties of the medium such as temperature and viscosity. It does not depend on the value of λ_2^0. Similarly λ_2^0 does not depend on the nature of the cation.

THE RELATION BETWEEN THE EQUIVALENT CONDUCTIVITY AND THE ABSOLUTE MOBILITY OF AN ION

The motion of an isolated body is governed by Newton's law that force = mass × acceleration, but in dealing with the motion of ions it is not usually necessary to consider the acceleration unless electrical fields of very high intensity or frequency are involved. Under normal conditions, the ions are almost instantaneously accelerated to the point where their motion is limited by the viscous drag of the solvent, and all the energy supplied by the electric field is dissipated by the viscous forces. The ions thus move with a constant limiting or terminal velocity, which for all reasonably small fields is directly proportional to the applied field. This is of course the reason for the validity of Ohm's law for electrolytes subjected to ordinary fields, and for the fact that the conductivities of ions have no simple relation to their masses. There is, for example, little difference between the ionic conductivities of chloride and iodide ions, though the latter has nearly four times the mass of the former.

In discussing motion against viscous forces it is convenient to define the mobility, u, of a body as the limiting velocity attained under unit force, *i.e.*,

$$u = v/F$$

The absolute mobility in the c.g.s. system is thus the velocity in cm per sec attained under a force of 1 dyn. It is, however, a common practice when dealing with ions to take as the unit of force a unit potential gradient acting on the ionic charge. Here we shall use u for the absolute mobility and u' for 'electrical mobility' defined as the velocity attained by the ion under unit potential gradient. Since 1 V (abs.) = 1/299·8 e.s.u. of potential and the protonic charge $e = 4{\cdot}802 \times 10^{-10}$ e.s.u. of charge, a field of 1 V/cm exerts on an ion of valency $|z|$ a force of $1{\cdot}602 \times 10^{-12}\,|z|$ dyn.

The equivalent ionic conductivity, λ, is simply related to the mobility. From the definition of specific conductivity, it follows that K_{sp} is the current flowing in a conductor of unit cross-section under unit potential gradient. The total ionic charge in unit volume is $\boldsymbol{F}c$ if c is measured in equivalents per unit volume, and this charge moving with velocity u' constitutes the current K_{sp}:

$$K_{sp} = \boldsymbol{F}cu'$$

or

$$\lambda = K_{sp}/c = \boldsymbol{F}u' \qquad \ldots.(2.45)$$

The absolute mobility is therefore:

$$u = u'/(|z|\,\boldsymbol{e}) = \boldsymbol{N}u'/(|z|\,\boldsymbol{F}) = \boldsymbol{N}\lambda/(|z|\,\boldsymbol{F}^2) \qquad \ldots.(2.46)$$

Because of these relations, one frequently finds the ionic equivalent conductivity λ referred to as the ionic mobility. In applying equation (2.46) some care is needed with the units; the usual ones give:

$$u/(\text{cm sec}^{-1}\,\text{dyn}^{-1}) = 6{\cdot}469 \times 10^6\,\lambda/(|z|\,\text{cm}^2\,\Omega^{-1}\,\text{equiv.}^{-1})$$
$$= 6{\cdot}466 \times 10^6\,\lambda/(|z|\,\text{cm}^2\,\text{int.}\,\Omega^{-1}\,\text{equiv.}^{-1})$$

THE RELATION BETWEEN THE SIZE AND MOBILITY OF IONS

For a particle of macroscopic dimensions moving in an ideal hydrodynamic continuum, it is possible to calculate the frictional resistance in terms of the dimensions of the particle and the viscosity (η) of the medium. For a spherical particle, the result was obtained by G. G. STOKES[2] as:

$$v = F/(6\pi\eta r) \qquad \ldots.(2.47)$$

where r is the radius of the sphere. If an ion can be considered to satisfy the conditions for Stokes' law motion, its radius is given by:

$$r = 1/(6\pi\eta u) \qquad \ldots.(2.48)$$

and since u is given in terms of the limiting equivalent conductivity by equation (2.46), we have:

$$r = |z|\,\boldsymbol{F}^2/(6\pi\boldsymbol{N}\eta^0\lambda^0) \qquad \ldots.(2.49)$$

If for convenience we express r in Å and η and λ in their usual units, this becomes:

$$r/\text{Å} = \frac{0{\cdot}820\,|z|}{\lambda/(\text{cm}^2\,\Omega^{-1}\,\text{equiv.}^{-1})\,.\,\eta/(\text{poise})} \qquad \ldots.(2.49\text{a})$$

For small ions the conditions for the validity of Stokes' law are not fulfilled; nevertheless equation (2.49) provides a useful starting point for discussing the dimensions of ions, and we shall refer to radii so calculated as 'Stokes' law radii'. For uncharged particles, the mobility can be calculated from the diffusion coefficient, D (see p. 46) as:

$$u = D/(\boldsymbol{k}T) \qquad \ldots.(2.50)$$

where $\boldsymbol{k}$ is Boltzmann's constant; this leads to the result:

$$r = \boldsymbol{k}T/(6\pi\eta D) \qquad \ldots.(2.51)$$

which is known as the Einstein-Stokes formula. Its validity is subject to the same kind of restrictions as equation (2.49).

TRANSPORT NUMBERS

The passage of electric current through an electrolyte solution is effected by the motion of ions of opposite charge moving in opposite directions under the applied potential.

Consider a tube of electrolyte solution 1 cm² in cross-section along which a potential gradient of 1 V/cm is set up, c being the concentration in equiv./l. All the positive ions at a distance u_1' on one side of an imaginary plane perpendicular to the gradient will cross that plane in one second. This number will be $\boldsymbol{N}cu_1'/(1000z_1)$ and the current will be $\boldsymbol{N}ecu_1'/1000$. The current set up by the motion of negative ions in the opposite direction will be $\boldsymbol{N}(-\,\boldsymbol{e})cu_2'/1000$ and the total current $\boldsymbol{N}ec(u_1' + u_2')/1000$. The fraction of the current carried by the positive ions is called the transport (or transference) number of the positive ion:

$$\left.\begin{aligned} t_1 &= u_1'/(u_1' + u_2') = \lambda_1/(\lambda_1 + \lambda_2) \\ \text{Similarly} \qquad t_2 &= u_2'/(u_1' + u_2') = \lambda_2/(\lambda_1 + \lambda_2) \end{aligned}\right\} \qquad \ldots.(2.52)$$

and

$$t_1 + t_2 = 1$$

We do not define the transport number in terms of absolute mobilities, u_1 and u_2, because in practice the oppositely charged ions are subjected not to the same force of one dyne but to the same potential gradient in terms of volts per centimetre, a gradient exerting on the ions a force which depends on their valency.

Considerable care has to be exercised in dealing with transport numbers of solutions of weak electrolytes or of electrolytes which form autocomplexes. These cases have been discussed by Spiro[3] and further reference will be made to them in Chapter 7.

DIFFUSION IN ELECTROLYTE SOLUTIONS

One of the most fundamental of irreversible processes is that of diffusion, by which a difference of concentration is reduced by the spontaneous flow of matter. In a solution containing a single solute, the solute moves from the region of higher to that of lower concentration, while the solvent moves in the opposite sense. From the point of view of molecular kinetics, no individual solute particle shows a preference for motion in any particular direction, but a definite fraction of the particles in an elementary unit volume may be considered to be moving in, say, the positive x-direction. In an adjacent volume-element, the same fraction may be considered as moving in the negative x-direction; now if the concentration in the first volume-element is greater than that in the second, this means that more particles will be leaving the first element for the second than will be re-entering from the second to the first, so there will be a resultant flow of solute in the direction of lower concentration. Further, one would naturally expect on the basis of this picture that the rate of flow would be at least approximately proportional to the concentration-difference existing between the two volume-elements.

While diffusion in everyday practical applications is often a two- or three-dimensional process, nothing essential to its understanding is lost by confining attention to the one-dimensional case, especially since most of the methods by which it is studied and measured involve a deliberate restriction to one-dimensional flow. In this case the following concepts and definitions are applicable:

The flux of matter, denoted by J, is defined as the amount (in moles, grams, *etc.*) of material crossing unit area of a plane perpendicular to the direction of flow in unit time.

The concentration gradient $\frac{\partial c}{\partial x}$ is the rate of increase of concentration with distance measured in the direction of the flow. It is usual to take the direction of flow as the positive direction of the distance x, to express the concentration c in the same units of moles, grams, *etc.*, as are used in defining the flux, and to take as the volume unit for the concentration c the cube of the unit of distance x. Thus if J is expressed in moles cm^{-2} sec^{-1}, and x in cm, c will be expressed in moles cm^{-3}.

The diffusion coefficient D is now defined by the equation:

$$J = -D\frac{\partial c}{\partial x} \qquad \ldots.(2.53)$$

the partial differential being necessary because c is, in general, dependent on time as well as on distance. The negative sign in equation (2.53) is introduced in order to make D a positive quantity, since $\frac{\partial c}{\partial x}$ is negative in virtue of our choice of sign for x, which increases as the concentration decreases. It will be seen that the diffusion coefficient D has dimensions $[L^2T^{-1}]$ and is independent

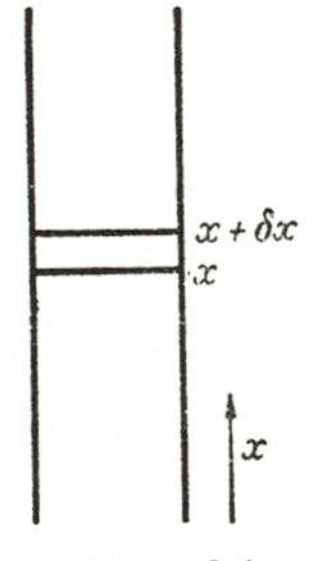

Figure 2.1

of the mass-units used provided that the same units are used in defining both J and c. With the c.g.s. units mentioned above, D will be in units of $cm^2\,sec^{-1}$. Although in practice experimental conditions are often chosen so that D is nearly constant, it is not defined as a constant, and the common practice of calling D the 'diffusion constant' is to be deplored, especially as it is frequently the variation of D with concentration in which we are interested.

Equation (2.53) is of importance in the study of diffusion by steady-state methods in which the concentration-gradient $\frac{\partial c}{\partial x}$ does not change with time. In many of the methods currently in use, however, the variation of c with both time and distance is of interest; for these cases (2.53) can be converted into a second-order partial differential equation connecting c, x and the time, t, as follows: consider (*Figure 2.1*) a tube of uniform unit cross-section intersected by two planes of unit area, normal to the x-axis, situated at x and $x + \delta x$ respectively. The amount of matter entering the volume-element between these planes through the plane at x in a time-interval δt is:

$$J \delta t = - \left(D \frac{\partial c}{\partial x} \right) \delta t$$

while the amount leaving through the plane at $(x + \delta x)$ is:

$$J' \delta t = - \left(D \frac{\partial c}{\partial x} \right)' \delta t$$

The difference between $\left(D \frac{\partial c}{\partial x} \right)'$ at the plane at $(x + \delta x)$ and $\left(D \frac{\partial c}{\partial x} \right)$ at the plane x may be expressed as:

$$\delta \left(D \frac{\partial c}{\partial x} \right) = \frac{\partial}{\partial x} \left(D \frac{\partial c}{\partial x} \right) \delta x$$

Hence the net amount of material accumulating in the volume-element considered in time δt is:

$$(J - J') \delta t = \frac{\partial}{\partial x} \left(D \frac{\partial c}{\partial x} \right) \delta x \, \delta t$$

This accumulation, since it occurs in an element of volume δx, gives rise to a concentration increase δc given by:

$$\delta c = \frac{\partial}{\partial x} \left(D \frac{\partial c}{\partial x} \right) \delta t$$

On proceeding to the limit, one obtains:

$$\frac{\partial c}{\partial t} = \frac{\partial}{\partial x} \left(D \frac{\partial c}{\partial x} \right) \qquad \ldots . (2.54)$$

which can be regarded as an alternative definition of the diffusion coefficient D. For diffusion in three dimensions, equations (2.53) and (2.54) take the forms:

$$J = - D \operatorname{grad} c$$

and

$$\frac{\partial c}{\partial t} = \operatorname{div} (D \operatorname{grad} c)$$

These two equations (2.53) and (2.54) are often loosely referred to as Fick's first and second laws of diffusion[4]. We have preferred here to develop them purely as equations defining the diffusion coefficient D; Fick's laws of diffusion may then be summed up by the statement that D, as defined by these equations, is a constant for a given system and temperature. This constancy is, however, only approximate, and the main importance of diffusion studies for electrolyte theory lies in the variation of the quantity D with concentration.

We have not yet stated how the distance x is to be determined. The obvious way is to measure from some arbitrary plane fixed with respect to the apparatus containing the diffusing system, and, indeed, when one is dealing with liquids it is difficult to see how any other experimental means of fixing the reference-plane could be found. In some cases, however, as when a liquid diffuses into a solid which swells as a result of the diffusion, it may be convenient to measure from the moving surface of the solid. Likewise, in discussing the theoretical aspects of liquid diffusion, it may be desirable to refer the measurement of distance to a plane so chosen that the amount of one component on one side of it remains constant; such a plane will, in general, move with respect to the apparatus. As long as equations (2.53) and (2.54) are regarded only as definitions of D, such procedures are quite legitimate; it must, however, be remembered that the value of D for a given system will depend on the method used to fix the reference-plane. HARTLEY and CRANK[(5)] have given a detailed account of the relations between diffusion coefficients defined with respect to various reference-planes.

REFERENCES

[1] GUGGENHEIM, E. A., "Thermodynamics: An advanced treatment for chemists and physicists", North-Holland Publishing Co., Amsterdam, 1949
[2] STOKES, G. G., *Trans. Camb. phil. Soc.*, VIII (1845) 287
[3] SPIRO, M., *J. chem. Educ.*, 33 (1956) 464
[4] FICK, A., *Pogg. Ann.*, 94 (1855) 59
[5] HARTLEY, G. S. and CRANK, J., *Trans. Faraday Soc.*, 45 (1949) 801

3

THE STATE OF THE SOLUTE IN ELECTROLYTE SOLUTIONS

CLASSIFICATION OF ELECTROLYTES

In our attempts to understand the complicated problems set by the study of electrolyte solutions, there is a key question which we must try to answer before embarking on detailed mathematical treatments: what are the actual kinetic entities in the solution? We have seen in the first chapter that in pure water, in spite of the existence of some considerable degree of short-range order, the only kinetically identifiable form of solvent particle is the single water molecule in equilibrium with minute amounts of hydrogen and hydroxyl ions. The introduction of a dissolved electrolyte complicates the position considerably. We may now have to cope with many kinds of solute entities: ions—solvated or unsolvated, electrostatically associated groups of ions, covalently bound molecules and complex ions.

In the face of this complexity, we are obliged to classify electrolyte solutions into several groups. The familiar division into 'strong' and 'weak' electrolytes, though convenient for elementary purposes, is not an entirely suitable basis for theoretical discussion. Instead we shall recognize two main classes, 'associated' and 'non-associated' electrolytes, which we shall define as follows.

Non-associated Electrolytes

A solute of this kind is believed to exist only in the form of the simple cation and anion, possibly solvated; there is no evidence for the presence of covalent molecules of the solute, or of any lasting association between oppositely charged ions. This class, although small in number, is of great importance in providing information with which we can make straightforward tests of the theory of electrolyte solutions. The archetype of this group is aqueous sodium chloride: with water as solvent the class comprises the alkali halides, the alkaline-earth halides and perchlorates, and some transition-metal halides and perchlorates. The chief criterion for placing an electrolyte in this class is the absence of valid evidence for any form of association. Since the validity of such evidence can be a matter of personal opinion (and this is particularly true of the evidence for ion association) there can be no general agreement; but our own inclination is to include also lithium nitrate, magnesium nitrate and possibly the rare-earth-metal halides. It is convenient to add to this class some electrolytes which show association only at extreme concentrations, in particular the halogen acids and perchloric acid. There is no well authenticated evidence that any electrolyte of this class can exist in non-aqueous solvents, with the possible exceptions of liquid hydrogen cyanide and some amides.

It may be convenient sometimes to refer to these electrolytes by the briefer title of 'strong electrolytes' but their essential characteristic is the absence of evidence of any lasting union between the ions.

Associated Electrolytes

The much more numerous associated electrolytes can conveniently be subdivided as follows:

1. We shall use the term 'weak electrolytes' to describe cases in which the solute can exist as undissociated (covalent) molecules as well as ions. All acids belong to this class; even the 'strong' halogen acids and perchloric acid are, strictly speaking, weak in terms of this definition, since there is no doubt that at high enough concentrations the molecular form does exist. In other solvents 'strong' acids are incompletely dissociated even at moderate concentrations: thus values of $pK_c = 1{\cdot}229$ have been found for hydrochloric acid in methanol[1] and 2·085 in ethanol[2] at 25°.

Bases are usually weak electrolytes, except for the alkali metal and the quaternary ammonium hydroxides. The class, however, contains very few aqueous salts, mercuric chloride being the chief example.

2. We shall use the term 'ion-pairing' in discussing a class of electrolytes in which association occurs as a result of purely electrostatic attraction between oppositely charged ions: this concept was introduced by Bjerrum shortly after the appearance of the Debye-Hückel theory, and has proved extremely useful in interpreting the behaviour of a large class of electrolyte solutions of which the bivalent metal sulphates in aqueous solution form an outstanding example. Almost all salts in non-aqueous solvents show evidence of this effect.

It must be emphasized that we adopt this classification of electrolytes on grounds of convenience rather than of logical rigour; there will be cases where a particular electrolyte cannot be clearly assigned to one of the above classes. Zinc iodide, for example, would be treated as a non-associated electrolyte only if the experiments were confined to a range of concentration below about 0·3 M. In more concentrated solution, however, there is good reason to believe that it forms ZnI_4^{--} ions, which may well be subject to ion-pair formation with Zn^{++} ions.

CHARACTERISTICS OF WEAK ELECTROLYTES

Since nearly all weak electrolytes are acids or bases, it is generally possible to recognize their weakness by conductivity or pH measurements or by potentiometric or conductimetric titration. Only in cases where the dissociation constant is of the order of 0·1 or greater is there any difficulty about determining whether the observed behaviour is due to the substance being a weak electrolyte or to interionic attraction effects. In these cases it may be necessary to use other methods for the detection of covalent molecules. If the solute exerts a detectable vapour pressure above the solution, we have strong reason to believe that covalent molecules of solute are present in the solution. The vapour form of the solute is certainly a true molecule and, however non-ideal the solution, at least some of the solute must also exist as molecules; ammonia solutions provide an obvious example. This criterion will, in some cases, force us to classify an electrolyte as strong in dilute solutions, but weak in concentrated solutions: such restrictions on the concentration-range are essential, for example, in discussing the cases of hydrochloric and sulphuric acids. The fact that hydrochloric acid solutions exert indetectable partial pressures of hydrogen chloride below about 3 N was indeed one of the classical anomalies that led to the development of the modern concept of strong electrolytes. At 10 N, however, the partial pressure of hydrogen chloride is of the order of 20 mm, in contrast with a value of about 7·6 atm. which would be estimated from the vapour pressure of pure liquid hydrogen chloride at 20° (41 atm.) on the basis of Raoult's law if the hydrochloric acid were completely undissociated and an ideal solute. While this figure is very rough, it indicates that an appreciable amount (of the order of 0·3 per cent) of the hydrochloric acid in a 10 N solution (the ordinary 'concentrated' acid) is in the form of covalent molecules; the acid at this concentration must therefore be regarded as a weak electrolyte, and an approximate value of its ionization constant can be calculated[3], of the order of 10^7.

In using such terms as 'indetectable vapour pressure' we are, in effect, admitting that no sharp boundary can be drawn between a strong and a weak electrolyte in these cases, since improvement in technique may result in the indetectable quantity of today becoming measurable within 0·1 per cent tomorrow. Certainly for solutes such as the halide acids, which can exist as liquids in their own right, it is only on grounds of experimental convenience that we can claim that they are in any circumstances true strong electrolytes. In practice we regard them as fully ionized when we believe that less than about 0·1 per cent of the solute is in molecular form. Had we any method of detecting covalent molecules in the presence of a large excess of ions as readily as we detect a few ions among a large excess of molecules (by the electrical conductance) we should no doubt hold different views on where the line should be drawn.

The vapour pressure criterion for the presence of undissociated solute molecules, valuable though it is, is applicable only where the solute is appreciably volatile in the pure state at the temperature of interest—generally room temperature; it can thus tell us nothing, for example, about sulphuric acid solutions. A criterion of much wider usefulness is the Raman spectrum. REDLICH[(4)] has emphasized the importance of this method, and remarks that 'there can be no doubt that the question of dissociation can be solved in any case in which complete knowledge of the vibration spectrum is available'. This method is based on the fact that the undissociated molecule necessarily has different symmetry properties from its ions. For simple molecules the general pattern, and in particular the number of lines, is predictable, though the frequencies and band widths may be modified by environmental factors such as concentration. Whenever the Raman spectrum of the molecule is detectable, we have clear evidence that the substance is not a strong electrolyte, but as Redlich points out, the absence of Raman lines cannot always be taken as final evidence that the substance is a strong electrolyte.

The scheme of classification proposed above will permit us to develop the fundamental theory and to test it for non-associated electrolytes; we then proceed to the discussion of weak electrolytes and ion-pair electrolytes, using the theoretical formulae for dealing with the free ions in these cases, and handling the associated part of the solute by suitable specific devices such as the introduction of finite dissociation constants. Some special cases such as the 'strong' acids and the transition metal halides will be discussed separately.

ION-SOLVENT INTERACTIONS

The reason for the ready solubility of so many electrolytes in water is the high dielectric constant of this solvent, which in turn is due to the polar nature of the water molecule and to the fact that its dimensions favour a tetrahedrally coordinated structure.

In a 'uniform' field a dipole experiences only a turning moment, but the field near an ion is highly divergent and therefore non-uniform, and a dipole near an ion is subject, in general, to both orienting and attractive forces. The mutual potential energy of a point charge $\boldsymbol{e}$ and a dipole of moment μ, *in vacuo* is:

$$\frac{\mu \boldsymbol{e} \cos \theta}{r^2}$$

where θ is the angle between the axis of the dipole and the radius vector passing through the ion (*Figure 3.1*). For an ion and a water

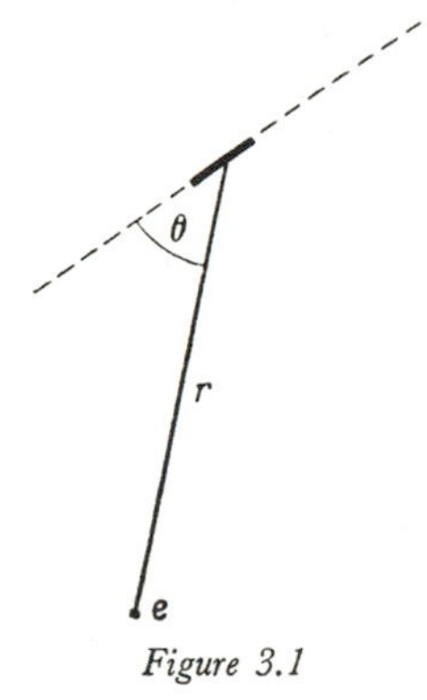

Figure 3.1

molecule, isolated *in vacuo*, the interaction energy could be calculated from this formula with confidence: putting $\mu = 1{\cdot}8 \times 10^{-18}$ e.s.u. and $\boldsymbol{e} = 4{\cdot}8 \times 10^{-10}$ e.s.u. and with r in Ångströms, we obtain $124 \cos \theta/r^2$ kcal/mole. In the completely oriented position, this energy would be greater than $\boldsymbol{RT}$ ($\sim 0{\cdot}6$ kcal/mole) up to about 14 Å. For an ion in liquid water, however, no such simple calculation is possible. For water molecules remote from the ion, a fair approximation could no doubt be made by inserting the dielectric constant ($\varepsilon \approx 80$) for the intervening water in the denominator of the energy expression: this immediately reduces the energy to the order of $1{\cdot}5 \cos \theta/r^2$ kcal/mole, a value which is much smaller than $\boldsymbol{RT}$ (even for favourably oriented molecules) for all r which could conceivably be regarded as 'remote'. For the first layer of molecules round an ion, however, the bulk dielectric constant of the medium is simply not relevant. It is not clear whether we should calculate the energy in this region with $\varepsilon = 1$, as for the charges *in vacuo*, or with $\varepsilon = 4$ or 5, the value estimated (see p. 20) from the atom and electron polarization of water; nor is it possible to calculate with any certainty the effect of other molecules in the first layer, unless simplifying assumptions are made about their number and orientation. However, it seems likely that all the water molecules in the first layer round all monatomic ions should have energies of interaction with it which are large compared with the thermal energy. Few ions, according to crystallographic measurements, are smaller than 0·8 Å in radius; and the radius of a water molecule may be taken as 1·4 Å, so that the least possible value of r^2 is about 5 Å^2. Water molecules in the second layer must be at least 2·8 Å further out, giving a minimum value of $r^2 = 25$ Å^2. Since they will also be less strongly oriented, the average value of $\cos\theta$ will be smaller also, and the effective dielectric constant must rise as we go out from the ion. For all these reasons it is clear that the second layer of water molecules will be much less strongly bound to the ion than the first. Indeed, it is probably only with polyvalent monatomic ions of small size that water molecules in the second layer have energies of interaction with the ion comparable to their thermal energy.

It is obviously of the utmost importance to our understanding of electrolyte solutions that we should know what the kinetic entity which we call an ion is, whether it is the bare ion, or whether it carries with it water molecules sufficiently firmly bound to be regarded as part of the ion, and if so, how many such molecules. It must be admitted that we do not know with any great certainty, though the importance of the problem has been realized for 50 years or more. One difficulty is that it is not possible to state quite unambiguously what we mean by a water molecule being 'bound to the ion'.

There are a few cases where the inner sheath of water molecules is permanent in a long-term sense and the water molecules are firmly attached, possibly by coordinate links. HUNT and TAUBE[(5)] have shown in a series of ingenious experiments using 0^{18} as tracer that the ion $[Cr(H_2O)_6]^{+++}$ exchanges its water with the solvent quite slowly, the half-life being about 40 h, whilst the single water molecule of the $[Co(NH_3)_5H_2O]^{+++}$ ion has a half-life of 24·5 h[(6)]. For other trivalent ions, however, the exchange is too rapid to detect, indicating a half-life of less than 3 min. The fact that the innermost layer of water molecules of the hydrated chromium ion is exceptionally firmly bound does not appear to have any particular effect on the electrolytic behaviour of the ion, which is quite similar in regard to both conductance and thermodynamic properties to other trivalent ions, probably because for all such ions there is a substantial second layer of water molecules which are also firmly enough held to form a part of the kinetic unit.

Chromium and cobalt are transition elements with a marked

tendency to form coordinate links. In the case of noble-gas type ions, and for any water molecules not in immediate contact with the ion, the forces involved are entirely electrostatic and no specific formula can be assigned to the aqueous ion. One may nevertheless hope to obtain an average value for the number of water molecules moving with the ion; such a value need not, of course, be integral, since the actual number per ion may vary from one ion to another and for the same ion from time to time. The hydration may be far from permanent in the everyday sense of the word; the permanence implied is, rather, relative to the time-scale of the Brownian motion.

We have already seen that the data for the entropy of solution of ions in water can be explained by assuming a layer of firmly bound water molecules around an ion, outside this layer the ion-solvent interaction being much weaker. The behaviour of the dielectric constant in the neighbourhood of an ion also fits into this picture. A somewhat similar conclusion was reached from some experiments on the vapour pressure of concentrated calcium nitrate solutions[(7)]. Although the solution of this electrolyte is saturated at 8·4 M at 25°, it readily supersaturates and on isothermal evaporation it passes into a transparent semi-solid gel without any discontinuity in the concentration-vapour pressure curve. It is only at about 21 M that the clear homogeneous gel breaks down into a striated form. Between about 9 M and 21 M it was found that the system could be treated as an adsorbent (calcium nitrate) and an adsorbate (water) and the BRUNAUER, EMMETT and TELLER adsorption isotherm[(8)] was applicable. This is a curious and perhaps unjustifiable extension of the original theory which was devised for gas adsorption on solid surfaces, but it has been used by PAULING[(9)] to explain the adsorption of water vapour, not only on fibrous proteins, but also on globular proteins. To carry it a stage further and apply it to the adsorption of water molecules from the liquid (or gel) phase on to single ions may be open to criticism, but it does lead to some interesting results. Calcium nitrate (probably, the $CaNO_3^+$ ion) is found to have 3·86 sites available for occupation by water molecules in the inner layer, each being held with an energy some 1300 cal/mole greater than the latent heat of evaporation of water, which is 10,480 cal/mole at 25°. Further adsorption can occur by building up outer layers, although the probability of such adsorption is less and also the energy of binding is less. Similar results are obtained for concentrated electrolyte solutions where no gel formation can be observed; the vapour pressures of very concentrated solutions of lithium chloride and bromide, hydrochloric and perchloric acid, zinc chloride and bromide and calcium chloride and bromide, can all be interpreted by assuming a reasonable value, between 3·5 and 7·1, for the number of water molecules which can be accommodated in the inner shell and a reasonable value, between 1000 and 3000 cal/mole over and above the latent heat of vaporization, for the energy of attachment in the solvation shell. In the theory of Brunauer, Emmett and Teller, it is assumed that any molecule in a layer other than the first is held with an energy equal to the heat of vaporization; if this assumption is made, a straightforward deduction of a comparatively simple isotherm ensues. ANDERSON[10] has produced a modification of this isotherm which permits layers beyond the first to have an energy not quite the same as the energy of vaporization of the adsorbate in bulk. This isotherm was developed for quite different purposes but it gives very plausible results with concentrated electrolyte solutions. The simpler Brunauer-Emmett-Teller isotherm gave the number of sites available for first layer adsorption round an ion: the number was close to either four or eight, depending on the valency of the cation. The number of sites should be integral, and assuming that this number is either four or eight, Anderson's theory enables us to calculate the energy of adsorption in the first layer and the somewhat smaller energy in subsequent layers. *Table 3.1* contains the results of some calculations made on this basis.

Table 3.1

Energy of 'Adsorption' of Water Molecules on Ions

Electrolyte	*No. of sites*	*Energy of adsorption in first layer* kcal/mole
Lithium chloride	4	12·1
Lithium bromide	4	12·7
Hydrochloric acid	4	12·1
Perchloric acid	4	12·9
Calcium nitrate	4	11·8
Zinc chloride	4	12·3
Zinc bromide	4	12·3
Calcium chloride	8	11·8
Calcium bromide	8	12·6

Calculations made from Anderson's equation:

$$\frac{ma_w}{55{\cdot}51[1 - a_w \exp(-\,d/\boldsymbol{R}T)]} = \frac{\exp(d/\boldsymbol{R}T)}{Cr} + \frac{C-1}{Cr}\,a_w$$

where r = number of sites available for occupation and $C \approx \exp\,(E - E_L)/\boldsymbol{R}T$;

where $(E - E_L)$ = excess of energy of adsorption in the first layer over the latent heat of vaporization of water;

and $(E_L - d)$ = energy of adsorption in outer layers.

The energy of adsorption in layers beyond the first is not listed in this table: calculation shows that it is about 100 cal/mole less than the latent heat of vaporization, whereas in the first layer the energy is between 1300 and 2400 cal higher. Evidently the attachment is much firmer in the first layer. The first four electrolytes with univalent cations require four molecules of water to complete the first layer; the last two electrolytes with bivalent cations require eight water molecules. Calcium nitrate, zinc chloride and zinc bromide may seem anomalous but there is evidence that the zinc salts in concentrated solution are more correctly represented by the formula $Zn^{++} [ZnX_4]^{--}$ and only half the zinc ions are available for hydration; calcium nitrate is subject to ion-pair formation and it is likely that in concentrated solution the predominant ion is the univalent $[CaNO_3]^+$. It is therefore reasonable to find that these three electrolytes require only four molecules of water in the inner shell. Hydrochloric acid solutions have also proved[11] amenable to this treatment over the temperature range 0° to 120°.

Until recently the principal way of estimating ionic hydration was Washburn's modification of the Hittorf method for measuring transport numbers. A supposedly inert non-electrolyte such as a carbohydrate is added to the electrolyte solution, and the concentration changes in the anode and cathode compartments are computed, (*a*) relative to the amount of water present and (*b*) relative to the concentration of non-electrolyte, which is assumed not to move in the electric field. From the difference between the transport numbers calculated on these two bases, it is possible to compute the difference between the numbers of molecules of water moving with the cation and with the anion. Hence, from a series of measurements with various electrolytes, hydration numbers can be allotted to each ion provided that that of one ion, say the chloride ion, is assumed. A great deal of work was done on this principle; a detailed account is given in a review by Bockris[12]. The method always leads to an unequivocal order of hydration values for the monovalent cations: $Li^+ > Na^+ > K^+ > Cs^+ > H^+$ but there is considerable disagreement over the actual values, which is due in part to different assumptions about the hydration of the chloride ion.

The basic assumption of this method is that the added non-electrolyte is inert. That it does not migrate in the electric field in the absence of the electrolyte can be demonstrated by measurements of the conductance. Recent very careful measurements by Longsworth[13], using a modification of the Tiselius electrophoresis apparatus, have, however, shown that when an electrolyte is present the non-electrolyte does migrate; as a result the calculated water transport per faraday of electricity passed depends on the particular non-electrolyte used. HALE and DE VRIES[14] have reached the same conclusion from a study of the transport numbers of tetra-alkylammonium iodides in the presence of various added non-electrolytes. The fact would seem to be that any non-electrolyte which is soluble enough in water to be useful in this method owes its solubility to the presence of polar groups, which are also responsible for interactions between the ions and the non-electrolyte molecules, so that in Gordon's words[15] 'added non-electrolytes are no more inert than the water molecules themselves'.

Another method is based on the principle that if the size of the solvated ion could be determined, it should be possible to calculate the number of water molecules involved in it. The most direct method available for determining the size of the kinetic unit would appear to be the measurement of its rate of motion under a known force, against the viscous drag of the solvent. Unfortunately for this method, we do not know with any great certainty the laws governing the motion of small molecules through a viscous medium. For large spherical molecules, the expression (2·47) derived by G. G. Stokes from classical hydrodynamics is known to be adequate.

This expression is used with success in the interpretation of diffusion and ultracentrifuge data for colloid molecules of approximately spherical shape, being used to derive the well-known Einstein-Stokes formula for the diffusion coefficient, $D = \boldsymbol{k}T/(6\pi\eta r)$. Clearly, if Stokes' law were valid for the motion of smaller molecules and ions, we should have a direct method for determining the sizes of ions. Unfortunately it is not valid, but a method of estimating the appropriate corrections to Stokes' law for small ions in water is proposed in Chapter 6.

There are many other methods by which the hydration of ions can be estimated; an account of these has been given by BOCKRIS[12]. He suggests that the term 'primary solvation' should refer to the comparatively firm attachment of solvent molecules to ions in such a way that an ion and its solvent molecules move as an entity in an electrolytic transport process, the solvent molecules having lost their own separate translational degrees of freedom. 'Secondary solvation' would designate all other ion-solvent interactions. It is, however, doubtful if any method has yet been devised to measure the primary solvation unequivocally. For example, even if the hydrodynamical theory of the flow of particles through a liquid medium could be developed so as to extend Stokes' law to particles of atomic dimensions, only a part of the difficulties would have been resolved. We could apply such a theory to the hypothetical case of an ion rigidly bound to a finite number of water molecules moving through a medium consisting of another set of water molecules whose only function would be to provide a medium through which the hydrated ion moved. This would not be the problem we are trying to solve; our problem is to set up and solve the equations of motion of an ion firmly bound to one set of water molecules, moving through a medium consisting of another set of water molecules, from which it is separated by a third set of water molecules which have neither the property of the first set of being 'permanently' (on the Brownian motion time-scale) bound to the ion nor the property of the far distant solvent molecules of being practically out of range of the ionic forces: instead they are subjected to a comparatively mild ion-solvent interaction. It is not surprising, therefore, that there is little concordance between the results of the many ingenious experimental methods which have been devised to measure the 'hydration number' of an ion; each method measures an average of the primary and secondary hydration but there are many ways of weighting an average. Some methods will tend to emphasize the secondary hydration and will therefore be more likely to give the upper limit to the hydration number. One such method consists in the distribution of a 'reference' substance between, first water and another immiscible solvent, and secondly an electrolyte solution and the other solvent. In general it is found that the addition of electrolyte to the aqueous layer drives the reference substance into the other layer and this is taken to mean that, by hydration, the electrolyte has withdrawn a certain amount of water from the state in which it can exert its solvent properties for the reference substance.

An extensive set of measurements was made by SUGDEN[16] who distributed acetic acid between aqueous salt solutions and amyl alcohol. He derived the following set of hydration numbers:

LiCl	10·5	NaCl	7·9	KCl	3·4
LiBr	9·0	NaBr	6·4	KBr	1·9
$LiNO_3$	4·4	$NaNO_3$	1·8	KNO_3	−2·7
$LiClO_3$	6·3	$NaClO_3$	3·7	$KClO_3$	−0·8
$LiBrO_3$	9·2	$NaBrO_3$	6·6	$KBrO_3$	2·1
$LiIO_3$	7·7	$NaIO_3$	5·1	KIO_3	0·6

These figures are of reasonable magnitude except for the negative hydration numbers of potassium nitrate and potassium chlorate—a strange result which Sugden explained by postulating a depolymerizing effect of these anions on the water structure. Otherwise, the results suggest that the method is giving hydration numbers of the right order.

Another method which probably also gives an estimate of the upper limit of the hydration number is described in Chapter 9. We are not now concerned with the details of the theory, but it is worthwhile anticipating some of the more important conclusions. In brief, the Debye-Hückel equation for the activity coefficient of an electrolyte, whose ions have finite size, is taken to apply to the solvated ions, since the ion sizes required by the theory are larger than those of the bare ions obtained from crystallographic data. It is found that these calculated activity coefficients do not agree with the experimentally determined coefficients which refer to the unhydrated solute, and the difference is ascribed to the formal thermodynamic effects of hydration. It is demonstrable that any allowance for hydration must result in a discrepancy between experiment and theory, but it is also known that there are a number of other effects operating, which should be included in the calculation of the activity coefficient. Unfortunately, whilst we know that these effects are important, theory has not advanced to the stage where we can calculate them; for want of a complete theory, all the difference between the experimental and the calculated activity coefficients is ascribed to hydration. It is true that the resulting equation is remarkably useful in describing observed results, but this should not obscure the fact that, whilst hydration is a very important factor, it is not the only one which determines the complicated equilibria in an electrolyte solution. To put it briefly, the secondary hydration has been stretched to cover a number of effects the quantitative nature of which requires a great deal of investigation.

In another part of this book (Chapter 11) there will be described a method of deriving hydration numbers, which depends on diffusion coefficient measurements. We think this method is weighted in favour of the firmly bound water molecules, *i.e.*, it should give a good approximation to the lower limit of the hydration number and measure the primary hydration. Unfortunately it requires diffusion measurements in a concentration range where theory cannot fully predict the influence of the viscosity factor. Again, there is a promising method which makes use of the compressibility of an electrolyte solution: it is assumed that the molecules of the solvent which are hydration molecules are compressed to their maximum extent by the intense electrical forces round the ion and that, on increasing the pressure, it is the remainder of the solvent which is compressed. This method[17] has been stimulated by improvements in technique for measuring the velocity of ultrasonic waves. A solution of volume V, containing a total number, n_A moles of water, of which n_h are attached to ions as hydrate molecules and do not contribute to the compressibility, would have a volume-pressure differential,

$$\left(\frac{\partial V}{\partial P}\right)_T = \frac{\partial}{\partial P}[\bar{V}_A^0(n_A - n_h)]_T$$

where $\bar{V}_A^0$ is the molal volume of pure water, and the measured compressibility can be equated to:

$$\beta = -\frac{n_A - n_h}{V}\frac{\partial \bar{V}_A^0}{\partial P} = (n_A - n_h)\frac{\bar{V}_A^0}{V}\beta_A^0$$

where β_A^0 is the compressibility of pure water. Thus if n_B is the number of moles of electrolyte in the solution, the hydration number is:

$$h = \frac{n_h}{n_B} = \frac{n_A}{n_B}\left(1 - \frac{\beta V}{\beta_A^0 n_A \bar{V}_A^0}\right)$$

which can be simplified to:

$$h = \frac{n_A}{n_B}\left(1 - \frac{\beta}{\beta_A^0}\right)$$

for dilute solutions. Hydration numbers of several salts have been measured in this way, and they can be compared with the numbers obtained from activity coefficient data and also with those calculated from diffusion data given in *Table 3.2.*

The compressibility method gives some unexpected results; those for the lithium salts are in good agreement with figures obtained by other methods, but the sodium salts give higher hydration numbers, whilst the potassium salts are shown as hydrated to a surprisingly high extent. The 2 : 1 salts have hydration numbers of the expected magnitude, except for barium chloride, whilst the 3 : 1 salts can be compared with lanthanum chloride for which a hydration number of 18·2 has been calculated from activity coefficient data.

The entropy change corresponding to the transfer of an ion from the gaseous state into solution has been attributed[18] to the entropy change of water molecules entering the hydration shell, a change which is assumed equal to that occurring when water molecules enter the solid state on freezing. All these methods determine an average—a weighted average for the strongly bound ionic hydration shell and those solvent molecules more distant but still under the attractive influence of the ion.

DEBYE[19] has proposed a theory which requires that an alternating potential, of the order of 10^{-6} V, should be set up between two electrodes in an electrolyte solution subjected to ultrasonic waves.

Table 3.2

Salt	*Hydration numbers*		
	From compressibility	*From activity*	*From diffusion*
LiBr	5–6	7·6	5·6
LiCl	6	7·1	6·3
KI	6–7	2·5	0·3
NaI	6–7	5·5	3·0
NaBr	6–7	4·2	2·8
KBr	6–7	2·1	0·3
KCl	7	1·9	0·6
NaCl	7	3·5	3·5
KF, NaF, $BeCl_2$	8–9	—	—
$Ca(ClO_4)_2$	11	17	—
$CdSO_4$	12	—	—
$Sr(ClO_4)_2$	13	15	—
$Pb(NO_3)_2$	13	—	—
$BaCl_2$	16–17	7·7	—
$MgCl_2$	16–17	13·7	—
$PrCl_3$	24	—	—
$LaCl_3$	—	18·2	—
$AlCl_3$	31–32	—	—

Data from BARNARTT, S., *Quart. Rev.*, 7 (1953) 84; PASSYNSKI, A., *Acta phys.-chim., U.S.S.R.*, 8 (1938) 385; GIACOMINI, A. and PESCE, B., *Ric. Sci.*, 11 (1940) 605; *Chem. Abstr.*, 33 (1939) 4494; *ibid.*, 35 (1941) 1292.

The potential should depend on the relative masses of the cations and the anions and should, therefore, measure the ionic hydration. Whilst the Debye effect has been detected[20], it is clear that there are many experimental difficulties to be overcome before quantitative results will emerge.

When we turn to other solvents than water, the problem of solvation becomes even more difficult to investigate. The hydrodynamic approach is frustrated at the outset by an almost total absence of measured transport numbers: ionic mobilities in most such solvents are at present known only in the form of sums for pairs of oppositely charged ions. Thermodynamic methods are complicated by the effects of the (usually) lower dielectric constants as compared with water, and by the relatively low accuracy of the experimental data. In the case of acids, it is reasonable to suppose that the proton is never a free entity, but is associated with at least one solvent molecule; the 'acid' ion NH_4^+ in liquid ammonia solutions is a familiar example. The striking fact that silver perchlorate forms an electrolytic solution in benzene is, similarly, attributable to the existence of a silver ion-benzene complex of an acid-base nature, which can be considered a form of solvated ion.

It is regrettable that while electrolyte solutions owe their very existence to ion-solvent interactions, we have so far been able to find out little of a quantitative nature about these interactions; the problem of ion-ion interactions has proved much more tractable.

THE FREE ENERGY AND ENTROPY OF IONS IN SOLUTION

Measurements of the vapour pressure, freezing and boiling points and osmotic pressure of solutions, which are essentially determinations of the chemical potential of the solvent, can be used to determine *via* the Gibbs-Duhem equation the change in chemical potential of the solute with concentration (see Chapter 8). These methods, alone, yield no information on the energy relations between the pure solute and the solution. The same is true of electromotive force measurements on concentration cells with transport. Electromotive force measurements on cells without transport, however, can be made to yield a little more: besides the change in chemical potential with concentration, they give the standard potential of the cell reaction. The details of the experimental methods and their numerical handling need not concern us here; they are discussed in Chapter 8. The point of interest here is the relation of the standard potential to the free energy change in the cell reaction. Consider the cell:

$$\mathrm{Zn}|\mathrm{ZnCl_2}(m)|\mathrm{AgCl, Ag}$$

containing as electrolyte a zinc chloride solution of molality m, in which the solute has the activity a_{ZnCl_2}. The cell reaction is:

$$\mathrm{Zn}\ (s) + 2\ \mathrm{AgCl}\ (s) \rightarrow 2\ \mathrm{Ag}\ (s) + \mathrm{ZnCl_2}\ (\text{solution of activity } a_{ZnCl_2})$$

and the free energy change per two faradays is:

$$\Delta G = \bar{G}^0_{ZnCl_2} + \boldsymbol{R}T \ln a_{ZnCl_2} - \bar{G}_{Zn} - 2\bar{G}_{AgCl} + 2\bar{G}_{Ag}$$

where $\bar{G}^0_{ZnCl_2}$ is the chemical potential of zinc chloride in the hypothetical mean molal solution which is the appropriate standard state. The potential is $E = -\dfrac{\Delta G}{2\boldsymbol{F}}$ and can be split up into two parts, a standard potential E^0 independent of the composition of the solution, and a composition-dependent term:

$$E = E^0 - \frac{RT}{2F} \ln a_{ZnCl_2}$$

Thus the standard potential E^0 is given by:

$$E^0 = -(\bar{G}^0_{ZnCl_2} - \bar{G}_{Zn} - 2\bar{G}_{AgCl} + 2\bar{G}_{Ag})/2F$$

so that in determining it we are finding a free energy difference between the ions in a standard state in solution, and certain related pure substances (here, the solids: silver chloride, zinc and silver). By differentiation of standard potentials with respect to temperature, it is likewise possible to obtain the partial molal entropy of an electrolyte in the standard state in solution, relative to those of the pure substances forming the electrodes; again the values obtained refer to the electrolyte as a whole, and are the sums of those for the constituent ions.

An alternative method of determining the same quantities involves the use of solubility data. Consider the process:

$$\text{KCl}\ (s) \rightarrow \text{K}^+ + \text{Cl}^- \text{ (in saturated solution)}$$

The free energy change of potassium chloride is zero, since the solid and saturated solution are in equilibrium; hence

$$\bar{G}_{(s)} = \bar{G}^0_{KCl} + RT \ln a_{KCl} \text{ (sat.)}$$

where a_{KCl} (sat.) is the activity of potassium chloride in the saturated solution (a quantity which can be determined by measurements of vapour pressure, *etc.*, extending up to the concentration of saturation) and $\bar{G}^0_{KCl}$ refers to the standard state on the appropriate scale. This constitutes a relation between the free energy of the solute as a solid and in its standard state in solution. The corresponding entropy change can be obtained by measurements of the heat of solution of the solid and employment of the relation $G = H - TS$. The heat content change of the solute for the process: solid $\rightarrow$ solution in the standard state, is the same as that for the process: solid $\rightarrow$ infinitely dilute solution, since the standard state is so defined that it has the same partial molal heat content as the infinitely dilute solution (see p. 35).

The third law of thermodynamics provides a basis for the calculation of absolute entropies of pure substances from heat capacity data extending down to low temperatures:

$$\bar{S} = \int_0^T \bar{C}_P \mathrm{d} \ln T$$

Allowance can also be made for any phase transitions occurring below the temperature of interest, so that the absolute entropies of the pure substances composing the electrodes of the cell discussed above, or of the solid potassium chloride are determinable. Their free energies are likewise obtainable, subject to the fixing of an energy zero for the pure substances. Entropy and free energy sums for anions and cations in solution in the standard state are therefore essentially determinable quantities. Considerable theoretical interest attaches to the question of how these sums should be divided between anions and cations, but before considering this a digression on the standard states is necessary.

We have considered above two specific solutes, potassium chloride and zinc chloride. For potassium chloride, we may choose the standard state of a hypothetical solution of unit mean molality and unit activity coefficient. This has the required property $a_{KCl} = m^2\gamma_{\pm}^2 = 1$. The molality of each ionic species, K^+ and Cl^-, is also unity, so that we need the additional assumption that not merely the mean activity coefficient, but that of each separate ion, is also unity, before we can write:

$$\bar{G}^0_{KCl} = \bar{G}^0_{K^+} + \bar{G}^0_{Cl^-}$$

In the case of zinc chloride, the relation

$$\bar{G}_{ZnCl_2} = \bar{G}^0_{ZnCl_2} + \boldsymbol{R}T \ln a_{ZnCl_2}$$

requires that in the standard state

$$a_{ZnCl_2} = 1 = 4m^3\gamma_{\pm}^3$$

The standard state is therefore again one of mean molality unity, and unit mean activity coefficient. The molality of zinc chloride in the standard state is, however, $4^{-1/3}$, that of the zinc ion is $4^{-1/3}$, and that of the chloride ion $2 \times 4^{-1/3}$. It thus appears that the molality of the chloride ion is different for the standard states of potassium chloride and of zinc chloride; this is undesirable as we obviously need a unique standard state for each ion (on a given concentration scale). This difficulty can be overcome as follows.

Imagine one mole of zinc ion to be concentrated from the hypothetical solution of molality $4^{-1/3}$ to a new one of molality 1, while two moles of chloride ion are diluted from a hypothetical molality $2 \times 4^{-1/3}$ to one of molality 1, the new solutions retaining the 'ideal' quality that $\gamma_{Zn^{++}} = \gamma_{Cl^-} = 1$. The free energy gained by the zinc ion will be:

$$\boldsymbol{R}T \ln 4^{1/3} = 1/3\boldsymbol{R}T \ln 4$$

while that gained by the chloride ion will be:

$$-2\boldsymbol{R}T\ln(2 \times 4^{-1/3}) = -\boldsymbol{R}T\ln 4 + \tfrac{2}{3}\boldsymbol{R}T\ln 4$$

The net free energy change for the whole process is therefore zero, and we can write:

$$\bar{G}^0_{ZnCl_2} = \bar{G}^0_{Zn^{++}} + 2\bar{G}^0_{Cl^-}$$

where $\bar{G}^0_{Zn^{++}}$ and $\bar{G}^0_{Cl^-}$ refer to standard states for the separate ions, which are hypothetical states of unit molality and unit activity coefficient. This logical difficulty which arises with salts of unsymmetrical valence type serves to emphasize the importance of choosing a hypothetical standard state; the corresponding steps could not be taken were we using an actual state of the solution as standard state.

Exactly similar considerations apply to the entropy in the standard state. For any electrolyte, therefore, we can write the free energy, entropy, *etc.*, in the standard state as the sum of values for the separate ions in their standard states, with the appropriate numerical weightings as indicated by the chemical formula of the electrolyte; and there is no inconsistency in defining the standard states as being hypothetical ones of mean molality unity for the electrolyte, and of molality unity for the separate ions. (This point is also of importance in defining the standard electrode potential for a half-reaction.)

If we arbitrarily assign to some particular ion in its standard state in solution a given partial molal entropy, the partial molal entropies of all other ions in the standard state can be calculated from the measurable entropy sums for anions and cations. The usual convention is to write $\bar{S}^0_{H^+} = 0$, which has the advantage of consistency with the usual definition of standard potentials relative to that of hydrogen; so that, to take a familiar example, the conventional ionic entropy of zinc ion may be readily computed from the temperature coefficient of the standard potential of zinc and the known entropies of zinc and hydrogen by considering the reaction:

$$Zn + 2H^+ \rightarrow Zn^{++} + H_2$$

A recent summary by POWELL and LATIMER[21] (*Table 3.3*) gives values of the ionic entropies of a number of simple ions computed from the best available data; a more extensive list is given by LATIMER, PITZER and SMITH[22]. Some obvious generalizations can be made about the ionic entropies in *Table 3.3.* First, for a given charge, the entropy increases with atomic weight; secondly, for approximately constant atomic weight, as in the ions Na^+, Mg^{++} Al^{+++}, the entropy decreases rapidly as the charge increases. Powell and Latimer point out that the values can be represented with very fair accuracy by the equation:

$$\bar{S}^0 = \tfrac{3}{2}\boldsymbol{R} \ln W + 37 - 270\,|z|/r_e^2$$

Table 3.3

Conventional Ionic Entropies at 25°C (298·16°K), *computed relative to* $\bar{S}^0_{H^+} = 0$ *in the hypothetical standard state of one gram-ion per kg of water.*

Ion	$\bar{S}^0$ cal deg⁻¹ mole⁻¹	*Ion*	$\bar{S}^0$ cal deg⁻¹ mole⁻¹	*Ion*	$\bar{S}^0$ cal deg⁻¹ mole⁻¹
H^+	(0·00)	Mg^{++}	− 28·2	Al^{+++}	− 74·9
Li^+	3·4	Ca^{++}	− 13·2	Cr^{+++}	− 73·5
Na^+	14·4	Sr^{++}	− 9·4	Fe^{+++}	− 70·1
K^+	24·5	Ba^{++}	3·0	Ga^{+++}	− 83
Rb^+	29·7	Mn^{++}	− 20	In^{+++}	− 62
Cs^+	31·8	Fe^{++}	− 27·1	Gd^{+++}	− 43
Tl^+	30·4	Cu^{++}	− 23·6	U^{+++}	− 36
Ag^+	17·67	Zn^{++}	− 25·45	Pu^{+++}	− 39
F^-	− 2·3	Cd^{++}	− 14·6	U^{++++}	− 78
Cl^-	13·17	Sn^{++}	− 5·9	Pu^{++++}	− 87
Br^-	19·25	Hg^{++}	− 5·4		
I^-	26·14	Pb^{++}	5·1		
OH^-	− 2·5	S^{--}	− 6·4		
SH^-	14·9				

Data from POWELL, R. E. and LATIMER, W. M., *J. chem. Phys.*, 19 (1951) 1139.

where W is the atomic weight, $|z|$ the valency treated as positive regardless of sign, and r_e an effective radius of the ion in solution, which is taken as 1·0 Å more than the crystal radius for anions and 2·0 Å more for cations.

In discussing the entropy and energy changes involved in the process of solution of ions it is desirable to eliminate any contributions due to the physical state of the pure solute. Thus, a comparison of the free energy changes when various substances dissolve from the solid state into the standard state in solution would involve the varying stability of the various crystals considered: to avoid this difficulty it is useful to compute the energies and entropies of hydration of ions, not from the solid state, but from a hypothetical gaseous state having the properties of an ideal gas.

The entropy of such an ideal gas can be computed by the methods of statistical mechanics. For monoatomic gases, the Sackur-Tetrode equation may be written[23]:

$$\bar{S}_{(g)} = 2{\cdot}303\boldsymbol{R}(\tfrac{3}{2}\log W + \tfrac{5}{2}\log T - \log P + \log Q_e - 0{\cdot}5055)$$

(W = atomic weight, P = pressure in atm., Q_e = the multiplicity of the ground state.) For atoms or ions of the inert-gas structure, $Q_e = 1$, and this expression reduces at 25°C and for one atmosphere pressure to:

$$\bar{S}^0_{(g)} = (6{\cdot}864 \log W + 26{\cdot}00) \text{ cal deg}^{-1} \text{ mole}^{-1}$$

The standard entropy of hydration of a given electrolyte, $\bar{S}^0_h$, may thus be computed as the sum of its standard ionic entropies in solution, less the corresponding sum calculated for the ideal gas at some specified pressure:

$$\bar{S}^0_h = \bar{S}^0_{\text{soln.}} - \bar{S}^0_g$$

The gas pressure may be taken as one atmosphere, though a pressure corresponding to a concentration of one mole per litre has something to recommend it since in this state the actual concentration of ions in the gas is close to that in the standard state in solution. Inter-conversion between these different standard states is a trivial matter, the entropy term being $\boldsymbol{R} \ln \left(\dfrac{22{\cdot}4T}{273}\right)$.

The heat content change in passing from the hypothetical gas state to the standard state in solution can be obtained as follows:

The crystal lattice energy $\Delta\bar{H}_{\text{cryst.}}$ is defined as the energy required to separate the ions of the crystal to infinite distance. Methods for its computation from thermal data are given by PAULING[24]. The heat of solution of the crystal to infinite dilution $\Delta\bar{H}_{\text{soln.}}$ is obtainable by calorimetric measurements; since the heat content of the solute in the standard state is the same as at infinite dilution, $\Delta\bar{H}_{\text{soln.}}$ also represents the heat of solution into the standard state. For a crystal MX we therefore have:

$MX(s) \rightarrow M^+ + X^-$ (in solution of infinite dilution) $\Delta H = \Delta\bar{H}_{\text{soln.}}$

$MX(s) \rightarrow M^+(g) + X^-(g)$ (infinite separation) $\quad \Delta H = \Delta\bar{H}_{\text{cryst.}}$

Hence for the process:

$$M^+(g) + X^-(g) \rightarrow M^+ + X^- \text{ (solution, standard state)}$$

we have the heat of hydration

$$\Delta\bar{H}_{(h)} = \Delta\bar{H}_{\text{soln.}} - \Delta\bar{H}_{\text{cryst.}}$$

Since we are assuming the gaseous ions to have ideal properties, which include the heat contents being independent of pressure, the quantity $\Delta\bar{H}_{(h)}$ so calculated is equal to the heat of hydration $\Delta\bar{H}^0_{(h)}$ between the chosen standard states. From the heat and entropy of hydration, we can then obtain the standard free energy of hydration $\Delta G^0_{(h)}$ by the relation $\Delta \bar{G}^0_{(h)} = \Delta \bar{H}^0_{(h)} - T\Delta S^0_{(h)}$. Free energies and entropies of hydration so calculated refer to sums for anions and cations. It is of great interest to attempt their separation into contributions for the separate ions. This involves, in the case of the entropy, the fixing of the absolute entropy of some one aqueous ion; the absolute entropies of other ions can then be obtained from their standard entropies (usually based on $S^0_{H^+} = 0$) and those of the gaseous ions can be obtained from the Sackur-Tetrode equation, giving the required individual $\Delta \bar{S}^0_{(h)}$ as a difference. In the case of the free energy, $\Delta \bar{G}^0_{(h)}$ for one ion must be fixed. LATIMER, PITZER and SLANSKY[25] and independently VERWEY[26], have considered this problem, and arrived at nearly identical conclusions. Both use effectively the same standard states (the former one mole per litre in the gas state, and the hypothetical molal solution; the latter equal low concentrations in the gas and the liquid) and both make use of Born's equation:

$$-\Delta G = \left(1 - \frac{1}{\varepsilon}\right)\frac{e^2}{2r}$$

for the energy of a sphere of charge e and radius r immersed in a medium of dielectric constant ε; Latimer *et al.* conclude that $\Delta \bar{G}^0_{(h)}$ for the chloride ion is $-84{\cdot}2$ kcal/mole and Verwey that it is -86; for the other alkali and halide ions the agreement is similarly good. Latimer *et al.* have also made calculations of the absolute entropies of separate ions, again making use of the Born expression; they conclude that the absolute entropy of the chloride ion in the hypothetical molal standard state is about 15 cal deg^{-1} mole^{-1}. Its standard entropy relative to hydrogen ion (*Table 3.3*) being 13 cal deg^{-1} mole^{-1}, it follows that the absolute entropy of the aqueous hydrogen ion is about 2 cal deg^{-1} mole^{-1}. Earlier, EASTMAN and YOUNG[27] estimated 18·1 cal deg^{-1} mole^{-1} for the absolute entropy of the chloride ion; thus it seems probable at least, that the absolute entropy of the aqueous hydrogen ion in the hypothetical molal solution is not far from zero and that, since most ionic entropies are in the range 10–100 cal deg^{-1} mole^{-1}, there will be only a small error involved in treating the standard entropies of *Table 3.3* as absolute entropies for the separate ions. The free energies and entropies of hydration computed by Latimer and his co-workers for the separate ions are given in *Table 3.4*.

VERWEY[26] emphasizes the fact that the hydration energies of the halide ions are substantially larger than those of cations of the same size. Fluoride ion, for example, has nearly the same crystal radius as potassium ion, yet the free energies of hydration are -114 and -73 kcal/mole respectively (see *Table 3.4*). This is consistent with the observed limiting mobilities $\lambda^0_{F^-} = 55{\cdot}4$ and $\lambda^0_{K^+} = 73{\cdot}5$ at 25°, which suggest a much stronger interaction between the fluoride ion and water molecules than between the potassium ion and water molecules; that, in fact, the fluoride ion is more 'hydrated' than the potassium ion. No other pair of ions in *Table 3.4* has such nearly equal radii as K^+ and F^-; the nearest approach to equality among the others being with $Cs^+(r_1 = 1{\cdot}69$ Å) and $Cl^-(r_2 = 1{\cdot}81$ Å). In spite of its larger size and therefore weaker surface field, chloride ion has a substantially higher hydration energy than caesium ion. In this case, however, the ionic mobilities are practically equal ($\lambda^0_{Cl^-} = 76{\cdot}35$, $\lambda_{Cs^+} = 77{\cdot}26$ at 25°). It seems likely that these ions are too large to be 'hydrated' in the sense of having a permanent hydration sheath, since they all

Table 3.4

Free Energies and Entropies of Hydration of Monovalent Ions at 25°C. *Standard states of one mole per litre for gaseous ions and hypothetical molal solution for aqueous ions.*

	$-\Delta\bar{G}_{(h)}$ kcal mole^{-1}	$-\Delta H^0_{(h)}$ kcal mole^{-1}	$-\Delta S^0_{(h)}$ cal deg^{-1} mole^{-1}	$r_{\text{cryst.}}$ Å (Pauling)
Li^+	114·6	121·2	22	0·60
Na^+	89·7	94·6	17	0·95
K^+	73·5	75·8	8	1·33
Rb^+	67·5	69·2	6	1·48
Cs^+	60·8	62·0	4	1·69
F^-	113·9	122·6	29	1·36
Cl^-	84·2	88·7	15	1·81
Br^-	78·0	81·4	12	1·95
I^-	70·0	72·1	7	2·16

Data from LATIMER, W. M., PITZER, K. S. and SLANSKY, C. M., *J. chem. Phys.*, 7 (1939) 108.

Note: The entropies of hydration given here differ from those in *Table 1.4* because of the adoption of different standard states and a different value of the absolute entropy of chloride ion. Frank and Evans used a standard state of one atmosphere for the gaseous ions, $N_B = 1$ for the solution, and 18·1 cal deg^{-1} mole^{-1} for the absolute entropy of chloride ion where Latimer *et al.* take 15.

These differences lead to Frank and Evans' values for the entropy of hydration of monovalent cations being less than Latimer's by $\boldsymbol{R} \ln \left(22{\cdot}4 \times \frac{298}{273} \times 55{\cdot}51\right) + 3{\cdot}1 = 17{\cdot}4$ cal deg^{-1} mole^{-1}, and for monovalent anions by

$$\boldsymbol{R} \ln \left(22{\cdot}4 \times \frac{298}{273} \times 55{\cdot}51\right) - 3{\cdot}1 = 11{\cdot}2 \text{ cal deg}^{-1} \text{ mole}^{-1}$$

The values of $r_{\text{cryst.}}$ are taken from PAULING, L., 'The Nature of the Chemical Bond', Cornell University Press (1940); see Appendix 3.1.

exhibit much the same rather high mobility ($\lambda^0 \sim 77$ at 25°). If negative ions comparable in size with lithium ($r = 0{\cdot}6$ Å) or sodium ($r = 0{\cdot}95$ Å) existed, it seems highly probable that they would be more strongly hydrated and possess lower mobilities than these cations.

Verwey has considered in detail the interaction of both positive and negative ions with water molecules and has succeeded in accounting rather well for the dependence of the hydration energy on the sign of the ionic charge. Thus for an ion of 1·36 Å radius he calculates 75–79 kcal/mole for a cation and 102–122 kcal/mole for an anion, which is clearly consistent with *Table 3.4.* The lower and upper limits quoted were obtained by the use of Verwey's models I and II for the charge distribution in the water molecule (see Chapter 1).

Powell and Latimer[21] have recently pointed out the curious fact that the electrostatic contribution to the entropy of aqueous ions is apparently proportional to the first power of the valency of the ion, rather than to its square as would be expected from Born's equation, and have suggested that such a result could arise from the inability of the water dipoles in the region close to the ion to rotate freely.

One must at present conclude that although the study of the energy and entropy of ions in solution gives some useful insight into the nature of ion-solvent interactions, it does not provide anything approaching a definite answer to the problem of identifying the kinetic entities in the solution. A good deal of the difficulty lies in the arbitrary nature of the division of free energy and entropy changes, measurable only for the electrolyte as a whole, into separate ionic values. This arises from the nature of thermodynamic arguments, which are essentially independent of the detailed molecular picture of the system. In electrical conductance, on the other hand, we encounter properties which are in principle and in practice determinable for the separate ions, viz., the limiting ionic mobilities; and it is by the fuller understanding of the hydrodynamics of small particles that progress towards a better picture of ionic solutions is likely to be made.

REFERENCES

1 Shedlovsky, T. and Kay, R. L., *J. phys. Chem.*, 60 (1956) 151

2 El-Aggan, A. M., Bradley, D. C. and Wardlaw, W., *J. chem. Soc.*, (1958) 2092

3 Wynne-Jones, W. F. K., *ibid.* (1930) 1064; Robinson, R. A., *Trans. Faraday Soc.*, 32 (1936) 743

4 Redlich, O., *Chem. Rev.*, 39 (1946) 333

5 Hunt, J. P. and Taube, H., *J. chem. Phys.*, 18 (1950) 757; 19 (1951) 602

[6] Rutenberg, A. C. and Taube, H., *ibid.*, 20 (1952) 825
[7] Stokes, R. H. and Robinson, R. A., *J. Amer. chem. Soc.*, 70 (1948) 1870
[8] Brunauer, S., Emmett, P. H. and Teller, E., *ibid.*, 60 (1938) 309
[9] Pauling, L., *ibid.*, 67 (1945) 555; see also Robinson, R. A., *J. chem. Soc.*, (1948) 1083; Green, R. W., *Proc. roy. Soc., New Zealand*, 77 (1948) 24, 313; Green, R. W. and Ang, K. P., *J. Amer. chem. Soc.*, 75 (1953) 2733
[10] Anderson, R. B., *ibid.*, 68 (1946) 686
[11] Feder, H. M., *ibid.*, 70 (1948) 3525
[12] Bockris, J. O'M., *Quart. Rev.*, 3 (1949) 173
[13] Longsworth, L. G., *J. Amer. chem. Soc.*, 69 (1947) 1288
[14] Hale, C. H. and de Vries, T., *ibid.*, 70 (1948) 2473
[15] Gordon, A. R., *Annu. Rev. phys. Chem.*, 1 (1950) 61
[16] Sugden, J. N., *J. chem. Soc.*, 129 (1926) 174
[17] Barnartt, S., *Quart. Rev.*, 7 (1953) 84; Passynski, A., *Acta phys.-chim., USSR*, 8 (1938) 385; Giacomini, A. and Pesce, B., *Ric. sci.*, 11 (1940) 605; *Chem. Abstr.*, 33 (1939) 4494; 35 (1941) 1292
[18] Ulich, H., *Z. Elektrochem.*, 36 (1930) 497
[19] Debye, P., *J. chem. Phys.*, 1 (1933) 13
[20] Yeager, E., Bugosh, J., Hovorka, F. and McCarthy, J., *ibid.*, 17 (1949) 411
[21] Powell, R. E. and Latimer, W. M., *ibid.*, 19 (1951) 1139
[22] Latimer, W. M., Pitzer, K. S. and Smith, W. V., *J. Amer. chem. Soc.*, 60 (1938) 1829
[23] Glasstone, S., 'Thermodynamics for Chemists', pp. 190–191, D. Van Nostrand Co. Inc. (1947)
[24] Pauling, L., 'The Nature of the Chemical Bond', Chap. X, Cornell University Press (1940)
[25] Latimer, W. M., Pitzer, K. S. and Slansky, C. M., *J. chem. Phys.*, 7 (1939) 108
[26] Verwey, E. J. W., *Rec. Trav. chim. Pays-Bas*, 61 (1942) 127
[27] Young, M. B., *Thesis*, University of California (1935)

4

IONIC DISTRIBUTION FUNCTIONS AND THE POTENTIAL

THE modern quantitative theory of electrolyte solutions is based on the concept of the interaction between the thermal motions of the ions and their electrical attractions and repulsions, and also involves in its higher refinements considerations of the physical dimensions of the ions and of their interactions with solvent molecules.

A fundamental idea in the theory of liquids in general is that of the 'distribution function' which gives the probability of finding a particle (molecule or ion) in a given position relative to another particle. In simple pure liquids the distribution-function has radial symmetry, *i.e.*, it depends only on the distance between the particles, and not on their mutual orientation. It shows a marked peak at the distance corresponding to the nearest neighbours, *i.e.*, to the first layer of molecules surrounding the central molecule; this is followed by one or two subsidiary peaks, and thereafter the distribution function flattens out to an effectively constant value, meaning that there are no preferred positions for molecules more than a few diameters distant from the central molecule considered. Such a radial distribution function is illustrated for the case of pure water in *Figure 1.3.* This situation is described as showing 'short-range order' and results from the fact that the intermolecular forces involved are of short range, *e.g.*, 'van der Waals' forces, dipole-dipole interactions, *etc.*, so that only at very small distances can they overcome the tendency of thermal motions to produce a purely random distribution.

The opposite extreme is found in the case of ionic crystals, where (except for minor lattice defects) the distribution function is characterized by a series of equal peaks at equal intervals of distance, and is, furthermore, strongly dependent on the direction chosen. In passing, it is interesting to note that very concentrated electrolyte solutions show definite traces of 'crystalline' structure, so that the long-range order characteristic of crystals can persist to some extent in such solutions[1].

In dilute non-electrolyte solutions, the distribution of solute is entirely random, subject only to the restriction that two particles cannot approach within a certain distance given by mutual contact.

Beyond this distance the distribution function is independent of both distance and direction. In an electrolyte solution, the distribution of ions results from competition between coulomb electrical forces, which are long-range forces in the statistical sense, and the thermal motions; this distribution is not random even at considerable distances.

If the distribution of ions is known, it is possible to calculate the electrical potential arising from this distribution; but the calculation of the distribution requires the use of the electrical potential. The first attempt to solve this problem was made by MILNER[(2)] in 1912 by a laborious method of numerical summation of interaction energies for all configurations of the system; this treatment is historically interesting, but has been superseded. The modern theory was founded by DEBYE and HÜCKEL[(3)] in 1923; refinements of their treatment of both equilibrium and transport properties have been made by numerous workers, chief among whom have been BJERRUM[(4)] (1926), ONSAGER[(5)] (1929) and FALKENHAGEN[(6)] (1952). Here we shall make no attempt to follow the historical development of the theory, but shall aim throughout at obtaining the modified forms which have proved most useful in treating the properties of electrolyte solutions at reasonable concentrations.

THE FUNDAMENTAL EQUATION FOR THE POTENTIAL

The essential feature of the Debye-Hückel theory is the calculation of the electrical potential ψ at a point in the solution in terms of the concentrations and charges of the ions and the properties of the solvent. This is achieved by the device of combining the Poisson equation of electrostatic theory with a statistical-mechanical distribution formula.

Poisson's equation is the most general expression of Coulomb's law of force between charged bodies and is written:

$$\nabla^2\psi = -\frac{4\pi}{\varepsilon}\rho \qquad \ldots.(4.1)$$

where ψ is the potential at a point where the charge density is ρ, ε being the dielectric constant of the medium in which the charges are immersed. The differential operator ∇^2, which may also be written (div grad), is given in Cartesian coordinates by

$$\left(\frac{\partial^2}{\partial x^2} + \frac{\partial^2}{\partial y^2} + \frac{\partial^2}{\partial z^2}\right)$$

In the special case of a distribution of charges possessing spherical symmetry about the origin, ψ depends only on the distance r of the point considered from the origin, and in this case the partial differential operator ∇^2 reduces to a total differential operator:

$$\nabla^2 = \frac{1}{r^2}\frac{\mathrm{d}}{\mathrm{d}r}\left(r^2\frac{\mathrm{d}}{\mathrm{d}r}\right) \text{ (for spherical symmetry)}$$

so that equation (4.1) becomes:

$$\frac{1}{r^2}\frac{\mathrm{d}}{\mathrm{d}r}\left(r^2\frac{\mathrm{d}\psi}{\mathrm{d}r}\right) = -\frac{4\pi}{\varepsilon}\rho \qquad \ldots.(4.2)$$

If a particular ion is chosen as the origin of coordinates and no external forces are acting on the ions, the time-average distribution of charge about that ion will obviously have spherical symmetry. Equation (4.2) is therefore taken to apply to the time-average values of the potential and the charge density ρ at distance r from the ion. Strictly speaking, Poisson's equation is valid for a system of charges at rest, but it is assumed that the time-averaging process will take care of any difficulties in this respect.

The average charge-density ρ at a point depends on the probabilities of an element of volume at that point being occupied by various kinds of ions. We denote the various ionic species by subscripts 1, 2 . . . s and their algebraic valencies by z_i, so that the ionic charge $z_i e$ is positive for a cation and negative for an anion. Since the solution as a whole is electrically neutral,

$$\sum_{i=1}^{s} n_i z_i = 0 \qquad \ldots.(4.3)$$

where n_i denotes the average number of i-ions per unit volume, *i.e.*, the bulk concentration. Now we select one particular ion, say a j-ion, as the centre of the coordinate system. The condition of electrical neutrality tells us that the net charge in the whole solution outside this ion is $-z_j e$. Furthermore, the average charge density at any point outside the central ion must be of opposite sign to the charge on the central ion. For example, if a cation is chosen as the centre of coordinates, in any spherical shell at distance r from it there will, on an average, be more anions than cations; the shell will therefore carry a net negative charge, and the totality of such shells, forming the whole solution outside the central ion, will carry a total negative charge equal to the positive charge of the cation. This may be expressed by the equation:

$$\int_a^{\infty} 4\pi r^2 \rho_j \, \mathrm{d}r = -z_j e \qquad \ldots.(4.4)$$

The distance a represents the limit within which no other ion can approach the central ion. Here the subscript j has been attached to ρ as a reminder that ρ is a quantity defined only in the coordinate system based on the j-ion as centre. It would be useless to talk about the time-average charge density at a fixed point in space (fixed, say, relative to the containing vessel), for such an average would obviously be zero everywhere.

The probability of an ion of species i being found in a volume-element $\mathrm{d}V$ at distance r will clearly be greatest when its electrical potential energy at that distance is lowest; it must also be proportional to the bulk concentration n_i of the i-ions, and to the volume $\mathrm{d}V$ of the element considered. At great distances from the central ion the electrical forces due to the central ion must be negligible, and the probability must approach $n_i\mathrm{d}V$ simply. Beyond these restrictions there is no absolute means of knowing what the distribution-function is. Debye and Hückel assumed the Boltzmann distribution law, according to which, since the electrical potential energy of an i-ion is $z_i e\psi_j$, the average local concentration n_i' of i-ions at the point in question is:

$$n_i' = n_i \exp\left(-\frac{z_i e\psi_j}{kT}\right) \qquad \ldots.(4.5)$$

Once more the subscript j attached to ψ reminds us that, like ρ, ψ is only meaningful in the (moving) coordinate system based on the j-ion as centre. Before discussing alternative distribution laws we shall proceed to evaluate ρ_j from (4.5), the better to see what effect the alternatives may have. Since each i-ion carries a charge $z_i e$, the net charge density at the point considered is, summing for all ionic species,

$$\rho_j = \sum_i n_i z_i e \exp\left(-\frac{z_i e\psi_j}{kT}\right) \qquad \ldots.(4.6)$$

According to equation (4.6), the Boltzmann distribution thus leads to an exponential relation between the charge density ρ and the potential ψ. However, a theorem of electrostatics, known as the principle of the linear superposition of fields, states that the potential due to two systems of charges in specified positions is the sum of the potentials due to each system separately. Thus, if all the ionic charges and therefore the charge density were doubled, the potential at any chosen point would according to this principle be doubled also. Yet according to equation (4.6) the potential would not be doubled, since (4.6) is an exponential and not a linear relation. This dilemma is of fundamental importance in electrolyte theory.

Its significance becomes clearer if the exponentials in (4.6) are expanded in the form:

$$e^{-x} = 1 - x + \frac{x^2}{2!} - \frac{x^3}{3!} + \ldots$$

when we obtain:

$$\rho_j = \Sigma n_i z_i e - \Sigma n_i z_i e \left(\frac{z_i e \psi_j}{kT}\right) + \Sigma \frac{n_i z_i e}{2!} \left(\frac{z_i e \psi_j}{kT}\right)^2 - \ldots \qquad \ldots . (4.7)$$

The first term on the right of (4.7) vanishes by the condition of electrical neutrality (4.3), and if $z_i e \psi_j \ll kT$, only the term linear in ψ is appreciable, giving the result:

$$\rho_j = -\sum_{i=1}^{s} \frac{n_i z_i^2 e^2 \psi_j}{kT} \qquad \ldots . (4.8)$$

The result in this approximate form is consistent with the superposition principle, since it states that ψ is directly proportional to ρ. The approximation is, however, valid only when the potential energy $z_i e \psi_j$ of the i-ion is small compared to its thermal energy kT; and though this may well be true for the majority of the i-ions in a dilute solution, which are at relatively great distances from the central j-ion, it is not true for those which are close to the j-ion. Furthermore, it is well known that even fairly dilute solutions of electrolytes show *large* deviations from ideal behaviour, and the reason for this is that the energy of the electrical interactions between ions is, in fact, not small compared to their thermal energy. Nevertheless, we shall use the Debye-Hückel expression (4.8) for the charge density, but must recognize that in so doing we are in fact rejecting the Boltzmann distribution (4.5) and replacing it by the linear relation:

$$n_i' = n_i \left(1 - \frac{z_i e \psi_j}{kT}\right) \qquad \ldots . (4.9)$$

There is one special case where this approximation is less drastic: when we are dealing with a solution of a single electrolyte of symmetrical valency type. Putting $z_1 = -z_2$ and $n_1 = n_2$ equation (4.7) for the charge density becomes:

$$\rho_j = 0 - 2n_1 z_1 e \left(\frac{z_1 e \psi_j}{kT}\right) + 0 - \tfrac{1}{3} n_1 z_1 e \left(\frac{z_1 e \psi_j}{kT}\right)^3 + 0 - \ldots \qquad \ldots . (4.10)$$

all the terms in even powers of ψ vanishing. In this special case, therefore, the approximation (4.8) involves no error of order

$\left(\frac{z_1 e\psi_j}{\boldsymbol{k}T}\right)^2$ as it does in the general case. This approximation is forced upon us by the nature of the coulomb forces, although it means abandoning the Boltzmann distribution, a well-established principle of statistical mechanics; therefore the resulting theory should work best in cases where the formula (4.8) for ρ_j involves the least departure from the Boltzmann distribution. Equation (4.10) shows that this applies for solutions of uni-univalent electrolytes, and we shall find in practice that the theory has its most pronounced quantitative success for these.

We therefore proceed with the derivation of the theory, using expression (4.8) corresponding to the distribution law (4.9), for the charge density ρ_j. Substituting this in the Poisson equation (4.2) for the case of radial symmetry, we have:

$$\frac{1}{r^2}\frac{\mathrm{d}}{\mathrm{d}r}\left(r^2\frac{\mathrm{d}\psi_j}{\mathrm{d}r}\right) = \frac{4\pi \boldsymbol{e}^2}{\varepsilon \boldsymbol{k}T}\sum_i n_i z_i^2 \psi_i = \kappa^2\psi_j \qquad \ldots.(4.11)$$

where κ is defined by:

$$\kappa^2 = \frac{4\pi \boldsymbol{e}^2 \sum_i n_j z_i^2}{\varepsilon \boldsymbol{k}T} \qquad \ldots.(4.12)$$

and is a function of concentration, ionic charge, temperature and the dielectric constant of the solvent, having the dimensions of a reciprocal length. Equation (4.11) is a *linear* second-order differential equation between ψ and r. It will be noted that had we retained the exact Boltzmann distribution in calculating ρ, (equation 4.5), we should have obtained a much more difficult *non-linear* differential equation for ψ; this difficulty applies equally to the Eigen-Wicke distribution function discussed below.

The substitution $u = \psi_j r$ reduces equation (4.11) to the standard form:

$$\frac{\mathrm{d}^2u}{\mathrm{d}r^2} = \kappa^2 u,$$

which has the general solution:

$$u = Ae^{-\kappa r} + Be^{\kappa r}$$

or

$$\psi_j = A\frac{e^{-\kappa r}}{r} + B\frac{e^{\kappa r}}{r}$$

where A and B are constants of integration to be determined from the physical conditions of the problem. Since the potential must remain finite at great values of r it is necessary that $B = 0$. The evaluation of A may be made by putting $\psi_j = A\dfrac{e^{-\kappa r}}{r}$ into the expression (4.8) for the charge density obtaining:

$$\rho_j = -A\frac{e^{-\kappa r}}{r}\sum_i \frac{n_i z_i^2 e^2}{kT} = -A\frac{\kappa^2\varepsilon}{4\pi}\cdot\frac{e^{-\kappa r}}{r}$$

Using this value for ρ_j in equation (4.4), which equation is in effect a statement that the solution as a whole is electrically neutral, we have:

$$A\kappa^2\varepsilon\int_a^\infty re^{-\kappa r}\,dr = z_j e,$$

from which on integrating by parts we obtain:

$$A = \frac{z_j e}{\varepsilon}\cdot\frac{e^{\kappa a}}{1+\kappa a},$$

so that the potential ψ is given by:

$$\psi_j = \frac{z_j e}{\varepsilon}\cdot\frac{e^{\kappa a}}{1+\kappa a}\cdot\frac{e^{-\kappa r}}{r} \qquad \ldots.(4.13)$$

Equation (4.13) is Debye and Hückel's fundamental expression for the time-average potential at a point at distance r from an ion of valency z_j in the absence of external forces; from it all the various manifestations of the interionic forces may be calculated. The quantity a has been introduced as the 'distance of closest approach' of the ions, *i.e.*, the sum of their effective radii in solution. However, it is implicitly assumed that a is the same for all pairs of ions, which means that they are all taken as spheres of diameter a. This is a rather drastic approximation in the case of electrolytes such as lanthanum chloride where there is strong reason to believe (*e.g.*, from a consideration of the ionic mobilities) that the sizes of the ions actually differ considerably. It must also be remembered that equation (4.13) has been derived on the basis of the linear distribution function (4.9), except in the case of symmetrical valency-types where it is consistent with a closer approximation to the Boltzmann distribution, viz.,

$$n_i' = n_i\left[1 - \left(\frac{z_i e\psi_j}{kT}\right) + \frac{1}{2}\left(\frac{z_i e\psi_j}{kT}\right)^2\right] \qquad \ldots.(4.14)$$

ALTERNATIVE DISTRIBUTION FUNCTIONS

The first essential modification to the Debye-Hückel evaluation of the potential given above was made by MÜLLER[7] and by GRONWALL,

LaMer and Sandved[8]. This consisted of accepting higher terms in the expansion of the exponential Boltzmann distribution function, resulting in a series expansion for the potential, the leading term of which was identical with that of Debye and Hückel. We shall not deal further with this approach, details of which may be found in the original papers.

Eigen and Wicke[9] have attacked the problem of the distribution function for ions, on a somewhat different basis. If one considers the Boltzmann distribution used by Debye and Hückel,

$$n_i' = n_i \exp\left(-\frac{z_i e \psi_j}{kT}\right)$$

one sees that for the case where the i-ion is of opposite sign to the central j-ion, the argument of the exponential is positive, so that n_i' is greater than n_i; *i.e.*, the concentration of anions around a given cation is greater than the average concentration of anions in the bulk. There is, however, a physical upper limit to the concentration of anions; this is reached when, because of the size of the anions, no more can be packed into a given volume-element. Eigen and Wicke therefore introduce a quantity, $\mathcal{N}_i$, the 'besetzungszahl' or number of sites available to i-ions in unit volume; this is the reciprocal of the effective volume v_i occupied by a single (hydrated) i-ion. They then modify the distribution function in such a way that it is impossible for n_i' to exceed $\mathcal{N}_i$. This is done by writing:

$$\frac{n_i'/(\mathcal{N}_i - n_i')}{n_i/(\mathcal{N}_i - n_i)} = \exp\left(-\frac{z_i e \psi_j}{kT}\right) \qquad \ldots.(4.15)$$

i.e., by replacing the actual concentrations n_i' and n_i in equation (4.5) by the ratios of these to the numbers of empty sites available per cubic centimetre for ions of the kind i. However, just as in the Debye-Hückel treatment the exponential Boltzmann function has to be approximated to a linear expression, so with this equation the approximation:

$$\frac{n_i'}{n_i} = 1 - \frac{z_i e \psi_j}{kT}\left(1 - \frac{n_i}{\mathcal{N}_i}\right) \qquad \ldots.(4.16)$$

is ultimately necessary, and it is on this distribution function that the theory of Eigen and Wicke is actually based. This should be compared with equation (4.9). The approximation by which equation (4.16) is obtained from equation (4.15) is, however, not altogether convincing.

From this distribution function there follows for the charge-density the result:

$$\rho_j = -\sum_i \left(\frac{n_i z_i^2 e^2 \psi_j}{kT}\right)\left(1 - \frac{n_i}{N_i}\right)$$

The calculation of the potential ψ_j then follows the same lines as those given on pp. 78–79 and leads to the result:

$$\psi_j = \frac{z_j e}{\varepsilon} \cdot \frac{e^{\kappa' a}}{1 + \kappa' a} \cdot \frac{e^{-\kappa' r}}{r}$$

where $$\kappa' = \left[\frac{4\pi e^2}{\varepsilon k T} \cdot \sum_i n_i z_i^2 \left(1 - \frac{n_i}{N_i}\right)\right]^{1/2}$$

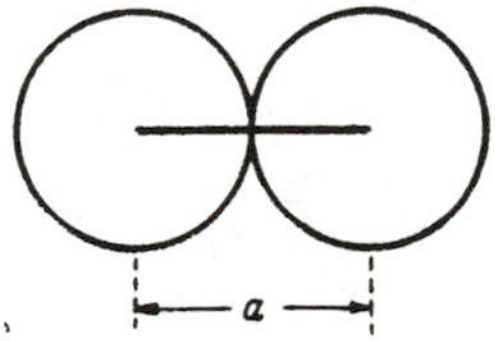

Figure 4.1

Defining an average number $\bar{N}$ of sites per 'molecule' of electrolyte by:

$$\frac{1}{\bar{N}} = \frac{\sum_i \nu_i z_i}{2\Sigma \nu_i} \sum \frac{1}{N_i}$$

and denoting the number of 'molecules' of electrolyte per cubic centimetre by n, one may write:

$$\kappa'^2 = \kappa^2 \left(1 - \frac{n}{\bar{N}}\right)$$

where κ is the ordinary quantity of the Debye-Hückel treatment.

Eigen and Wicke further propose that the mean effective volume of an ion should be calculated in terms of the parameter a, the distance of closest approach of the ions, by the relation,

$$\frac{4}{3}\pi a^3 = \frac{v_1 + v_2}{2}$$

This is, however, palpably inconsistent with the potential derivation since if a is the distance of closest approach it is the *sum* of the effective ionic radii, not their mean (*Figure 4.1*). If the ions have radii $a/2$ one has:

$$v_1 = v_2 = \frac{4}{3}\pi\left(\frac{a}{2}\right)^3$$

so that:

$$\tfrac{1}{2}(v_1 + v_2) = \tfrac{1}{6}\pi a^3$$

It would thus appear that the calculations of Eigen and Wicke are based on the use of effective volumes which are some eight times too large. Their justification for this course is that the value of a which affects the potential is the distance of closest approach of oppositely charged ions for which mutual partial penetration of the hydration shells may occur, while the value limiting the local concentrations of ions of the same sign (and thus giving N_i) is larger since such penetration will not occur.

In point of fact, it is easily shown that the simpler Boltzmann distribution adopted by Debye and Hückel can never lead to physically impossible high values of the local concentration of ions of one kind, when account is taken of the dimensions of the ions. The argument runs as follows:

It is clear that the maximum physically possible value of n_i' will occur when the i-ion is of opposite charge to the 'central' j-ion, when the concentration of the solution is as high as possible, and when the ions are as small as possible. For a fully dissociated 1 : 1 electrolyte, the minimum diameter of the ions is $a = \frac{1}{2}\frac{e^2}{\varepsilon kT}$; for ions smaller than this, Bjerrum has shown that ion-pair formation will occur so that the electrolyte can no longer be regarded as fully dissociated (see Chapter 14). Now the maximum theoretically possible concentration of spheres of diameter a will occur when they are in contact, with close-packing, and is given by:

$$n_{(\max)} = \frac{\sqrt{2}}{a^3} \text{ spheres per cubic centimetre}$$

Since these spheres must be half anions and half cations, we have as the maximum bulk concentrations attainable:

$$n_{1(\max)} = n_{2(\max)} = \frac{\sqrt{2}}{2a^3}$$

The maximum value of κ is therefore given by:

$$\kappa^2{}_{(\max)} = \frac{4\pi e^2}{\varepsilon kT}\cdot\frac{\sqrt{2}}{a^3}$$

and the maximum value of (κa) by:

$$(\kappa^2 a^2)_{(\max)} = \frac{4\pi e^2}{\varepsilon kT}\frac{\sqrt{2}}{a}$$

Now putting $a = \frac{1}{2}\frac{e^2}{\varepsilon kT}$ we have,

$$(\kappa^2 a^2)_{(\max)} = 8\pi\sqrt{2} = 35{\cdot}54, \text{ or } (\kappa a)_{(\max)} = 5{\cdot}96$$

This is the highest theoretically attainable value of (κa) for a fully ionized 1 : 1 electrolyte: in aqueous solutions at 25° it demands a concentration of 26 mole/litre, which is never obtained in practice owing to limitations of solubility.

The Debye-Hückel expression for the potential at distance r from a cation in a 1 : 1 electrolyte:

$$\psi = \frac{e}{\varepsilon}\frac{e^{\kappa a}}{1 + \kappa a}\frac{e^{-\kappa r}}{r}$$

reaches a maximum, as far as neighbouring ions are concerned, at the minimum physically possible value of r, $r = a = \frac{1}{2}\frac{e^2}{\varepsilon kT}$, where:

$$\psi_{(\max)} = \frac{e}{\varepsilon}\frac{1}{a(1 + \kappa a)}$$

At this distance, *i.e.*, in contact with the central ion, which we will take to be a cation for the present illustration, the concentration of anions is at its maximum given by:

$$n'_{2(\max)} = n_{2(\max)} \exp\left[\frac{e^2}{\varepsilon kT} \cdot \frac{1}{a(1 + \kappa a)}\right]$$

In this expression, we now put $a = \frac{1}{2}\frac{e^2}{\varepsilon kT}$, and $\kappa a \approx 6$, when the result becomes:

$$n'_{2(\max)} = n_{2(\max)}\, e^{2/7} = 1{\cdot}33 n_{2(\max)} = 0{\cdot}67 n_{(\max)}$$

If instead of the exact Boltzmann distribution we assume the approximate distribution law (4.9), the corresponding result is $n'_{2(\max)} = 1{\cdot}28 n_{2(\max)}$.

The important conclusion to be drawn from these figures is that even when the most extreme conditions of concentration and small ion size are assumed, the Debye-Hückel formulae do not lead to impossibly high values for the local concentration of ions: for the figures show that at the worst only two-thirds of the available 'sites' near the central ion need be filled by ions of one kind in order to satisfy the equations of the Debye-Hückel theory. At lower concentrations than the extreme one considered here, ψ_j will be greater and n'_2/n_2 will therefore be greater than 1·33; but this does not involve any difficulty since n_i will be much smaller and n'_i is therefore unable to exceed the physically possible limit.

It is also of some interest to estimate the maximum value of the quantity $\left(\frac{z_i e \psi_j}{kT}\right)$ which measures the ratio of the electrical energy of an i-ion to its thermal energy. According to the Debye-Hückel equation for ψ_i this quantity is given by:

$$\frac{z_i e \psi_j}{kT} = \frac{z_i z_j e^2}{\varepsilon kT} \frac{e^{\kappa a}}{1 + \kappa a} \frac{e^{-\kappa r}}{r} \qquad \ldots .(4.17)$$

and is greatest when the i-ion is as close as is physically possible to the central j-ion, *i.e.*, at $r = a$. If we take as the minimum a value of interest the Bjerrum critical distance (cf. eq. 14.1):

$$a = q = \frac{|z_1 z_2| e^2}{2\varepsilon kT}$$

this gives:

$$\left|\frac{z_i e \psi_j}{kT}\right|_{(\max)} = \frac{2}{1 + \kappa a}$$

Thus the electrical energy of the i-ion does not exceed $2kT$, and diminishes as the concentration increases. The maximum value is of course only attained by ions in actual contact with the central ion. To evaluate $\frac{z_i e \psi_j}{kT}$ for greater distances, we may consider distances, a, $2a$, $3a$, *etc.*: at the distance $r = pa$ we have by equation (4.17):

$$\left(\frac{z_i e \psi_j}{kT}\right)_{r=pa} = \frac{z_i z_j e^2}{\varepsilon kT} \cdot \frac{e^{\kappa(1-p)a}}{pa(1 + \kappa a)}$$

This quantity is shown in graphical form for several values of (κa) in *Figure 4.2.* It will be seen that the ratio of the electrical to the thermal energy cannot be regarded as small compared to unity in dilute solutions until considerable distances from the central ion are reached; but that, rather unexpectedly, this ratio is smaller and more rapidly decreasing with increasing distance in more concentrated solutions. At distances up to a few ionic diameters, however, the approximation:

$$\exp\left(-\frac{z_i e \psi_j}{kT}\right) \approx 1 - \frac{z_i e \psi_j}{kT}$$

made in the Debye-Hückel treatment cannot be justified on the ground that $z_i e \psi_j$ is small compared to kT as is usually claimed: instead it must be justified on grounds of mathematical expediency in order to obtain a distribution function consistent with the principle of the linear superposition of fields. Unfortunately, we now see that this linear approximation can lead to absurd results in

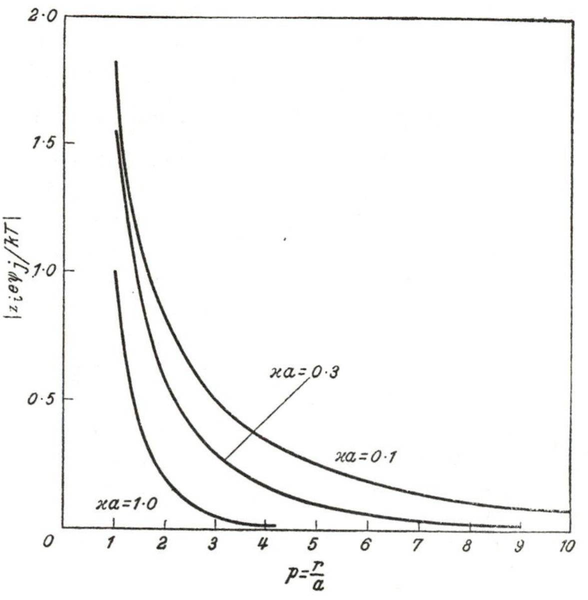

Figure 4.2. Variation of the quantity $|z_i e\psi_j/(kT)|$ *with concentration and with distance from the central j-ion. The unit of distance is taken as the Bjerrum critical distance for ion-pair formation,* $a = \frac{1}{2}\frac{|z_1 z_2|e^2}{\varepsilon kT}$. *For* 1 : 1 *electrolytes in water at* 25°, $a = 3{\cdot}57$ Å, *and the curves for* $\kappa a = 0{\cdot}1$, $0{\cdot}3$ *and* $1{\cdot}0$ *thus correspond to concentrations of* 0·0073 N, 0·0653 N *and* 0·726 N *respectively.*

certain cases; for the local concentration of ions of the *same* species as the central ion is given by:

$$n_j' = n_j\left(1 - \frac{z_j e\psi_j}{kT}\right)$$

and if, as it now appears, there are regions of distance and concentration where $z_j e\psi_j > kT$, the local concentration of j-ions becomes negative! This absurdity is, however, remedied if we consider the next term of the expansion of the Boltzmann exponential expression and write:

$$n_j' = n_j\left[1 - \frac{z_j e\psi_j}{kT} + \frac{1}{2}\left(\frac{z_j e\psi_j}{kT}\right)^2\right] \qquad \ldots .(4.18)$$

a course which, as we have shown above, is justified for symmetrical valency-type electrolytes where it does not violate the requirement of linear superposition. For since outside the Bjerrum region $z_j e\psi_j < 2\mathbf{k}T$, n'_j cannot by equation (4.18) fall below zero. Thus once more we reach the conclusion that the theory is only really adequate for symmetrical electrolytes. The use of the modified distribution function proposed by Eigen and Wicke offers no solution of this particular difficulty, for as *Figure 4.2* shows, the greatest values of the vital quantity $\frac{z_i e\psi_j}{\mathbf{k}T}$ are attained in dilute solutions, where the extra factors introduced by Eigen and Wicke make no material difference. We may, however, conclude from a study of *Figure 4.2* that, since the Debye-Hückel treatment of the potential problem is known from experiment to be successful for dilute solutions of fully dissociated symmetrical electrolytes, it should, if anything, be more so in more concentrated solutions; and for 1 : 1 electrolytes, at least, it does not lead to any physically absurd distributions such as negative concentrations of ions, or concentrations too high to be consistent with the known sizes of the ions. We therefore intend to use the Debye-Hückel expression (4.13) for the potential as a basis for all our theoretical calculations, along with the distribution function (4.14) for symmetrical electrolytes and with the less adequate (4.9) for unsymmetrical ones.

The alternative distribution functions which we have discussed offer no improvement in respect of the self-consistency of the resulting theory, but have the disadvantage of adding considerably to the complexity of the formulae.

REFERENCES

1. BECK, J., *Phys. Z.*, 40 (1939) 474
2. MILNER, S. R., *Phil. Mag.*, 23 (1912) 551; 25 (1913) 742
3. DEBYE, P. and HÜCKEL, E., *Phys. Z.*, 24 (1923) 185
4. BJERRUM, N., *K. danske vidensk. Selsk.*, 7 (1926) No. 9
5. ONSAGER, L., *Phys. Z.*, 28 (1927) 277
6. FALKENHAGEN, H., LEIST, M. and KELBG, G., *Ann. Phys. Lpz.*, 11 (1952) 51
7. MÜLLER, H., *Phys. Z.*, 28 (1927) 324; 29 (1928) 78
8. GRONWALL, T. H., LAMER, V. K. and SANDVED, K., *ibid.*, 29 (1928) 358
9. WICKE, E. and EIGEN, M., *Naturwissenschaften*, 38 (1951) 453; 39 (1952) 545; *Z. Elecktrochem.*, 56 (1952) 551; 57 (1953) 319; *Z. Naturf.*, 8a (1953) 161

THE MEASUREMENT OF CONDUCTIVITIES AND TRANSPORT NUMBERS

METHODS OF MEASURING ELECTROLYTIC CONDUCTIVITY

THE requirements for precise measurement of electrolytic conductivity may be summed up as (*a*) accurate temperature control, (*b*) avoidance of polarization at the electrodes, and (*c*) accuracy in the electrical measurements themselves.

As regards temperature control, the majority of aqueous electrolytic solutions have a temperature coefficient of conductivity close to 2 per cent per degree at 25°. That of the hydrogen ion is appreciably lower, about 1·4 per cent per degree at 25°. If an accuracy of 0·01 per cent is sought, the thermostat should therefore be capable of controlling the temperature to $\pm$ 0·005° or better throughout the measurements. When, as is frequently the case, the solution studied has a temperature coefficient similar to that of the standard potassium chloride solution used for the cell calibration, it is less important to know the exact temperature with the same accuracy provided that it is constant, for a constant error of a few hundredths of a degree will largely be compensated by a corresponding change in the conductivity of the standard. This, of course, does not apply when the temperature coefficients differ appreciably or when the measurements and the cell calibration are made at different temperatures; in such cases sensitive and recently-calibrated thermometers, or preferably a platinum resistance thermometer, are desirable. The use of water as a thermostat liquid should be avoided owing to undesirable capacity effects across the cell walls in a.c. measurements and to the risk of electrical leakage currents in d.c. measurements. A light paraffin such as kerosene is a satisfactory thermostat medium at ordinary temperatures. The errors caused by the use of water as a thermostat liquid were thoroughly investigated by JONES and JOSEPHS[1], who found differences of up to 0·5 per cent between the resistance in an oil- and a water-filled thermostat. The errors varied in a complicated manner with the cell design, the conductivity of the thermostat water, and the resistance of the cell being measured; they were greater at higher frequencies and with high cell resistances, indicating that they arose mainly from capacitance by-paths through the cell walls and the thermostat water. Earthing the thermostat tank changed the sign of the errors, and reduced their magnitude, but by no means eliminated them. Rather surprisingly, they were smaller when the conductivity of the thermostat water was increased by the addition of potassium chloride.

Polarization errors are usually minimized by the use of audio frequency alternating current for the measurements and by coating the electrodes with a heavy deposit of platinum black, a system initiated by KOHLRAUSCH(2). While this procedure correctly applied is undoubtedly effective, the use of alternating current enormously complicates the electrical technique required for high-accuracy measurements, owing to the need for compensation of capacitative and inductive effects in the circuit. The only alternative is, however, the use of electrodes which are truly reversible to one of the ions in the solution; this permits the use of the simpler direct current measuring techniques. Though the d.c. method has received increasing attention in recent years the conventional a.c. method seems likely to remain the standard one for general applications.

ALTERNATING CURRENT CONDUCTANCE MEASUREMENTS

In the simple Wheatstone bridge (*Figure 5.1*) used for d.c. resistance measurements, the galvanometer shows no deflection at balance; the potentials at A and B are therefore equal, whence $R_1/R_2 = R_3/R_4$. In the a.c. bridge (*Figure 5.2*) the battery is replaced by a sinusoidal alternating potential from an oscillator, and the galvanometer by a suitable detector. The condition for balance (*i.e.*, no signal in the detector) is that the alternating potentials at A and B are of equal amplitude and exactly in phase, which leads to the relation $Z_1/Z_2 = Z_3/Z_4$, where the impedance Z is the a.c. analogue of resistance.

Impedance is conveniently represented as a complex quantity having the following properties: (*i*) impedances combine like resistances, *i.e.*, impedances in series add, while if in parallel their reciprocals (admittances) add; (*ii*) a pure resistance R has an impedance $Z = R$ which is entirely a real quantity; (*iii*) a perfect condenser of capacity C has impedance $Z = 1/(\mathrm{j}\omega C)$ where ω is angular frequency and j is an operator having the mathematical properties of $\sqrt{-1}$, which represents a phase displacement of 90° between current and potential; (*iv*) a pure inductance L has impedance $Z = \mathrm{j}\omega L$. This representation of impedances by complex numbers

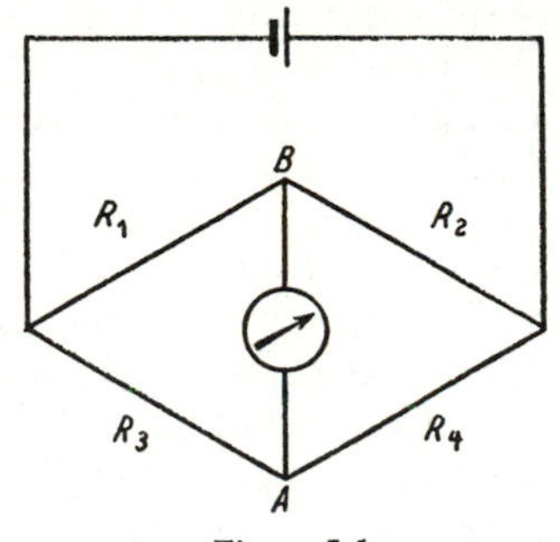

Figure 5.1

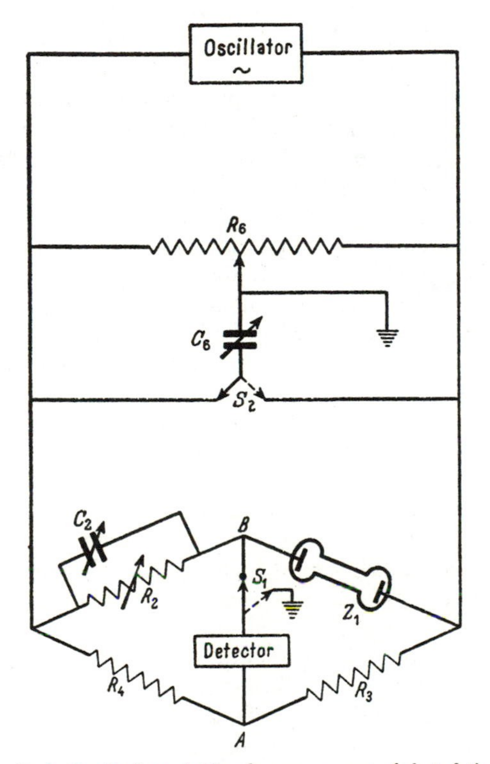

Figure 5.2. Basic circuit of a.c. bridge for measurement of electrolytic conductance

has considerable utility since equality of two impedances demands equality of both real and imaginary parts, so that amplitudes and phases in any part of an a.c. network can in principle be calculated by methods formally similar to those used in d.c. work.

The design and construction of high-precision conductance bridges was studied intensively by GRINNELL JONES[3–9] and his colleagues and also by SHEDLOVSKY[10] and the principles they laid down are the basis of today's designs. Referring to *Figure 5.2*, the ratio-arms R_3 and R_4 are made equal (usually 1000 ohms) and of identical construction so that the residual capacities between turns and between the coils and nearby objects are exactly equal. The usual device of winding with doubled wire ensures that the inductances of all resistance coils in the bridge are negligible at audio-frequencies. The measuring-arm R_2 may be a separate resistance-box but is usually built into the same box as R_3 and R_4. In parallel with R_2 is a variable capacitor C_2 of maximum capacity 0·001 microfarad; this is necessary to obtain a sharp balance-point since the cell will in general have an impedance Z_1 which is not purely resistive. A further valuable aid to a sharp balance-point is the 'Wagner earth' comprising R_6 and C_6; the object of these components, which are used in conjunction with the earthing-switch S_1, is to ensure that at balance the potentials at A and B are not merely equal but are actually earth-potential so that pick-up of hum and stray noise by the detector is minimized. A rough balance is first obtained with S_1 in the position shown by the full line; then S_1 is turned to earth and R_6 and C_6 are adjusted to give a minimum signal in the detector; S_1 is then restored to its original position and a final balance is obtained. IVES, PRYOR and FEATES[11] obtain an equivalent effect by the use of two 1000 Ω radio potentiometers connected across the output of the oscillator, the moveable tappings being taken to the bridge input.

The oscillator and detector are important auxiliaries. The oscillator should give a good sinusoidal wave-form at all frequencies from 500 c/s to several thousand. The amplitude should be variable from a few volts down to very low values and both the output terminals should be isolated from earth. It is best to isolate the bridge from both the oscillator and the detector by good quality transformers as otherwise the Wagner earth will not function properly. The detector consists first of one or two stages of amplification, in which an automatic gain control may be incorporated to limit the maximum signal when the bridge is far off balance. The amplifier may be followed by a telephone headset, which is remarkably sensitive around 1000 c/s, but the tendency today is to use a cathode ray oscilloscope which is much more versatile and less nerve-racking. The horizontal deflection plates are fed from the oscillator and the vertical ones from the amplifier; the out-of-balance trace is an ellipse which becomes a horizontal straight line at balance. An outfit of this kind can readily detect a variation of a few parts in a million in the bridge setting.

Cell design. The object of the measurements is to determine the pure ohmic resistance R_1 of the solution between the electrodes. If the cell impedance Z_1 consisted only of this resistance, R_1 would equal R_2 at all frequencies and the capacitor C_2 would be required only to compensate for capacity between the cell leads and for the small capacity in parallel with the cell due to the action of its electrodes as a capacitor with the solution as dielectric. In practice there are several other sources of impedance which cause R_2 and C_2 to vary appreciably with frequency. Some of these can be avoided by proper design and the others are inherent in the electrode processes. The former comprise the Parker effect and the effect of a conducting thermostat medium discussed on p. 87. The Parker effect arises when the cell leads pass near to the cell solution giving the effect of a capacitor connected between one end of R_1 and some point in the middle of R_1; it can be avoided by spacing the leads well away from parts of the cell containing solution as in the designs shown in *Figure 5.3.* Mercury-filled lead-in tubes are often used but are a nuisance: we replace them by heavy silver wires welded to the outer ends of the electrodes beyond the seals.

Designs for electrode assemblies which can be used as dipping electrodes in containers of any size are given by BRODY and FUOSS[12].

The effects associated with the electrode processes themselves are of interest and their understanding is necessary in order to eliminate them from the measurements. KOHLRAUSCH[2] showed that this could be largely achieved by coating the electrodes with platinum-black, when R_2 becomes practically independent of frequency; this course is, however, not always possible since platinum-black may catalyse unwanted reactions and may in dilute solutions adsorb appreciable quantities of solute, making necessary emptying and refilling of the cell until a constant reading is obtained. The platinizing solution recommended by JONES and BOLLINGER[8] is 0·025N hydrochloric acid containing 0·3 per cent of platinic chloride and 0·025 per cent of lead acetate; the lead acetate improves the adherence of the deposit. The platinizing current should be 10 mA /cm^2, the polarity being reversed every ten seconds. Even a barely visible deposit greatly reduces the frequency dependence and a deposit corresponding to a few coulombs/cm^2 is ample.

Electrode effects can also be eliminated by eliminating the electrodes. This is done physically in the transformer bridge due to CALVERT[13] *et al.*, in which a loop of solution links a current transformer with a voltage transformer; and virtually by the double cell

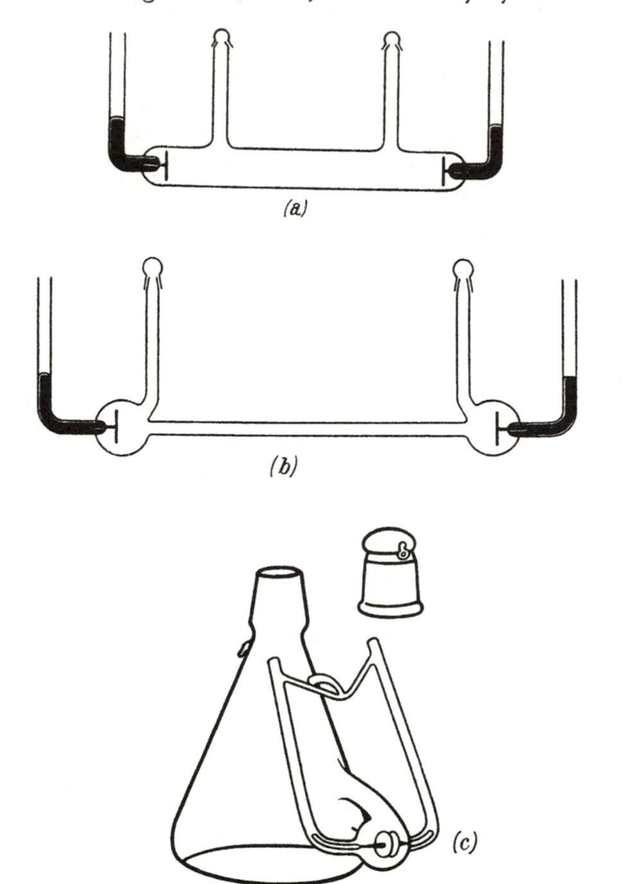

Figure 5.3. Typical conductivity cell designs for (a) moderate, (b) high, and (c) low concentrations; the last after DAGGETT, H. M., BAIR, E. J. and KRAUS, C. A., *J. Amer. chem Soc.*, 73 (1951) 799

design of FEATES, IVES and PRYOR[11] who use two cells with identical electrodes but with different lengths of solution between them and measure the difference in the resistances of the two cells, this difference showing only a small residual frequency dependence. They also employ two independent leads to each electrode so that lead resistances can be completely eliminated by the 'four leads' method used in platinum resistance thermometry. By these means they have attained remarkable accuracy in the conductometric measurement of ionization constants.[14]

The elimination of electrode effects from ordinary cells, without platinization, is still of great importance. The modern theory of electrode processes leads to a schematic representation of the conductance cell shown in *Figure 5.4*, which is that proposed by IVES' school[11] with the addition of the 'Warburg impedance' –W–. In this

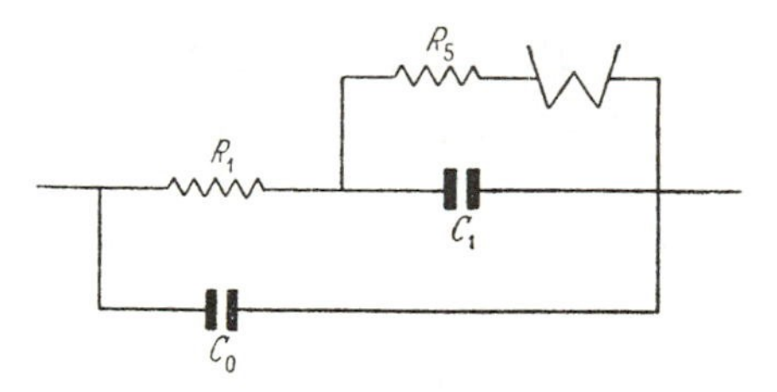

Figure 5.4. Network electrically equivalent to conductance cell

figure R_1 is the true ohmic resistance of the electrolyte, which is to be determined. This is independent of frequency at audio frequencies, because the Falkenhagen effect, associated with the relaxation of the ionic atmospheres, does not become appreciable until radio frequencies are attained. In series with R_1 is the capacity C_1 of the double layer of ions at the electrode surfaces: this is also expected to be independent of frequency. Because of the small thickness of the double layer, this capacity is surprisingly large, often amounting to several microfarads per sq. cm. of electrode surface. The current through R_1 is transported across the double layer mainly by virtue of this capacity without any actual discharge or formation of ions, for the cell gives definite resistance readings when the potential across it is only a few millivolts, far too little to cause electrolysis of most solutions at bright platinum electrodes. However, some electrolysis will normally occur simultaneously, perhaps through the depolarizing action of dissolved oxygen and the discharge of ions of the solvent, and in some cases through reversible discharge of ions of the electrolyte, for example in a cell with silver electrodes in a solution of silver nitrate. The electrolysis process is represented as a 'faradaic leakage' in parallel with the double layer. In general, as shown by GRAHAME[15] and by RANDLES,[16] it will consist of two parts: a pure resistance R_5, independent of frequency, and a 'Warburg impedance' at the electrodes. For the full theory of the Warburg impedance, the original literature should be consulted: here we merely note that it can be regarded as equivalent to a resistance and a capacity in series, the impedance of both being the same at any one frequency but both varying inversely as $\omega^{\frac{1}{2}}$. They can therefore be represented together by:

$$-\text{W}- = k(1 - \text{j})/\sqrt{\omega}$$

where k is a constant of dimensions (resistance $\times$ time$^{-\frac{1}{2}}$). Solution of the balance conditions leads to the result that, if the impedance of the cell arm *with the omission of* C_0 is denoted by Z, then:

$$1/R_2 = \text{real part of } Z^{-1}$$

from which R_2 may be determined in terms of R_1, C_1, R_5 and k. The following special cases are of interest.

(*i*) R_5 infinite: the electrodes are ideally polarized and:

$$R_2 = R_1 + (\omega^2 C_1^2 R_1)^{-1} \qquad \ldots.(5.1)$$

This is unlikely to arise in practice, as pointed out by Ives, Pryor and Feates, because of the depolarizing action of dissolved oxygen: it might perhaps be expected in solvents of very low self-dissociation and with an electrolyte whose ions have high discharge potentials.

(*ii*) If the Warburg impedance is negligible compared to R_5,

$$R_2 = R_1 + \frac{R_1 R_5 + R_5^2}{R_1(1 + \omega^2 C_1^2 R_5^2) + R_5}$$

which, since in any ordinary conductance cell $R_1 >> R_5$, approximates well to:

$$R_2 = R_1 + R_5/(1 + \omega^2 C_1^2 R_5^2) \qquad \ldots.(5.2)$$

This is the model proposed by Feates, Ives and Pryor as applicable to a conductance cell with grey platinized electrodes; it also corresponds to the behaviour of a cell with bright platinum electrodes in an aqueous solution. Although equation (5.2) is ill-adapted to graphical extrapolation to infinite frequency, it is found in practice[17] that R_1 as obtained by solving it for three frequencies agrees well with the value obtained by a *linear* extrapolation of R_2 against ω^{-1}.

(*iii*) If C_1 is very large, so that its impedance is small compared with that of the faradaic leakage, $R_2 = R_1$ at all frequencies. A very near approach to this behaviour is found with heavily blacked platinum electrodes.

(*iv*) Where the Warburg impedance is large compared to R_5 but small compared to R_1, one obtains solutions approximating to:

$$R_2 = R_1 + k/\sqrt{\omega} \qquad \ldots.(5.3)$$

This corresponds to the behaviour found by JONES and CHRISTIAN[7] in their studies of electrode polarization, more particularly at silver electrodes in silver nitrate solution or at platinum electrodes in acid solutions. It does not appear to be as general as was thought earlier for bright platinum electrodes: Jones and Christian themselves found marked curvature of the R_2 vs. $1/\sqrt{\omega}$ curves for such electrodes in potassium chloride solution and BRODY and FUOSS[12] report a number of cases in which R_2 is quadratic in $k/\sqrt{\omega}$. These are presumably intermediate between (*ii*) and (*iv*).

To sum up, it is necessary when using bright platinum electrodes to measure at a number of frequencies, preferably including higher frequencies than the usual 2 Kc/s., and to extrapolate to infinite frequency according to the kind of frequency dependence observed.

Standards of specific conductance—The actual measurements are of the resistance between two electrodes of fixed shape and size in a cell filled with the solution. This resistance naturally depends on the geometry of the cell as well as on the dimensions and separation of the electrodes; it is therefore the invariable practice to calibrate the cell by means of a solution of known specific resistance. It is usual to define a cell constant a by

$$K_{sp} = a/R,$$

R being the measured resistance with a solution of specific conductivity, K_{sp}, in the cell. To provide such a standard, a great deal of careful work has been expended on determinations of the specific resistance of potassium chloride solutions. The standards generally accepted today are those of JONES and BRADSHAW[6], and in view of their fundamental importance the methods used in obtaining them will now be described.

At the time the work was done, the accepted unit of electrical resistance was the international ohm, defined as the d.c. resistance at the ice-point of a uniform column of mercury 106·300 cm in length and of 14·4521 g mass. Thus a conductivity cell could be calibrated in international ohm units by measuring its resistance when filled with mercury at 0°C. Jones and Bradshaw prepared cells having resistances of approximately one ohm when filled with mercury at 0°C, and measured their resistance (to direct current) on a Kelvin bridge in terms of the international ohm. However, it was not possible to use these cells directly for determining the specific conductivity of standard potassium chloride solutions, since a one molar solution at 0° would have had a resistance of the order of 100,000 Ω, which is too high for accurate determination on an a.c. bridge. They therefore measured the conductivity at 0° of a relatively concentrated (6 N) sulphuric acid solution, which was then used to calibrate smaller cells. In these smaller cells, the specific conductivity of one demal solutions (see below) was measured at 0°, 18° and 25°, with allowance for the thermal expansion of the cell at the higher temperatures. The one demal potassium chloride solution was then used to calibrate still smaller cells in which the conductivity of 0·1 demal potassium chloride was measured and finally the 0·1 demal solution was used to calibrate even smaller cells in which the conductivity of 0·01 demal potassium chloride was found. Thus the series of steps by which the standard values for potassium chloride were obtained may be represented:

$$\text{International ohm} \rightarrow \text{Hg} \rightarrow H_2SO_4 \rightarrow 1\ \text{D KCl} \rightarrow 0{\cdot}1\ \text{D KCl} \rightarrow 0{\cdot}01\ \text{D KCl}.$$

In these measurements, the exact concentration of the sulphuric acid was not required as it served only as an intermediate standard; but the concentration of the potassium chloride was very carefully defined in terms of the weight of salt in 1,000 g of solution. Appendix 5.1 gives the compositions and specific conductivities of the three standard solutions determined by Jones and Bradshaw.

The term 'demal' was introduced by PARKER and PARKER[18] in an earlier determination of standards; it is not in general use as a measure of concentration, but the name was retained by Jones and Bradshaw as a convenient label for their standard compositions. It will be noted that Jones and Bradshaw's standards, being defined only in terms of weights *in vacuo*, are independent of volume standards and of atomic weights, changes in both of which have caused considerable confusion ever since the first standards were proposed by Kohlrausch. It is particularly unfortunate that the extensive compilation of electrolytic conductivities in the International Critical Tables is in terms of the earlier Parker and Parker standards, which are now generally considered to be unsatisfactory. Practically all recent work, in English-speaking countries at least, has, however, been based on the Jones and Bradshaw standards (Appendix 5.1), and for the sake of consistency they should be retained even if future work shows them to be slightly in error. Already one change has occurred which emphasizes the difficulty of defining a standard to a high degree of accuracy: the international ohm is no longer the recommended unit of resistance, having been replaced by the absolute ohm which is defined in terms of the fundamental units of the c.g.s. electromagnetic system. The relation between the absolute and international ohm is:

$$1 \text{ int. ohm} = 1{\cdot}00050 \text{ abs. ohm.}$$

It follows that the measure of a given resistance in absolute ohms is 0·050 per cent larger than its measure in international ohms; and the measure of a given specific conductivity in (abs. ohm)$^{-1}$ cm^{-1} 0·050 per cent smaller than in (int. ohm)$^{-1}$ cm^{-1}.

However, there seems to be no point in revising all the literature values of conductivities of electrolytes into (abs. ohm)$^{-1}$ cm^{-1}, as we are far more concerned with the *variation* of conductivity with concentration than with its exact value to five significant figures; theory is unable to predict conductivities *a priori* to even two significant figures, though it can do much better with the *change* of conductivity with concentration. The fact that resistance bridges are now calibrated in absolute ohms need not cause any difficulty, as the usual experimental determination is not that of an actual specific conductance, but its ratio to the specific conductance of one of the standard solutions, *via* the cell constant. Hence as long as the standard is defined in international units, the conductivity of the substance studied will be in the same units. It must be noted that the standard specific conductivities recorded in Appendix 5.1 are corrected for the specific conductivity of the water used in preparing them. As this is usually of the order of $1 \times 10^{-6}\ \Omega^{-1}\ \text{cm}^{-1}$, it will not be significant for the 1 demal standard, but must be allowed for when the 0·1 D and especially the 0·01 D solutions are used for cell calibration; in the last it can make a difference of the order of 0·1 per cent to the cell constant.

Variation of the Cell-constant with Temperature

The standard solutions specified by Jones and Bradshaw have accurately known conductivities at 0°, 18°, and 25°. For work at other temperatures, the cell constant measured at one of the standard temperatures must be adjusted slightly to allow for expansion of the glass and the platinum electrodes. The naïve expectation that the correction factor would be the same regardless of the geometry of the cell is disappointed, as the following argument shows.

Treating the cell as consisting of a number of regions in each of which the current density is uniform, the cell constant is given by

$$a = \Sigma \frac{l}{A}$$

where l is the length of each region and A its cross-sectional area normal to the current. Two extreme types of cell design may now be considered:

(*i*) A long narrow tubular cell with large electrodes in bulbs at the ends. Here nearly all the resistance is contributed by the narrow tube. If the glass has a linear expansion coefficient α_g, the relative change in a with temperature (t) is given by:

$$\frac{1}{a}\frac{\mathrm{d}a}{\mathrm{d}t} \approx \frac{1}{l}\frac{\mathrm{d}l}{\mathrm{d}t} - \frac{1}{A}\frac{\mathrm{d}A}{\mathrm{d}t}$$

$$= \alpha_g - 2\alpha_g = -\alpha_g$$

(*ii*) A cell consisting of two large electrodes of area A separated by a small distance l. The electrodes are supported by platinum wires sealed into the cell at points distant S apart,

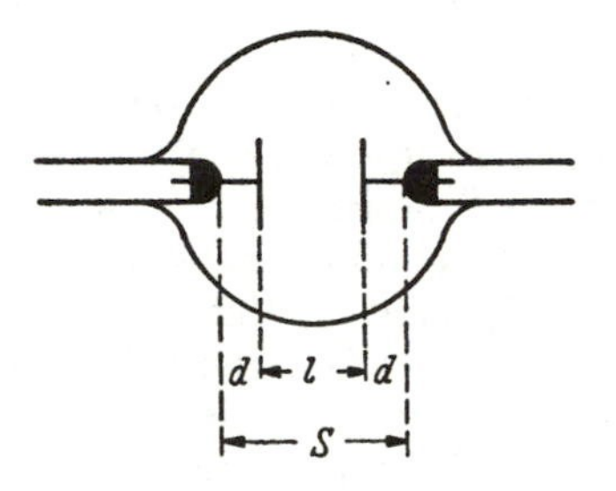

the length of wire between the electrode and the seal being d in each case so that the separation of the electrodes is $l = S - 2d$. Thermal expansion has three distinct effects:

(*a*) The area A of the electrodes is increased.
(*b*) The distance S is increased.
(*c*) The distances d are increased.

The cell constant is given approximately by:

$$a = \frac{l}{A} = \frac{S - 2d}{A}$$

By logarithmic differentiation with respect to temperature one obtains the temperature coefficient:

$$\frac{1}{a}\frac{\mathrm{d}a}{\mathrm{d}t} = \alpha_g \frac{S}{S - 2d} - 2\alpha_{Pt}\frac{S - d}{S - 2d}$$

If the expansion coefficients of the glass and the platinum are equal, as is approximately the case with a soda-glass cell, we have

$$\frac{1}{a}\frac{\mathrm{d}a}{\mathrm{d}t} = -\alpha_g$$

just as for the long tubular cell. In this case therefore the expansion correction is the same for both, and the result clearly generalizes to cells of any shape. But if Pyrex glass is used, differences can arise. Taking

$$\alpha_g = 3{\cdot}6 \times 10^{-6} \text{ deg C}^{-1}, \text{ and } \alpha_{Pt} = 9 \times 10^{-6} \text{ deg C}^{-1}$$

we have for the two cases:

(*i*) $\dfrac{1}{a}\dfrac{\mathrm{d}a}{\mathrm{d}t} = -3{\cdot}6 \times 10^{-6} \text{ deg C}^{-1}$

(*ii*) Putting $S = 10$ mm and $d = 2$ mm

$$\frac{1}{a}\frac{\mathrm{d}a}{\mathrm{d}t} = -18 \times 10^{-6} \text{ deg C}^{-1}$$

so that for a temperature change of 100° the cell constants would change by 0·04% and 0·18% respectively.

A very large temperature coefficient may occur if the electrodes are very close together and supported by rather long wires sealed into Pyrex glass: thus for example if $S = 20$ mm and $d = 9$ mm, giving electrodes only 2 mm apart supported on 9 mm lengths of wire:

$$\frac{1}{a}\frac{\mathrm{d}a}{\mathrm{d}t} = 63 \times 10^{-6} \text{ deg C}^{-1}$$

corresponding to a change of 0·63 per cent in the cell constant for 100° change.

Direct Current Conductivity Measurements

It will be clear from the foregoing account that the alternating current method for conductivity measurements, though capable of extreme accuracy, introduces a great many new complications due to capacity effects in the circuit; the compensating advantages are the elimination of polarization, the fact that electronic amplifiers are easily incorporated in the detector circuit and that thermo-electric effects and contact potentials in the resistance box are unimportant.

The direct current method is therefore simpler in principle, requiring only the passage of a steady current through the solution and through a standard resistance in series, and the comparison of the potential developed between two fixed points in the solution with that across the standard resistance. Since potentiometric measurements can be made with an accuracy of 0·001 per cent the method should be capable of accuracy comparable with that of the best a.c. technique. It is, however, essential that the potential measurements in the solution be made between strictly reversible

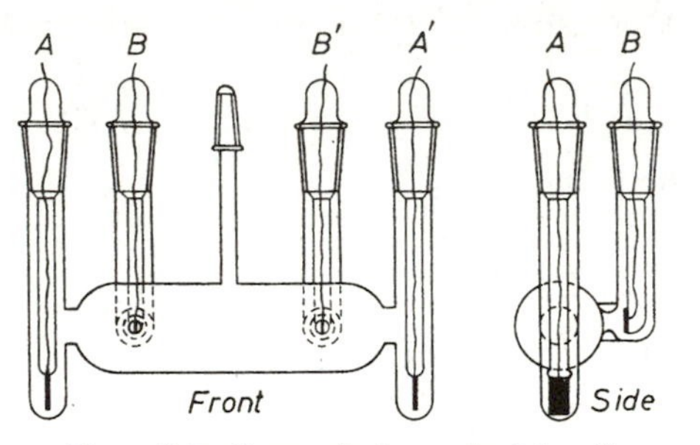

Figure 5.5. GORDON's *d.c. conductivity cell*

electrodes. The greatest success with the d.c. method to date has been achieved by GORDON and his collaborators[19] on dilute halide solutions in water and in methanol.

A slightly modified form of their cell, described by ELIAS and SCHIFF[19a], is shown in *Figure 5.5.*

A cylindrical Pyrex tube, about 20 cm in length and 5 cm in diameter, has two side-tubes about 10 cm apart to hold the probe-electrodes, *B*, *B'*, each of which is made of an 8 mm platinum disc, so mounted that its position is not subject to appreciable variation in successive experiments. Each electrode is covered with fused glass except for a narrow slip 1 × 6 mm in size which is silver-plated and chloridized (or bromidized). The electrodes, *A*, *A'*, which introduce the current are inserted in narrow collimating tubes at each end: they are of heavy silver-plated platinum, dipped in fused silver chloride (or bromide). Through this cell, with a calibrated 500 Ω resistance in series, current is passed by means of the constant current circuit used in transport number work. The potential across the 500 Ω resistance is measured first, then the potential across the probe-electrodes; to eliminate any bias, it is advisable to reverse the current and measure the potential again, with a further measurement across the standard resistance as a check.

In Gordon's design the fact that there is a potential gradient near the potential-measuring electrodes makes it essential that these be small and reproducibly located, so that only silver-silver halide electrodes are really suitable. IVES and SWAROOPA[19b] used a cell in which the solution being measured was connected to two quinhydrone electrodes via arms so placed that no potential gradient occurred in them; this made possible the use of a liquid junction between the quinhydrone electrode and the solution, increasing the scope of the method. ELIAS and SCHIFF[19a] have developed a more precise apparatus on the same principle. Their cell is that of *Figure 5.5*, but the tubes B and B' carrying the potential-measuring electrodes are replaced by the components shown in *Figure 5.6*.

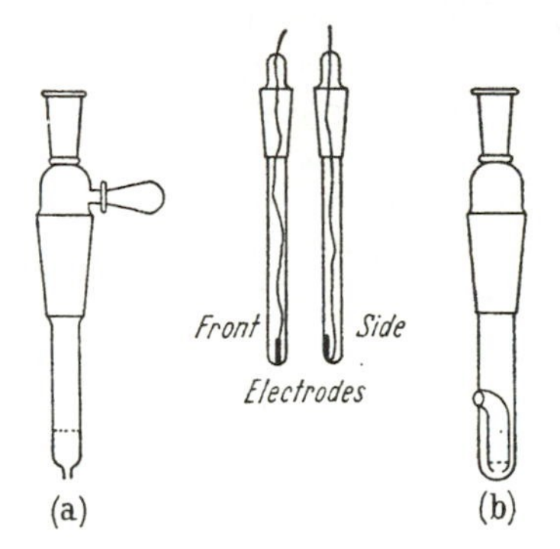

Figure 5.6. Liquid junction electrode vessels. After ELIAS, L. *and* SCHIFF, H. I., *J. phys. Chem.*, 60 (1956) 595

Silver-silver halide electrodes mounted in the small central tubes are surrounded by alkali halide solution in tubes *a* or *b*, this solution making a liquid junction with the main cell solution at the position shown by the dotted lines. Type *a* is used when the cell solution is denser than the probe solution and type *b* in the opposite case. Since no electrolysis current flows through the solution in the side arm, the potential between the probe electrodes is proportional to the resistance of the cell solution. The current carrying electrodes are silver plated platinum; no electrode is needed reversible to either ion of the electrolyte being studied and the method has general application.

Radio Frequency Measurement of Conductivity

The audio frequency a.c. bridge technique and the d.c. methods discussed above are high precision methods designed to meet the exacting requirements of the physical chemist whose concern is with the nature of electrolyte solutions and the mathematical interpretation of their behaviour. Conductivity measurement is, however, also an everyday analytical tool of immense value; for purposes such as conductimetric analysis it is seldom necessary to adopt all the refinements of technique described above. Many commercial instruments for conductimetric titration, for example, operate at the rather low frequency ($\sim$ 50 c/sec) of the electric mains, with an accuracy of one or two per cent, which is quite adequate for the purpose. A very interesting development of recent years, with great possibilities for analytical and process-control applications, is the use of radio-frequency methods for conductivity measurements. The great virtue of these methods is that the electrodes need not be in contact with the solution; polarization errors are therefore completely absent. The cell vessel may be a simple test-tube or flask which is placed either within a coil or between the plates of a condenser which form the elements of an oscillatory circuit. The presence of the electrolytic resistance alters the frequency of the oscillations, or in another method, the coupling between two oscillatory circuits, and the change is measured by suitable meters.

MEASUREMENT OF TRANSPORT NUMBERS

The experimental methods available for measuring transport numbers fall into three categories: (*i*) the Hittorf method, (*ii*) the moving boundary method, and (*iii*) a method depending on concentration cells with a liquid junction.

The first of these is so familiar that it requires little comment. Devised in 1853, it was the instrument for an outstandingly comprehensive study lasting over half a century and, although it has been superseded by other techniques, the value of this one man's contribution should be recognized. It is especially remarkable that many of these measurements were made before the Arrhenius ionic theory was developed. Several modifications of the apparatus have been made, but all consist essentially of an anode compartment, a cathode compartment and a third intervening compartment. Current, in amount measured by a coulometer, is passed and the change in composition of each section determined analytically; assuming that the current is not passed so long that the composition of the central compartment changes, then the loss in either the anode or the cathode compartment gives one of the two transport numbers.

The application of the Hittorf method is limited by two main factors: at least one and preferably both the electrodes must be reversible and extreme accuracy is needed in the analysis of the solution before and after electrolysis. There are few electrodes through which it is possible to pass a substantial quantity of electricity, *e.g.*, 20 coulombs, without gassing or other unwanted side-reactions. Well annealed, very pure silver behaves satisfactorily in aqueous silver nitrate solution provided oxygen is excluded; silver-silver halide electrodes are suitable for aqueous chloride and bromide solutions. The possibility of electrode reactions with the solvent must be considered. The analytical problem can be reduced in importance by passing sufficient electricity to produce a large change in concentration, but this involves either long runs with the danger of diffusion of the anode and cathode solutions into one another, or large currents with consequent overheating of the solution, resulting in turbulent mixing.

Figure 5.7 shows a modern form of Hittorf apparatus developed by STEEL and STOKES[20] for use with alkali bromide solutions in

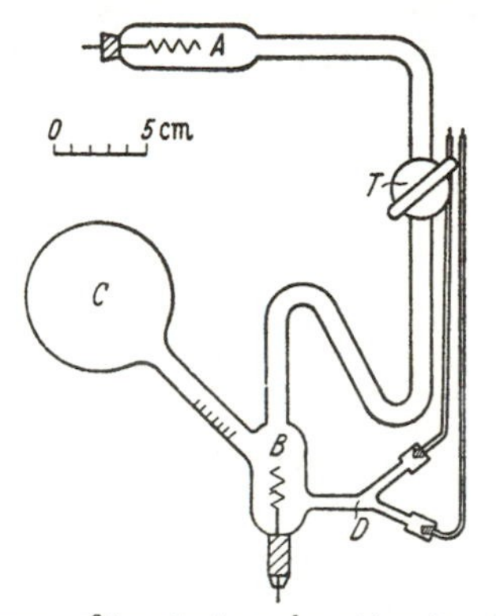

Figure 5.7. Diagram of transport number apparatus after STEEL, B. J *and* STOKES, R. H., *J. phys. Chem.*, 62 (1958) 450

mixed solvents. The important feature of this apparatus is the built-in conductance-cell which makes possible the analysis of the cathode solution without removal from the apparatus. After an initial measurement of the conductance, current is passed through the silver-silver bromide electrodes to produce a 10 to 20 per cent change in concentration of the solution near the electrodes. The tap is then closed, the whole of the cathode solution is thoroughly mixed in the bulb and returned to the position shown for a second conductance measurement. As a check on the reversibility of the electrode-reactions, the tap is then opened, the anode and cathode solutions are remixed in the bulb, and a final conductance reading—which should agree with the first—is taken. The volume of the cathode solution is read off from the calibrated stem of the mixing-bulb; to ensure proper drainage, the bulb is coated with a water-repellent silicone film. The 'apparent' transport number of the cation is $v\Delta c\boldsymbol{F}/Q$, where v is the volume of the cathode solution, Δc is the change in its concentration in equivalents per unit volume and Q

is the quantity of electricity passed; this is converted to the Hittorf frame of reference, (motion relative to the solvent), by the use of density data. Some results obtained by this method will be considered in Chapter 11; while not quite equal in precision to the moving-boundary method, it offers some prospect of success in non-aqueous solutions where transport number data are urgently needed.

The Moving Boundary Method

The considerations upon which this method is based are simple: let solutions of two salts (having a common anion, X^-), form a boundary *ab* and let current be passed so that the cations move up the tube and the anions down the tube (*Figure 5.8*). If the conditions are chosen properly the boundary will remain distinct but will move

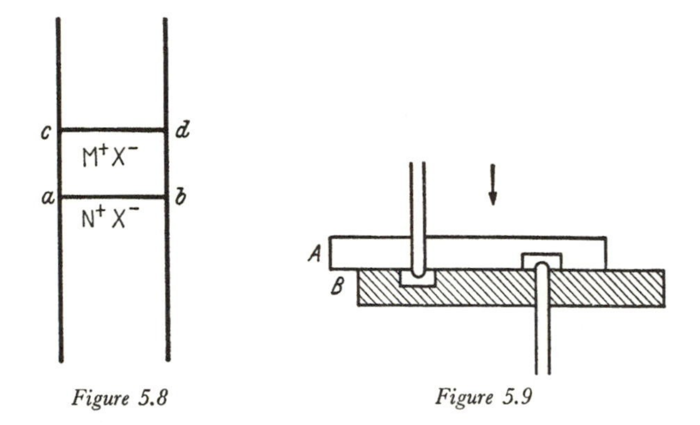

Figure 5.8 *Figure 5.9*

up the tube. After a certain time let it be at the position *cd*. In this interval all the cations, M^+, in the volume V between *cd* and *ab* must have crossed a plane at *cd*. If the amount of electricity passing be Q coulombs, then the amount moving upwards is t_1Q coulombs. If V is the volume between *ab* and *cd* and the concentration of the solution of M^+X^- is c ion equivalents of M^+ per unit volume, the amount of electricity moving upwards must be $V[c]\boldsymbol{F}$ whence:

$$t_1 = Vc\boldsymbol{F}/Q$$

All variants of this method depend fundamentally on measuring this volume for unit amount of electricity passed and the successful application of the method depends on three factors: (1) the construction of an apparatus capable of producing a sharp boundary, (2) the use of a suitable salt N^+X^- (called the indicator) and its use at the proper concentration, and (3) a small correction for changes in the position of the boundary due to volume changes. The sharp boundary can be produced by one of three methods. The first was developed by MacInnes and Brighton[21] and is called the 'sheared boundary, method. In its simplest form (*Figure 5.9*) the tube containing the two solutions is divided with the upper half held in a hole in a disc *A* over a depression in another disc so that a drop of the solution hangs from the end of the tube. The other half is held in the disc *B* immediately under a cut or indent in the disc *A* and is filled with the other solution until a drop protrudes at the end.

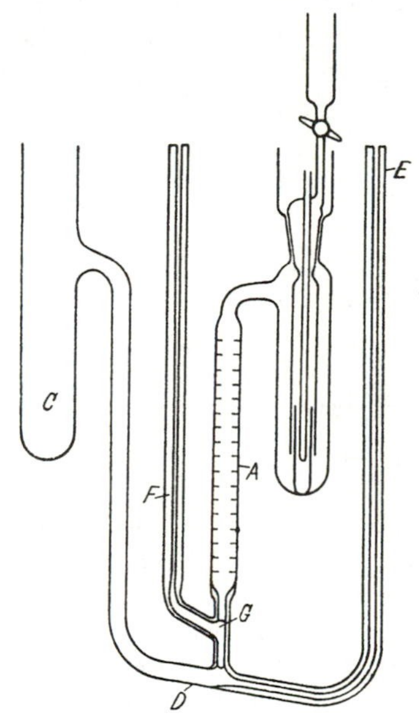

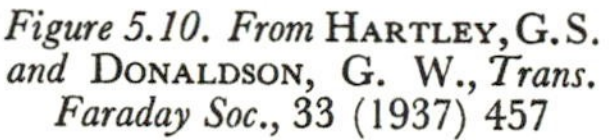

Figure 5.10. From Hartley, G. S. *and* Donaldson, G. W., *Trans. Faraday Soc.*, 33 (1937) 457

If the two discs have plane surfaces and the discs are moved over one another so that the two tubes are adjacent the excess of each liquid is 'sheared off' and a sharp boundary is formed. The second method is the 'autogenic' method of Franklin and Cady[22] in which the indicator solution is formed by making the anode at the bottom of the tube of a metal such as cadmium, with a solution of potassium chloride over it: on passing a current, a solution of cadmium chloride is formed and a boundary is produced between this solution and the potassium chloride solution. Again, an anode of silver can be used to give a boundary between a silver nitrate solution and a potassium nitrate solution. The third method is the 'air bubble' method[23]. The apparatus is shown in *Figure 5.10.* To the top of the capillary tube *F*, two pinchcocks are attached, one closed and one half open. At the beginning of the experiment, the whole apparatus is filled with the leading solution and by closing one pinchcock an air bubble is formed by compressing the air in the capillary *F*; this expands into *A* at *G* separating the solution into two parts. That in *C*, *D* and *E* is removed, these portions of the apparatus are rinsed several times with water and filled with indicator solution. By unscrewing the pinchcock the air bubble can now be withdrawn into *F* and a boundary formed at *G*.

If the boundary moves up the tube, it is necessary to have the indicator solution of greater density than the leading solution and, conversely, if the indicator solution is to be on the top of the leading

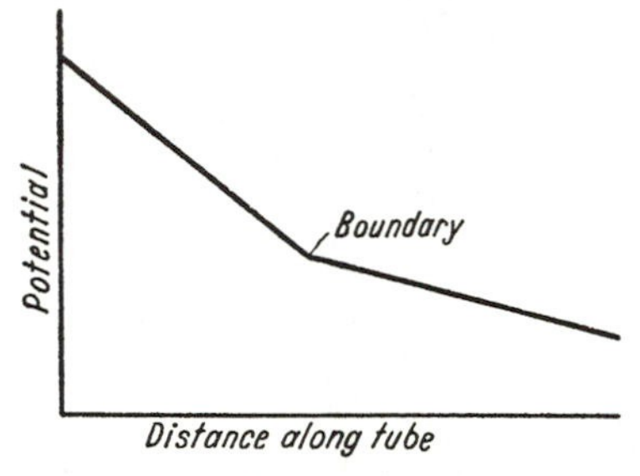

Figure 5.11

solution and the boundary is to move downwards, then the indicator solution should be lighter. Moreover, in whichever direction the boundary moves, the ion of the indicator solution must have a mobility lower than the ion of the leading solution. Fortunately, there is a kind of self-regulating effect which restores the sharpness of the boundary if it is for some reason diminished. Let us consider what will be the concentration of the indicator ion N^+ behind the moving boundary. We have already seen that during the passage of an amount, Q coulombs, of electricity all the M^+ ions between *cd* and *ab* cross the boundary *cd*. Let a further Q coulombs pass: then all the N^+ ions between *ab* and *cd* must cross the plane at *cd* and

$$t_1' = \frac{Vc_{N^+}F}{Q}$$

where t_1' is the transport number of the N^+ ion at the concentration c_{N^+} at which it is present in the volume V behind the boundary.

Then $$t_1/t_1' = c_{M^+}/c_{N^+}$$

This is sometimes called Kohlrausch's regulating function. Now t_1' must be less than t_1 and therefore c_{N^+} must be less than c_{M^+}. Thus if we consider the fall in potential along the tube (*Figure 5.11*), both the lower mobility and the lower concentration of the indicator ion will combine to produce a sharper drop in potential in the indicator solution than in the leading solution. If by any mischance a leading ion diffuses into the indicator solution its comparatively high mobility shoots it forward again down the comparatively large potential gradient; conversely should an indicator ion find itself too far ahead, its lower mobility combined with the lower gradient slows it up: in either event, there is a mechanism at work to restore the sharpness of the boundary. The same mechanism adjusts the concentration c_{N^+} of the indicator solution: if c_{N^+} is initially too large, the potential gradient is lower than that demanded by the condition $t_1/t_1' = c_{M^+}/c_{N^+}$ and the N^+ ion travels more slowly until the correct value of c_{N^+} is reached, and conversely, if the initial value of c_{N^+} is too small. Naturally it cannot be expected that this 'self-regulating mechanism' can cope with wide departures from ideal conditions. MacInnes has found by experiment, however, that considerable tolerance is permitted so that it is sufficient to adjust the concentration of the indicator solution to within only 5 to 10 per cent of the required value.

During the passage of the current there may be volume changes resulting from electrode reactions, *etc.* Except in concentrated solutions, the correction for this effect is small but it may be considered here in some detail because of its bearing on the distinction between the Hittorf and the moving boundary transport numbers, a matter appreciated very clearly by LEWIS[24] as long ago as 1910. The moving boundary method gives an ionic mobility relative to the fixed glass tube in which the measurements are made. The Hittorf experiment measures the number of ions crossing a plane fixed relative to a hypothetical plane in the solvent which of course moves if the solvent moves. MACINNES and LONGSWORTH[25] illustrate this by reference to an experiment (*Figure 5.12*) in which potassium chloride forms the leading solution, barium chloride the indicator solution and the tube is sealed at the bottom by a silver electrode; x marks the position of a water molecule of a hypothetical type, hypothetical since it is presumed free of Brownian motion. cd is a plane in a region not subject to change of concentration. On passing a faraday of electricity the following changes occur:

1. t_1 equivalents of K^+ cross the plane cd in an upward direction giving a volume decrease $t_1\bar{V}^{KCl}_{K^+}$, *i.e.*, t_1 times the partial molal volume of the potassium ion in potassium chloride solution.

2. t_2 equivalents of Cl^- cross cd in a downward direction with a volume increase $t_2\bar{V}^{KCl}_{Cl^-}$.

3. An equivalent of metallic silver is lost with volume decrease $\bar{V}_{Ag}$.
4. An equivalent of silver chloride is formed with volume increase $\bar{V}_{AgCl}$.
5. An equivalent of Cl^- is lost by reaction with silver with volume decrease $\bar{V}_{Cl^-}^{BaCl_2}$.
6. Since the number of Ba^{++} ions below the boundary remains unchanged the loss in process 5 above must be exactly compensated by a transfer of an equivalent of Cl^- downward across the boundary with a volume increase $\bar{V}_{Cl^-}^{BaCl_2} - \bar{V}_{Cl^-}^{KCl}$.

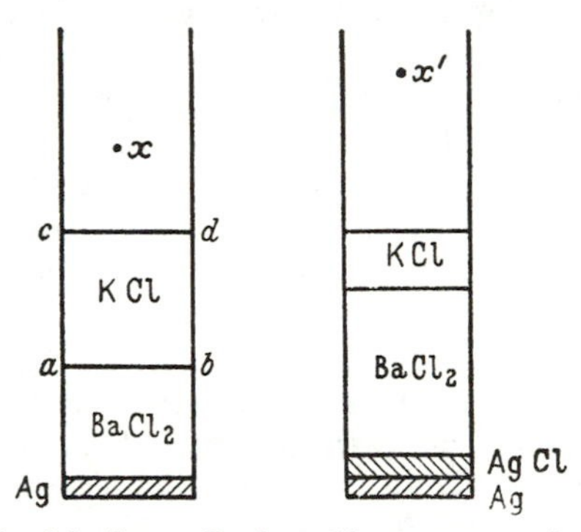

Figure 5.12. After MACINNES, D. A. *and* LONGSWORTH, L. G., *Chem. Rev.*, 11 (1932) 204

The net volume increase is:

$$\Delta V = \bar{V}_{AgCl} - \bar{V}_{Ag} - t_1 \bar{V}_{K^+}^{KCl} - (1 - t_2)\bar{V}_{Cl^-}^{KCl}$$
$$= \bar{V}_{AgCl} - \bar{V}_{Ag} - t_1 \bar{V}_{KCl}$$

The moving boundary transference number is $t_1 = Vc_{K^+}$ but because of the increase in volume, ΔV, the water molecule at x is raised to x'. Relative to this molecule, the boundary has not moved so much and the Hittorf transport number is:

$$t_1 = (V - \Delta V)c_{K^+}$$

There is a further correction to be applied if the solvent conducts an appreciable fraction of the current, which has been shown[26] to take the form:

$$(1 + K_{sp}\ \text{solvent}/K_{sp}\ \text{solution})$$

This becomes important with very dilute solutions.

Although microcoulometers have been used to measure the current[27], the usual practice is to hold a known current as steady as possible for a measured time interval. A somewhat elaborate device is used by the Rockefeller Institute workers[28] although others[23, 29] have preferred a simpler mechanism.

The accuracy of moving-boundary measurements depends to a large extent on the precision with which the boundary position can be determined at any time. The usual method makes use of the

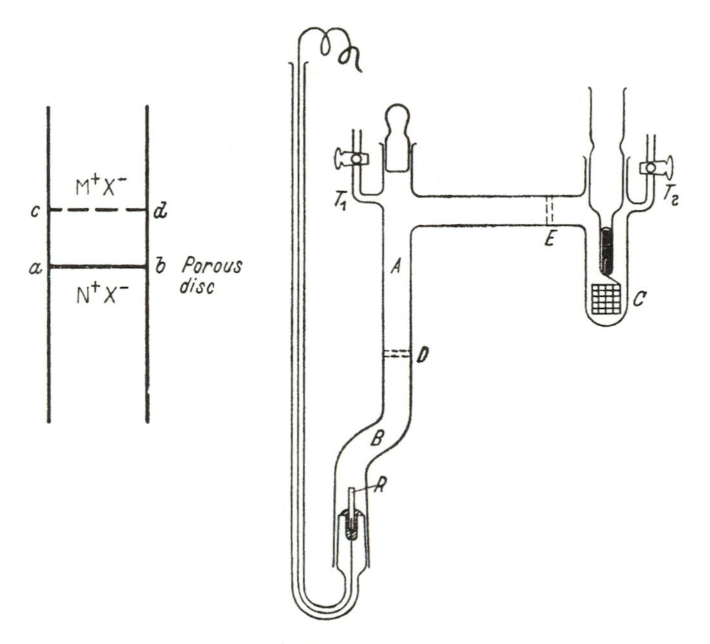

Figure 5.13. After SPIRO, M. *and* PARTON, H. N. *Trans. Faraday Soc.*, 48 (1952) 265

difference in refractive index between the leading and indicator solutions, a schlieren image of the boundary being formed by a lens system. It is not however always possible to secure a clearly visible boundary while meeting the other requirements of the method, and GORDON and his co-workers[30] have reported a technique of detecting the boundary by means of the abrupt change in conductance which occurs as it passes a pair of micro-conductance electrodes sealed into the walls of the tube.

Transport Numbers by the Analytical Boundary Method

In this variant of the moving boundary method[31], a tube is divided (*Figure 5.13*) into two compartments by a sintered glass disc; one compartment contains the solution under investigation and the other the indicator solution. After electricity in amount Q has passed, the boundary, originally at the disc, has moved to cd through a volume V given by

$$FV = \frac{Qt_1}{c_{M^+}} = \frac{Qt_1'}{c_{N^+}}$$

where the primed quantity refers to N^+ and the unprimed to M^+. Thus a quantity $c_{N^+}V$ of the ionic species N^+ passes through the disc; this can be measured by ordinary analytical methods and is equal to Qt_1'/F. The upper solution is now the indicator and the method gives the transport number of the ion in the following solution. The apparatus used by Spiro and Parton is shown in *Figure 5.13.* The indicator solution containing sodium or potassium nitrate is in the tube A of 20·5 mm diameter, the disc D has a pore diameter of 20–30 μ and the cathode, C, is a piece of platinum gauze in ferric nitrate solution, the ferric nitrate being introduced to reduce the amount of gas evolution. E is a coarse porous disc designed to prevent diffusion of ferric nitrate into A, and T_1 and T_2 are filling taps. The non-gassing anode R is a rod of silver and the compartment B contains 0·1 M silver nitrate. The amount of silver ion migrating into the compartment A is determined by a careful potentiometric titration. For the details of the current regulator the original paper should be consulted. Spiro and Parton found, using potassium nitrate as indicator, that there was a range of indicator concentration around 0·11 M (the Kohlrausch concentration for 0·1 M $AgNO_3$) for which the transport number was independent of current, time and indicator concentration. The value found was 0·4676 compared with the accepted value of 0·4682. Using sodium nitrate as indicator, *i.e.*, using an indicator cation which moves more slowly than the silver ion, there appeared to be no range of indicator concentration over which the transport number remained constant. The correct result was obtained not at the Kohlrausch concentration but at that concentration of indicator where the specific conductivities of the two solutions were equal. This may have been coincidental and further work is needed on this point. The paper by Brady is interesting in that in one set of experiments he used radioactive tracers and made the analysis by counter methods. He developed the method for colloidal electrolytes which do not lend themselves to the moving boundary method and in a subsequent paper[32] he has described the determination of the transport numbers of four surface active agents.

Transport Numbers from Cell Measurements

The potential, E, of the cell:

$$\text{Ag, AgCl}|\text{MCl}(m')|M|\text{MCl}(m)|\text{Ag, AgCl} \qquad \text{Cell I}$$

can be combined with the potential, E_t, of the cell:

$$\text{Ag, AgCl}|\text{MCl}(m')|\text{MCl}(m)|\text{Ag, AgCl} \qquad \text{Cell II}$$

to give the transport number of the cation, t_1, as:

$$t_1 = E_t/E$$

This is the well-known Helmholtz relation and leaves undecided, without a more detailed study of the theory, the concentration to

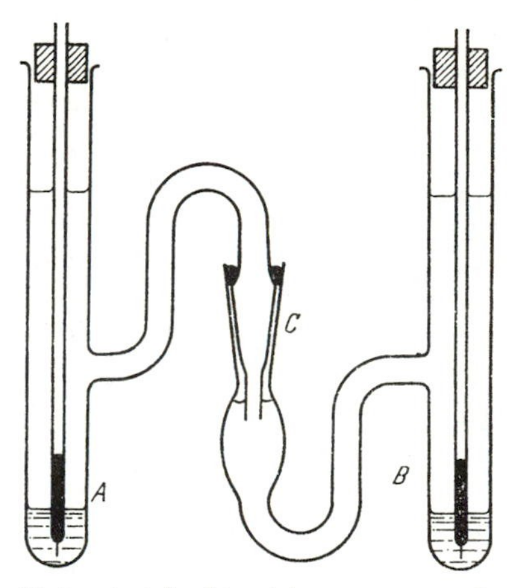

Figure 5.14. Cell with transport for determining transport numbers in zinc perchlorate solutions. Stokes, R. H. *and* Levien, B. J., *J. Amer. chem. Soc.*, 68 (1946) 333

which t_1 refers. By considering the case where m' is held fixed and m is varied, measurements of E and E_t being made for a range of m values, it is readily shown that $t_1 = \mathrm{d}E_t/\mathrm{d}E$. Systems similar to those designated Cell II have been used by MacInnes and Shedlovsky and also by Gordon to determine activity coefficients, *i.e.*, finding Cell I not particularly amenable to measurement because of experimental difficulties with the electrode M, they prefer to determine the transport number by one of the other methods described in this chapter and combine the result with the potential of Cell II to give an activity coefficient. Very valuable work has been done in this way, particularly with solutions less than 0·1 M concentration, to which reference will be made in Chapter 8. For the present we shall content ourselves with a description of the use of Cell II for the measurement of a transport number, for example, that of zinc in zinc perchlorate, by this method[33]. Whilst the zinc amalgam electrode is known to work very well and could replace the silver-silver chloride electrode mentioned in Cell I above, there is no electrode known to be reversible with respect to the perchlorate ion which could take the place of the electrode *M*. However, using the isopiestic vapour pressure method, it was possible to determine the activity coefficient of zinc perchlorate over a wide concentration range, from which the potential of the hypothetical cell:

$$Zn_xHg|Zn(ClO_4)_2(m')|X|Zn(ClO_4)_2(m)|Zn_xHg$$

where *X* is an electrode reversible to the perchlorate ion, can be calculated. The cell with transport is shown in *Figure 5.14*. The vessel *A* was first filled with the more dilute solution and warm liquid 5 per cent zinc amalgam run in. The solution had previously been degassed and the dissolved air replaced by hydrogen. The other vessel, *B*, was filled with the more concentrated solution to the mark shown, amalgam run in and the two parts of the cell united at the ground glass joint *C*. The electromotive force became steady after an hour and remained steady within 0·03 mV for a day. It was found that the potential of the cell with transport E_t, was related to that of the cell without transport, *E*, at the same concentrations, by the equation:

$$E_t = aE + bE^2 + cE^3$$

so that $t_2 = dE_t/dE$ could be obtained easily. A similar study[34] of the transport number of zinc iodide failed to reveal any such simple relation between *E* and E_t, and a method of calculating the differential due to RUTLEDGE[35] was used.

The transport number given by these amalgam cells is not that of either the zinc ion or the halide ion since allowance has to be made for the formation of autocomplexes. For example, if we suppose that a concentrated solution of zinc iodide consists only of zinc and complex ZnI_4^{--} ions in equal amount we find, on considering the details of the cell reactions, that:

$$t_1 \text{ (observed)} = 1 - dE_t/dE = 1 - 2t_{ZnI_4^{--}} = t_{Zn^{++}} - t_{ZnI_4^{--}}$$

KERKER and ESPENSCHIED[36] given an interesting discussion of the cells:

$$Hg, Hg_2HPO_4|H_3PO_4(m')|H_2\text{–}Pt\text{–}H_2|H_3PO_4(m'') \; Hg_2HPO_4, Hg \quad \text{I}$$

and

$$Hg, Hg_2HPO_4|H_3PO_4(m')||H_3PO_4(m'')|Hg_2HPO_4, Hg \quad \text{II}$$

the mercury-mercurous hydrogen phosphate electrode being reversible not directly to the $H_2PO_4^-$ ion, which is the only anionic species present in any amount, but to the HPO_4^{--} ion which is in equilibrium with it. At the concentrations used, very small amounts of HPO_4^{--} and PO_4^{---} ions are present and the current is carried by H^+ and $H_2PO_4^-$ ions.

The passage of two faradays through cell I ($m'' < m'$) corresponds to the reaction:

$$Hg_2HPO_4 + H_2 \rightarrow H_3PO_4\ (m'') + 2Hg$$

in the right hand half cell and

$$H_3PO_4(m') + 2Hg \rightarrow Hg_2HPO_4 + H_2$$

in the other half cell, so that the net reaction is:

$$H_3PO_4(m') \rightarrow H_3PO_4(m'')$$

and the e.m.f. of the cell is given by:

$$-2E\boldsymbol{F} = \Delta G = \boldsymbol{R}T \ln a''/a'$$

There are three processes to be considered when two faradays pass through cell II:

(*a*) $$2Hg + HPO_4^{--} \rightarrow Hg_2HPO_4 + 2e^-$$

i.e., the loss of a mole of HPO_4^{--} in the left hand compartment;

(*b*) a corresponding gain in the right hand compartment;

(*c*) a transfer of $2t_{H^+} = 2\lambda_{H^+}/(\lambda_{H^+} + \lambda_{H_2PO_4^-})$ moles of hydrogen ion from left to right across the liquid junction and a transfer of $2t_{H_2PO_4^-} = 2\lambda_{H_2PO_4^-}/(\lambda_{H^+} + \lambda_{H_2PO_4^-})$ dihydrogen phosphate ions in the opposite direction.

Some reaction will proceed in each compartment to maintain the various ionic species in equilibrium but since $t_{H^+} + t_{H_2PO_4^-} = 1$, the stoichiometric result is the transfer of $(2t_{H^+} - 1)$ moles of phosphoric acid from the left to the right compartment so that:

$$-E_t\,\mathbf{F} = (2t_{H^+} - 1)\ \boldsymbol{R}T \ln \frac{a''}{a'}$$

and in the limit when m' and m$''$ differ only infinitesimally:

$$\begin{aligned} t_{\text{observed}} &= dE_t/dE \\ &= (2t_{H^+} - 1) \equiv (1 - 2t_{HPO_4^-}) \\ &= t_{H^+} - t_{H_2PO_4^-} \end{aligned}$$

Thus the 'observed transport number' is neither of the true transport numbers but their difference. It seems desirable when using the cell method to work out the details of the cell reaction for each case along the above lines rather than to depend blindly on the formula $t = dE_t/dE$, which is clearly of limited application.

Transport Numbers from Centrifugal Cells

The first measurements on gravity and centrifugal cells were made by DES COUDRES[37]. The effect in a gravity cell is of the order of a few microvolts per metre but by means of a specially constructed potentiometer circuit GRINNELL and KOENIG[38] have increased the accuracy of measurement and obtained 0·4900 and 0·4893 for the cation transport number of 0·975 and 0·712 M potassium iodide at 20°. TOLMAN[39] made experiments with a powerful centrifuge corresponding to a gravity cell some 1,200 m in height and his potentials were of the order of several millivolts.

MACINNES[40] has devoted much attention to cells of this type in recent years. In a centrifugal cell such as: $Pt|I_2$ in $MI|Pt$ where M is a cation, two identical iodine-iodide electrodes situated at distances r_1, r_2 from the point about which the cell is rotated develop a potential E. If current passes inside the cell from the outer to the inner electrode, the cell reaction is:

$$I^- \rightarrow \tfrac{1}{2}I_2 + e^- \text{ at the outer electrode,}$$

$$\tfrac{1}{2}I_2 + e^- \rightarrow I^- \text{ at the inner electrode.}$$

At the same time, for each faraday of electricity which passes, t_1 equivalents of the cation pass from the region around the outer electrode to the region around the inner electrode and t_2 equivalents of iodide ion pass in the opposite direction. The net result is the transport of one equivalent of iodine from the inner to the outer electrode and t_1 equivalents of the salt MI in the opposite direction.

MacInnes and Ray have given a rigid deduction of the equation for the potential of such a cell. The equation can be derived less rigidly as follows:

The kinetic energy due to the rotation of the equivalent of iodine at the outer electrode is $2\pi^2r_2^2\omega^2 W_I$, ω being the number of revolutions per second and W_I the atomic weight of iodine. The increase in kinetic energy of the iodine on transferring an equivalent from the inner to the outer electrode is therefore: $2\pi^2\omega^2(r_2^2 - r_1^2) W_I$ and there must be a similar term for the t_1 equivalents of salt transferred in the opposite direction. But the volumes occupied by the salt and by iodine may not be the same and therefore there may be a movement of solution as a whole to compensate, involving a transfer of

$\rho(\bar{V}_I - t_1\bar{V}_{MI})$ grams of solution where ρ is the density of the solution. By equating the electrical work to the net change in kinetic energy we get:

$$EF = 2\pi^2\omega^2(r_2^2 - r_1^2)[(W_I - t_1W_{MI}) - \rho(\bar{V}_I - t_1\bar{V}_{MI})] \quad \ldots .(5.4)$$

The apparatus used by MacInnes and his colleagues is shown in *Figure 5.15.* The rotor R is a magnesium disc 23 cm diameter and 5 cm thick, turned by the pressure of the disc D on the plate P

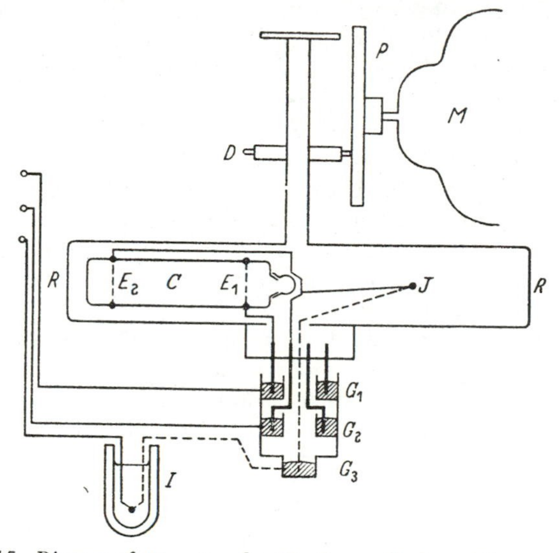

Figure 5.15. Diagram of apparatus after MacInnes, D. A. *and* Dayhoff, M. O. *J. chem. Phys.*, 20 (1952) 1035

which is rotated by a synchronous motor M. The potential is measured through the mercury wells G_1 and G_2 whilst the wells G_2, G_3 are used to measure the temperature of the rotor with a copper-constantan junction J and an external ice-bath I. Radial temperature gradients are eliminated as far as possible by maintaining a vacuum of 10 μ around the rotor (thus avoiding gas friction) and circulating cooling water at the vacuum bearing. Another important feature of the technique arose from the presence of minute suspended particles in the cell solution, which all precautions failed to eliminate and which gave rise to erratic potentials. This error can be avoided by sealing the electrodes E_1, E_2 in the form of platinum rings several millimetres from the ends of the cell C. The centrifugal force then drives the suspended particles clear of the electrodes and they collect harmlessly at the base of the cell. Rotor speeds between 40 and 120 rev/sec are used. It is important to realize that the cell potential originates from a difference of centrifugal potential at the two electrodes and not from a concentration gradient set up in the solution as a result of centrifugal force. The theory assumes uniformity of concentration. On continued centrifuging a concentration gradient should be set up sufficient to reduce the cell potential to zero, and at high speeds of the rotor MacInnes did observe a slow fall in potential if the experiment was prolonged. PEDERSEN[(41)] has obtained similar sedimentation with some salts using an ultracentrifuge at much higher speeds.

The work of MacInnes is still at the stage where very fine technique is being developed: so far the experiments have yielded a transport number $t_{Na^+} = 0{\cdot}3827$ for 0·1911 N sodium iodide and $t_{K^+} = 0{\cdot}4873$ for 0·1941 N potassium iodide compared with the value, 0·4887, found by LONGSWORTH[(42)]. More recently[(43)], the transport numbers of lithium, rubidium and caesium iodide have been measured. The method is being developed for application to non-aqueous solutions where other methods encounter difficulties due to the joule heat.

It should be added that the experimental data had to be interpreted in the light of a further complication, the formation of complex iodide ion, and that the assumption of the formula I_3^- for this complex ion was sufficient to reconcile the, at first sight, confusing results obtained with varying iodide concentrations.

REFERENCES

1 JONES, G. and JOSEPHS, R, C., *J. Amer. chem. Soc.*, 50 (1928) 1049
2 KOHLRAUSCH, F., *Wied. Ann.*, 60 (1897) 315
3 JONES, G. and BOLLINGER, G. M., *J. Amer. chem. Soc.*, 51 (1929) 2407
4 JONES, G. and BOLLINGER, G. M., *ibid.*, 53 (1931) 411
5 JONES, G. and BOLLINGER, G. M., *ibid.*, 53 (1931) 1207
6 JONES, G. and BRADSHAW, B. C., *ibid.*, 55 (1933) 1780
7 JONES, G. and CHRISTIAN, S. M., *ibid.*, 57 (1935) 272
8 JONES, G. and BOLLINGER, D. M., *ibid.*, 57 (1935) 280
9 JONES, G. and PRENDERGAST, M. J., *ibid.*, 59 (1937) 731
10 SHEDLOVSKY, T., *ibid.*, 52 (1930) 1793
11 FEATES, F. S., IVES, D. J. G. and PRYOR, J. H., *J. electrochem. Soc.* 103 (1956) 580
12 BRODY, O. V. and FUOSS, R. M., *J. phys. Chem.*, 60 (1956) 177
13 CALVERT, R., CORNELIUS, J. A., GRIFFITHS, V. S. and STOCK, I. D., *ibid.*, 62 (1958) 47
14 IVES, D. J. G. and PRYOR, J. H., *J. chem. Soc.*, (1955) 2104; FEATES, F. S. and IVES, D. J. G., *ibid.*, (1956) 2798
15 GRAHAME, D. C., *Ann. Rev. phys. Chem.*, 6 (1955) 346
16 RANDLES, J. E. B., *Disc. Faraday Soc.*, 1 (1947) 11
17 STEEL, B. J. and STOKES, R. H., Unpublished work (1958)
18 PARKER, H. C. and PARKER, E. W., *J. Amer. chem. Soc.*, 46 (1924) 312

[19] GUNNING, H. E. and GORDON, A. R., *J. chem. Phys.*, 10 (1942) 126; 11 (1943) 18; BENSON, G. C. and GORDON, A. R., *ibid.*, 13 (1945) 470; JERVIS, R. E., MUIR, D. R., BUTLER, J. P. and GORDON, A. R., *J. Amer. chem. Soc.*, 75 (1953) 2855
[19a] ELIAS, L. and SCHIFF, H. I., *J. phys. Chem.*, 60 (1956) 595
[19b] IVES, D. J. G. and SWAROOPA, S., *Trans. Faraday Soc.*, 49 (1953) 788
[20] STEEL, B. J. and STOKES, R. H., *J. phys. Chem.*, 62 (1958) 450
[21] MACINNES, D. A. and BRIGHTON, T. B., *J. Amer. chem. Soc.*, 47 (1925) 994
[22] FRANKLIN, E. C. and CADY, H. P., *ibid.*, 26 (1904) 499; CADY, H. P. and LONGSWORTH, L. G., *ibid.*, 51 (1929) 1656
[23] HARTLEY, G. S. and DONALDSON, G. W., *Trans. Faraday Soc.*, 33 (1937) 457
[24] LEWIS, G. N., *J. Amer. chem. Soc.*, 32 (1910) 862
[25] MACINNES, D. A. and LONGSWORTH, L. G., *Chem. Rev.*, 11 (1932) 171
[26] LONGSWORTH, L. G., *J. Amer. chem. Soc.*, 54 (1932) 2741
[27] REEVELEY, W. O. and GORDON, A. R., *Trans. Electrochem. Soc.*, 63 (1933) 167
[28] MACINNES, D. A., COWPERTHWAITE, I. A. and BLANCHARD, K. C., *J. Amer. chem. Soc.*, 48 (1926) 1909
[29] LE ROY, D. J. and GORDON, A. R., *J. chem. Phys.*, 6 (1938) 398; see also HOPKINS, D. T. and COVINGTON, A. K., *J. sci. Instr.*, 34 (1957) 20
[30] LORIMER, J. W., GRAHAM, J. R. and GORDON, A. R., *J. Amer. chem. Soc.*, 79 (1957) 2347
[31] BRADY, A. P., *ibid.*, 70 (1948) 911; SPIRO, M. and PARTON, H. N., *Trans. Faraday Soc.*, 48 (1952) 263
[32] BRADY, A. P. and SALLEY, D. J., *J. Amer. chem. Soc.*, 70 (1948) 914
[33] STOKES, R. H. and LEVIEN, B. J., *ibid.*, 68 (1946) 333
[34] STOKES, R. H. and LEVIEN, B. J., *ibid.*, 68 (1946) 1852
[35] RUTLEDGE, G., *Phys. Rev.*, 40 (1932) 262
[36] KERKER, M. and ESPENSCHIED, W. F., *J. Amer. chem. Soc.*, 80 (1958) 776
[37] DESCOUDRES, T., *Ann. Phys.*, 49 (1893) 284; 55 (1895) 213; 57 (1896) 232
[38] GRINNELL, S. W. and KOENING, F. O., *J. Amer. chem. Soc.*, 64 (1942) 682
[39] TOLMAN, R. C., *ibid.*, 33 (1911) 121
[40] MACINNES, D. A. and RAY, B. R., *ibid.*, 71 (1949) 2987; MACINNES, D. A. and DAYHOFF, M. O., *Symposium on Electrochemical Constants*, Washington (1951); *J. chem. Phys.*, 20 (1952) 1034
[41] PEDERSEN, K. O., *Z. phys. Chem.*, 170A (1934) 41
[42] LONGSWORTH, L. G., *J. Amer. chem. Soc.*, 57 (1935) 1185
[43] RAY, B. R., BEESON, D. M. and CRANDALL, H. F., *ibid.*, 80 (1958) 1029

6

THE LIMITING MOBILITIES OF IONS

THE transport of electricity through electrolytes differs fundamentally from metallic conduction in that the carriers are ions, the dimensions and masses of which are much larger than those of the electrons responsible for metallic conduction. The ions of course share in the general Brownian motion of the liquid, and may be expected to have randomly-directed instantaneous velocities of the order of 10^4 cm sec^{-1}, though of course with the extremely short mean free path characteristic of the liquid state. In the absence of an external field or a concentration-gradient, the Brownian movement is entirely random, and does not lead to a drift of ions in any one direction. The presence of an electric field, as in conductance, or of a concentration-gradient, as in diffusion, has the effect of biasing the Brownian movement in a particular direction. In a field of 1 V/cm the average velocity of the ions in the direction of the field is of the order of 10^{-3} to 10^{-4} cm sec^{-1}, and hence represents only a very small perturbation of the random ionic motions. The actual path of an ion under an electric field of ordinary intensity is thus extremely erratic, bearing very little resemblance to that of a billiard-ball sinking in water. Nevertheless, the drastic simplification of substituting for the actual chaotic motion a steady progress of all the ions of one kind with equal velocities in one direction of the field is extraordinarily successful: the Brownian motion needs to be considered only in regard to its effect on the interionic forces.

Experimental data on conductivity are fortunately extremely plentiful and of high accuracy, at least for low concentrations; in the best work agreement to one part in 10,000 between different workers is not uncommon. Especially in non-aqueous solutions and in mixed solvents, conductivity measurements are far more easily made than those of activities, and provide the greater part of our knowledge of the behaviour of electrolytes in such solutions. Furthermore, the measurements can be carried to extraordinarily low concentrations provided proper precautions are taken. Whereas the measurement of the electromotive forces of cells usually becomes unreliable at concentrations below about 0·001 M even in the most favourable cases, accurate conductivity measurements can be made at concentrations down to about 0·00003 M. The experimental techniques have been discussed in Chapter 5; here we are concerned with the theoretical interpretation of the results.

The equivalent conductivity of strong electrolytes at low concentrations is found to be accurately a linear function of the square root of concentration, decreasing as the concentration increases. Extrapolation to zero concentration yields the limiting equivalent conductivity Λ^0, and the equivalent conductivity Λ, at *very low concentrations* can therefore be represented by the equation:

$$\Lambda = \Lambda^0 - A\sqrt{c} \qquad \ldots.(6.1)$$

as was observed by Kohlrausch.

A complete theory of electrolytic conduction should therefore be capable of (*a*) predicting the value of Λ^0 from the dimensions, charges and other properties of the ions and the solvent molecules, (*b*) predicting the value of the constant A in equation (6.1), (*c*) accounting quantitatively for deviations from equation (6.1) at higher concentrations. Of these three problems, the first is farthest from solution, the second is solved and the position with regard to the third has recently been greatly improved.

THE LIMITING VALUES OF EQUIVALENT CONDUCTIVITY

In the state of infinite dilution to which Λ^0 refers, the motion of an ion is limited solely by its interactions with the surrounding solvent molecules, there being no other ions within a finite distance. In these circumstances, the validity of Kohlrausch's law of the independent migration of ions is almost axiomatic; according to this law each species of ion present contributes at infinite dilution a definite amount to the total equivalent conductivity, regardless of the nature of the other ions present. Thus for an electrolyte giving two kinds of ions,

$$\Lambda^0 = \lambda_1^0 + \lambda_2^0 \qquad \ldots.(6.2)$$

The values of λ_1^0 and λ_2^0 may be determined by measurements of transport numbers t, which may also be extrapolated linearly to infinite dilution against the square root of concentration. Thus

$$\begin{aligned} \lambda_1^0 &= t_1^0\Lambda^0 \\ \lambda_2^0 &= t_2^0\Lambda^0 \end{aligned} \qquad \ldots.(6.3)$$

The accuracy with which such measurements confirm Kohlrausch's law of the independent migration of ions may be seen from the data in *Table 6.1* for aqueous potassium and sodium chloride compiled from papers by GORDON and his collaborators[(1)]. These measurements represent probably the most accurate test yet made of the Kohlrausch principle; it will be noted that even at 45° where the experimental difficulties are most marked, the two independent values of $\lambda^0_{Cl^-}$ agree within 0·04 per cent. Thus the step of resolving Λ^0 for a salt into values of λ^0 for its separate ions can be taken with complete confidence provided that accurate values of the transport numbers are available at concentrations low enough to permit extrapolation to zero concentration. To account for the observed

Table 6.1

Test of Kohlrausch's Law of the Independent Migration of Ions

Temp. °C	15°	25°	35°	45°
Λ^0 KCl	121·07	149·85	180·42	212·41
t_2^0 (KCl)	0·5072	0·5095	0·5111	0·5128
λ_2^0 (KCl)	*61·41*	*76·35*	*92·21*	*108·92*
Λ^0 NaCl	101·18	126·45	153·75	182·65
t_2^0 (NaCl)	0·6071	0·6038	0·5998	0·5961
λ_2^0 (NaCl)	*61·43*	*76·35*	*92·22*	*108·88*

Λ^0 in (cm² Int. Ω^{-1} equiv.$^{-1}$)

values of λ^0 in terms of other properties of the ions is, however, a much more difficult problem, of which at present only a qualitative treatment can be given.

The limiting equivalent conductivities of a number of ions at 25° in water are compiled in Appendix 6.1. These have been obtained as follows: the best available data for Λ^0 for various salts have been selected from the literature as indicated by the references quoted. In the case of chlorides, the cation mobility λ_1^0 has been computed as $\lambda_1^0 = \Lambda^0 - \lambda^0_{Cl^-}$ using the value $\lambda^0_{Cl^-} = 76·35$ obtained from Gordon's data (*Table 6.1*). The values for other anions have then been computed as $\lambda_2^0 = \Lambda^0 - \lambda_1^0$ using wherever possible Λ^0 values for potassium or sodium salts and the tabulated values for $\lambda^0_{Na^+}$ or $\lambda^0_{K^+}$. This is done to ensure the self-consistency of the tabulated values, but means that in some cases the value given for λ^0 is not quite that decided upon by the workers referred to, owing to a different choice of the limiting transport numbers. The table will, however, permit the calculation of Λ^0 from the constituent λ^0 values within the experimental error.

THE INTERPRETATION OF THE LIMITING EQUIVALENT CONDUCTIVITIES OF IONS

The most striking feature of the ionic conductivities compiled in Appendix 6.1 is the extremely high mobility of the hydrogen ion, which clearly suggests that a special mechanism is involved in its motion. It is scarcely possible to imagine that the bare proton could be moving freely through the solution, for this would lead to an almost infinite mobility. Nor is it possible to regard the moving entity as the H_3O^+ ion, (though this formula is often written for the aqueous hydrogen ion) since this ion would have dimensions similar to those of a water molecule, and the mobility of the water molecule is known from experiments on the self-diffusion of water to be similar to that of simple ions such as K^+ and Cl^-.† A reasonable explanation has been found[2] in terms of a 'proton jump' mechanism, by which a proton passes from one water molecule to a favourably oriented neighbouring one, but in doing so leaves these molecules unfavourably oriented for another jump.

At any one time, only a few of the protons in a solution will be indulging in these 'jumps'. The majority will be definitely associated with one water molecule or another, and to this extent it is legitimate to write the hydrogen ion as H_3O^+. However, it is believed that this ion can fit into the normal coordinated structure of water almost as well as can an ordinary water molecule, so the charged molecule may become the centre of a rather firmly associated group of water molecules; it may, in fact, become further hydrated. This would explain the remarkable similarity between the activity coefficients of lithium chloride, bromide, iodide and perchlorate and those of the corresponding acids, which implies that from a thermodynamic point of view the hydrated lithium ion and the hydrogen ion are of nearly the same size and involve about the same number of water molecules, while the proton-jump mechanism accounts for the fact that the mobility of the hydrogen ion under an applied electrical field is some ten times that of the lithium ion. The suggested proton jump mechanism can be represented diagrammatically (after GLASSTONE, LAIDLER and EYRING[2]) as follows:

$$\begin{array}{c} \mathrm{H} \\ | \\ \mathrm{H{-}O{-}H} \\ + \end{array} + \begin{array}{c} \mathrm{H} \\ | \\ \mathrm{O{-}H} \\ \end{array} \rightarrow \begin{array}{c} \mathrm{H} \\ | \\ \mathrm{H{-}O} \\ \end{array} + \begin{array}{c} \mathrm{H} \\ | \\ \mathrm{H{-}O{-}H} \\ + \end{array}$$

The abnormally high mobility of the aqueous hydroxide ion, which is second only to that of hydrogen ion, may be similarly accounted for by the proton-transfer process:

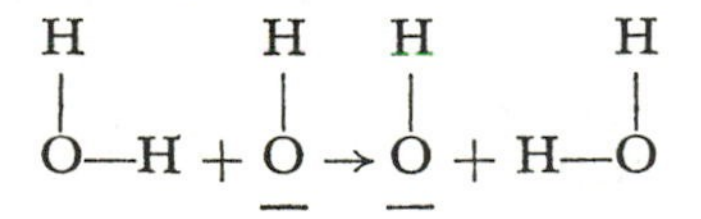

† The mobility of the H_3O^+ ion as a unit may be approximately calculated in units of equivalent conductivity as $D^* = \boldsymbol{R}T\lambda/(|z|\boldsymbol{F}^2)$ where D^* is the self-diffusion coefficient of water ($\sim 2.4 \times 10^{-5}$ cm^2 sec^{-1} at 25°). This gives $\lambda_{H_3O^+} \approx 90$ cm^2 Ω^{-1} equiv.$^{-1}$.

If $(\Lambda^0_{HCl} - \Lambda^0_{KCl})/\Lambda^0_{KCl}$ is taken as a measure of the abnormal hydrogen ion mobility, the data in Appendix 6.2 give 2.26, 1.84 and 1.07 for this ratio at 0°, 25° and 100° respectively, suggesting that the breaking of the water structure reduces the abnormal mobility. A pressure of 3,000 atm.,[2a] however, increases the abnormal mobility, the ratio being 2.15 compared with 1.84 at 1 atm.

The equivalent conductivity of hydrochloric acid which is 426·1 in water at 25° is only 198·5 $cm^2\ \Omega^{-1}\ equiv^{-1}$ in methanol whilst it has a minimum value in a water-methanol mixture containing about 10 per cent by weight of water[3]. In this mixed solvent its conductivity is similar to that of sodium chloride. It is evident that the proton jump via the $CH_3OH_2^+$ complex is less effective than it is via the H_3O^+ complex and that the abnormal mobility is absent in 90 per cent methanol.

Having eliminated these two exceptional cases, we find that some further interesting generalizations emerge from an inspection of Appendix 6.1. The maximum mobility of monovalent ions (at 25° in water) is about 75 equivalent conductivity units; the mobilities of K^+, Tl^+, NH_4^+, Cl^-, Br^-, I^-, NO_3^-, ClO_4^- all cluster closely about this value. It appears that these ions lie in a critical range of size: if they were smaller (in terms of crystallographic radius) they would acquire a permanent hydration sheath and end up larger and with lower mobility, as do sodium, lithium and fluoride ions; if they were larger in crystallographic radius, they would not hydrate, but would be slower-moving merely on account of their size like, for example, the carboxylic acid anions.

The order of the mobilities of the alkali-metal cations is the inverse order of their crystallographic size, which is of course in accordance with the expectation that ions of the greatest surface charge will be most strongly hydrated. The same order holds for the bivalent cations, though the practically identical values for Ca^{++} and Sr^{++} suggest that these two hydrated ions have very similar dimensions. (This similarity is not so marked in the activity coefficients of calcium and strontium salts.) The mobilities of the bivalent cations cover only a small range, about 53–63 units; this may well be because they all have one firmly attached layer of water molecules and only a few in a second layer. Among the few bivalent anions for which data are available the symmetrical tetrahedral sulphate ion shows a substantially higher mobility than the others and even than the bivalent cations, suggesting that it is sufficiently 'padded' with oxygen atoms to prevent any extensive hydration. Comparison of the structurally rather similar sulphate and perchlorate ions on the basis of Stokes' law radius,

$r = 0{\cdot}820|z|/(\lambda^0\eta^0)$, however, indicates that the sulphate ion has a substantially larger 'radius', about 70 per cent greater than that of ClO_4^-.

The trivalent cations of the rare earths, as might be expected, all show very similar mobilities, close to $\lambda^0 = 70$; the ions are evidently all hydrated to much the same large extent. This conclusion is confirmed by the fact that the activity coefficient data for their chlorides all require values of the ion size parameter *a* which lie in the range 5·6–6·0 Å. Their mobilities are strikingly lower than those of the trivalent complex ions $Co(NH_3)_6^{+++}$ and $Fe(CN)_6^{---}$, which are both close to 100 units; in these ions the place of the first layer of water molecules is taken by NH_3 and CN^- respectively, and water molecules do not appear to attach themselves to these 'foreign' groups as readily as to other water molecules. The various polyphosphate ions, which have been carefully investigated by DAVIES and MONK[(4)], provide interesting examples of anions with high negative charges, and attention should also be drawn to JAMES' study[(5)] of the sexavalent cation $[Co_2\ \mathit{trien}_3]^{6+}$ of *tris*-triethylenetetraminecobaltic chloride, a quadridentate compound containing two cobalt atoms and three triethylenetetramine

$$(NH_2{\cdot}CH_2{\cdot}CH_2{\cdot}NH{\cdot}CH_2{\cdot}CH_2{\cdot}NH{\cdot}CH_2{\cdot}CH_2{\cdot}NH_2)$$

molecules.

The tetra-alkyl ammonium ions[(6)] are of great theoretical interest because they combine large size and symmetrical shape with low charge, and furthermore, some of their salts are soluble in many solvents besides water. In the other solvents, however, the individual ionic mobilities are less certainly known because (*a*) the conductivity measurements are in general less easily extrapolated to infinite dilution and (*b*) the limiting transport numbers are seldom known experimentally but have to be guessed on some reasonable basis. The λ^0 values used in compiling Appendix 6.1 for these ions in water at 25° represent the latest values given by KRAUS and his collaborators[(7)], obtained from measurements by the most fastidious techniques extending to concentrations as low as 10^{-4} molar, and are almost certainly to be preferred to the numerous earlier values to be found in the literature. They are of great value as a test of the validity of Stokes' law for ions in aqueous solutions. There is strong reason to believe, from an examination of the temperature-dependence of ionic mobility (see pp. 128–129) that for ions which are (*a*) intrinsically large and of low surface charge, or (*b*) of sufficiently large surface charge to form firmly hydrated entities, Stokes' law is of the correct form though the numerical constant may not be 6π. For these ions the product $\lambda^0\eta^0$ is very nearly constant over a fair range of temperatures in water. The possibility therefore presents itself that we might use the mobilities of the tetra-substituted ammonium ions to calculate correction factors for Stokes' law in water, and then by using these factors, calculate the size of the strongly hydrated ions from their mobilities. In order to do this it is of course necessary to know the sizes of the substituted ammonium ions. A fair approximation to their sizes may be obtained as follows:

1. The effective radius of the $N(CH_3)_4^+$ ion can be estimated from the N–C internuclear distance of 1·47 Å to which is added Pauling's value of 2·0 Å for the Van der Waals radius of the methyl group as a whole, giving 3·47 Å.

2. For the ion $N(C_2H_5)_4^+$ a similar calculation from bond-lengths and angles indicates a maximum radius of about 4·2 Å, while a scale model (using 'Catalin' atomic models) suggests an average radius of about 4·0 Å; the value is somewhat dependent on the configuration given to the C—C—H linkages. The latter value is probably preferable.

3. For the higher homologues, it is not easy to estimate a radius from bond lengths or models, as too many configurations exist. The following rather tentative method may be tried: the first two members of the series are structurally very similar to the symmetrical paraffins $C(CH_3)_4$ and $C(C_2H_5)_4$ which have molal volumes of approximately 120 cm³ and 170 cm³ respectively. One would expect the radii to be directly proportional to the cube roots of the molecular volumes, and one finds, in fact, that the empirical relation:

$$r \approx 0{\cdot}72\bar{V}^{1/3}$$

(with r in Å and $\bar{V}$ in cm³ per mole) gives for the first two members $r = 3{\cdot}55$ Å and $r = 3{\cdot}99$ Å in adequate agreement with the values given above. One may then estimate approximate radii for the higher members by this formula, assuming for the density of the corresponding paraffins the value of 0·75 which is typical of the higher paraffins. The radii of the ions calculated in this way are given in the column headed r in *Table 6.2*. The Stokes' law radii obtained from the limiting mobilities of Appendix 6.1 are given in the column headed r_S; since the viscosity of water at 25° is 0·008903 poise, equation 2.49 becomes $r_S = 92{\cdot}1/\lambda^0$ for monovalent ions. The ratio r/r_S can be regarded as a correction factor for Stokes' law in water and the table suggests that the law is applicable for particles greater than ~ 5 Å in radius, but gives radii which are

Table 6.2

Ion	r (Å)	r_S (Å)	r/r_S
$N(CH_3)_4^+$	3·47	2·05	1·69
$N(C_2H_5)_4^+$	4·00	2·82	1·42
$N(C_3H_7)_4^+$	4·52	3·93	1·15
$N(C_4H_9)_4^+$	4·94	4·73	1·04
$N(C_5H_{11})_4^+$	5·29	5·27	1·00

r = radius estimated from molecular volumes or models.

r_S = radius calculated from the limiting mobility by Stokes' law.

considerably too small when applied to particles smaller than this. The correction factor is plotted against the Stokes' law radius in *Figure 6.1.* We may now very tentatively use this graph to estimate

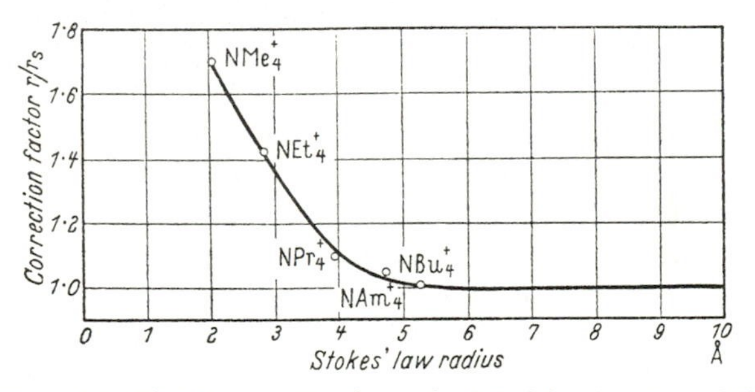

Figure 6.1. Tentative correction factors for Stokes' law in water at 25°

the radii of heavily hydrated ions from their limiting mobilities, assuming these corrections to apply. The results for a number of ions are given in *Table 6.3*; the calculation is of course confined to cases where the ion is of symmetrical shape and has a Stokes' law radius, $r_S = 0{\cdot}820|z|/(\lambda^0\eta^0)$, in the range above 2·0 Å.

The 'corrected Stokes' law radius' of the hydrated ion can then be used to estimate its volume, and since the volume of the bare ion itself is negligible compared to the resulting values, a rough estimate can then be made of the average number of water molecules involved in the hydrated entity by neglecting the electrostriction of these molecules, and ascribing to them their ordinary liquid volume of 30 Å^3. The hydration numbers 'h' so obtained are given in the last column of *Table 6.3*, and it must be admitted that they are eminently reasonable.

Table 6.3

Estimates of the Radii of Hydrated Ions from Modified Stokes' Law:

$$r = \frac{0 \cdot 820|z|}{\lambda^0 \eta^0} \left(\frac{r}{r_s}\right) \text{ in Å}$$

Ion	λ^0	r_s	r	r (*crystal-lographic*)	$\frac{4}{3}\pi r^3$ (cu Å)	h
Na^+	50·10	1·83	3·3	0·97	150	5
Li^+	38·68	2·37	3·7	0·60	210	7
Be^{++}	45	4·08	4·6	—	410	13–14
Mg^{++}	53·05	3·46	4·4	0·65	360	12
Ca^{++}	59·50	3·09	4·2	0·99	310	10
Sr^{++}	59·45	3·09	4·2	1·13	310	10
Ba^{++}	63·63	2·88	4·1	1·35	290	9–10
Zn^{++}	53·0	3·46	4·4	0·74	360	12
La^{+++}	69·75	3·95	4·6	1·15	410	13–14

(The correction factor $\frac{r}{r_s}$ is read from *Figure 6.1* for the value of r_s in the third column.)

THE VARIATION OF LIMITING IONIC CONDUCTIVITIES WITH TEMPERATURE

The data given in Appendix 6.1 are mainly confined to cases where accurate values of the limiting conductances of salts of the ions have been determined at 25° by means of measurements extending down to very low concentrations, *e.g.*, 10^{-3} to 10^{-4} N, making possible reliable extrapolations for Λ^0. A few less reliable values, *e.g.*, that for Be^{++}, have been included for completeness. A much more extensive tabulation, including data for other temperatures, is given by WALDEN[8]. There is an extremely large body of conductivity data for 18°, which was the standard temperature for many physico-chemical studies in Britain and Europe until the 1920's, when the American practice of using 25° as the standard temperature became general. Other temperatures for which data are fairly plentiful are 0° and 100°. However, the use of the precise moving boundary method for determining transport numbers has been mainly confined to 25°, with the result that transport numbers for other temperatures are less certainly known.

The most accurate information we have on the variation of ionic mobilities with temperature comes from the work of GORDON and his collaborators[1, 9], who have measured the conductivities and transport numbers of potassium chloride, sodium chloride and calcium chloride and the conductivity of potassium bromide at 15°,

25°, 35° and 45°. Their measurements were carried to concentrations low enough to permit reliable extrapolation to zero concentration.

Their transport numbers for the chloride ion in potassium chloride solution at infinite dilution, $t^0_{Cl^-}$, are plotted against the temperature in *Figure 6.2.* (The results do not confirm the claim of some earlier workers that transport numbers tend to approach the value 0·5 as the temperature is raised.) From this curve a value

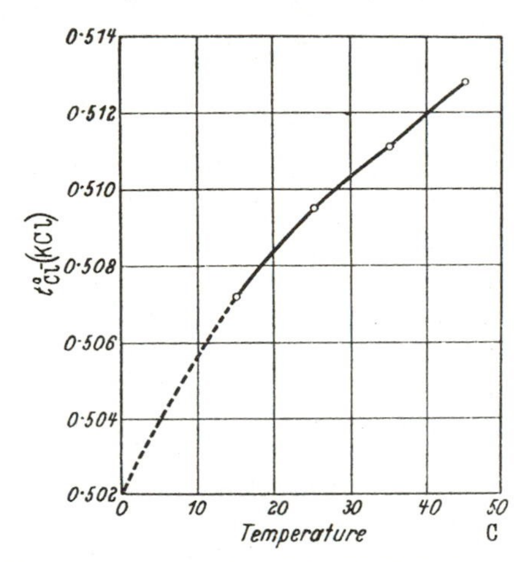

Figure 6.2. Limiting transport number of the chloride ion in aqueous potassium chloride as a function of temperature

of $t^0_{Cl^-} = 0{\cdot}5079$ may be interpolated at 18°. The four points on this graph suggest that an extrapolation to 0° would give $t^0_{Cl^-} \approx 0{\cdot}504$. Walden, before Gordon's data were available, estimated a value of 0·507. OWEN(10) has fitted the conductivity data for a number of electrolytes to a cubic equation in the temperature; if his equation is used for extrapolation, it appears that at low temperatures the transport number decreases more rapidly with decreasing temperature than the results between 15° and 45° suggest. On this basis values of 0·502 and 0·504 at 0° and 5° respectively would seem reasonable and will be used for further calculations. At high temperatures the position is much less satisfactory. To extrapolate Gordon's data for more than 10° or so beyond 45° would be risky; but it seems likely that even at 100° t_{Cl^-} should lie between 0·51 and 0·53. Owen's equations lead to 0·522; Walden in his compilation in Landolt-Börnstein's 'Tabellen' adopts the value 0·509; here we shall assume that $t^0_{Cl^-}$ (100°) = 0·52;

the difference is unimportant for the purpose of examining trends in ionic mobility with temperature, but our figure seems more in

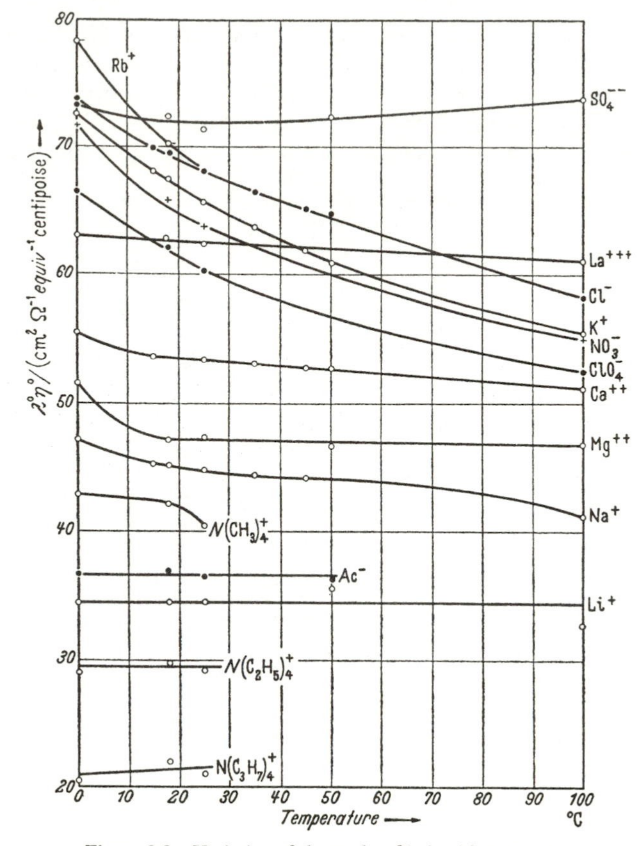

Figure 6.3. Variation of the product $\lambda^\circ\eta^\circ$ with temperature

keeping with the slow increase observed by Gordon between 15° and 45°.

Having settled upon these values for the transport number of chloride ion in potassium chloride, we can now use them in conjunction with the limiting conductivities of potassium chloride to calculate values for the limiting conductivity of the chloride ion at various temperatures.

With these values for the chloride ion as a basis, the limiting conductivities of many other single ions may be computed from the Λ^0 values for various salts. A representative selection is given in

Appendix 6.2 which has been compiled from the sources given in the footnotes.

The temperature variation of ionic conductivity is large, involving a five- or six-fold change over the range 0° to 100°. There is no doubt that the increasing mobility is closely related to the increasing fluidity of water; this can be shown by plotting the product $\lambda^0\eta^0$ (η^0 = the viscosity of water) against the temperature as in *Figure 6.3.* It is noteworthy that the ions for which $\lambda^0\eta^0$ is most nearly constant are those of large size, whether this large size is due to their being polyatomic (*e.g.*, acetate and substituted ammonium ions) or to extensive hydration (*e.g.*, Li^+, Ca^{++}, La^{+++}). This observation does a good deal to justify the arguments by which we estimated the sizes of highly hydrated ions in terms of those of the substituted ammonium ions (see pp. 124–126).

The monatomic ions, K^+, Rb^+, Cl^-, Br^-, I^-, and ClO_4^-, NO_3^- are of similar mobility, and show a similar variation of the product $\lambda^0\eta^0$ with temperature; this behaviour shows up clearly in *Figure 6.3* in contrast to the approximate constancy of $\lambda^0\eta^0$ for the larger ions. It should be noted, however, that even with these ions the variation of $\lambda^0\eta^0$ with temperature is only of the order of 30 per cent over the range 0° to 100°. This suggests that ordinary viscous forces account for most of the resistance to the motion of these ions in water, though there is evidently some other effect operative as well, which is important enough to render useless any attempt to estimate the sizes of these ions on the basis of Stokes' law.

IONIC MOBILITIES IN NON-AQUEOUS SOLVENTS

The measurement of the conductivities of non-aqueous solutions is a straightforward matter, the main requirements being careful attention to the purity of materials and the exclusion of atmospheric moisture. The work of KRAUS and his school[(11)] may be quoted as examples of the most precise techniques. It is, however, much more difficult to obtain from the experimental results accurate values for *limiting* ionic conductivities. First, the low dielectric constants of most non-aqueous solvents result in a much more pronounced decrease of the equivalent conductivity with concentration than is the case for aqueous solutions; and the theory required in extrapolating the conductivity to zero concentration is complicated by the effects of ion-pair formation. These difficulties can, however, be overcome, partly by carrying the measurements to very low concentrations, and partly by the introduction of a finite dissociation-constant into the conductivity formulae. The latter method has been highly developed by Fuoss, and is discussed in Chapter 14.

The second and more serious difficulty is that at present there are practically no accurate transport number data available for non-aqueous electrolytes. The measurement of transport numbers in aqueous solutions has been developed to a high pitch of precision, the moving-boundary method of Longsworth being the standard method for dilute solutions. GORDON and his co-workers[12] have successfully applied this method to sodium and potassium chlorides in pure methanol and in methanol–water mixtures, (see Chapter 7) and HARNED and DREBY[13] have derived the transport numbers of hydrochloric acid in dioxane–water mixtures from electromotive force studies. Measurements in mixed solvents, interesting though they are, present new theoretical problems connected with the preferential solvation of ions by one component of the solvent, and are therefore of less direct interest than would be the corresponding data in single non-aqueous solvents.

Many attempts have been made to estimate separate ionic conductivities from those of salts on a hydrodynamic basis. WALDEN[14] found that the limiting equivalent conductivity of tetra-ethyl-ammonium picrate in a wide variety of solvents including water conformed closely with the formula:

$$\Lambda^0\eta^0 = \text{constant}$$

which is derivable from Stokes' law and is known as Walden's rule. Walden's data showing the constancy of the product $\Lambda^0\eta^0$ for this particular salt at various temperatures and in various solvents are quite striking. A similar constancy was found for tetramethyl-ammonium picrate, but the higher homologues, *e.g.*, tetra-*iso*-amyl ammonium picrate, showed considerably larger variations of the $\Lambda^0\eta^0$ product. Walden considered that the constancy found for tetraethyl ammonium picrate justified the assumption that the picrate ion separately would have constant values of the product $\Lambda^0\eta^0$; on this basis, he deduced λ^0 values for the picrate ion in non-aqueous solvents from the known value of λ^0 in water, the only solvent for which accurate transport number measurements are available: for water, the value is found to be $\lambda^0\eta^0 = 0{\cdot}270$ cm^2 Ω^{-1} equiv^{-1} poise. Limiting conductivities of other cations can then be found by subtraction of the appropriate value for the picrate ion from observed value of Λ^0 for various picrates in other solvents.

Walden's assumption of a constant value of $\lambda^0\eta^0$ for the picrate ion has been challenged by KRAUS[6], who prefers to estimate transport numbers on the basis that the large tetra-*n*-butylammonium and triphenylborofluoride ions should have equal mobilities in all solvents: this permits the estimation of separate ionic equivalent conductivities which are believed to be reliable within 5 per cent. The ionic conductivity values so obtained for the picrate and tetra-ethylammonium ions do not exhibit the constancy of the product $\lambda^0\eta^0$ assumed by Walden, that for the picrate ion varying from $\lambda^0\eta^0 \approx 0{\cdot}24$ in ethylene dichloride to 0·30 in pyridine. Furthermore, Kraus cites data which show that the product $\Lambda^0\eta^0$ for tetra-ethylammonium picrate itself is by no means as constant as Walden claimed: this argument is independent of any arbitrary choice of transport numbers. In pyridine, in particular, the conductivity is abnormally high. While Walden's rule gives a useful guide to the conductivity to be expected, it cannot be considered quantitatively reliable. In a limited way, it is of value in interpreting the variation of ionic conductivities with temperature in one solvent: this aspect has been considered for aqueous ions on p. 124. The obtaining of accurate experimental transport number data for non-aqueous solutions is clearly the key to further progress in understanding the interactions of ions with these solvents.

The conductivities of electrolytes in methanol and hydrogen cyanide and some amides as solvents are discussed in the next chapter and concentrated sulphuric acid as a solvent for electrolytes is considered in Chapter 13. In solvents of lower dielectric constant, electrolytes readily form ion-pairs; this is discussed in Chapter 14, but mention may now be made of Appendix 14.2, which gives the limiting equivalent conductivities and dissociation constants of a number of electrolytes in non-aqueous solvents.

REFERENCES

[1] ALLGOOD, R. W., LEROY, D. J. and GORDON, A. R., *J. chem. Phys.*, 8 (1940) 418; ALLGOOD, R. W. and GORDON, A. R., *ibid.*, 10 (1942) 124; BENSON, G. C. and GORDON, A. R., *ibid.*, 13 (1945) 473

[2] BERNAL, J. D. and FOWLER, R. H., *ibid.*, 1 (1939) 515; see also GLASSTONE, S., LAIDLER, K. J. and EYRING, H., 'The Theory of Rate Processes', Chap. X, McGraw-Hill Book Co. Inc. (1941)

[2a] HAMANN, S. D., 'Physico-Chemical Effects of Pressure', p. 123, Butterworths Scientific Publications, London, (1957)

[3] SHEDLOVSKY, T. and KAY, R. L., *J. phys. Chem.*, 60 (1956) 151; ERDEY-GRÚZ, T., KUGLER, E. and REICH, A., *Magyar Kém. Folyóirat*, 63 (1957) 242; ERDEY-GRÚZ, T. and MAJTHÉNYI, L., *ibid.*, 64 (1958) 212; TOURKY, A. R. and MIKHAIL, S. Z., *Egypt. J. Chem.*, 1 (1958) 1, 13, 187

[4] DAVIES, C. W. and MONK, C. B., *J. chem. Soc.*, (1949) 413; MONK, C. B., *ibid.*, (1949) 423, 427

[5] JAMES, J. C., *Trans. Faraday Soc.*, 47 (1951) 392

[6] KRAUS, C. A., *Ann. N.Y. Acad. Sci.*, 51 (1949) 789

[7] DAGGETT, H. M., BAIR, E. J. and KRAUS, C. A., *J. Amer. chem. Soc.*, 73 (1951) 799
[8] WALDEN, P., LANDOLT-BÖRNSTEIN, 'Tabellen', Eg. III, p. 2059, Julius Springer, Berlin (1936)
[9] KEENAN, A. G., MCLEOD, H. G. and GORDON, A. R., *J. chem. Phys.*, 13 (1945) 466
[10] OWEN, B. B., *J. Chim. phys.*, 49 (1952) C 72
[11] HNIZDA, V. F. and KRAUS, C. A., *J. Amer. chem. Soc.*, 71 (1949) 1565
[12] DAVIES, J. A., KAY, R. L. and GORDON, A. R., *J. chem. Phys.*, 19 (1951) 749
[13] HARNED, H. S. and DREBY, E. C., *J. Amer. chem. Soc.*, 61 (1939) 3113
[14] WALDEN, P., ULICH, H. and BUSCH, G., *Z. phys. Chem.*, 123 (1926) 429; WALDEN, P. and BIRR, E. J., *ibid.*, 153 A (1931) 1

7

THE VARIATION OF CONDUCTIVITIES AND TRANSPORT NUMBERS WITH CONCENTRATION

In the last chapter we have considered the equivalent and ionic conductivities at infinite dilution, *i.e.*, in a state where the ions are far enough apart to be without influence on one another. We now take up the question of the variation of conductivity with concentration, a problem which calls on all the resources of ionic interaction theory. There are two main effects of the interaction between the electric charges of the ions: these are the *electrophoretic effect* and the *relaxation effect.*

THE ELECTROPHORETIC EFFECT

The electrophoretic effect arises in the following way. When an ion moves through a viscous medium it tends to drag along with it the solution in its vicinity. Neighbouring ions therefore have to move not in a stationary medium but with or against the stream according as they are moving in the same direction as the first ion or oppositely. The effect will clearly be concentration-dependent, falling to zero at infinite dilution, and its computation will require the use of the distribution function, since it involves the distances between ions. For the equilibrium case where no external forces such as electric fields or concentration-gradients are acting on the solution, we have been obliged to adopt the distribution functions (4.9) for unsymmetrical electrolytes, and (4.14) for symmetrical electrolytes; these conform to the Boltzmann distribution law as nearly as is permitted by the principle of the linear superposition of fields, and are mathematically consistent with the expression (4.13) for the potential ψ_j. If the ions are moving under the influence of external forces, these distributions will in general be disturbed. In the case of the diffusion of a single electrolyte, however, all the ions must move with the same velocity, and the symmetry of the distribution is not affected. In this case, therefore, the electrophoretic effect may legitimately be computed from these distribution functions. In electrical conduction the symmetry will be disturbed; this gives rise to the relaxation effect which will be discussed later, but we shall neglect the effect of the dissymmetry when calculating the electrophoretic effect in conduction. Also, for the sake of generality, we shall use the Boltzmann distribution law (4.5) rather than forms (4.9) or (4.14), in order to facilitate the investigation of questions of convergence; the results corresponding to the distributions (4.9) and (4.14) can then be obtained as special cases of the general formula. However, we retain the simple expression (4.13) for the potential ψ_j. The treatment of electrophoresis given here is essentially that of ONSAGER and FUOSS[1], but employs the general term of the Boltzmann distribution law. We shall also confine ourselves to solutions containing only a single electrolyte, subscripts 1 and 2 denoting cations and anions respectively, and a subscript A the solvent.

In the cases considered here, bulk motion of the solution as a whole is irrelevant; it follows that the forces k_1 and k_2 acting on the ions must be balanced by other forces k_A acting on the solvent molecules; and denoting the respective bulk concentrations by n_1, n_2 and n_A, we have:

$$n_A k_A = -n_1 k_1 - n_2 k_2 \qquad \ldots.(7.1)$$

At a distance r from a chosen cation the local concentrations of ions are given by the Boltzmann expression (4.5). A spherical shell of radius r and thickness $\mathrm{d}r$ is subject to a resultant force given by:

$$(n_1' k_1 + n_2' k_2 + n_A k_A) 4\pi r^2 \,\mathrm{d}r$$

Provided that we neglect any variation in n_A at this point from its bulk value (a course which is safe for dilute solutions), we can eliminate $n_A k_A$ by means of (7.1) obtaining for the resultant force on the shell:

$$[(n_1' - n_1)k_1 + (n_2' - n_2)k_2] 4\pi r^2 \,\mathrm{d}r$$

This force is assumed to cause the shell and all points within it to move with a velocity obtained, according to Stokes' law, by dividing the force by $6\pi\eta r$. Each shell thus contributes an electrophoretic increment to the velocity of the central ion, and the whole increment, Δv_1, is obtained by integrating over all the shells, beginning at $r = a$ (the distance within which no other ions can penetrate and within which the electrophoretic velocity remains constant). This gives:

$$\Delta v_1 = \frac{2}{3\eta} \int_{r=a}^{\infty} [(n_1' - n_1)k_1 + (n_2' - n_2)k_2] r \,\mathrm{d}r \qquad \ldots.(7.2)$$

If we take n_1' and n_2' to be given by the Boltzmann distribution (4.5), we have on expanding the exponentials:

$$\left.\begin{aligned} n_1' - n_1 &= n_1 \sum_{n=1}^{\infty} \frac{(-1)^n}{n!}\left(\frac{z_1 e\psi}{kT}\right)^n \\ n_2' - n_2 &= n_2 \sum_{n=1}^{\infty} \frac{(-1)^n}{n!}\left(\frac{z_2 e\psi}{kT}\right)^n \end{aligned}\right\} \qquad \ldots.(7.3)$$

It will prove convenient to express the concentrations n_1 and n_2 in these formulae in terms of the quantity (κa). From equation (4.12) we have, using the electrical neutrality condition, $n_1 z_1 + n_2 z_2 = 0$,

$$n_1 = \frac{(\kappa a)^2}{4\pi a^2}\left(\frac{e^2}{\varepsilon kT}\right)^{-1} \frac{1}{z_1(z_1 - z_2)}$$

$$n_2 = \frac{(\kappa a)^2}{4\pi a^2}\left(\frac{e^2}{\varepsilon kT}\right)^{-1} \frac{1}{z_2(z_2 - z_1)}$$

Using these values in (7.3) and taking the potential ψ to be given by equation (4.13), we obtain from equation (7.2):

$$\Delta v_1 = \frac{1}{6\pi\eta}\frac{(\kappa a)^2}{a^2}\left(\frac{e^2}{\varepsilon kT}\right)^{-1} \sum_{n=1}^{\infty}\left[z_1^n \frac{z_1^{n-1}k_1 - z_2^{n-1}k_2}{z_1 - z_2}\cdot\frac{(-1)^n}{n!} \cdot\left(\frac{e^2}{\varepsilon kT}\right)^n \left(\frac{e^{\kappa a}}{1+\kappa a}\right)^n \int_a^\infty \frac{e^{-n\kappa r}}{r^{n-1}}\,\mathrm{d}r\right] \qquad \ldots.(7.4)$$

The integral occurring in (7.4) can always be evaluated as:*

$$\int_a^\infty \frac{e^{-n\kappa r}}{r^{n-1}}\,\mathrm{d}r = \frac{S_n(\kappa a)}{a^{n-2}} \qquad \ldots.(7.5)$$

where $S_n(\kappa a)$ is a function of (κa) only. Equation (7.4) may therefore be more briefly expressed as:

$$\Delta v_1 = \sum_{n=1}^{\infty} \frac{(-1)^n}{6\pi\eta n!}\left(\frac{e^2}{\varepsilon kT}\right)^{n-1} \frac{z_1^n}{a^n}\,\frac{z_1^{n-1}k_1 - z_2^{n-1}k_2}{z_1 - z_2}\,\phi_n(\kappa a) \qquad \ldots.(7.6)$$

where the function $\phi_n(\kappa a)$ is a function of (κa) only, and is defined by:

$$\phi_n(\kappa a) = (\kappa a)^2\left(\frac{e^{\kappa a}}{1+\kappa a}\right)^n S_n(\kappa a)$$

* See appendix to this chapter, p. 170.

The corresponding equation for the electrophoretic increment to the velocity of an anion, Δv_2, is obtained by merely interchanging the subscripts 1 and 2 throughout equation (7.6). The further abbreviation:

$$\Delta v_1 = \sum_{n=1}^{\infty} A_n \cdot \frac{z_1^n(z_1^{n-1}k_1 - z_2^{n-1}k_2)}{a^n(z_1 - z_2)} \qquad \ldots.(7.7)$$

is useful where A_n depends only on (κa), temperature, and solvent properties and is given by:

$$A_n = \frac{(-1)^n}{n!6\pi\eta}\left(\frac{e^2}{\varepsilon kT}\right)^{n-1} \phi_n(\kappa a) \qquad \ldots.(7.8)$$

Equation (7.7) contains the (as yet) unspecified forces k_1 and k_2 which act on the ions. In conductance, these forces are given by the product of the field intensity and the ionic charge; in diffusion, they are the combination of a virtual force produced by the gradient of chemical potential, and an electrical force due to the 'diffusion potential' which results from the electrical attraction of the faster moving for the slower moving ionic species. The application of equation (7.7) to these phenomena will be discussed later.

THE 'RELAXATION EFFECT' IN CONDUCTIVITY

In general, the motion of ions under the influence of external forces will disturb the symmetrical distribution of the ions, and one would therefore expect that this disturbance would tend to decrease the velocity of the ions. In the solution in equilibrium, the 'ionic atmosphere' (which is a convenient description of the whole assemblage of ions outside the central one chosen) is on a time-average distributed with spherical symmetry, and therefore exerts no resultant force on the central ion. The central ion may then be pictured as moving to an off-centre position and experiencing a restoring force, which, however, rapidly dies away as the 'atmosphere' is rearranged by the thermal motions of its constituent ions. The molecular picture thus involves the concept of the 'relaxation of the ionic atmosphere' and the average restoring force experienced by the ion is called the relaxation effect. The external force acting on the ion may, in the conductivity problem, be taken as a field of intensity X acting in the x-direction; the 'relaxation field' will clearly act in the same direction but in the opposite sense, and will be denoted by ΔX. The computation of ΔX involves a combination of the ideas of the interionic attraction theory with the equation of continuity of hydrodynamics, and is mathematically the most difficult part of electrolyte theory. Because of the unavoidable complexity of the treatment we shall not give the complete derivation here, but shall merely give a statement of the main results.

The first attack on the problem of the relaxation effect was made by DEBYE and HÜCKEL[2]; a more successful approach, however, was that of ONSAGER[3], who obtained the following limiting law for the relaxation effect on the conductivity of an extremely dilute solution of a single electrolyte dissociating into ions 1 and 2:

$$\frac{\Delta X}{X} = \frac{z_1 z_2 e^2}{3\varepsilon kT} \cdot \frac{q\kappa}{1 + \sqrt{q}} \quad \ldots.(7.9)$$

Here the quantity q is defined by:

$$\begin{aligned} q &= \frac{|z_1 z_2|}{|z_1| + |z_2|} \cdot \frac{\lambda_1^0 + \lambda_2^0}{|z_2|\lambda_1^0 + |z_1|\lambda_2^0} \\ &= \frac{|z_1 z_2|}{(|z_1| + |z_2|)(|z_2|t_1^0 + |z_1|t_2^0)} \\ &= \tfrac{1}{2} \text{ for symmetrical electrolytes where } |z_1| = |z_2| \quad \ldots.(7.10) \end{aligned}$$

The total electric force acting on the ion is thus given by $Xz_je \left(1 + \frac{\Delta X}{X}\right)$ and produces a velocity (relative to the solvent) of:

$$v_j' = Xz_jeu_j^0 \left(1 + \frac{\Delta X}{X}\right) \quad \ldots.(7.11)$$

where u_j^0 is the absolute mobility of the ion. At infinite dilution, the velocity produced by the field X is:

$$v_j^0 = Xz_jeu_j^0 \quad \ldots.(7.12)$$

Hence introducing (7.9) we have:

$$v_j' = v_j^0 \left(1 + \frac{z_1 z_2 e^2}{3\varepsilon kT} \cdot \frac{q\kappa}{1 + \sqrt{q}}\right)$$

as Onsager's expression for the velocity of the ion, corrected for the relaxation effect. Before calculating the further correction required to take the electrophoretic effect into account, we shall consider later developments in the theory of the relaxation effect. In Onsager's treatment, several approximations are made: (*a*) The potential ψ_j is taken as given by the expression:

$$\psi_j = \frac{z_je}{\varepsilon} \frac{e^{-\kappa r}}{r}$$

i.e., the factor $\frac{e^{\kappa a}}{1 + \kappa a}$ is omitted from equation (4.13). This means that the resulting expression is valid only at great dilutions where κa is small compared with unity. (*b*) Various other approximations involving the relation $\Delta X \ll X$ are made: these also will be admissible at great dilutions where the relaxation effect is small.

For some twenty-five years after the appearance of Onsager's theory no major progress was made with the relaxation-effect. From 1952 onwards however a number of extensions and refinements have appeared, in all of which the most important new feature is the introduction of the ion size parameter a, making possible a considerable increase in the range of validity of the theory.

The 1952 paper of FALKENHAGEN, LEIST and KELBG[4] employed the Eigen-Wicke distribution function mentioned in Chapter 4, in place of the usual Boltzmann one; this makes only a minor difference in the meaning of the quantity κ and need not be considered further here. They obtained for the relaxation-effect, allowing for finite ion size, the expression:

$$\frac{\Delta X}{X} = \frac{z_1 z_2 e^2}{3\varepsilon kT} \cdot \frac{q}{1 - q} \cdot \frac{\kappa}{(1 + \kappa a)\kappa a} [e^{\kappa a(1-\sqrt{q})} - 1] \qquad \ldots.(7.13)$$

Expanding the exponential in (7.13) as far as the first power of (κa) gives:

$$\frac{\Delta X}{X} = \frac{z_1 z_2 e^2}{3\varepsilon kT} \cdot \frac{q}{1 + \sqrt{q}} \cdot \frac{\kappa}{1 + \kappa a} \qquad \ldots.(7.14)$$

which differs from Onsager's result (7.9) only by the factor $(1 + \kappa a)$ in the denominator. Thus they found the effect of finite ion size on the relaxation-effect to be of the same form as its effects on the free energy and on the electrophoretic effect (*cf.* equation 9.5 and 7.27).

Almost simultaneously, PITTS[5] investigated the conductance of symmetrical electrolytes; his equation is compared with others in reference (6). His result, on separating out the part dealing with the relaxation-effect, may be written:

$$-\frac{\Delta X}{X} = \frac{z^2 e^2}{3\varepsilon kT} \cdot \frac{q}{1 + \sqrt{q}} \cdot \frac{\kappa}{(1 + \kappa a)(1 + \kappa a\sqrt{q})} + \left(\frac{z^2 e^2 \kappa}{\varepsilon kT}\right)^2 \frac{S_1}{3} \qquad \ldots.(7.15)$$

The second term on the right arises from the 'higher terms', S_1 being given as a function of (κa) in a table in the original paper. Ignoring this term, which will not appear if we adopt the 'self-consistent' treatment of the potential problem as discussed in Chapter 4, we see that (7.15) differs from (7.14) by the presence of a further factor $(1 + \kappa a\sqrt{q})$ in the denominator.

In a subsequent paper[7], Falkenhagen's school introduced another boundary condition, that the normal component of the relative motion of two ions must vanish at the surface of ions in contact since they are treated as hard spheres. On this basis they evaluated the relaxation-effect as:

$$\frac{\Delta X}{X} = -\frac{e^2}{3\varepsilon kT} \cdot \frac{q}{1 + \sqrt{q}} \cdot \frac{\kappa}{(1 + \kappa a)\,[1 + \kappa a\sqrt{q} + \kappa^2 a^2/6]} \quad \ldots\ldots(7.16)$$

for 1:1 electrolytes, with $q = 0.5$. This equation is identical with the first-order term of Pitt's result (7.15), except for the further term $(\kappa a)^2/6$ in the denominator.

In 1953 MIRTSKHULAVA[8] also published a treatment of the problem along the same lines as Pitts. Her result for the relaxation-effect is given as a complicated power series (equation 38 of ref. 8), involving also a term containing the exponential integral function. The latter is of especial interest since at very low concentrations it gives rise to a term in $(c \log c)$, as had been anticipated by ONSAGER and FUOSS[1].

FUOSS and ONSAGER[9] have recently given the most comprehensive treatment, including numerical tables of certain transcendental functions related to the exponential integral functions. Their final result, giving the contributions of both the relaxation effect and the electrophoretic effect, is expressed as Onsager's original limiting-law result together with a very complicated function of κ and a, for which the original papers must be consulted. They demonstrate that the transcendental functions involved lead to a term of order $(c \log c)$, although the approximations giving this form are valid only at extremely low concentrations. This contribution from the transcendental terms is quite small, but its relative importance increases at low concentrations, and Fuoss and Onsager show that neglect of it can lead to small errors in the extrapolation of conductance data. They emphasize that the approximation of their transcendental terms to the form $(c \log c)$ at low concentrations does not justify the use of such terms with *arbitrary* coefficients for the representation of conductance data at higher concentrations, a practice which has been followed by many workers in the past. An arduous but worthwhile task for an enthusiastic algebraist would be to determine to what extent the formulae of Fuoss and Onsager agree with those of Mirtskhulava, which as far as we know have not yet been tested against experimental results.

We do not consider it practicable to present the Fuoss-Onsager treatment in detail; even the original papers give only a condensed account of the development of the formulae. We shall therefore use for illustrating the general form of the theory, Falkenhagen's expression (7.16) for the relaxation effect.

THE EFFECT OF ELECTROPHORESIS ON THE CONDUCTIVITY

The general equation (7.7) for the electrophoretic increment to the ionic velocity may now be specialized for the case of conductivity by replacing the forces k_1 and k_2, which act on the ions, by the sum of the forces produced by the external field X and the relaxation-field ΔX giving:

$$k_1 = (X + \Delta X)z_1 e, \quad k_2 = (X + \Delta X)z_2 e \qquad \ldots.(7.17)$$

Equation (7.7) then becomes:

$$\left.\begin{aligned} \Delta v_1 &= (X + \Delta X)e \sum_n A_n \frac{z_1^{2n} - z_1^n z_2^n}{a^n(z_1 - z_2)} \\ \Delta v_2 &= (X + \Delta X)e \sum_n A_n \frac{z_1^n z_2^n - z_2^{2n}}{a^n(z_1 - z_2)} \end{aligned}\right\} \qquad \ldots.(7.18)$$

Hence the final velocity of the ions, corrected for both electrophoretic and relaxation effects, is given by combining equations (7.11) and (7.18):

$$\begin{aligned} v_1 &= v_1' + \Delta v_1 \\ &= (X + \Delta X)z_1 e u_1^0 + (X + \Delta X)e \sum_n A_n \frac{z_1^{2n} - z_1^n z_2^n}{a^n(z_1 - z_2)} \end{aligned} \qquad \ldots.(7.19)$$

But the absolute mobility u_1^0 is also given by equation (7.12) in terms of the velocity v_1^0 produced by the field X at infinite dilution:

$$v_1^0 = X z_1 e u_1^0 \qquad \ldots.(7.20)$$

Dividing equation (7.19) by (7.20) gives:

$$\frac{v_1}{v_1^0} = \left(1 + \frac{\Delta X}{X}\right)\left[1 + \frac{1}{z_1 u_1^0} \sum_n A_n \frac{z_1^{2n} - z_1^n z_2^n}{a^n(z_1 - z_2)}\right] \qquad \ldots.(7.21)$$

Since the velocities v_1 and v_1^0 are those attained under the same external field X in the actual solution and in the infinitely dilute solution respectively, the ratio v_1/v_1^0 may be replaced by the ratio of the equivalent ionic conductivities λ_1/λ_1^0; and the factor $\dfrac{1}{z_1 u_1^0}$ preceding the summation on the right may be put in terms of λ_1^0, by the relation (cf. 2.46):

$$u_1^0 = \mathbf{N}\lambda_1^0/(\mathbf{F}^2|z_1|)$$

so that (7.21) becomes:

$$\lambda_1 = \left(\lambda_1^0 + \frac{\mathbf{F}^2}{\mathbf{N}} \sum_n A_n \frac{z_1^{2n} - z_1^n z_2^n}{a^n(|z_1| + |z_2|)}\right)\left(1 + \frac{\Delta X}{X}\right) \quad \ldots.(7.22)$$

the relaxation term $\dfrac{\Delta X}{X}$ being given by equations (7.9), (7.14) or (7.16) according to the degree of approximation desired. The corresponding expression for the anion is:

$$\lambda_2 = \left(\lambda_2^0 + \frac{\mathbf{F}^2}{\mathbf{N}} \sum_n A_n \frac{z_2^{2n} - z_1^n z_2^n}{a^n(|z_1| + |z_2|)}\right)\left(1 + \frac{\Delta X}{X}\right) \quad \ldots.(7.23)$$

The equivalent conductivity of the electrolyte $\Lambda = \lambda_1 + \lambda_2$, is therefore:

$$\Lambda = \left(\Lambda^0 + \frac{\mathbf{F}^2}{\mathbf{N}} \sum_n A_n \frac{(z_1^n - z_2^n)^2}{a^n(|z_1| + |z_2|)}\right)\left(1 + \frac{\Delta X}{X}\right) \quad \ldots.(7.24)$$

Though we have retained the general expression[10] for the electrophoretic terms in developing these expressions, it will be recalled that the Boltzmann distribution on which this expression is based is not mathematically consistent with the Poisson equation, and that for consistency the series can be taken only as far as the first term for unsymmetrical valency types, and the second for symmetrical types. Furthermore, it is obvious from equations (7.22), (7.23) and (7.24) that in the case of symmetrical electrolytes, ($z_1 = -z_2$), the second-order electrophoretic term ($n = 2$) vanishes identically. Hence, in all cases the first-order term alone need really be considered though examination of the convergence of higher-order terms may throw useful light on the validity of the approximation made to the Boltzmann distribution.

The first-order electrophoretic term in the conductivity equation

Taking only the term for $n = 1$ and using the definitions of A_n, ϕ_n and S_n, given on pp. 135–136, we have:

$$A_1 = -\frac{1}{6\pi\eta} \cdot \frac{\kappa a}{1 + \kappa a}$$

Equations (7.22), (7.23) and (7.24) therefore become:

$$\lambda_1 = \left(\lambda_1^0 - \frac{\mathbf{F}^2}{6\pi\eta\mathbf{N}} |z_1| \frac{\kappa}{1 + \kappa a}\right)\left(1 + \frac{\Delta \mathrm{X}}{\mathrm{X}}\right) \quad \ldots.(7.25)$$

$$\lambda_2 = \left(\lambda_2^0 - \frac{\mathbf{F}^2}{6\pi\eta\mathbf{N}} |z_2| \frac{\kappa}{1 + \kappa a}\right)\left(1 + \frac{\Delta \mathrm{X}}{\mathrm{X}}\right) \quad \ldots.(7.26)$$

$$\Lambda = \left(\Lambda^0 - \frac{\mathbf{F}^2}{6\pi\eta\mathbf{N}} (|z_1| + |z_2|) \frac{\kappa}{1 + \kappa a}\right)\left(1 + \frac{\Delta \mathrm{X}}{\mathrm{X}}\right) \quad \ldots.(7.27)$$

THE ONSAGER LIMITING LAW FOR THE CONDUCTIVITY

In Onsager's treatment, the further approximation of writing $(1 + \kappa a) \approx 1$ in the denominator of the first-order electrophoretic correction is made, and the relaxation term $\frac{\Delta \mathrm{X}}{\mathrm{X}}$ is expressed by the limiting equation (7.9). Further, in evaluating the electrophoretic effect the forces k_1 and k_2 are taken as $\mathrm{X} z_1 \mathbf{e}$ and $\mathrm{X} z_2 \mathbf{e}$ rather than as $(\mathrm{X} + \Delta\mathrm{X}) z_1 \mathbf{e}$ and $(\mathrm{X} + \Delta\mathrm{X}) z_2 \mathbf{e}$; this is equivalent to neglecting the cross-product of the electrophoretic and relaxation terms in (7.25)–(7.27). All these approximations are of course quite justified for the purpose of finding the limiting law, but it is clear that the resulting expression will apply only at extreme dilutions, since (κa) is far from negligible compared to unity at ordinary concentrations, and diminishes only as $\sqrt{c}$. The Onsager limiting law is thus:

$$\Lambda = \Lambda^0 - \frac{|z_1 z_2| \mathbf{e}^2}{3\varepsilon \boldsymbol{k} T} \cdot \frac{\Lambda^\circ q \kappa}{1 + \sqrt{q}} - \frac{\mathbf{F}^2}{6\pi\eta\mathbf{N}} (|z_1| + |z_2|)\kappa \quad \ldots.(7.28)$$

Since κ is given by equation (4.12) it may be written as:

$$\kappa = \left(\frac{8\pi\mathbf{N}\mathbf{e}^2}{1000\varepsilon \boldsymbol{k} T}\right)^{\frac{1}{2}} \sqrt{I}$$

where I is the 'ionic strength' defined by:

$$I = \frac{c}{2}(\nu_1 z_1^2 + \nu_2 z_2^2)$$

with the usual convention that c implies the mole/l scale. Equation (7.28) then becomes, on inserting the values of the physical constants:

$$\Lambda = \Lambda^0 - \left[\frac{2.801 \times 10^6 |z_1 z_2| q \Lambda^0}{(\varepsilon \mathrm{T})^{3/2}(1 + \sqrt{q})} + \frac{41.25(|z_1| + |z_2|)}{\eta(\varepsilon \mathrm{T})^{\frac{1}{2}}}\right]\sqrt{I} \quad \ldots .(7.29)$$

in which η must be expressed in poise and T in °K.

This limiting law is of the form:

$$\Lambda = \Lambda^0 - \mathrm{A}\sqrt{c} \quad \ldots .(7.30)$$

which was found by Kohlrausch to describe the variation of equivalent conductivity with concentration in dilute solutions. For aqueous solutions at 25° it reduces on putting $\varepsilon = 78{\cdot}30$, $\mathrm{T} = 298{\cdot}16$°K and $\eta = 0{\cdot}008903$ poise, with Λ expressed as (cm² Ω^{-1} equiv^{-1}), to:

$$\Lambda = \Lambda^0 - \left[0{\cdot}7852|z_1 z_2| \frac{q\Lambda^0}{1 + \sqrt{q}} + 30{\cdot}32\,(|z_1| + z_2|)\right]\sqrt{I} \quad \ldots .(7.31)$$

CONDUCTIVITY EQUATIONS FOR HIGHER CONCENTRATIONS

For many years equation (7.29) was employed with added terms in c, $c^{3/2}$, $c \log c$, *etc.*, to represent data at concentrations above 0·001N, where Onsager's limiting form is no longer adequate. For purposes of extrapolation, an equation proposed by SHEDLOVSKY[11] has been widely used: equation (7.29) may be written:

$$\Lambda = \Lambda^0 - (B_1 \Lambda^0 + B_2)\sqrt{c}$$

where B_1 and B_2 are parameters given by the theory. Rearranging this to:

$$\Lambda^0 = (\Lambda + B_2\sqrt{c})/(1 - B_1\sqrt{c}), \quad \ldots .(7.32)$$

Shedlovsky observed that for strong aqueous 1:1 electrolytes the quantity on the right of (7.32) is not constant, as it would be if equation (7.29) were obeyed exactly, but varies almost linearly with c up to concentrations of about 0.1N. He therefore defined an extrapolation function, $\Lambda^{0\prime}$, by:

$$\Lambda^{0\prime} = (\Lambda + B_2\sqrt{c})/(1 - B_1\sqrt{c}) \qquad \ldots.(7.33)$$

which when plotted against c would yield on extrapolation to $c = 0$, the true limiting conductivity Λ^0. This implies that data up to 0.1N can be fitted by the expression:

$$\Lambda = \Lambda^0 - (B_1\Lambda^0 + B_2)\sqrt{c} + b\, c(1 - B_1\sqrt{c}) \quad \ldots.(7.34)$$

where the coefficient b is chosen to fit the data. Useful as this device is, its weakness lies in its empirical nature, no simple meaning being attached to the b coefficient. FUOSS and ONSAGER[9] have shown that the approximate constancy of b can be accounted for as a fortuitous consequence of the numerical values of certain terms in their complete theory.

Still denoting the coefficients of the relaxation and electrophoretic terms in equation (7.29) by B_1 and B_2 respectively, we can combine Falkenhagen's equation (7.16) with equation (7.27) to obtain:

$$\Lambda = \left(\Lambda^0 - \frac{B_2\sqrt{c}}{1 + \kappa a}\right)\left(1 - \frac{B_1\sqrt{c}}{(1 + \kappa a)\,(1 + \kappa a\sqrt{q} + \kappa^2 a^2/6)}\right) \quad \ldots.(7.35)$$

If we expand this product in powers of $\sqrt{c}$, putting $\kappa a = Ba\sqrt{c}$, we obtain:

$$\Lambda = \Lambda^0 - (B_1\Lambda^0 + B_2)\sqrt{c} + c(aBB_2 + B_1B_2 + 1{\cdot}707\Lambda^0\, a\, BB_1)$$
$$- 2{\cdot}707\, a\, BB_1B_2\, c^{3/2} + \ldots.$$

Since for most aqueous 1:1 electrolytes $Ba \approx 1$ (mole$^{-\frac{1}{2}}$ litre$^{\frac{1}{2}}$), whilst $B_1 \approx 0{\cdot}2$, $B_2 \approx 60$ and $\Lambda \approx 100$, the coefficients of the terms in c and $B_1c^{3/2}$ are of the same magnitude; to this extent the expression provides some justification for the form of Shedlovsky's function.

Another useful approximation is that proposed by the authors[12] shortly after the appearance of Falkenhagen's earlier equation (7.13). Combining (7.14) with (7.27) and neglecting the cross-product of the relaxation and electrophoretic terms, gives:

$$\Lambda = \Lambda^0 - \frac{B_1\Lambda^0 + B_2}{1 + \kappa a}\sqrt{c} \qquad \ldots.(7.36)$$

i.e., we have merely to divide the square-root term of Onsager's original limiting law by $(1 + \kappa a)$ in order to allow for the finite ionic size. Equation (7.36) gives a very fair account of the conductances of aqueous 1:1 electrolytes up to 0·05 or 0.1N. It may be rearranged as:

$$\Lambda^0 = \Lambda + \frac{B_1\Lambda + B_2}{1 + (Ba - B_1)\sqrt{c}}\sqrt{c} \qquad \ldots .(7.37)$$

Table 7.1

Values of the function $\phi_n(\kappa a) = (\kappa a)^2 \left(\frac{e^{\kappa a}}{1 + \kappa a}\right)^n S_n(\kappa a)$

at round (κa) *where* $S_n(\kappa a) = a^{n-2}\int_a^{\infty} \frac{e^{-n\kappa r}}{r^{n-1}}\,\mathrm{d}r$

κa	$100\phi_1(\kappa a)$	$100\phi_2(\kappa a)$	$100\phi_3(\kappa a)$	$100\phi_4(\kappa a)$	$100\phi_5(\kappa a)$
0·0	0	0	0	0	0
0·05	4·762	0·4566	0·1609	0·0884	0·0584
0·1	9·091	1·235	0·4761	0·2624	0·1693
0·2	16·67	2·911	1·166	0·6192	0·3764
0·3	23·08	4·405	1·734	0·876	0·499
0·5	33·33	6·628	2·425	1·100	0·552
0·7	41·18	7·987	2·686	1·096	0·491
1·0	50·00	9·032	2·678	0·938	0·37
1·2	54·55	9·327			
1·4	58·33	9·434			
1·5	60·00	9·429			
1·6	61·54	9·411			
1·8	64·29	9·316			
2·0	66·67	9·170			
2·5	71·43	8·692			
3·0	75·00	8·172			
3·5	77·78	7·662			
4·0	80·00	7·187			
4·5	81·82	6·753			
5·0	83·33	6·359			
5·5	84·62	6·002			
6·0	85·71	5·681			

From STOKES, R. H., *J. Amer. chem. Soc.*, 75 (1953) 4563.

where $\kappa = B\sqrt{c}$, which is a useful form for determining Λ^0. Its advantage over Shedlovsky's function is that the parameter a has a simple physical meaning and can be expected to lie in the range 3 – 5·5Å; for fully dissociated 1:1 electrolytes it has been found to be nearly independent of temperature for any given electrolyte (see *Table 7.3*).

To each of these equations (7.29, 7.35, 7.36) for Λ there of course corresponds a pair of equations for λ_1 and λ_2 separately; these are obtained from equations (7.25) and (7.26) by the same method, and differ only in that λ_1 or λ_2 replaces Λ, λ_1^0 or λ_2^0 replaces Λ^0, and in the electrophoretic term $|z_1|$ or $|z_2|$ replaces the sum $(|z_1| + |z_2|)$. The relaxation term for the separate ions in the equations corresponding to (7.33) and (7.34) is exactly the same as in these equations for Λ.

CONVERGENCE OF THE ELECTROPHORETIC TERMS

We now return to equation (7.24) where the electrophoretic contribution appears as the series:

$$\frac{\mathbf{F}^2}{\mathbf{N}} \sum_n A_n \frac{(z_1^n - z_2^n)^2}{\mathrm{a}^n(|z_1| + |z_2|)}$$

In this series the quantities A_n are given by the equation:

$$A_n = \frac{(-1)^n}{n!6\pi\eta} \left(\frac{\boldsymbol{e}^2}{\varepsilon \boldsymbol{k} T}\right)^{n-1} \phi_n(\kappa a)$$

The dimensions of A_n are those of (viscosity^{-1} length^{n-1}). The dimensionless function ϕ_n (κa) for values of n up to 5 is given in *Table 7.1*. In deriving the conductivity equations, we have used only the first-order electrophoretic term, obtained by putting $n = 1$

Table 7.2

Values of the valency factor $\dfrac{(z_1{}^n - z_2{}^n)^2}{a^n(|z_1| + |z_2|)}$

Valency type	$n = 1$	$n = 2$	$n = 3$	$n = 4$	$n = 5$
1 : 1	$2/a$	0	$2/a^3$	0	$2/a^5$
2 : 2	$4/a$	0	$64/a^3$	0	$1024/a^5$
1 : 2 and 2 : 1	$3/a$	$3/a^2$	$27/a^3$	$75/a^4$	$363/a^5$

in the above formula, a course adopted in the interest of self-consistency. The question as to whether the series does converge rapidly enough to make this expedient successful has been investigated by STOKES[10] who showed that for aqueous solutions at 25° convergence depends on the factor:

$$\frac{(z_1^n - z_2^n)^2}{a^n\ (|z_1| + |z_2|)}$$

rather than the quantity A_n. *Table 7.2* shows how this factor behaves for various valency types.

Bearing in mind that the ion-size parameter a is in the vicinity of 4Å for most simple ions, it is apparent from *Table 7.2* that the electrophoretic terms given by the series in (7.24) will converge quite satisfactorily for 1 : 1 electrolytes in water, so that the formulae proposed for the conductivity, in which only the term $n = 1$ is accepted, should be adequate. For 2 : 2 electrolytes, in spite of the vanishing of the even-order terms, the convergence is unsatisfactory since the third- and fifth-order terms will be of comparable magnitude to the first. For unsymmetrical valency types, none of the terms vanish and convergence will be slow, though somewhat assisted by the alternation of signs. In non-aqueous solvents the dielectric constant is usually lower than that of water (hydrocyanic acid is one exception); therefore the factor $\frac{e^2}{\varepsilon kT}$ which appears to the $(n - 1)th$ power in the expression for A_n, will be larger and will militate further against satisfactory convergence. We cannot, therefore, really expect the present treatment to be quantitatively successful except for 1 : 1 electrolytes in water, unless there are good reasons for ascribing a large effective size a to the ions; this foreboding is, in fact, fulfilled. Merely to include the higher-order electrophoretic terms is not a satisfactory solution, since the formulae from which these terms have been computed are self-consistent only as far as the first order for unsymmetrical electrolytes and as far as the second order for symmetrical ones. The formulae obtained above for the higher-order terms are therefore helpful only when these are negligible; in other cases they serve to demonstrate the inadequacy of the treatment rather than to provide an adequate one.

EXPERIMENTAL TESTS OF THE THEORY OF CONDUCTIVITY

The Onsager limiting law (7.29) has been exhaustively tested by extremely precise experimental studies, and its validity, for the conditions assumed in its derivation, has been conclusively demonstrated. These conditions may be summed up by the requirements that the dimensionless parameter (κa) should be very small compared to unity, and that the electrolyte should be fully dissociated into ions. For aqueous solutions at ordinary temperatures, κ is approximately $0{\cdot}3 \times 10^8 \sqrt{I}$ (see Appendix 7.1); and the mean ionic diameter a is $3\text{–}5 \times 10^{-8}$ cm; hence κa is of the same order of magnitude as $\sqrt{I}$. At a concentration where the ionic strength I is 0·001, κa is about 0·03, and the approximation of neglecting it in the factor $(1 + \kappa a)$ therefore involves an error of about 3 per cent in the value of $(\Lambda^0 - \Lambda)$. At this ionic strength, $(\Lambda^0 - \Lambda)$ is about three equivalent conductivity units for 1 : 1 electrolytes, and rather more for higher valency types, so that the approximation is of the order of 0·1 units in Λ. This is several times the experimental error of the best work and shows that a concentration of one-thousandth normal must be regarded as a theoretical upper bound to the range of validity of the limiting law, even for aqueous 1 : 1 electrolytes. In other solvents and for higher valency-types the

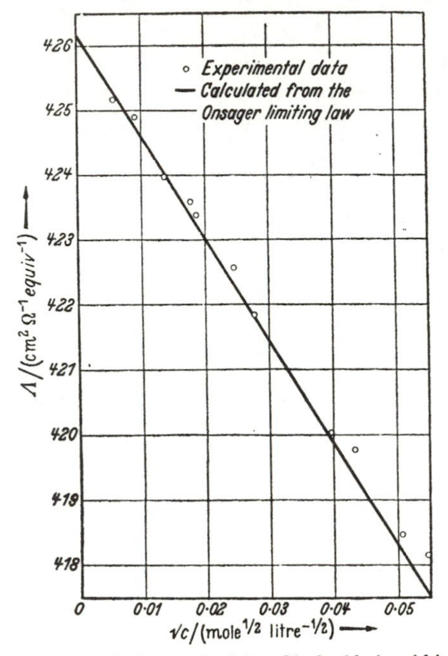

Figure 7.1. Equivalent conductivity of hydrochloric acid in very dilute aqueous solution at 25°

bound is even lower. However, conductivity measurements can in favourable cases be carried out accurately at concentrations as low as 0·00003 N: although the measured values at such concentrations are often close to the limiting value for infinite dilution, the experimental precision is such that the difference still makes a significant test of the limiting law. The very careful measurements of SHEDLOVSKY *et al.*(13) have shown that in the concentration-range from 0·00003 to 0·001 N the Onsager formula (7.29) is obeyed within experimental error by aqueous sodium chloride, potassium chloride, hydrochloric acid, silver nitrate, calcium chloride and lanthanum chloride.

Silver nitrate actually conforms to the limiting law up to substantially higher concentrations: this is due to the effect of ion-pair formation, which, though negligible below 0·001 N, reduces the conductivity at higher concentrations by amounts comparable to those involved by neglecting the factor $(1 + \kappa a)$. *Figure 7.1* illustrates the concordance between the experimental data of Shedlovsky for aqueous hydrochloric acid and the predictions of the Onsager

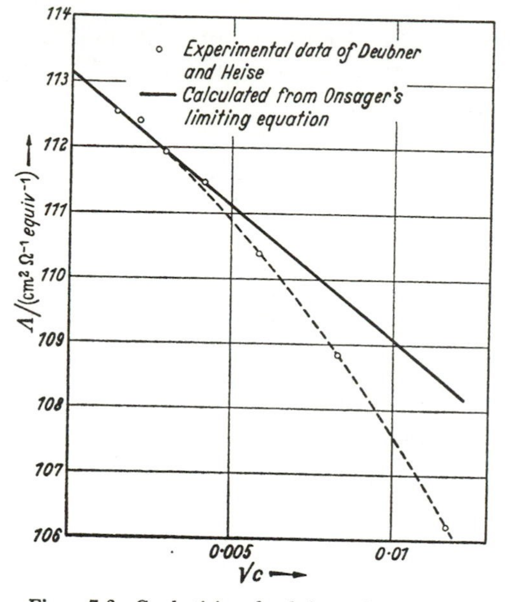

Figure 7.2. Conductivity of cadmium sulphate at 18°

limiting law at concentrations below $c = 0{\cdot}003$. For 2 : 2 electrolytes and higher valency types, however, the limiting law is obeyed only at extraordinarily low concentrations, the formation of ion-pairs being appreciable even at high dilutions. It is only recently that evidence has been advanced to show that the limiting Onsager equation is valid for a 2 : 2 electrolyte. By taking extraordinary precautions, DEUBNER and HEISE(14) have been able to measure the conductivity of cadmium sulphate solutions at concentrations as low as $c = 2 \times 10^{-6}$. *Figure 7.2* shows how the seven values of the conductivity determined by Deubner and Heise do agree with the values predicted by the limiting law:

$$\Lambda = 113{\cdot}15 - 408{\cdot}1\sqrt{c}$$

It is clear that with increasing dilution the experimental values approach more nearly to the theoretical and are in good agreement for the four most dilute solutions. At higher concentrations the experimental values are lower than the predicted; as we shall see in Chapter 14 this is characteristic of salts subject to ion-pair formation. The deviation is in the opposite direction with non-associated salts, and *Figure 7.3* illustrates this by using the data of Shedlovsky for sodium chloride up to comparatively high concentrations.

Valuable though the confirmation of the limiting law at these extreme dilutions is, there is little practical use for a theory dealing

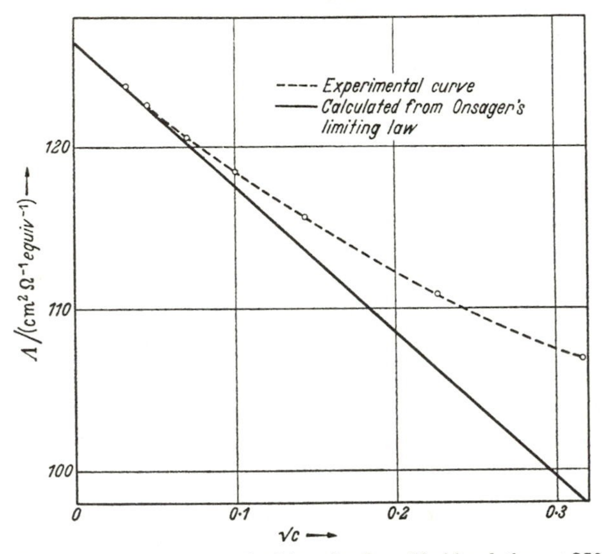

Figure 7.3. Equivalent conductivity of sodium chloride solutions at 25°

only with such solutions. More interest therefore attaches to testing the more complete equations for higher concentrations in which the factor $(1 + \kappa a)$ is not omitted; since these all necessarily reduce to the Onsager form when $\kappa a \ll 1$, it is clear that if they hold, the Onsager limiting law must also do so when only the concentration-range below about 0·001 N is considered.

The most careful experimental work has largely been confined to solutions less than 0·1 N in concentration, and the very precise direct current method of Gordon has been employed only up to 0·01 N. Below 0·01 N the best measurements by different observers often agree within about 0·03 in Λ, but at higher concentrations uncertainties of several tenths of a unit exist. For example, the

Table 7.3

Tests of Equation (7.36) $\Lambda^0 = \Lambda + \dfrac{(B_1\Lambda + B_2)\sqrt{c}}{1 + (Ba - B_1)\sqrt{c}}$

Temp. °C	Λ^0 eq. (7.36)	*Mean* δ per cent	*Max.* δ per cent	*Number of points*	*Range (molar)*	*Observer*	Λ^0 (S.E.F.)
			HCl. a = 4·3 Å *at all temperatures*				
5°	297·61	0·03	0·09	12	0·001–0·083	O & S	297·6
15°	361·89	0·04	0·09	11	0·001–0·082	O & S	362·0
25°	425·98	0·03	0·06	12	0·002–0·086	O & S	426·2
25°	426·10	0·03	0·05	11	0·00003–0·003	S	426·16
35°	489·02	0·02	0·06	14	0·001–0·062	O & S	489·2
45°	550·18	0·02	0·05	11	0·002–0·090	O & S	550·3
55°	609·34	0·02	0·05	11	0·002–0·070	O & S	609·5
65°	666·64	0·02	0·06	12	0·001–0·072	O & S	666·8

O & S: Owen, B. B. and Sweeton, F. H., *J. Amer. chem. Soc.*, 63 (1941) 2811
S: Shedlovsky, T. (Λ^0 converted to Jones and Bradshaw 0·1 demal standard) *ibid.* 54 (1932) 1411
Λ^0 (S.E.F.) denotes the values obtained by the observers named, using the Shedlovsky extrapolation function (7.33)

From Robinson, R. A. and Stokes, R. H., *J. Amer. chem. Soc.*, 76 (1954) 1991 (where similar data are given for four 1:1 salts at various temperatures) equivalent conductivity of 0·1 N potassium bromide solution at 25° is given as 131·19 by JONES and BICKFORD(15) and as 131·39 by LONGSWORTH(16). However, since the significant quantity for the theory is $(\Lambda^0 - \Lambda)$ and this is larger at higher concentrations, such discrepancies in the data are not serious; one should, however, be prepared to tolerate deviations of a few tenths of a unit in Λ at the higher concentrations.

All of the proposed equations can be expressed in the form:

$$\Lambda = \Lambda^0 + f(c, a)$$

so that the problem is to find the best values of the two constants Λ^0 and a. To illustrate the precision with which their equation will reproduce the experimental results for a strong 1 : 1 electrolyte, FUOSS and ONSAGER(9) quote details for aqueous potassium bromide at 25°. The experimental data used are those of OWEN and ZELDES(17) which cover the concentration range 0·0014–0·0072 N. By a method of successive approximations, the best values are found to be $\Lambda^0 = 151{\cdot}75$, $a = 3{\cdot}6$ Å, and the corresponding equation fits the data within 0·01 cm² ohm⁻¹ equiv.⁻¹.

Shedlovsky's empirical equation (7.34) will represent the same data with only slightly inferior accuracy but requires $\Lambda^0 = 151{\cdot}68$. Our equation (7.36) gives a similar fit with $\Lambda^0 = 151{\cdot}67$ and $a = 3{\cdot}2$ Å. Equation (7.35), which is that of Falkenhagen's school, or that of Pitts without the 'higher terms,' requires an ion size of 2·0 Å and gives $\Lambda^0 = 151{\cdot}71$. In *Table 7.4* the predictions of these various equations are compared with the experimental results. Whilst the table shows that the Fuoss-Onsager treatment gives an almost perfect fit to this set of experimental results, it must be emphasized that the other theoretically less exact equations give deviations of at the worst only 0·02 cm² ohm⁻¹ equiv⁻¹ or 0·014 per cent. The important question is therefore whether the Λ^0 value given by the Fuoss–Onsager theory is more correct than the others, which are 0·04 to 0·08 lower. There is no method of determining Λ^0 absolutely; it must always be found by extrapolation and the Fuoss-Onsager theory demands that the simpler extrapolation functions should curve upwards slightly in very dilute solutions. Whether this actually occurs is obscured by the increasing importance of experimental errors at high dilutions. Probably the most effective test can be made by using data for very dilute hydrochloric acid; here the solvent corrections are less important than for other electrolytes, and the data in the dilute region should be correspondingly more reliable. Application of the Fuoss-Onsager formulae to SHEDLOVSKY's(18) data for the range up to 0·003 N, gives $\Lambda^0 = 426{\cdot}27$, compared with the values 426·16 obtained by Shedlovsky, and 426·10 using equation 7·36 with $a = 4{\cdot}3$ Å, from the same data. It seems safe therefore to estimate that Λ^0 values obtained by the SHEDLOVSKY[11] or ROBINSON-STOKES[12] methods from data in the range 0·001 – 0·01 N will not be more than 0·05 per cent lower than the Fuoss-Onsager values.

Bearing in mind that unless quite exceptional precautions are taken, the experimental data will not be of better than 0·02 per cent

Table 7.4

Equivalent conductance of potassium bromide solutions at 25°

cm² Int. Ω^{-1} equiv^{-1}

10^4c mole/l.	$\Lambda_{exp.}$	Λ_{F-O}	Λ_S	Λ_{R-S}	Λ_{P-F}
13·949	148·27	148·27	148·26	148·26	148·28
27·881	146·91	146·91	146·92	146·92	146·93
42·183	145·88	145·89	145·89	145·90	145·90
59·269	144·90	144·91	144·90	144·91	144·89
71·696	144·30	144·30	144·28	144·29	144·28
Λ^0	—	151·75	151·68	151·67	151·71
	—	$a = 3{\cdot}6$ Å	$b = 91$	$a = 3{\cdot}2$ Å	$a = 2{\cdot}0$ Å

Λ_{F-O}—calculated by Fuoss-Onsager theory including transcendental terms (ref. 9).

Λ_S —calculated by Shedlovsky's function, equation (7.34).

Λ_{R-S}—calculated by Robinson-Stokes equation, (7.36).

Λ_{P-F}—calculated by Falkenhagen or Pitt's equations, ignoring 'higher terms,' equation (7.35).

accuracy, it is clear that the simpler equations will be adequate for most work, and the arithmetical labour of applying the Fuoss-Onsager theory will be undertaken only when extreme precision is needed. To illustrate the reality of experimental errors, we compare in *Fig. 7.4* the results for potassium bromide at 25° obtained by three investigators, OWEN and ZELDES[17], BENSON and GORDON[19], and JONES and BICKFORD[15]. The quantity plotted is the arbitrary deviation function $(\Lambda + 81\sqrt{c})$. It is evident that the differences between various investigators are as great as the differences between Λ^0 obtained by the various equations tested in *Table 7.4.*

Equation 7.36 is particularly convenient for representing conductances up to 0·1 N, though the best value of the parameter a and consequently of Λ^0 depend slightly on the concentration range fitted. *Table 7.5* compares the experimental conductances of sodium chloride with the predictions of this equation using $a = 4$ Å; the deviations are less than 0·05 per cent up to 0·05 N. Since the equation involves little more calculation than does the limiting law (7.29),

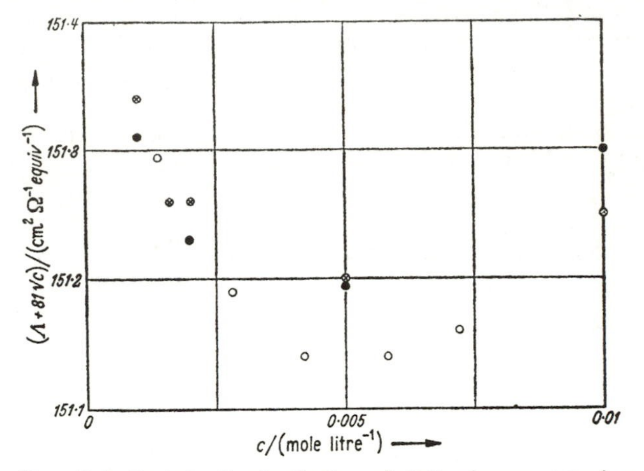

Figure 7.4. Deviation function for the conductivity of aqueous potassium bromide solutions at 25° according to various investigators

⊙ *Owen and Zeldes.* ● *Benson and Gordon.* ⊗ *Jones and Bickford*

its slight theoretical inadequacy compared with the more exact equation of Fuoss and Onsager may well be forgiven.

Table 7.5

Conductivities of Sodium Chloride Solutions at 25°

c mole/l.	Λ_{obs}	Λ (Eq. 7.36)	$\Lambda_{L.L.}$ (Eq. 7.29)
0	(126·45)	(126·45)	(126·45)
0·0005	124·51	124·51	124·45
0·001	123·74	123·75	123·63
0·002	122·66	122·68	122·46
0·005	120·64	120·68	120·14
0·01	118·53	118·57	117·53
0·02	115·76	115·81	113·83
0·05	111·06	111·03	106·50
0·1	106·74	106·52	98·23

Λ (Eq. 7.36) calculated with $a = 4$Å;
$\Lambda_{L.L.}$ by Onsager limiting law.

LIMITATIONS OF CONDUCTIVITY EQUATIONS

In deriving the various equations for the conductivity discussed above, the following assumptions are made, each resulting in some restriction on the applicability of the final equations. (*a*) Complete ionization is assumed; the formulae will however apply to the ionized part of weak electrolytes and to the 'non-paired' part of electrolytes in which ion-association occurs. In practice, only a small number of 1 : 1 electrolytes in water and possibly in some other solvents of high dielectric constant can be treated as completely ionized. Fuoss and Onsager comment that the mere fact of introducing the ion size parameter a implies some degree of ion-association, since it involves the idea that some ions do approach to mutual contact and whilst in mutual contact they will not contribute to the conductance. (This appears to us to be not quite the case, for even whilst the ions are in contact they can contribute something to the conductance by moving round each other.) (*b*) The treatments of Fuoss and Onsager and of Falkenhagen are based on the Debye-Hückel expression (4.13) for the potential in the absence of an external field. The limitations of this have been discussed in Chapter 4; it is an approximation which accords well with thermodynamic data (*see* Chapter 9) and is most accurate for ions of low charge in media of high dielectric constant. The more complicated potential expression employed by Pitts and by Mirtskhulava is a doubtful improvement in view of the departure from mathematical self-consistency involved.

The theory is weaker at every point for unsymmetrical electrolytes: the potential expression is less exact since the term in ψ^2 does not vanish from equation (4.7); the convergence for the series for the electrophoretic effect is unsatisfactory; and the theory of the relaxation effect has not been properly developed beyond the first approximation given by equation (7.9). There is therefore little justification for using any theoretical treatment except the Onsager limiting law (7.29) in such cases; empirical terms in c, $c \ln c$, $c^{3/2}$ *etc*., may be added but only for convenience of representation. Mere division of $\sqrt{I}$ in equation 7.29 by $(1 + \kappa a)$ does indeed give a rough fit, and with reasonable ion size parameters, but the discrepancies between its predictions and the measurements are much greater than experimental error. Two examples are given in *Table 7.6*.

Even for aqueous 1 : 1 electrolytes, one is straining the mathematics to the limit in applying theory to solutions as strong as 0·1 N, and only approximate treatments can be given at higher concentrations. Some of these will be discussed in Chapter 11.

THE VARIATION OF TRANSPORT NUMBERS WITH CONCENTRATION

Experimental results show that transport numbers are in general concentration-dependent, and the interpretation of this observation provides a useful test of the theory. In the case of non-associated uni-univalent electrolytes the form of the concentration dependence is as follows:

(*a*) If the cation transport number is close to 0·5, it scarcely varies with concentration; this is seen in the case of potassium chloride.

Table 7.6

Conductivity of Calcium and Lanthanum Chloride Solutions at 25°

$CaCl_2$ (*i*)				$LaCl_3$ (*ii*)			
c	$\Lambda_{obs.}$	$\Lambda_{calc.}$	$\Lambda_{L.L.}$	c	$\Lambda_{obs.}$	$\Lambda_{calc.}$	$\Lambda_{L.L.}$
0	(135·85)	—	—	0	(145·9)	—	—
0·00025	131·90	132·02	131·88	0·000167	139·6	139·9	139·6
0·0005	130·32	130·52	130·23	0·000333	137·0	137·6	137·0
0·0010	128·20	128·47	127·91	0·00167	127·5	128·8	126·0
0·0015	126·61	126·95	126·12	0·00333	121·8	123·0	117·8
0·0025	124·23	124·65	123·29	0·00667	115·3	115·9	106·2
0·0035	122·47	122·85	120·99	0·0167	106·2	104·3	83·1
0·0050	120·36	120·69	118·09	0·0333	99·1	94·3	57·1
0·01	115·65	115·65	110·73				
0·025	108·47	107·19	96·14				
0·05	102·46	99·52	79·68				
(a = 4·31 Å)				(a = 4·9 Å)			

(*i*) Data up to c = 0·005 from BENSON, G. C. and GORDON, A. R., *J. chem. Phys.*, 13 (1945) 470; above c = 0·005 from SHEDLOVSKY, T. and BROWN, A. S., *J. Amer. chem. Soc.*, 56 (1934) 1066

(*ii*) JONES, G. and BICKFORD, C. F., *J. Amer. chem. Soc.*, 56 (1934) 602; LONGSWORTH, L. G. and MACINNES, D. A., *ibid.*, 60 (1938) 3070

(*b*) If the cation transport number is less than 0·5 as with lithium chloride, it decreases further with increasing concentration.

(*c*) If the cation transport number is greater than 0·5, it increases with concentration; this occurs for example with hydrochloric acid.

These findings are completely and quantitatively explained by the interionic attraction theory[20]. According to equations (7.25), (7.27) and (7.36) the transport number t_1 of the cation is given by:

$$t_1 = \frac{\lambda_1}{\Lambda} = \frac{\lambda_1^0 - \frac{1}{2}|z_1|B_2\sqrt{I}/(1+\kappa a)}{\Lambda^0 - \frac{1}{2}(|z_1|+|z_2|)B_2\sqrt{I}/(1+\kappa a)} \quad \ldots.(7.39)$$

where

$$B_2 = \frac{82 \cdot 5}{\eta(\varepsilon T)^{1/2}}, \quad \text{and} \quad \kappa a = Ba\sqrt{I}$$

For 1 : 1 electrolytes this simplifies to:

$$t_1 = \frac{\lambda_1^0 - \frac{1}{2}B_2\sqrt{c}/(1 + \kappa a)}{\Lambda^0 - B_2\sqrt{c}/(1 + \kappa a)} \qquad \ldots .(7.40)$$

Thus the transport number expression contains only electrophoretic terms, the relaxation factor having cancelled out from equations (7.25) and (7.27). It is clear from (7.39) that if the limiting transport number $t_1^0 = \dfrac{\lambda_1^0}{\Lambda^0}$ is 0·5 exactly, t_1 will not vary from this value, and that the behaviour described under (*b*) and (*c*) above is also accounted for. Equation (7.39) gives an excellent quantitative account of the observed transport numbers; for aqueous 1 : 1 electrolytes at 25°, $B_2 = 60 \cdot 65$ and $\kappa a = 0 \cdot 3291\, a\sqrt{c}$. *Table 7.7* gives some observed and calculated values of t_1. The high degree of agreement with theory exhibited in *Table 7.7* is striking evidence for the soundness of the treatment of the electrophoretic effect for 1 : 1 electrolytes. The values of the ion size parameter a needed to account for the transport numbers are very reasonable, and similar to those found from consideration of the activity coefficient data. The ions of potassium chloride, for example, appear from their mobilities to be very nearly of the same effective size, and the crystal radius of the chloride ion is 1·8 Å; the value $a = 3 \cdot 7$ Å is thus just about what would be expected. For the fairly strongly hydrated sodium and lithium ions we have estimated in Chapter 6 radii of 3·3 and 3·7 Å respectively by using a modified Stokes' law formula: combining these with 1·8 Å for the chloride ion, we have $a = 5 \cdot 1$ and 5·5 Å for sodium chloride and lithium chloride respectively, as compared with the value $a = 5 \cdot 2$ Å found adequate for both salts in *Table 7.7*. A Stokes' law estimate cannot be used for hydrogen ion because of the abnormal transport mechanism involved, but the value $a = 4 \cdot 4$ Å for hydrochloric acid compares very well with $a = 4 \cdot 47$ Å required for the activity coefficient data (Chapter 9).

Very few transport number data are available for solvents other than water, and such as there are have mostly been obtained in mixed solvents, *e.g.*, the transport numbers of hydrochloric acid have been measured by HARNED and DREBY[21] in a number of dioxane-water mixtures, and GORDON and his collaborators[22] have used the moving boundary method for sodium and potassium chlorides in equimolar methanol-water mixtures. These results for

Table 7.7

Observed and Calculated Cation Transport Numbers of Aqueous 1 : 1 *Electrolytes at* 25°: *Tests of Equation* (7.39)

	HCl*†		LiCl*		NaCl‡§		NaAc‡		KCl*§		KAc‖	
	t_1 obs.	t_1 calc.	t_1 obs.	t_1 calc.	t_1 obs.	t_1 calc.	t_1 obs.	t_1 calc.	t_1 obs.	t_1 calc.	t_1 obs.	t_1 calc.
0	(0·8209)	0·8209	(0·3363)	0·3363	(0·3962)	0·3962	(0·5506)	0·5506	(0·4905)	0·4905	(0·6425)	0·6425
0·01	0·8251	0·8249	0·3289	0·3285	0·3918	0·3918	0·5537	0·5538	0·4902	0·4901	0·6498	0·6495
0·02	0·8266	0·8263	0·3261	0·3258	0·3902	0·3902	0·5550	0·5550	0·4901	0·4900	0·6523	0·6521
0·05	0·8292	0·8287	0·3211	0·3211	0·3876	0·3875	0·5573	0·5573	0·4899	0·4898	0·6569	0·6570
0·1	0·8314	0·8310	0·3168	0·3165	0·3854	0·3849	0·5594	0·5596	0·4898	0·4895	0·6609	0·6619
0·2	0·8337	0·8337	0·3112	0·3112	0·3821	0·3819	0·5610	0·5626	0·4894	0·4892	—	—
0·5	0·838	0·838	0·303	0·301	—	—	—	—	0·4888	0·4887	—	—
1·0	0·841	0·841	0·297	0·287	—	—	—	—	0·4882	0·4883	—	—
2·0	0·843	0·843	—	—	—	—	—	—	—	—	—	—
3·0	0·843	0·845	—	—	—	—	—	—	—	—	—	—
a (Å)	4·4		5·2		5·2		3·7		3·7		3·7	

* LONGSWORTH, L. G., *J. Amer. chem. Soc.*, 54 (1932) 2741.
† HARNED, H. S. and DREBY, E. C., *ibid.*, 61 (1939) 3113.
‡ LONGSWORTH, L. G., *ibid.*, 57 (1935) 1185.
§ ALLGOOD, R. W., LEROY, D. J. and GORDON, A. R., *J. chem. Phys.*, 8 (1940) 418 and 10 (1942) 124.
‖ LEROY, D. J. and GORDON, A. R., *ibid.*, 6 (1938) 398.

Note: Experimental values given to four figures are by moving boundary method; those given to three figures are by the e.m.f. method for hydrochloric acid and by the Hittorf method for lithium chloride.

hydrochloric acid appear to conform reasonably well with the requirements of the theory at low concentrations, but not so well at higher concentrations. This is perhaps to be expected since the convergence of the theoretical formulae for the electrophoretic effect is less satisfactory in media of low dielectric constant.

For higher valency types, even in water, the theory is also found to be inadequate; with calcium chloride, for example, equation (7.39) gives results which are nearer to the observed values than are those predicted by the limiting law, but are still substantially low. In 0·05 M calcium chloride at 25° the observed cation transport number is 0·4070; equation (7.39) with $a = 5$ Å gives 0·3952; and the limiting law equation (*i.e.*, equation (7.39) with $a = 0$) gives 0·3545. Thus while the ion-size correction gives a useful improvement, it does not lead to the quantitative agreement which can be obtained with 1 : 1 electrolytes. This is probably attributable to the lower degree of self-consistency of the interionic attraction theory for unsymmetrical electrolytes.

For bi-bivalent electrolytes (*i.e.*, of the zinc sulphate type), the theory should be applicable; but here the difficulty arises that a large proportion of the ions are present as closely associated ion-pairs; this effect is quite important at the lowest concentration accessible to transport number experiments, *viz.*, about 0·005 M. Unfortunately no moving-boundary measurements have been made on 2 : 2 electrolytes, and we therefore have to rely on older and less precise measurements by the Hittorf method. The measurements on cadmium sulphate at 18° by JAHN and his co-workers[23] seem to be the best available, and internal evidence suggests a reliability of a few units in the third decimal place of the transport number. The curve of t_1 against $\sqrt{c}$ also bears a strong resemblance to that found by the electromotive force method[24] for zinc sulphate, except in the most dilute region, where the Hittorf method may well be more reliable.

The transport number of the cadmium ion is found to fall almost linearly in $\sqrt{c}$ from $t_1^0 = 0{\cdot}396$ at $c = 0$ to $t_1 = 0{\cdot}254$ at $c = 1$ mole per litre; deviations from this straight line scarcely exceed the experimental error. Taking the limiting equivalent conductivities at 18° as $\lambda^0_{\mathrm{Cd}^{++}} = 44{\cdot}8$, $\lambda^0_{\mathrm{SO}_4^{--}} = 68{\cdot}4$, and inserting the appropriate numerical values for 18° in equation (7.39), we have:

$$t_1 = \frac{44{\cdot}8 - 101{\cdot}7\sqrt{c}/(1 + 0{\cdot}6546 \times 10^8\, a\sqrt{c})}{113{\cdot}2 - 203{\cdot}4\sqrt{c}/(1 + 0{\cdot}6546 \times 10^8\, a\sqrt{c})} \quad \ldots.(7.41)$$

With the value $a = 3{\cdot}5$ Å, equation (7.41) gives very fair agreement with experiment as shown in *Table 7.8.* The formation of ion-pairs in a symmetrical salt does not result in the appearance of any new ionic species and the effect on the transport number should therefore be merely that of 'diluting' the solution by removing some of the ions to form electrically neutral pairs, the 'dilution'

Table 7.8

Cation Transport Number in Cadmium Sulphate at 18°

c (mole/litre)	0	0·01	0·09	0·25	0·49	1·00
t_1 obs.	(0·396)	0·384	0·353	0·323	0·295	0·254
t_1 calc.	0·396	0·378	0·347	0·321	0·299	0·270

factor being α, the degree of dissociation of the ion-pairs. The success of equation (7.41) in which the ion pairing is ignored is apparently due to this 'dilution factor' being approximately compensated by the use of a rather small value (3·5 Å) for the ion-size parameter.

Negative cation transport numbers

The behaviour of the transport numbers of calcium chloride or zinc perchlorate may be regarded as normal for 2 : 1 electrolytes in spite of the inability of theory to cope with it. Many of the

Table 7.9

Cation Transport Numbers of Aqueous 2 : 1 *Electrolytes at* 25°, *showing Effect of Autocomplex Formation in Zinc Halides*

m	$Zn(ClO_4)_2$*	ZnI_2†	$ZnBr_2$‡	$ZnCl_2$§
0	(0·440)	(0·408$_5$)	(0·404$_1$)	(0·409$_7$)
0·05	—	0·382	0·366	0·365
0·1	0·409	0·363	0·349	0·350
0·2	0·389	0·345	0·331	0·335
0·5	0·361	0·320	0·306	0·331
1·0	0·335	0·291	0·286	0·171
2·0	0·303	0·178	0·181	0·000
3·0	0·281	0·056	− 0·059	− 0·137
4·0	0·271	− 0·050	− 0·151	− 0·256
5·0	—	− 0·190	− 0·233	− 0·364
8·0	—	− 0·444	− 0·445	− 0·562
10·0	—	− 0·550	− 0·563	− 0·559

m = mole salt per kg water

* STOKES, R. H. and LEVIEN, B. J., *J. Amer. chem. Soc.*, 68 (1946) 333
† STOKES, R. H. and LEVIEN, B. J., *ibid.*, 68 (1946) 1852
‡ PARTON, H. N. and MITCHELL, J. W., *Trans. Faraday Soc.*, 35 (1939) 758
§ HARRIS, A. C. and PARTON, H. N., *ibid.*, 36 (1940) 1139

transition-metal halides[25], however, show very different behaviour, which is illustrated in *Table 7.9*; at high concentrations the cation transport number decreases rapidly to zero, then becomes negative. This is in marked contrast to the behaviour of the transport number of zinc in its perchlorate[26] which may be taken as typical behaviour for a normal 2 : 1 electrolyte at higher concentrations; the explanation of the anomaly is that the metal ion is to a large extent present as a complex negative ion, believed to be mainly an ion of type ZnX_4^{--}, a view which is supported by measurements[27] of the vapour pressures of ZnX_2—KX mixtures. Though no quantitative treatment of the effect can be given, it would appear (p. 112) that to give a negative apparent transport number for the metal ion, the complex negative ion must have a higher mobility than the normal (hydrated) metal ion. This is quite possible since the bivalent cations are known to be strongly hydrated.

CONDUCTIVITIES IN NON-AQUEOUS SOLVENTS

There is a substantial body of experimental data for conductivities in non-aqueous solvents. Such solvents frequently have a much lower conductivity than pure water, with the result that measurements can be made at lower concentrations without serious loss of accuracy; on the other hand they are more difficult to purify, and may require careful protection from atmospheric moisture, while simple salts are often only slightly soluble in them, with a consequent restriction of the concentration-range which can be studied. The theoretical interpretation of the results is at present hampered by a severe shortage of reliable transport numbers in non-aqueous solvents. It is to be hoped that the centrifugal cell method of MacInnes will soon be developed to a point where this difficulty is overcome; in the meantime a valuable start has been made by Gordon and his collaborators who have obtained accurate moving-boundary transport number measurements in anhydrous methanol solutions of sodium and potassium chlorides. These, together with their direct-current conductivity measurements in the same solvent[28], provide the most precise information we have on the transport properties of ions in non-aqueous solutions. The results of their measurements are summarized in *Table 7.10*. Both cation and anion transport numbers were measured in several cases, the sum being within 0·0003 of unity; this provides a valuable check on the results.

Gordon's school[29] have also made measurement of transport numbers and conductances for lithium, sodium and potassium chlorides in anhydrous ethanol. Owing to the low solubilities, (the maximum concentration used was 0·0025 N), and to the occurrence of ion-association, only their limiting ion conductivities are given in *Table 7.10.*

First, examining the limiting ionic conductivities, we see that for both anions and cations the λ^0 values increase as the crystal radii

Table 7.10

Transport Numbers in Methanol at 25°

c	t_1 (NaCl)	t_1 (KCl)
0	(0·4633)	(0·5001)
0·003	0·4603	—
0·005	0·4595	0·5007
0·007	0·4588	0·5009
0·01	0·4582	0·5013
0·02	—	0·5012

c in mole/litre

From DAVIES, J. A., KAY, R. L. and GORDON, A. R., *J. chem. Phys.*, 19 (1951) 749

Equivalent Conductivities, Λ, *in Methanol at* 25°

$c \times 10^4$ mole/l	LiCl	NaCl	NaBr	KCl	KBr	KI
0	(92·20)	(97·61)	(101·76)	(104·78)	(108·95)	(115·15)
1	89·74	—	99·19	—	106·34	112·52
2	88·70	94·11	98·11	101·16	105·26	111·43
5	86·65	92·09	96·04	99·07	103·10	109·29
10	84·52	89·87	93·80	96·72	100·71	106·94
20	81·74	86·91	90·86	93·56	97·51	103·74
30	79·73	84·84	88·80	91·24	95·19	101·50
50	76·73	81·80	85·66	87·79	91·80	98·16
70	—	79·43	—	85·28	—	—
100	—	76·71	—	82·32	—	—

BUTLER, J. P., SCHIFF, H. I. and GORDON, A. R., *J. chem. Phys.*, 19 (1951) 752; JERVIS, R. E., MUIR, D. R., BUTLER, J. P. and GORDON, A. R., *J. Amer. chem. Soc.*, 75 (1953) 2855

Limiting Ionic Conductivities in Methanol and Ethanol at 25°

Ion	Li^+	Na^+	K^+	Cl^-	Br^-	I^-
λ^0(MeOH)	39·82	45·22	52·40	52·38	56·55	62·75
λ^0(EtOH)	17·05	20·31	23·55	21·85	—	—
r(Å)	0·60	0·95	1·33	1·81	1·95	2·16

increase, though the anion and cation values do not fall on the same curve. This is a slightly more regular situation than prevails for the same ions in water, where for anions the order of increasing mobilities bears no relation to the crystal radii. It is unlikely that any of the anions carries a permanent solvation sheath in methanol, but the increase in mobility with size may be due to decreased interaction of the larger ion with the solvent dipoles. Since the molal volume of methanol is about 41 cm^3, the 'radius' of a methanol molecule must be substantially greater than that of a water molecule, so that the conditions for the validity of Stokes' law are far from realization with these small ions. One may perhaps picture the ion as causing some rotation of solvent dipoles as the ion passes the molecule, which will result in a dissipation of energy and will therefore increase the effective resistance; this interaction will increase rapidly with reduction of the ion-to-dipole distance, as required by the observations. Such an effect might well be more important in methanol than in water, where the 'structure' of the solvent is more definite. The possible existence of this effect requires further investigation.

The variation with concentration of the transport number of potassium chloride in methanol is scarcely significant, as is required by the theory for cases where the limiting transport numbers are nearly 0·5. That of sodium chloride may profitably be examined by the theory. For methanol at 25°, the viscosity is 0·005445 poise and the dielectric constant 31·52*; these give for the values of the constants in equation (7.36):

$B = 0{\cdot}5188 \times 10^8$, $B_1 = 0{\cdot}9004$, $B_2 = 156{\cdot}2$; or in the transport number equation (7.39), the constant $B_2 = 156{\cdot}2$ and $\kappa a = 0{\cdot}5188 \times 10^8 \, a\sqrt{c}$. The equation:

$$t_1 = \frac{45{\cdot}22 - 78{\cdot}1\sqrt{c}/(1 + 3{\cdot}06\sqrt{c})}{97{\cdot}61 - 156{\cdot}2\sqrt{c}/(1 + 3{\cdot}06\sqrt{c})}$$

reproduces the observed transport numbers up to 0·01 N within 0·0001. The ion size corresponding to $\kappa a = 3{\cdot}06\sqrt{c}$ is $a = 5{\cdot}9$ Å, a rather large value unless the sodium ion at least is solvated, but about the value one would expect if the ions approach until separated by one methanol molecule only.

In this medium, of dielectric constant 31·52, a considerable amount of ion-pair formation is to be expected even with 1 : 1 electrolytes, since the Bjerrum critical distance is 8·9 Å. We have remarked before that ion-pair formation in a symmetrical electrolyte will affect the transport numbers only by a sort of 'dilution' effect. The equivalent conductivities on the other hand, will be reduced in nearly direct proportion to the amount of ion-pairing; they are much more sensitive to it than are transport numbers. It is therefore

* A more recent value is 32·63.

not surprising to find that the conductivities of the alkali halides in methanol conform rather too closely to the Onsager limiting law. For sodium chloride, for example, the limiting law becomes:

$$\Lambda = 97{\cdot}61 - 244{\cdot}1\sqrt{c}$$

which at 0·001 N gives $\Lambda = 89{\cdot}89$ in agreement with the observed value $\Lambda = 89{\cdot}87$. Since the value of (κa) must be at least 0·05, this agreement is too close for a fully ionized electrolyte. In general, the conductivities at higher concentrations lie above the limiting law values, but not as much as they should for fully dissociated electrolytes. The values cannot be accurately represented by equation (7.36); a value of $a = 3{\cdot}2$ Å for sodium chloride gives a rough fit, within about 0·5 in Λ, but this value of the ion size is inconsistent with that needed for the transport number equation.

Liquid hydrogen cyanide is of great interest as an electrolytic solvent, having a dielectric constant of about 160 at 0° and 120 at 18°, so that ion-pair formation should be less than in water. It has also a much lower viscosity than that of water, the values at 0° being 0·00232 poise for hydrogen cyanide and 0·01787 poise for water. COATES and TAYLOR[30] studied a number of alkali-metal salts in this solvent at 18°, and LANGE, BERGÅ and KONOPIK[31] have made measurements at 0° on some potassium and some tetra-substituted ammonium salts. The 18° measurements were all at low concentrations (0·0001–0·0025 mole per litre) and conform to relations of the type:

$$\Lambda = \Lambda^0 - A\sqrt{c}$$

with Λ^0 and A values given in *Table 7.11*. These linear relations hold over the whole of the concentration-range studied (up to 0·002 or 0·003 N in most cases) except for lithium chloride, nitrate and thiocyanate and sodium nitrate. These show a downward curvature, most marked in the case of lithium thiocyanate, and probably indicative of ion-pair formation. The limiting conductivities are reasonably consistent with the Kohlrausch principle, as shown by the nearly constant differences:

$$\Lambda^0_{\mathrm{LiX}} - \Lambda^0_{\mathrm{NaX}} \approx 3$$
$$\Lambda^0_{\mathrm{KX}} - \Lambda^0_{\mathrm{NaX}} \approx 19{\cdot}6$$

The range of Λ^0 values is noticeably more restricted than in water, and there are indications that there is less actual solvation of ions than occurs in water. The values of the slopes (A) of the Λ versus $\sqrt{c}$ curves do not agree any too well with the theoretical limiting result, which for hydrogen cyanide at 18° becomes:

$$\Lambda = \Lambda^0 - [0{\cdot}1271\Lambda^0 + 233]\sqrt{c}$$

Since the electrophoretic term is here much larger than the relaxation term, all the theoretical slopes are much the same, lying in the range $A = 259$–269. In the cases where the curvature of the plots suggests ion-pair formation, the observed slope in the most dilute

Table 7.11

Conductivities of Salts in Hydrogen Cyanide Solution at 18°

$$\Lambda = \Lambda^0 - A\sqrt{c}$$

	Λ^0	A		Λ^0	A
LiCl	345·4	335	Na Picrate	266·9	195
LiBr	346·9	270	KCl	363·4	280
LiI	348·0	258	KBr	363·2	248
$LiNO_3$	336·6	402	KI	363·9	235
$LiClO_4$	336·9	230	KNO_3	353·9	253
LiCNS	340·6	400	$KClO_4$	353·3	275
NaBr	343·8	243	KCNS	358·0	243
NaI	344·9	238	RbCl	363·2	195
$NaNO_3$	333·8	250	CsCl	368·2	200
$NaClO_4$	335·5	235	$N(Et)_4$ Picrate	282·3	215
NaCNS	337·7	230			

The A values given are the experimentally observed slopes; theoretical values of A lie between 259 and 269.

Coates, J. E. and Taylor, E. G., *J. chem. Soc.* (1936) 1245

region is considerably steeper than the theoretical, which is reasonable; but the remaining presumably 'normal' salts appear to give straight lines lying *above* the theoretical slope. This is most noticeable with the picrates and rubidium and caesium chlorides, which have large ions. For the salts such as sodium bromide, where the points lie only slightly above the limiting-law lines, the introduction of the factor $(1 + \kappa a)$ into the denominator, as required by the more complete theory, gives a satisfactory account of the results. In this solvent at 18°, $\kappa = 0{\cdot}2703 \times 10^8\sqrt{c}$ and for sodium bromide the value $a = 5{\cdot}1$ Å in the equation:

$$\Lambda = \Lambda^0 - \frac{(0{\cdot}1271\Lambda^0 + 223)}{(1 + \kappa a)}\sqrt{c}$$

is adequate and reasonable. Caesium chloride, however, requires $a = 28$ Å, which is quite absurd in view of the not greatly different limiting conductivities, which imply ions of comparable size to those of sodium bromide, but the concentrations are rather too low to permit an accurate evaluation of the parameter a.

The measurements at 0° in hydrogen cyanide are rather more interesting, as they extend to concentrations high enough to make a significant test of equation (7.36) involving the ion-size correction. Data for potassium iodide solutions[31], interpolated to round concentrations, are given in *Table 7.12*. For this solvent at 0°C, the viscosity (η) is 0·00232 poise, and the dielectric constant (ε) is 161. (The latter figure is by no means well established, however.) These values give the following values of the constants in equation (7.36):

$$B_1 = 0{\cdot}0890,\ B_2 = 169{\cdot}6,\ B = 0{\cdot}240 \times 10^8$$

Now if the ions of potassium iodide are unsolvated in hydrogen cyanide solution, the distance of closest approach may be estimated

Table 7.12

Conductivities of Potassium Iodide in Hydrogen Cyanide at 0°

c	$\Lambda_{obs.}$	$\Lambda_{calc.}$
0	(310·3)	(310·3)
0·001	304·4	304·2
0·002	301·8	301·8
0·003	300·2	300·0
0·005	297·3	297·1
0·007	294·9	294·9
0·010	292·1	292·1
0·015	288·1	288·4
0·02	284·9	285·4
0·05	269·6	273·2
0·10	252·2	261·0

c in mole/litre $a = 3{\cdot}5$ Å

Data from Lange, J., Bergå, J. and Konopik, N., *Monatsh.*, 80 (1949) 708

from the crystal radii as $a \approx 1{\cdot}33 + 2{\cdot}16 \approx 3{\cdot}5$ Å. Taking the limiting value of the conductivity as $\Lambda^0 = 310{\cdot}3$, we then have for this solution from equation (7.35):

$$\Lambda_{calc.} = 310{\cdot}3 - \frac{197{\cdot}2\sqrt{c}}{1 + 0{\cdot}84\sqrt{c}}$$

The values of $\Lambda_{calc.}$ given by this equation are included in *Table 7.12*; the agreement is quantitative up to 0·01 N and satisfactory up to 0·02 N, after which the calculated values are increasingly high.

On the whole, the conductivity measurements in hydrogen

Table 7.13

Limiting Conductances in Amide Solvents. Λ^0 *in* $cm^2\ \Omega^{-1}\ equiv^{-1}$; η *in poise*

Solvent	Temp. °C	η	ε	HI	NaI	KI	CsI	Me_4NI	Et_4NI	Bu_4NI
Formamide (32)	25	0·0330	109·5	27·4	26·8	29·3	—	—	—	23·5
N-methyl formamide (33)	25	0·0165	182·4	—	44·4	45·0	47·2	—	49·0	—
N-N-dimethyl formamide (33, 34)	25	0·00796	36·7	—	82·0	82·6	—	91·0	87·5	77·7
N-methyl acetamide (35, 36)	40	0·0302	165·5	23·7	22·8	23·0	(LiI 21·2)	26·6	26·2	22·4
N-N-dimethyl acetamide (37)	25	0·00919	37·8	—	67·6	67·1	—	—	74·5	64·6
N-methyl propionamide (30°–60°) (38)	30	0·0457	164·3	—	13·4	13·7	—	—	—	—
N-methyl butyramide (30°–60°) (38)	30	0·0747	124·7	—	(NaCl 6·3)	—	—	—	—	—

Table 7.13—Contd.

Solvent	Temp. °C	Me_3PhNI	KCl	KBr	KPi (picrate)	KSCN	KNO_3	$KPhSO_3$	$KClO_4$
Formamide (32)	25	27·3	29·8	—	—	—	—	23·1	—
N-methyl form-amide (33)	25	—	41·9	43·7	35·3	—	—	—	—
N-N-dimethyl form-amide (33, 34)	25	—	—	84·1	—	90·3	88·1	—	82·8
N-methyl acetamide (35, 36)	40	24·8	19·9	21·2	20·2	24·5	22·9	18·7	25·2
N-N-dimethyl acetamide (37)	25	70·1	—	68·5	56·8	74·1	71·6	56·3	68·1
N-methyl propion-amide (30°–60°) (38)	30	—	11·6	12·4	—	—	—	—	—
N-methyl butyramide (30°–60°) (38)	30	—	6·5	—	—	—	—	—	—

cyanide thus tend to support the theory; but there are some anomalies which clearly call for further investigation. Determinations of transport numbers, verification of the dielectric constant, and studies of the effect of solutes on the viscosity would all be of great value.

The amides of the lower aliphatic acids and their N-methyl derivatives form a class of liquids with extremely high dielectric constants (Appendix 1.2), the most striking case being that of N-methyl formamide $H.CO.NH(CH_3)$ which has $\varepsilon = 182{\cdot}4$ at 25°. Comprehensive studies of conductances in these solvents have recently been made by Sears and Dawson and co-workers and by French and Glover. In the case of formamide, an approximate measurement of the limiting transport number by the Hittorf method has also been made, giving $t_{K^+}^{KCl} = 0{\cdot}406$ at 25°, so that individual ionic mobilities in this solvent are known. In the other solvents of this class no measured transport numbers are yet available, but reasonable estimates have been made on the basis of the behaviour of very large ions in relation to solvent viscosity. The original data are too extensive to present in detail, but *Table 7.13* gives a summary of the major results. For compactness we have given Λ^0 values only of iodide ion with various cations and of potassium ion with various anions; these do not necessarily represent salts actually studied in the original work, but in some cases have been obtained by the application of the Kohlrausch principle to measurements on related salts.

A noteworthy feature is that the Λ^0 values for strong acids are similar to those for salts; evidently the hydrogen ion has no special transport mechanism available in these solvents, as it has in water and the lower alcohols.

The general pattern of concentration-dependence of the conductance is, as might be expected, one of approach to the Onsager limiting law from above; but in the di-N-methyl amides, which have dielectric constants about half that of water, the results lie close to the limiting-law curve, and with some salts actually fall below it, indicating a slight degree of ion-association.

Most of the other commonly used non-aqueous solvents have lower dielectric constants than methanol, and the conductivity of solutions in these appears to be so strongly influenced by ion association that little progress can be made by attempting to treat them as strong electrolytes. The extensive researches of Kraus and his collaborators have done much towards elucidating the behaviour of ion-aggregates in these solutions, and are discussed more fully in Chapter 14.

APPENDIX TO THE THEORY OF THE ELECTROPHORETIC EFFECT

EVALUATION OF THE INTEGRAL $S_n(\kappa a)$ OF EQUATION (7.5)

This integral takes an elementary form only for the case $n = 1$, when it becomes:

$$S_1(\kappa a) = \frac{1}{a}\int_a^\infty e^{-\kappa r}\,\mathrm{d}r = \frac{e^{-\kappa a}}{\kappa a}$$

For $n \geqslant 2$ it involves the exponential integral function $Ei(x)$ defined by:

$$Ei(x) = \int_x^\infty e^{-y}y^{-1}\,\mathrm{d}y$$

(y being merely a variable of integration). This function is available from tables[39] for various values of x.

For $n = 2$

$$S_2(\kappa a) = \int_a^\infty e^{-2\kappa r}r^{-1}\,\mathrm{d}r = \int_a^\infty e^{-2\kappa r}(2\kappa r)^{-1}\mathrm{d}(2\kappa r) = Ei(2\kappa a)$$

For $n > 2$ it is necessary to perform successive integrations by parts until the integral reduces to an exponential integral function. Thus one obtains:

$$S_3(\kappa a) = e^{-3\kappa a} - 3\kappa a Ei(3\kappa a)$$

$$S_4(\kappa a) = e^{-4\kappa a}(\tfrac{1}{2} - 2\kappa a) + 8(\kappa a)^2 Ei(4\kappa a)$$

and in general for $n > 2$:

$$S_n(\kappa a) = e^{-n\kappa a}\left[\frac{1}{n-2} + \frac{(-n\kappa a)}{(n-2)(n-3)} + \frac{(-n\kappa a)^2}{(n-2)(n-3)(n-4)} + \ldots + \frac{(-n\kappa a)^{n-3}}{(n-2)!}\right] + \frac{(-n\kappa a)^{n-2}}{(n-2)!}Ei(n\kappa a)$$

there being $(n - 2)$ terms in the series enclosed in the square bracket. If the theory is taken only as far as the 'self-consistent' approximation, only $S_1(\kappa a)$ and $S_2(\kappa a)$ are involved, and the latter appears only in the theory of diffusion for symmetrical electrolytes. The higher-order terms have been computed to facilitate the investigation of questions of convergence.

SUMMARY OF EQUATIONS FOR CONDUCTIVITY AND TRANSPORT NUMBERS

1 = cation, 2 = anion

$$q = \frac{|z_1 z_2|}{(|z_1| + |z_2|)(|z_1|t_2^0 + |z_2|t_1^0)} = \frac{1}{2} \text{ for symmetrical valence-types} \quad \ldots.(7.10)$$

$$\kappa^2 = \frac{4\pi e^2}{\varepsilon kT}(n_1 z_1^2 + n_2 z_2^2) = \frac{8\pi N e^2}{1000\varepsilon kT} I \quad \ldots.(4.12)$$

where I = ionic strength $= \frac{1}{2}\Sigma c_i z_i^2$ (c in mole/litre.)

$$\kappa = 50{\cdot}29 \times 10^8(\varepsilon T)^{-1/2}\sqrt{I}$$

Equivalent Conductivities

$$\Lambda = \Lambda^0 - \left[\frac{2{\cdot}801 \times 10^6|z_1 z_2|q\Lambda^0}{(\varepsilon T)^{3/2}(1+\sqrt{q})} + \frac{41{\cdot}25(|z_1| + |z_2|)}{\eta(\varepsilon T)^{1/2}}\right]\sqrt{I} \quad \ldots.(7.29)$$

(Onsager limiting law for extreme dilutions).

$$\Lambda = \Lambda^0 - \left[\frac{2{\cdot}801 \times 10^6|z_1 z_2|q\Lambda^0}{(\varepsilon T)^{3/2}(1+\sqrt{q})} + \frac{41{\cdot}25(|z_1| + |z_2|)}{\eta(\varepsilon T)^{1/2}}\right]\frac{\sqrt{I}}{1+\kappa a} \quad \ldots.(7.36)$$

(valid for moderate concentrations with suitable choice of a, especially for 1 : 1 electrolytes).

Formulae reduced for the case of 1 : 1 electrolytes:

$$\Lambda = \Lambda^0 - (B_1\Lambda^0 + B_2)\sqrt{c} \text{ (Onsager limiting law.)}$$

$$\left.\begin{aligned} \Lambda &= \Lambda^0 - (B_1\Lambda^0 + B_2)\sqrt{c}/(1 + Ba\sqrt{c}) \\ \text{or} \quad \Lambda^0 &= \Lambda + \frac{(B_1\Lambda + B_2)\sqrt{c}}{1 + (Ba - B_1)\sqrt{c}} \end{aligned}\right\} \quad \ldots.(7.36)$$

(for moderate concentrations).

For values of B, B_1 and B_2 for aqueous solutions, see Appendix 7.1. In other solvents:

$$B = 50{\cdot}29(\varepsilon T)^{-1/2} \times 10^8$$

$$B_1 = 8{\cdot}204 \times 10^5(\varepsilon T)^{-3/2}$$

$$B_2 = 82{\cdot}5/[\eta(\varepsilon T)^{1/2}] \text{ with } \eta \text{ in poise, } T \text{ in deg. K}$$

Transport Numbers (using first order electrophoretic terms only):

$$t_1 = \frac{\lambda_1^0 - \frac{1}{2}|z_1|B_2\sqrt{I}/(1+\kappa a)}{\Lambda^0 - \frac{1}{2}(|z_1| + |z_2|)B_2\sqrt{I}/(1+\kappa a)} \quad \ldots.(7.39)$$

which reduces for 1 : 1 electrolytes to:

$$t_1 = \frac{\lambda_1^0 - \frac{1}{2}B_2\sqrt{c}/(1 + Ba\sqrt{c})}{\Lambda^0 - B_2\sqrt{c}/(1 + Ba\sqrt{c})} \qquad \ldots .(7.40)$$

with B, B_2 as given above.

Limiting law for transport numbers, valid at extreme dilution:

$$t_1 = t_1^0 + \frac{B_2}{2\Lambda^0}[(|z_1| + |z_2|)t_1^0 - |z_1|]\sqrt{I}$$

REFERENCES

1 ONSAGER, L. and FUOSS, R. M., *J. phys. Chem.*, 36 (1932) 2689
2 DEBYE, P. and HÜCKEL, E., *Phys. Z.*, 24 (1923) 305
3 ONSAGER, L., *ibid.*, 28 (1927) 277
4 FALKENHAGEN, H., LEIST, M. and KELBG, G., *Ann. Phys., Lpz.* [6], 11 (1952) 51
5 PITTS, E., *Proc, Roy. Soc.*, 217A (1953) 43
6 PITTS, E., TABOR, B. E. and DALY, J. *Trans. Faraday Soc.* 65 (1969) 849
7 KELBG, G., DISS., *Rostock* (1954); FALKENHAGEN, H. and KELBG, G., *Z. Elektrochem.*, 58 (1954) 653
8 MIRTSKHULAVA, I. A., *J. phys. Chem.*, (*U.S.S.R.*) 27 (1953) 840
9 FUOSS, R. M. and ONSAGER, L., *J. phys. Chem.*. 61 (1957) 668
10 STOKES, R. H., *J. Amer. chem. Soc.*, 75 (1953) 4563
11 SHEDLOVSKY, T., *ibid.*, 54 (1932) 1405
12 ROBINSON, R. A. and STOKES, R. H., *ibid.*, 76 (1954) 1991
13 SHEDLOVSKY, T., BROWN, A. S. and MACINNES, D. A., *Trans. electrochem. Soc.*, 66 (1934) 165; SHEDLOVSKY, T., *J. Amer. chem. Soc.*, 54 (1932) 1411; SHEDLOVSKY, T. and BROWN, A. S., *ibid.*, 56 (1934) 1066
14 DEUBNER, A. and HEISE, R., *Ann. Phys. Lpz.*, [6] 9 (1951) 213
15 JONES, G. and BICKFORD, C. F., *J. Amer. chem. Soc.*, 56 (1934) 602
16 LONGSWORTH, L. G., *ibid.*, 57 (1935) 1186
17 OWEN, B. B. and ZELDES, H., *J. chem. Phys.*, 18 (1950) 1083
18 SHEDLOVSKY, T., *J. Amer. chem. Soc.*, 54 (1932) 1411
19 BENSON, G. C. and GORDON, A. R., *J. chem. Phys.*, 13 (1945) 473
20 STOKES, R. H., *J. Amer. chem. Soc.*, 76 (1954) 1988
21 HARNED, H. S. and DREBY, E. C., *ibid.*, 61 (1939) 3113
22 SHEMILT, L. W., DAVIES, J. A. and GORDON, A. R., *J. chem. Phys.*, 16 (1948) 340
23 JAHN, H., *Z. phys. Chem.*, 58 (1907) 641
24 PURSER, E. P. and STOKES, R. H., *J. Amer. chem. Soc.*, 73 (1951) 5650
25 PARTON, H. N. and MITCHELL, J. W., *Trans. Faraday Soc.*, 35 (1939) 758; HARRIS, A. C. and PARTON, H. N., *ibid.*, 36 (1940) 1139; STOKES, R. H. and LEVIEN, B. J., *J. Amer. chem. Soc.*, 68 (1946) 1852
26 STOKES, R. H. and LEVIEN, B. J., *ibid.*, 68 (1946) 333
27 STOKES, R. H., *Trans. Faraday Soc.*, 44 (1948) 137

[28] BUTLER, J. P., SCHIFF, H. I. and GORDON, A.R., *J. chem. Phys.*, 19 (1951) 752; JERVIS, R. E., MUIR, D. R., BUTLER, J. P. and GORDON, A. R., *J. Amer. chem. Soc.*, 75 (1953) 2855; DAVIES, J. A., KAY, R. L. and GORDON, A. R., *J. chem. Phys.*, 19 (1951) 749

[29] GRAHAM, J. R. and GORDON, A. R., *J. Amer. chem. Soc.*, 79 (1957) 2350; GRAHAM, J. R., KELL, G. S. and GORDON, A. R., *ibid.*, 79 (1957) 2352; see also SMISKO, J. and DAWSON, L. R., *J. phys. Chem.*, 59 (1955) 84

[30] COATES, J. E. and TAYLOR, E. G., *J. chem. Soc.*, (1936) 1245

[31] LANGE, J., BERGÅ, J. and KONOPIK, N., *Monatsh.*, 80 (1949) 708

[32] DAWSON, L. R., WILHOIT, E. D. and SEARS, P. G., *J. Amer. chem. Soc.*, 79 (1957) 5906; DAWSON, L. R. and BERGER, C., *ibid.*, 79 (1957) 4269 (transport numbers); DAWSON, L. R., NEWELL, T. M. and MCCREARY, W. J., *ibid.*, 76 (1954) 6024

[33] FRENCH, C. M. and GLOVER, K. H., *Trans. Faraday Soc.*, 51 (1955) 1418; data at 15° also are given

[34] AMES, D. P. and SEARS, P. G., *J. phys. Chem.*, 59 (1955) 16; SEARS, P. G., WILHOIT, E. D. and DAWSON, L. R., *ibid.*, 59 (1955) 373

[35] DAWSON, L. R., WILHOIT, E. D., HOLMES, R. R. and SEARS, P. G., *J. Amer. chem. Soc.*, 79 (1957) 3004

[36] FRENCH, C. M. and GLOVER, K. H., *Trans. Faraday Soc.*, 51 (1955) 1427; extensive data for 35° and 45° in N-methyl-acetamide are given

[37] LESTER, G. R., GOVER, T. A. and SEARS, P. G., *J. phys. Chem.*, 60 (1956) 1076

[38] DAWSON, L. R., GRAVES, R. H. and SEARS, P. G., *J. Amer. chem. Soc.*, 79 (1957) 298

[39] JAHNKE, E. and EMDE, F., 'Tables of Functions', Dover Publications, New York (1943). Note that the function called Ei(x) in our appendix is the — Ei(— x) function of Jahnke and Emde

8

THE MEASUREMENT OF CHEMICAL POTENTIALS

THE determination of the chemical potentials of the components of an electrolyte solution usually resolves itself either into a measurement of the activity of the solvent and the calculation of the activity coefficient of the solute by using the Gibbs-Duhem equation or *vice versa*. The methods in general use are therefore conveniently discussed under two headings:

1. Methods depending on measuring the activity of the solvent.
    - (*A*) Vapour pressure methods.
        - (*i*) The static method
        - (*ii*) The dynamic method
        - (*iii*) The isopiestic method.
    - (*B*) Determination of the depression of the freezing point. The elevation of the boiling point is similar in principle but has not been studied to the same extent.
2. Methods which measure the activity of the solute, usually by measuring the potentials of suitable cells with or without liquid junction.

In addition there are some methods which, because of difficulties of technique or for reason of limited application, have not come into widespread use:

- (*a*) Osmotic pressure measurements
- (*b*) Solubility measurements
- (*c*) Measurement of the solute vapour pressure
- (*d*) Distribution of solute between two solvents
- (*e*) Sedimentation in an ultracentrifuge.

THE MEASUREMENT OF VAPOUR PRESSURE BY THE DIRECT STATIC METHOD

In its essentials this method is a direct manometric measurement. *Figure 8.1* shows an apparatus due to GIBSON and ADAMS[(1)], simple in construction but capable of high accuracy if a few precautions are observed. One of the features of their apparatus is the use of *n*-butyl phthalate as the manometer liquid; its vapour pressure is even lower than that of mercury, but as its density is 1·0418 at 25°, its use results in a displacement in the manometer arms many times that which would be given by mercury and consequently greatly increased accuracy of measurement. However, some precautions have to be observed in its use and SHANKMAN and GORDON[2] prefer Cenco Hyvac pump oil (density at 25°: 0·895) as the manometer

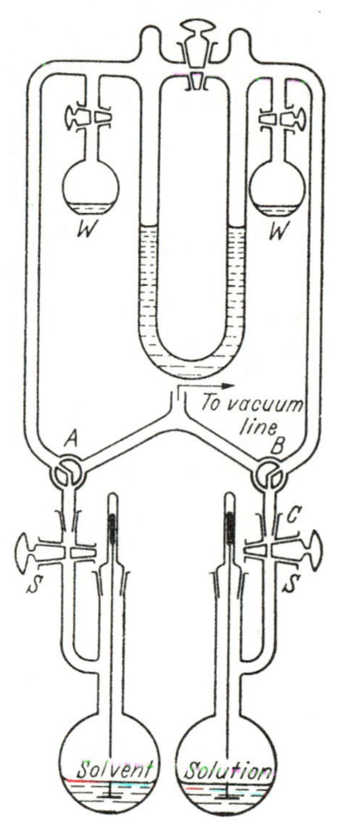

Figure 8.1. From GIBSON, R. E., *and* ADAMS, L. H., *J. Amer. chem. Soc.*, 55 (1933) 2679

liquid. Thorough outgassing of the solution is essential and this is accomplished by repeated solidification and melting whilst the flask is evacuated through the stopcock *S*. The flask is then connected to the manometer set up at *C*. The solvent is treated in the same way but, once outgassed and connected to the manometer, is left permanently in position. The solvent is then connected by the three-way stopcock *A* to one arm of the manometer, the other arm being connected to the vacuum line by stopcock *B*. The resulting displacement of the manometer fluid gives the vapour pressure of the solvent. By turning both stopcocks, the solution can be connected to one arm of the manometer, the other being connected to the vacuum line so that the vapour pressure of the solution can be measured. Finally, suitable manipulation of the stopcocks connects both solution and solvent to the manometer so that the difference in vapour pressure between solution and solvent can be measured. When stopcock B is opened the manometer arm is filled with vapour by evaporation from the solution; this results in a slight cooling and the equilibrium pressure is only reached slowly. To overcome this trouble, the subsidiary flask W containing solvent is provided. Opening its stopcock for a short time fills the manometer arm with vapour after which stopcock B can be opened and only a small amount of vapour is condensed on the solution before equilibrium is reached. Because of the loss of solvent from the solution during the outgassing process, it is necessary to analyse the solution after completing the measurements.

Although minor fluctuations in temperature may result in appreciable changes in the vapour pressure of solvent and solution, this apparatus gives remarkably concordant values of the water activity. In one experiment Shankman and Gordon quote values $p^0 = 361{\cdot}1$, $p = 207{\cdot}45$ and $\Delta p = 153{\cdot}75$ mm of pump oil for the vapour pressure of the solvent, the solution and the differential lowering respectively, so that three values of the water activity $p/p^0 = 0{\cdot}5745$, $\dfrac{p^0 - \Delta p}{p^0} = 0{\cdot}5742$ and $\dfrac{p}{p + \Delta p} = 0{\cdot}5743$ can be derived depending on which two of the three measurements are used in the calculation. Twenty-four hours later they recorded $p^0 = 360{\cdot}1$, $p = 206{\cdot}7$, $\Delta p = 153{\cdot}4$ mm giving a value of 0·5740 for the water activity. Therefore, although the individual readings changed by about one part in three hundred, the water activity changed by only three parts in fifty-seven hundred.

THE MEASUREMENT OF VAPOUR PRESSURE BY THE DYNAMIC METHOD

In principle this method is extremely simple: if a dry inert gas is passed in succession through (1) water, (2) a desiccant to absorb water, (3) an aqueous solution, and (4) a second desiccant then, if the proper experimental conditions are observed, the amount of water absorbed in the first desiccant is proportional to the vapour pressure of the solvent and the amount absorbed by the second is proportional to the vapour pressure of the solution. A modern apparatus constructed by Bechtold and Newton[3] uses successive layers of barium perchlorate and magnesium perchlorate as desiccants, air is passed at a rate kept constant by a manostat relay, being bubbled through five saturators and then passed over the liquid surface in a final saturator in order to equilibrate the air stream with the solvent; after absorption of the water vapour by the desiccants, the air stream is saturated with water vapour at the pressure over the solution by passage through a similar set of saturators containing the solution. The total pressure over the pure solvent is somewhat greater than that over the solution because there is a decrease in pressure owing to the resistance offered by the packed desiccants; it can easily be shown that if the total pressures at the outlet ends of the two series of saturators are P^0 and P and the water vapour pressures are p^0 and p, then:

$$\frac{p}{p^0} = \frac{wP}{w^0P^0 - w^0p^0 + wp^0}$$

when w^0 and w are the weights of water vapour absorbed by the two desiccants. It is clearly desirable to maintain constant the pressures P^0 and P in order to avoid a series of tedious pressure readings during the course of an experiment with the consequent errors introduced by a process of averaging. For this reason a second manostat is introduced at the point where the air stream leaves the final saturator. From the data given by Bechtold and Newton for solutions of calcium chloride and barium chloride, the method seems to give water activities with a probable error of the order of 0·0001 in a_w.

THE MEASUREMENT OF VAPOUR PRESSURE BY THE ISOPIESTIC METHOD

Introduced by BOUSFIELD(4) in 1918 and improved by SINCLAIR(5), this is a comparative method depending on the principle that two solutions of non-volatile solutes will distil from one to the other until their concentrations are such that the solutions have equal vapour pressure. The comparative nature of the method is a disadvantage in that the vapour pressure—concentration curve of some one 'reference' electrolyte must be known with accuracy but, apart from this drawback, the method is one which gives results rapidly and with an accuracy limited only by the accuracy with which the data for the reference electrolyte are known.

Let X and Y be two solutions initially at the same temperature, the vapour pressure of X being initially greater than that of Y and let them be connected by a path through which vapour can pass. Then solvent will distil from solution X to solution Y, resulting in a cooling of X and a heating of Y from the heat of vaporization generated during the process. Because of these temperature changes, the vapour pressure of X decreases and that of Y increases and, if perfect thermal insulation could be maintained between the two solutions, a steady state would be set up with a temperature difference between the two solutions sufficient to equalize the vapour pressures. For example, 4 M solutions of sodium and potassium chlorides differ in vapour pressure by 0·4442 mm. Hg at 25° and a temperature difference of about 0·32° would equalize the vapour pressures. A method based on this principle will be described later: we are now concerned with the extreme case when perfect thermal contact is offered between the solutions and heat can flow back from solution Y to X. The distillation of solvent can now continue with a concentration of X and a dilution of Y, the vapour pressure of X decreasing and that of Y increasing as a result, not of a temperature difference, but of a concentration difference. Equilibrium will occur when this concentration difference suffices to equalize the vapour pressure. For example, starting with two solutions each containing one gram of water and sufficient sodium and potassium chloride respectively to make each solution 4 M, the distillation of 61 mg of water will concentrate the potassium chloride solution to 4·260 M and dilute the sodium chloride to 3·770 M at which concentrations the vapour pressures are equal.

The attainment of equilibrium is greatly accelerated by evacuation of the container to the vapour pressure of the solutions; another critical feature of the experiment is the thermal communication between the solutions. This is secured by containing the solutions in metal dishes of high thermal conductivity such as silver, although platinum or stainless steel dishes can be used with corrosive solutions. Seamless spun circular dishes about 4 cm in diameter, with hinged lids, are convenient. The dishes rest on a thick copper block (about 2·5 cm thick) and the upper surface of this block and the base of each dish should be as flat and smooth as possible. Thermal contact is further improved by a film of solution between each dish and the copper block. If it were desired to measure the vapour pressure of a sodium chloride solution with respect to a potassium chloride solution, sodium chloride would be weighed accurately into each of two dishes in amount sufficient, with between 1 and 2 ml of water (which need not be known with any accuracy) to give approximately the concentration of sodium chloride at which it is desired to study the vapour pressure. Alternatively, between 1 and 2 ml of sodium chloride solution could be weighed out, provided that its concentration was known accurately. In a similar way, a potassium chloride solution is introduced into each of another pair of dishes. The four dishes are placed on the copper block which rests in a glass desiccator which is then evacuated by a good filter-pump. The desiccator is placed in a thermostat and rocked slowly to agitate the solutions gently. The time required for equilibrium to be attained depends on the concentration of the solutions. Generally speaking, for solutions above 1 M, twenty-four hours should suffice; below this concentration, the time required increases and at 0·1 M three or four days may be necessary. Equilibrium having been attained, the dishes are reweighed and the concentrations of sodium chloride and potassium chloride calculated. These solutions have equal vapour pressure and are called 'isopiestic'. The solutions can now be diluted, the experiment repeated and the concentrations of another pair of isopiestic solutions found. Alternatively, the inclusion of a fifth dish containing a more concentrated solution will give a pair of isopiestic solutions higher in the concentration scale. It is possible to have a simple wire device attached to the inlet tube of the desiccator so that, at the end of a run, the lids of the dishes can be closed before air is admitted to the desiccator, thus diminishing error due to evaporation of the solutions or introduction of grease particles. It is also possible, with a slight modification of the apparatus[(6)], to introduce the solutions out of contact with air; measurements can then be made on electrolytes such as ferrous chloride which are readily oxidized on exposure to the atmosphere. The apparatus has been modified to permit the microdetermination of molecular weights using three to seven milligram samples[(6a)].

Measurements are made more easily in concentrated solution and the only limit is the saturation of one of the solutions. At the other end of the concentration scale, about 0·1 M is the lower limit at which measurements are practicable although, by taking extreme precautions, GORDON[(7)] has used the method down to about 0·03 M. From a series of measurements at different concentrations we can construct a curve of the isopiestic ratio against the molality of either electrolyte. The isopiestic ratio is defined by:

$$R = \frac{\nu_B m_B}{\nu_C m_C} \qquad \ldots.(8.1)$$

where m_B is the molality of electrolyte B in solution X and m_C the molality of electrolyte C in solution Y. B is the reference electrolyte, the vapour pressures of whose solutions are known over the necessary concentration range. It is usually convenient to plot R against m_C. The condition of equal vapour pressure is given by:

$$\nu_B m_B \phi_B = \nu_C m_C \phi_C$$

or

$$\phi_C = R\phi_B \qquad \ldots.(8.2)$$

Thus ϕ_C can be derived from R and ϕ_B.

The accuracy of the method therefore depends on two factors: (*a*) R depends on the accuracy of weighing the dishes and can easily be measured with an accuracy of 0·1 per cent; with care the error can be reduced even further; (*b*) assuming that ϕ_B is also known within 0·1 per cent (and this is a problem to which we shall have to return later) then ϕ_C can be determined to this degree of accuracy.

From ϕ_C the activity coefficient γ_C can be calculated by some modification of the Gibbs-Duhem equation[8] as, for example equation (2.27):

$$-\ln \gamma_C = h_C + \int_0^{m_C} h_C \, \mathrm{d} \ln m_C \qquad \ldots.(8.3)$$

where $h_C = (1 - \phi_C)$. The alternate forms of the integral, $2\int h_C/\sqrt{m_C} \, . \, \mathrm{d}\sqrt{m_C}$ and $\int h_C/m_C \, . \, \mathrm{d}m_C$ can be used in many instances. If the activity coefficient of the reference salt is known then another method of calculation is available. For:

$$-55{\cdot}51 \, \mathrm{d} \ln a_w = \nu_B m_B \, \mathrm{d} \ln m_B \gamma_B = \nu_C m_C \, \mathrm{d} \ln m_C \gamma_C$$

where the subscript B refers to the reference electrolyte and the subscript C to the electrolyte whose activity coefficient is being determined.

Then

$$\nu_B m_B \, \mathrm{d} \ln \gamma_B + \nu_B m_B \, \mathrm{d} \ln m_B = \nu_C m_C \, \mathrm{d} \ln \gamma_C + \nu_C m_C \, \mathrm{d} \ln m_C$$

$$\mathrm{d} \ln \gamma_C + \mathrm{d} \ln m_C = R \, \mathrm{d} \ln \gamma_B + R \, \mathrm{d} \ln m_B$$

$$= \mathrm{d} \ln \gamma_B + \mathrm{d} \ln m_B + (R - 1) \mathrm{d} \ln \gamma_B m_B$$

and

$$\ln \gamma_C = \ln \gamma_B + \int_0^{m_B} \mathrm{d} \ln m_B/m_C + \int_0^{m_B} (R - 1) \mathrm{d} \ln \gamma_B m_B$$

whence remembering that:

$$\mathop{\mathrm{Lt}}_{m_B \to 0} \; m_B/m_C = \nu_C/\nu_B$$

$$\ln \gamma_C = \ln \gamma_B + \ln \mathrm{R} + 2 \int_0^{m_B} \frac{R - 1}{\sqrt{a_B}} \, \mathrm{d}\sqrt{a_B} \qquad \ldots.(8.4)$$

and the last term can be evaluated either graphically or by tabulation, the equivalent form $\int (R - 1)/a_B . \mathrm{d}a_B$ being sometimes easier for numerical computation, especially for very concentrated solutions. Isopiestic measurements do not usually extend below 0·1 M;

for many 1 : 1 electrolytes the curve used in evaluating the integral in equation (8.4) can be extrapolated to zero with considerable confidence, provided that the reference salt B is also a 1 : 1 electrolyte. In such cases the method gives the absolute value of γ_C, *i.e.*, values relative to $\gamma_C = 1$ at $m = 0$. With higher valency types, however, the extrapolation is longer and less certain since the curve of R versus m often has a minimum below the experimental lower limit of 0·1 M, and a variety of methods have been tried to fix the values of γ at 0·1 M. GUGGENHEIM and STOKES[8a] have recently proposed a method for 2 : 1 and 1 : 2 electrolytes based on the fact that the isopiestic method gives absolute values of the osmotic coefficient φ. If γ is given by a Debye-Hückel expression (assumed valid to at least 0·3 M):

$$-\ln \gamma = \frac{\alpha\sqrt{m}}{1+\beta\sqrt{m}} - 2\,bm \qquad \ldots.(8.4a)$$

the corresponding expression for φ is:

$$\varphi = 1 - \frac{\alpha\sqrt{m}}{3}\,\sigma(\beta\sqrt{m}) + bm \qquad \ldots.(8.4b)$$

(*See* p. 34.) The function:

$$\varphi^0 = 1 - \frac{\alpha\sqrt{m}}{3}\,\sigma(\beta\sqrt{m})$$

is tabulated for several values of the parameter in Appendix 2.3. A value of β is chosen such that $(\varphi - \varphi^0)$ is directly proportional to m for $m = 0{\cdot}1$ to 0·3, the proportionality factor being the other parameter b. Insertion of these parameters in Equation (8.4a) for $m = 0{\cdot}1$ gives the required value of $\gamma_{0\cdot1}$. The γ values for 2 : 1 and 1 : 2 electrolytes in Appendix 8.10 have been adjusted to this new basis wherever practicable.

THE MEASUREMENT OF VAPOUR PRESSURE BY THE METHOD OF 'BITHERMAL EQUILIBRATION'

We have already mentioned that if two solutions are connected by a vapour path but are thermally insulated, a steady state is set up in which the initial difference in vapour pressure between the two solutions is eliminated by the creation of a temperature difference. STOKES[9] has described a method depending essentially on this principle. Water is maintained at a fixed temperature, t, in vapour contact with a solution at 25°; distillation continues until the concentration of the solution is such that its vapour pressure at 25° is equal to that of water at the lower temperature *t*. A knowledge of the vapour pressure of water at *t*, and an analysis of the solution when the steady state has been set up suffice to give the vapour pressure at 25° of a solution at a known concentration. Thus, with the water at a temperature 9·972° below that of a solution of sodium hydroxide at 25°, it was found that the latter changed in concentration until it reached 9·150 M. The vapour pressure of water at 25°

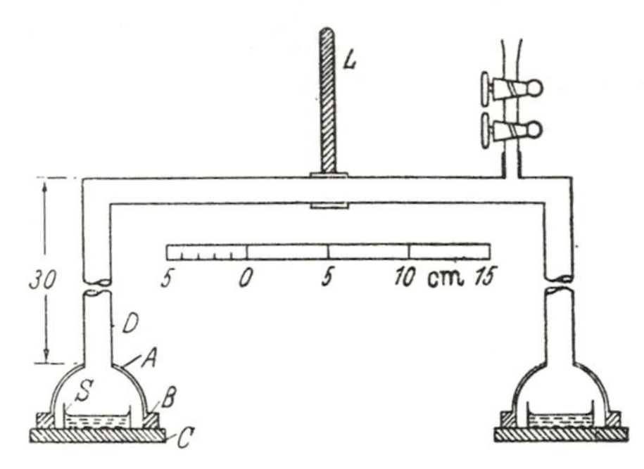

Figure 8.2. From Stokes, R. H., *J. Amer. chem. Soc.*, 69 (1947) 1291

is 23·753 mm and at 15·028° it is 12·807 mm whence it follows that the water activity of the solution is 12·807/23·753 = 0·5391. If the temperature difference were 9·977° the vapour pressure would have been 12·803 mm and the water activity of the solution 0·5390 so that to secure an accuracy of $\pm$ 0·0002 in the water activity, the temperature difference between the two liquids must be controlled to within $\pm$ 0·005°. The apparatus used by Stokes is shown in *Figure 8.2.* The copper domes *A* were soldered to brass rings *B*, the lower faces of which were turned and lapped to fit the flat copper plates *C*. The resulting 'bells' were connected to the thin-walled copper tube *D* so that each leg could be put in a separate thermostat. The horizontal part of this tube carried a side tube for evacuation and a lever *L*, by means of which the apparatus could be rocked. The interior of the apparatus was heavily silver plated. The thermostats were equipped with special thermoregulators[10] designed to control the temperature within $\pm$ 0·001°. The temperature difference between the two thermostats was measured by a 100-junction copper-constantan thermocouple. To start a run, a silver dish, similar to that used in the isopiestic method and containing solution, was placed on one of the plates *C*, and another dish containing water was placed on the other plate. The 'bells' were sealed on with stopcock grease and the apparatus evacuated for an hour on a filter pump followed by ten minutes evacuation with a Hyvac pump with a phosphorus pentoxide tube between the apparatus and the pump. A minimum of twenty-four hours of gentle rocking in the thermostats was necessary, after which air was admitted to the apparatus and the solution removed and analysed. The method is capable of high accuracy, but is a cumbersome one; it was developed for the specific purpose of making possible a choice between two alternative sets of data for the vapour pressure of sulphuric acid, one obtained by the static manometric method by SHANKMAN and GORDON[2], and the other derived from electromotive force measurements by HARNED and HAMER[11]. The resulting independent measurements led to the adoption of a standard set of vapour pressure data for sulphuric acid solutions in close agreement with those of Shankman and Gordon; these have been used as a basis for isopiestic measurements on numerous other concentrated electrolyte solutions.

THE DEPRESSION OF THE FREEZING POINT

The condition for ice to be in equilibrium with pure liquid water at the freezing point T_0 is that the molal free energy shall be the same in each phase:

$$\bar{G}_{\text{ice}(T_0)} = \bar{G}^0_{A(T_0)}$$

A solution has a lower freezing point, T_F (we are dealing with solutions which freeze out the pure solvent phase and not a solid solution) and the condition for equilibrium is now given by:

$$\bar{G}_{\text{ice}(T_F)} = \bar{G}_{A(T_F)} = \bar{G}^0_{A(T_F)} + \boldsymbol{R}T_F \ln a_A$$

$[\bar{G}^0_{A(T_F)} - \bar{G}_{\text{ice}(T_F)}]$ is the increase in free energy on the fusion of a mole of ice to pure liquid water at T_F. Call it $\Delta\bar{G}_{T_F}$: it will be a function of T. From the Gibbs-Helmholtz equation:

$$\frac{\partial}{\partial T}\left(\frac{\Delta\bar{G}}{T}\right) = -\frac{\bar{L}}{T^2}$$

where $\bar{L}$ is the latent heat of fusion of a mole of ice, it follows that:

$$-\boldsymbol{R} \ln a_A = \frac{\Delta\bar{G}_{T_F}}{T_F} = -\int_{T_0}^{T_F} \frac{\bar{L}}{T^2} \mathrm{d}T$$

since $\Delta\bar{G} = 0$ at T_0.

The latent heat of fusion can be written as a function of the temperature:

$$\bar{L} = \bar{L}_0 + \bar{J}(T_F - T_0)$$

where $\bar{L}_0$ is the latent heat of fusion at T_0 and $\bar{J}$ is the difference of the molal heat capacities of liquid water and ice. In most work, $\bar{J}$ can be assumed to be independent of temperature. Then:

$$-\ln a_A = \frac{1}{R}(\bar{L}_0 - \bar{J}T_0)\left(\frac{1}{T_F} - \frac{1}{T_0}\right) + \frac{\bar{J}}{R}\ln\frac{T_0}{T_F} \quad \ldots .(8.5)$$

It is convenient to eliminate T_F by introducing the lowering of the freezing point, $\theta = (T_0 - T_F)$, when equation (8.5) approximates to:

$$-\ln a_A = \frac{\bar{L}_0}{RT_0^2}\theta + \left[\frac{\bar{L}_0}{RT_0} - \frac{\bar{J}}{2R}\right]\frac{\theta^2}{T_0^2} \quad \ldots .(8.6)$$

At 0° the latent heat of fusion of ice is 1435·5 cal mole^{-1} whilst the heat capacities of ice and liquid water are 0·5026 and 1·0081 cal deg^{-1} gram^{-1} respectively[11a]. For aqueous solutions equation (8 6) becomes:

$$-\log a_A = 0{\cdot}004207\theta + 2{\cdot}1 \times 10^{-6}\theta^2$$

In making this approximation, expansions in powers of θ/T_0 have been introduced and taken only as far as the second power of θ/T_0. In very accurate work it may be necessary to consider higher terms, but, if this is done, consideration should also be given to the possible variation of $\bar{J}$ with temperature. The water activity a_A obtained from these formulae is of course that at the temperature T_F. An alternative way of evaluating a_A, which is especially convenient for concentrated solutions, is as follows: values of the vapour pressures of ice and supercooled liquid water are available at various temperatures below 0°C. Then since at the temperature T_F the solution is in equilibrium with ice, its vapour pressure is p_{ice} and the water activity of the solution is therefore:

$$a_A(T_F) = p_{\text{ice}(T_F)}/p_{\text{water}(T_F)}$$

For example, if the freezing point of the solution is − 10°, its water activity at − 10° is $a_A = 1{\cdot}950/2{\cdot}149 = 0{\cdot}9074$.

As the solution becomes more dilute:

$$a_A \rightarrow N_A\ (= 1 - \nu N_B) \text{ and } -\ln a_A \approx \nu N_B$$

so that we can write:

$$\underset{m\to 0}{\mathrm{Lt}}\left(\frac{\theta}{m}\right) = \nu\,\frac{RT_0^2}{\bar{L}_0}\,\frac{W_A}{1000} = \nu\lambda \qquad \ldots .(8.7)$$

The quantity $\lambda = \dfrac{RT_0^2 W_A}{1000\,\bar{L}_0}$ is called the molal lowering of the freezing point. For water as solvent it has the value of 1·860.

CALCULATION OF THE ACTIVITY COEFFICIENT FROM FREEZING POINT RESULTS

If a_A is independent of temperature, the calculation of the activity coefficient of the solute is a simple matter because, by the Gibbs-Duhem equation:

$$\mathrm{d}\ln a_B = \frac{1000\bar{L}_0}{W_A RT_0^2}\cdot\frac{\mathrm{d}\theta}{m} + \frac{2000}{W_A RT_0^2}\left[\frac{\bar{L}_0}{T_0} - \frac{\bar{J}}{2}\right]\frac{\theta\mathrm{d}\theta}{m}$$

$$= \frac{\mathrm{d}\theta}{\lambda m} + \xi\,\frac{\theta\mathrm{d}\theta}{m}$$

where ξ is a parameter independent of θ. The relative magnitudes of these terms can be seen by substituting the values for water as solvent, giving: $\xi = 0{\cdot}00054$ whilst $\lambda = 1{\cdot}860$ so that the first term is by far the more important.* The integration is facilitated by introducing a function defined by:

$$j = 1 - \frac{\theta}{\nu m\lambda}$$

Then

$$\frac{\mathrm{d}\theta}{\nu m\lambda} = -\,\mathrm{d}j + (1-j)\mathrm{d}\ln m$$

$$\mathrm{d}\ln\gamma = -\,\mathrm{d}j - j\mathrm{d}\ln m + \xi\,\frac{\theta\mathrm{d}\theta}{\nu m}$$

$$-\ln\gamma = j + \int_0^m j\mathrm{d}\ln m - \xi\int_0^\theta \frac{\theta\mathrm{d}\theta}{\nu m} \qquad \ldots .(8.8)$$

The last term is equivalent to $\xi\lambda\int_0^\theta (1-j)\mathrm{d}\theta$, the integration being made over the range of θ corresponding to the range of m from zero to the molality in question. The j function is therefore used in much the same way as the h function in computing vapour pressure

* ξ is very sensitive to the values selected for the heat capacities of ice and liquid water: we have used the data of WASHBURN, E. W., quoted by DORSEY[(11a)].

results; indeed, they would be identical if the ξ term were zero. For most 1 : 1 electrolytes, at least, this ξ term contributes only a few units in the fourth decimal place of log γ.

Table 8.1

Calculation of the Activity Coefficient of Sodium Chloride at the Freezing Point

m	j	*First term*	*Second term*	*Third term*	$-\log\gamma$
0·1	0·0663	0·0288	0·0753	0·0002	0·1039
0·2	0·0785	0·0341	0·0971	0·0003	0·1309
0·3	0·0843	0·0366	0·1114	0·0004	0·1476
0·4	0·0876	0·0380	0·1222	0·0006	0·1596
0·5	0·0895	0·0389	0·1308	0·0007	0·1690
0·6	0·0904	0·0393	0·1380	0·0009	0·1764
0·7	0·0907	0·0394	0·1442	0·0010	0·1824
0·8	0·0902	0·0392	0·1491	0·0011	0·1872
0·9	0·0896	0·0389	0·1539	0·0013	0·1915
1·0	0·0884	0·0384	0·1578	0·0014	0·1948

First term $= 0{\cdot}4343j$. Second term $= \int_0^m j\,\mathrm{d}\log m$.

Third term $= 0{\cdot}4343\,\xi\int_0^\theta \dfrac{\theta\mathrm{d}\theta}{\nu m}$

Table 8.1 illustrates the calculation of the activity coefficient of sodium chloride from freezing point measurements[12]. From this table we can see that the third term is almost negligible, and that it is the second term which dominates. Care must therefore be exercised in the tabular or graphical evaluation of this integral especially in the region of low concentrations. Equation (8.6) is written for aqueous solutions:

$$\nu m\,\lambda\varphi = (1 + 4{\cdot}9 \times 10^{-4}\theta)\theta$$

and then using a procedure similar to that outlined on p. 181, we get:

$$(1 + 4{\cdot}9 \times 10^{-4}\theta)\theta - \nu\lambda m\varphi^0 = \nu\lambda b m^2$$

so that a graph of the left hand side against m^2 should give a straight line whose slope determines the parameter b and hence the activity coefficient at 0·1 M by an equation similar to (8.4a). GUGGENHEIM and TURGEON[13] have made such calculations for a number of 1 : 1 electrolytes with $\beta = 1$ in equation (8.4b).

CALCULATION OF ACTIVITY COEFFICIENTS AT TEMPERATURES OTHER THAN THE FREEZING POINT

If, as is usually the case, a_A does vary with the temperature, the correction from the freezing point T_F to some other temperature

T_S is a more complicated matter. It will often be necessary to calculate an activity coefficient at $T_S = 298{\cdot}16°$, a temperature at which measurements by other methods are more frequently made. For this purpose we write:

$$\bar{L}_A = \bar{L}_{A(T_S)} + \bar{J}_A(T - T_S)$$

$\bar{L}_A$ being the relative partial molal heat content at the variable temperature T, $\bar{L}_{A(T_S)}$ its value at the fixed temperature T_S, and $\bar{J}_A$ the relative partial molal heat capacity of the solvent, which can usually be assumed independent of the temperature. $\bar{L}_A$ and $\bar{J}_A$ are to be distinguished from the $\bar{L}_0$ and $\bar{J}$ terms used before; unlike the latent heat of fusion, $\bar{L}_A$ and $\bar{J}_A$ are partial molal properties of the solution and are concentration dependent. By the Gibbs-Helmholtz equation:

$$\left(\frac{\partial \ln a_A}{\partial T}\right) = -\frac{\bar{L}_A}{\boldsymbol{R}T^2} \qquad \ldots.(2.34)$$

whence:

$$\ln \frac{a_{A(T_S)}}{a_{A(T_F)}} = -\int_{T_F}^{T_S} \frac{\bar{L}_A}{\boldsymbol{R}T^2} \mathrm{d}T = -\bar{L}_{A(T_S)}\left(\frac{T_S - T_F}{\boldsymbol{R}T_S T_F}\right)$$
$$+ \bar{J}_A\left(\frac{T_S}{\boldsymbol{R}} \cdot \frac{T_S - T_F}{T_S T_F} - \frac{1}{\boldsymbol{R}} \ln \frac{T_S}{T_F}\right)$$

or

$$\log \frac{a_{A(T_S)}}{a_{A(T_F)}} = -\bar{L}_{A(T_S)}y + \bar{J}_A z = x_A$$

where $y = \dfrac{T_S - T_F}{2{\cdot}303\boldsymbol{R}T_S T_F}$ and $z = T_S y - \dfrac{1}{\boldsymbol{R}} \log \dfrac{T_S}{T_F}$

The functions y and z have been tabulated for a range of T_F values[11b, 14] and the calculation of $a_{A(T_S)}$ at, say, 25° from its value at the freezing point is not difficult. Since:

$$\nu m\, \partial \ln \frac{\gamma_{T_S}}{\gamma_{T_F}} = -\frac{1000}{W_A}\, \partial \ln \frac{a_{A(T_S)}}{a_{A(T_F)}}$$

it follows that:

$$\log \gamma_{T_S} = \log \gamma_{T_F} - \frac{1000}{\nu W_A} \int_0^m \frac{\mathrm{d}x}{m} \qquad \ldots.(8.9)$$

the integration to be carried out over the range of y and z values corresponding to the molality range from zero to the value at which

γ_{T_S} is to be calculated. γ_{T_F} is obtained by the methods already outlined. Equation (8.9) can easily be transformed into:

$$\log \gamma_{T_S} = \log \gamma_{T_F} + \frac{x_B}{\nu}$$

where x_B is defined in terms of the relative partial molal heat content and heat capacity of the solute:

$$x_B = - \bar{L}_B y + \bar{J}_B z$$

We can illustrate this by reference to sodium chloride (*Table 8.2*), the activity coefficient of which has been determined at the freezing point by Scatchard and Prentiss.

Table 8.2

Calculation of the Activity Coefficient of Sodium Chloride at 25° from Freezing Point Data

m	$-\log \gamma_{T_F}$	$\bar{L}_B$	$\bar{J}_B$	$y\bar{L}_B$	$z\bar{J}_B$	$-\log \gamma_{25°}$
0·1	0·1039	102	5·0	0·0069	0·0055	0·1051
0·2	0·1309	90	7·0	0·0062	0·0064	0·1308
0·3	0·1476	62	8·7	0·0043	0·0082	0·1456
0·4	0·1596	28	10·0	0·0020	0·0097	0·1557
0·5	0·1690	− 10	11·1	− 0·0007	0·0111	0·1631
0·6	0·1764	− 48	12·2	− 0·0035	0·0124	0·1684
0·7	0·1824	− 85	13·2	− 0·0063	0·0139	0·1723
0·8	0·1872	− 120	14·1	− 0·0090	0·0152	0·1751
0·9	0·1915	− 156	14·9	− 0·0119	0·0164	0·1773
1·0	0·1948	− 188	15·8	− 0·0145	0·0179	0·1786

The heat content and heat capacity data are from the paper of GULBRANSEN, E. A. and ROBINSON, A. L., *J. Amer. chem. Soc.*, 56 (1934) 2637, the interpolation of the heat content data having been made by HARNED, H. S., and OWEN, B. B., 'The Physical Chemistry of Electrolytic Solutions,' Reinhold Publishing Corp. (1950) p. 541; the heat capacity data are represented by $\bar{J}_B = 15{\cdot}8\sqrt{m}$.

The determination of freezing points with the necessary accuracy is no easy matter; equation (8.8) shows that, if $\log \gamma$ is to be determined within 0·0001, then j must be known within 0·0002; to secure this at a concentration of 1 M, the depression of the freezing point must be measured within $\pm$ 0·0007°. The permissible error decreases proportionally to the molality. Scatchard, after a careful consideration of the accuracy attainable with modern thermocouple technique, concluded that freezing point depressions could be measured within about two hundred-thousandths of a degree and that a concentration of 0·001 M was about the lowest at which measurements could profitably be made. Thus at 1 M concentration the thermometric errors are negligible, but at 0·001 M an error of 2×10^{-5} in the temperature measurement corresponds to an error of about 0·005 in j and about 0·002 in log γ. Successful measurements therefore call for highly skilled experimental work, and Scatchard and his co-workers have obtained valuable results on 26 salts. Their data are given in Appendix 8.7 as values of the activity coefficients quoted to three significant figures; the original papers give log γ to four significant figures and should be consulted if four-figure activity coefficients are required.

In their work all temperature measurements were differences between the temperature of ice in equilibrium with pure water and ice in equilibrium with a solution. Two gold plated silver containers, 8 cm in diameter and 20 cm deep were used. They were divided into three compartments, the centre one being 4 cm wide and the two outer ones comparatively smaller. The central compartment was used to hold the ice and, by means of a pumping system, the solution was forced through the ice from each of the outer compartments. Silvered Dewar flasks were used to contain the vessels in order to secure as nearly as possible adiabatic conditions; since nitrogen has only half the solubility of oxygen in water, dissolved air which would affect the freezing point was removed by passing a stream of nitrogen through the solutions. The temperatures were measured by a 48-junction copper-constantan thermocouple and the concentration of the solution after coming to equilibrium with the ice was determined by finding the specific conductivity at 10° of an aliquot removed from the equilibrium mixture.

THE ELEVATION OF THE BOILING POINT

The theory of this effect is very similar to that of the freezing point depression, but the molal elevation of the boiling-point, if water is the solvent, is only 0·513°, about one-quarter the molal depression of the freezing point; thus boiling points must be measured with nearly four times the accuracy of freezing points to give activity coefficients of the same accuracy. Moreover, the experimental difficulties seem to be much greater. This is unfortunate because the boiling point elevation could give most useful information at temperatures where other methods fail. Unlike the freezing point, the boiling point is markedly susceptible to the pressure and, by making experiments at a series of reduced pressures, data over a temperature range could be acquired. Very little attention has been given to this method in recent years apart from the outstanding contribution of Smith[15] whose paper may well be read by anyone considering developing the technique: Smith considers that his results are consistent within $\pm$ 0·0002°: this corresponds to an accuracy of $\pm$ 0·0001 in the osmotic coefficient at a concentration of 2 M—NaCl, but only $\pm$ 0·004 at 0·05 M. Results, summarized in Appendix 8.8, have been obtained for sodium chloride and potassium bromide between 60° and 100°.

ACTIVITY COEFFICIENTS FROM CONCENTRATION CELLS WITHOUT TRANSPORT

If a faraday of electricity passes through the cell:

$$\text{Ag, AgCl}|\text{HCl}\ (m')|\text{H}_2(\text{Pt}) - (\text{Pt})\text{H}_2|\text{HCl}\ (m)|\text{AgCl, Ag}$$

(the positive current flowing from right to left through the potentiometer circuit outside the cell), the cell reactions are:

$$\text{AgCl} + e^- \rightarrow \text{Ag} + \text{Cl}^-(m)$$
$$\tfrac{1}{2}\text{H}_2 \rightarrow \text{H}^+(m) + e^-$$
$$\text{H}^+(m') + e^- \rightarrow \tfrac{1}{2}\text{H}_2$$
$$\text{Ag} + \text{Cl}^-(m') \rightarrow \text{AgCl} + e^-$$

The net reaction is:

$$\text{HCl}(m') \rightarrow \text{HCl}(m)$$

and the increase in free energy is:

$$\Delta\bar{G} = \bar{G}_{\text{HCl}(m)} - \bar{G}_{\text{HCl}(m')} = 2\boldsymbol{R}T \ln \frac{\gamma m}{\gamma' m'}$$

where γ', γ are the mean ionic activity coefficients at m', m respectively. The (reversible) potential of the cell (assuming that the hydrogen gas is at the same pressure at each electrode) is given by:

$$E\boldsymbol{F} = -\Delta\bar{G}$$

or

$$E = 2\frac{2{\cdot}303\boldsymbol{R}T}{\boldsymbol{F}} \log \frac{\gamma' m'}{\gamma m} = 2k \log \frac{\gamma' m'}{\gamma m} \qquad \ldots .(8.10)$$

The expression, $2{\cdot}303\boldsymbol{R}T/\boldsymbol{F}$, occurs so frequently that it is convenient to abbreviate it to the symbol k; this is not likely to be confused with Boltzmann's constant. Values of $2{\cdot}303\boldsymbol{R}T/\boldsymbol{F}$ are given in Appendix 8.1 for temperatures between 0° and 100°. From equation (8.10) we can determine the activity coefficient at one concentration relative to that at another. In practice it is found easier to measure the potential, E, of the half cell:

$$(\text{Pt})\text{H}_2|\text{HCl}(m)|\text{AgCl, Ag}$$

Let E^0 be the (standard) potential of the half cell

$$(Pt)H_2|HCl|AgCl, Ag$$

in which the acid is present at unit activity in its standard state. Then

$$E = E^0 - 2k \log \gamma m \qquad \ldots .(8.11)$$

and the problem reduces to one of finding the standard potential E^0. The simplest way is to plot the quantity $E' = [E + 2k \log m]$

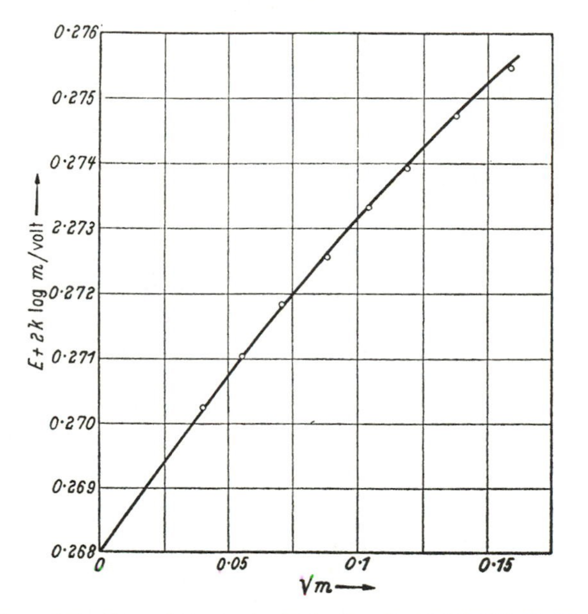

Figure 8.3. Extrapolation to give E^0 for the cell: $H_2|HCl|HgCl, Hg$

against some function of the concentration, say the square root. Then the limiting value of E' as $m \to 0$ is E^0. This extrapolation is shown in *Figure 8.3* for the potential of the analogous cell:

$$H_2|HCl|HgCl, Hg$$

on which very careful measurements have been made recently by HILLS and IVES[16]. It is easy to see that E^0 is not far from 0.2680, but the accuracy of the work justifies something better than this. We therefore seek a deviation function and find (see Chapter 9) that the Debye-Hückel theory gives for the activity coefficient in very dilute solutions:

$$\log \gamma \approx -A\sqrt{(md_0)}$$

where $A \approx 0{\cdot}5$ mole$^{-\frac{1}{2}}$ l$^{\frac{1}{2}}$ and the density of water, d_0, is introduced because the theory gives the activity coefficient expressed in terms

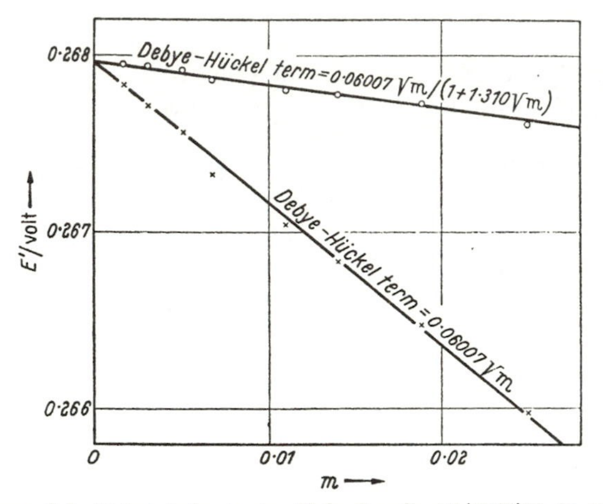

Figure 8.4. Extrapolation to give E^0 *for the cell:* $H_2|HCl|HgCl, Hg$

of molarities (strictly the activity coefficient on the mole fraction scale, but the difference is inappreciable in dilute solution). Then we can write:

$$E' = E + 2k \log m - 2kA\sqrt{(md_0)}$$

and plot this against the molality. This is shown in the lower curve of *Figure 8.4.* Over the same concentration range, the extrapolation function now covers only a range of 0·002 V as against 0·007 V in *Figure 8.3.* This is sometimes called the Hitchcock method[17]. The abscissa in *Figure 8.3* is $\sqrt{m}$; but in *Figure 8.4* it is m, because the $\sqrt{m}$ term has been incorporated in the extrapolation function, and deviations from the Debye-Hückel formula should be approximately proportional to m.

An even easier extrapolation can be made by using a fuller form of the Debye-Hückel equation, corresponding to equation (9.7):

$$\log \gamma \approx -\frac{A\sqrt{(md_0)}}{1 + Ba\sqrt{(md_0)}}$$

i.e., we plot the function:

$$E' = E + 2k \log m - 2kA\sqrt{(md_0)}[1 + Ba\sqrt{(md_0)}]$$

against the molality, selecting a reasonable value of a, the mean diameter of the ions. The upper curve of *Figure 8.4* shows this, using $a = 4$ Å. The function covers a range of only 0·0003 V and no doubt this could be reduced still further by using an a value somewhat higher. As it is, the extrapolation can be made easily, giving $E^0 = 0{\cdot}26796$ V. Once E^0 is known, a potential measurement at a given molality suffices to give the activity coefficient at that molality by equation (8.11).

An extensive series of measurements of the cell: $H_2|HCl|AgCl, Ag$ covering the concentration range 0·003 to 4 M and the temperature range 0° to 90° has been made[18]. Another study[19] has extended the measurements up to about 16 M over the range 0° to 50°. The analogous cell:

$$H_2|HBr|AgBr, Ag$$

has been measured[20] between 0·0001 and 0·004 M at 25° and between 0·001 and 1 M over the temperature range 0° to 60° to give the activity coefficient of hydrobromic acid[21]. An independent check[22] of the standard potential of this cell has been made.

Another method of arriving at the standard cell potential is due to Owen[23]. It will be shown in Chapter 12 that the potential, E_1, of the cell:

$$H_2|HA(m), NaA(m), KCl(m)|AgCl, Ag \qquad \text{Cell I}$$

where HA is a very weak acid (in this work, boric acid) and, for simplicity, the molalities of the three components have been put equal to one another, is:

$$E_1 = E^0_{HCl} - k \log K - k \log \frac{\gamma_{Cl^-}\gamma_{HA}}{\gamma_{A^-}} - k \log m$$

K being the ionization constant of the acid; γ_{Cl^-}, γ_{A^-}, ionic activity coefficients; and γ_{HA} the activity coefficient of the *undissociated* acid. Cells of this type have been used extensively to determine the dissociation constants of weak acids, the standard potential, E^0_{HCl}, of the cell:

$$H_2|HCl|AgCl, Ag$$

being known. But there is no reason why the procedure should not be reversed: if K is known, then this cell could be used to determine E^0_{HCl}. This particular standard potential is already well known but that of the cell containing hydriodic acid is not, so that the cell:

$$H_2|HA(m), NaA(m), KI(m)|AgI, Ag \qquad \text{Cell II}$$

where the last activity coefficient term may be slightly different from the former, can be measured at a series of values of m and the quantity $E + k \log (mK)$ extrapolated to infinite dilution to give E^0_{HI}.

It is even easier to make parallel measurements on both cells I and II together, when:

$$E_1 - E_2 = E^0_{HCl} - E^0_{HI}$$

an equation which will be accurate except for the slight difference in the activity coefficient terms. By taking the concentrations low enough, say $m = 0{\cdot}003$, this term will be less than 0·01 mV and therefore beyond the experimental error. Thus measurements are made in solutions dilute enough for the activity coefficient term to be negligible and yet in a buffered solution of sufficient concentration to yield stable potentials.

By using flowing amalgam cells of the type:

$$\text{Ag, AgX}|\text{MX}(m')|\text{M}_x\text{Hg}|\text{MX}(m)|\text{AgX, Ag}$$

the activity coefficients of a number of alkali halides have been determined. Among the salts studied are lithium chloride[24], sodium chloride[27], potassium chloride[28], caesium chloride[26], lithium bromide[24], sodium bromide[29], potassium bromide[24], sodium iodide[25] and potassium iodide[25].

A combination of hydrogen and amalgam electrodes will give the activity coefficient of the hydroxides:

$$(\text{Pt})\text{H}_2|\text{MOH}(m')|\text{M}_x\text{Hg}|\text{MOH}(m)|\text{H}_2(\text{Pt})$$

The theory of the cell is slightly more complicated because the solvent takes part in the cell reaction:

$$\text{MOH}(m') + \text{H}_2\text{O} \rightarrow \text{MOH}(m) + \text{H}_2\text{O}$$

with the distinction that the water of the left-hand side of the equation disappears from the right half of the cell and reappears as water in the left half and allowance has to be made for the change in water activity so that:

$$E = 2k \log \frac{\gamma' m'}{\gamma m} + k \log \frac{a_w}{a'_w}$$

We shall deal with the general question of cells in which the solvent participates in the cell reaction on pp. 196–197. Cells of this type have been used with lithium[30], sodium[31], potassium[32, 33] and caesium[26] hydroxide as electrolyte.

If the electrolyte is polyvalent, allowance must be made for the multiple charge on the ions. For example, the potential of the cell:

$$\mathrm{In}|\mathrm{In_2(SO_4)_3}(m)|\mathrm{Hg_2SO_4, Hg}$$

has been measured[34]. The reversible working of the cell involves the reactions:

$$2\mathrm{In} \rightarrow 2\mathrm{In}^{+++} + 6e^-$$

$$3\mathrm{Hg_2SO_4} + 6e^- \rightarrow 6\mathrm{Hg} + 3\mathrm{SO_4^{--}}$$

The electrical work per mole of indium sulphate would be $6EF$ and

$$E = E^0 - \frac{RT}{6F} \ln a_{\mathrm{In_2(SO_4)_3}}$$

$$= E^0 - \frac{5RT}{6F} \ln \sqrt[5]{108}\, m\gamma_{\pm}$$

because: $$a_{\mathrm{In_2(SO_4)_3}} = \gamma_{\mathrm{In}}^2 \gamma_{\mathrm{SO_4}}^3 m_{\mathrm{In}}^2 m_{\mathrm{SO_4}}^3$$

$$= \gamma_{\pm}^5 (2m)^2 (3m)^3 = 108 m^5 \gamma_{\pm}^5$$

$\gamma_{\pm}$ being the mean ionic activity coefficient. In general, for an electrolyte dissociating into ν_1 positive and ν_2 negative ions, where $(\nu_1 + \nu_2) = \nu$, and n electrons are involved at the electrodes for each molecule reacting and m is the stoichiometric molality of the electrolyte,

$$E = E^0 - \frac{\nu RT}{nF} \ln [(\nu_1^{\nu_1} \nu_2^{\nu_2})^{1/\nu} m\gamma_{\pm}]$$

The cells:

$$\text{Zn amalgam}|\mathrm{ZnSO_4}|\mathrm{Hg_2SO_4, Hg}$$

and $$\text{Cd amalgam}|\mathrm{CdSO_4}|\mathrm{Hg_2SO_4, Hg}$$

are two examples of systems which give reproducible potentials, stable over a long interval of time; if the amalgams and solutions are saturated, the cells are the standard Clark and Weston cells. The former has been measured[35] over a concentration range whilst a variant[36] of the second cell:

$$\mathrm{Cd_xHg}|\mathrm{CdSO_4}|\mathrm{PbSO_4, Pb_xHg}$$

has been used to give the activity coefficient of cadmium sulphate. The chlorides, bromides and iodides of both zinc and cadmium have been studied by combining zinc or cadmium amalgam electrodes with the appropriate silver–silver halide electrode[37]. The barium amalgam electrode seems to work satisfactorily in solutions of barium chloride[38] or barium hydroxide[39] and the strontium amalgam electrode in solutions of strontium chloride[40] but it is doubtful if the calcium amalgam electrode gives true reversible potentials. The cell:

$$Na_xHg|Na_2SO_4|PbSO_4, Pb_xHg$$

is suitable for determining the activity coefficient of sodium sulphate[41] and a similar cell has been used with lithium and potassium sulphate[42]. Finally, mention should be made of two cells[11, 43]:

$$(Pt)H_2|H_2SO_4|Hg_2SO_4, Hg$$

and

$$(Pt)H_2|H_2SO_4|PbSO_4, PbO_2(Pt)$$

which give the activity coefficient of sulphuric acid. The cell reaction of the latter is:

$$H_2 + PbO_2 + H_2SO_4 \rightarrow PbSO_4 + 2H_2O$$

so that the formula for the cell potential will include a term for the water activity, and we may now consider the generalized treatment of such cells[44]. A complete concentration cell can be written:

Electrode A|Solution (m_{ref})|Electrode B|Solution (m)|Electrode A

and the cell reaction as four processes:

(*a*) A loss of one molecule of electrolyte at concentration m from the right-hand solution.

(*b*) A gain of one molecule of electrolyte at concentration m_{ref} in the left-hand solution.

(*c*) A loss of r molecules of water from the left-hand solution.

(*d*) A gain of r molecules of water in the right-hand solution.

The increment in free energy per mole of electrolyte reacting is:

$$\Delta\bar{G} = [\bar{G}_{B(\text{ref})} - \bar{G}_{B(m)}] + r[\bar{G}_{w(m)} - \bar{G}_{w(\text{ref})}]$$

where the subscript B refers to the solute. If $\boldsymbol{n}$ electrons are involved in the reaction:

$$\boldsymbol{n}E\boldsymbol{F} = -\Delta\bar{G} = [\bar{G}_{B(m)} - \bar{G}_{B(\text{ref})}] + r[\bar{G}_{w(\text{ref})} - \bar{G}_{w(m)}]$$

or

$$\begin{aligned} \boldsymbol{nF}\mathrm{d}E &= \mathrm{d}\bar{G}_{B(m)} - r\mathrm{d}\bar{G}_{w(m)} \\ &= \boldsymbol{R}T\mathrm{d}\ln a_B - r\boldsymbol{R}T\mathrm{d}\ln a_w \\ &= -\boldsymbol{R}T\frac{55{\cdot}51}{m}\,\mathrm{d}\ln a_w - r\boldsymbol{R}T\mathrm{d}\ln a_w \\ &= -\boldsymbol{R}T\frac{55{\cdot}51 + rm}{m}\,\mathrm{d}\ln a_w \end{aligned}$$

therefore

$$\ln \frac{a_w}{a_{w(\text{ref})}} = -\frac{\boldsymbol{F}}{RT}\int_{m_{\text{ref}}}^{m}\left(\frac{nm}{55{\cdot}51 + rm}\right)\text{d}E$$

Defining $m' = \dfrac{nm}{55{\cdot}51 + rm}$,

$$\log \frac{a_w}{a_{w(\text{ref})}} = -\frac{\boldsymbol{F}}{2{\cdot}303\boldsymbol{R}T}\int_{m_{\text{ref}}}^{m} m'\text{d}E$$

In every case so far examined it has been found that a simple deviation function, $x = E + f(m')$ can be defined, in such a way that x varies by only a few millivolts over a concentration range where E varies by several hundred millivolts. The form of $f(m')$ is decided by trial: a logarithmic form, $x = E + a \log m'$, is usually applicable below 1 M, whilst at higher concentrations the form $x = E + b\sqrt{m'}$ or $x = E + cm'$ may be more suitable. In any case:

$$\int m'\,\text{d}E = \int m'\,\text{d}x - \int m'[\text{d}f(m')/\text{d}m']\text{d}m'$$

The second term on the right is a simple analytical integral and the first term may be obtained by graphical or tabular integration. Since the first term contributes only a few per cent to the total value of $\int m'\,\text{d}E$, it is readily evaluated with all the accuracy inherent in the electromotive force determinations. This accuracy is not obtained if the direct integration of m' with respect to E or of m to $\log \gamma$ is attempted. Alternatively the Gibbs-Duhem equation can be used to eliminate $\bar{G}_w$ instead of $\bar{G}_B$, and the activity coefficients of the solute computed without successive approximations.

Concentration Cells without Transport in Non-Aqueous Solvents

Many measurements have been made of the potentials of cells containing hydrochloric acid in non-aqueous media or in mixed solvents of which water is one component. Harned and his co-workers[45] have made an intensive study of water-dioxan mixtures, that containing 82 per cent by weight dioxan having a dielectric constant of about 10. Measurements have also been made in solvents such as pure methanol[46, 47], ethanol[48, 49, 50], and formic and acetic acid[51] and in aqueous solvents to which were added methanol[47, 52], ethanol[49, 53, 53a], n-propanol[54], iso-propanol[53, 54a], acetone[55], glycerol[50, 55a], glycols[56, 56a], glucose[57], fructose[57a] or sucrose[58].

EXPERIMENTAL MEASUREMENTS

The $H_2|HCl|AgCl$, Ag cell usually takes the form of an H-tube, one arm of which holds the platinum electrode around which bubbles hydrogen. The hydrogen may be obtained from a cylinder, in which case it should be freed from any traces of oxygen by passing it over heated copper; it can also be made by electrolysis of a strong solution of sodium hydroxide. The gas should be passed through a saturator containing the same solution as the cell so that the passage of gas through the cell does not result in a change of concentration by evaporation. The platinum electrodes may conveniently be of 0·5 cm by 2·5 cm size, coated with platinum black by electrolysis of a chloroplatinic acid solution (a solution containing 0·5 g of platinum per 100 ml has been recommended with a current density of about 200 mA/cm² for 10 min.: the amount of platinum black should be reduced by plating for shorter times, say one minute, if the electrode is to be used in very dilute acid solution, because heavily plated electrodes are sluggish in coming to equilibrium in very dilute solution).

The normal potentials as tabulated assume that the hydrogen is at a partial pressure of one atmosphere. In practice there will be a small correction because of barometric variations and the vapour pressure of the solution in the cell:

In the cell:

$$H_2 \text{ (pressure } \overline{P-p})|HCl|H_2 \text{ (1 atm.)}$$

where P is the total pressure and p the vapour pressure, the reaction is:

$$H_2 \text{ (pressure } \overline{P-p}) \rightarrow H_2 \text{ (1 atm.)}$$

and the free energy change per mole is:

$$-\boldsymbol{R}T \ln (P-p)$$

so that the potential of the cell is $\dfrac{\boldsymbol{R}T}{2\boldsymbol{F}} \ln (P-p)$ and the observed potential is to be corrected by subtracting $\frac{1}{2}k \log (P-p)$.

With proper precautions, the glass electrode gives results as accurate as the hydrogen electrode: Covington and Prue[59] have used cells with and without transport to get precise activity coefficients and transport numbers of hydrochloric, perchloric and nitric acid. An important study [60] has been made of the glass electrode in methanol-water mixtures from which it is concluded that accurate pH measurements can be made provided that (*i*) the electrode is stored in and equilibrated with solvent of the composition which is to be used in the pH measurement, (*ii*) the electrode is standardized with a buffer in the same solvent and (*iii*) a small correction is made for the liquid junction potential.

Silver-silver chloride electrodes of several types have been tried. What is sometimes called the Carmody type[61] is a piece of platinum gauze about 1 cm square, plated with silver by electrolysis in a solution of potassium silver cyanide. It is important not to use the excess of potassium cyanide which is common in ordinary silver plating; instead, a salt twice recrystallized from water is used. A current of 8 mA for 8 h has been recommended. The electrode is washed thoroughly for several days in running water and is then chloridized in a hydrochloric acid solution for one hour at 3 mA. It is well to do these operations with the electrode protected from direct lighting.

A variant of this electrode is used[62] in cells with transport. The dimensions are very much reduced by using a 1 cm length of platinum wire, 0·045 cm in diameter, silver-plated by electrolysis for 2–6 h at 2–0·5 mA in a solution from which excess cyanide has been removed by adding a small amount of silver nitrate until opalescence occurs. After washing, the electrode is chloridized in 0·1 N hydrochloric acid for half an hour at 2 mA.

GÜNTELBERG[63] used a platinum wire spiral filled with silver oxide, the oxide being converted to metal by heating to 450°–500° and the spiral then immersed in crystalline silver chloride made by evaporating an ammoniacal silver chloride solution over sulphuric acid. In a third type[64] the oxide is converted to metal as before but the chloride layer is formed by electrolysis in normal hydrochloric acid solution at 2 mA/cm^2 for 2 h. It is well to avoid rubber stoppers, sulphur compounds in which cause the formation of silver sulphide and dissolved air should be removed from the solutions, especially if they are dilute.

The calomel or mercury-mercurous chloride electrode has received more attention lately after being unfashionable for many years. HILLS and IVES[65] prepared their calomel electrolytically and coated their electrode vessels with a hydrophobic reagent (Dow-Corning Silicone Fluid No. 200 deposited from 1 per cent carbon tetrachloride solution). The excellent consistency of the results they obtained with solutions of hydrochloric acid as dilute as 0·0016 M is shown by the graph in *Figure 8.4*.

Silver–silver bromide electrodes are prepared rather more easily. A mixture of 90 per cent silver oxide and 10 per cent silver bromate is ground in an agate mortar, made into a paste with water, fed into a platinum spiral and heated at 650° for about 7 min[20].

For studies on alkali halide solutions it is not possible to use the pure alkali metal as one electrode, owing to its irreversible reaction with water. Instead a very dilute (~ 0·01 per cent) alkali metal

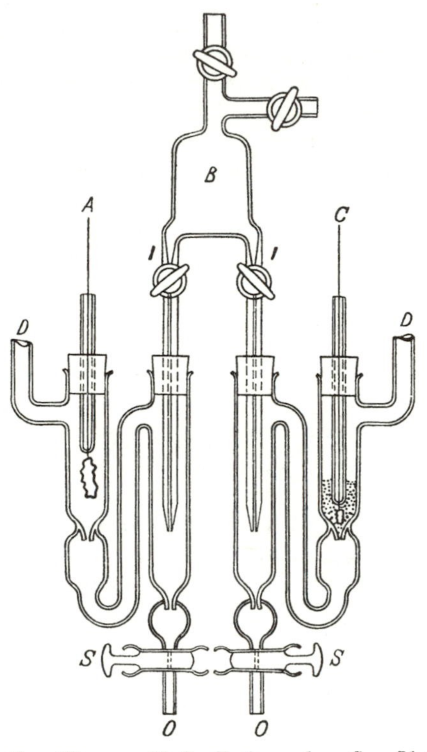

Figure 8.5. From HARNED, H. S., *J. Amer. chem. Soc.*, 51 (1929) 417

amalgam is used; in order to avoid the problems introduced by variation of the amalgam composition, a complete concentration cell such as:

$$\text{Ag, AgCl}|\text{KCl}(m')|\text{K}_x\text{Hg} - \text{K}_x\text{Hg}|\text{KCl}(m)|\text{AgCl, Ag}$$

is employed, the amalgam being dropped in a fine stream through the solutions from a common reservoir. Cells of this type have been perfected and studied extensively by Harned and his collaborators.

Figure 8.5 shows a simple design of cell[24] for alkali metal halide solutions. *A* and *C* are silver–silver halide electrodes, *DD* are inlet tubes through which solutions could be fed into the cell compartments, the solutions having been previously boiled *in vacuo*. The amalgam is made by electrolysis of an hydroxide solution to give about 0·1 per cent amalgam; the amalgam is dried in vacuo and allowed to stand until the impurities have risen to the surface, after which the clean amalgam is run off through an outlet in the base of the container into mercury and diluted to about 0·01 per cent. Using vacuum technique the amalgam is introduced into *B* and by manipulating the stopcocks *I–I* the amalgam flows through the capillary tubes of the dropper *B*, through the solutions at a rate of about 1 cm^3 in 20 sec. During the flow of amalgam, the stopcocks *S–S* are manipulated to remove the amalgam through *O–O* and as many potentiometer readings as possible are made whilst the supply of amalgam is flowing. The apparatus is designed so that a new dropper can be introduced, solutions rejected through *O–O* and the cells filled with fresh solutions. The elimination of oxygen is essential to the proper working of these cells.

ACTIVITY COEFFICIENTS FROM CONCENTRATION CELLS WITH TRANSPORT

A paper from the Rockefeller Institute in 1935 forms a good introduction to this subject[66]; this work was a natural corollary of the study of transport numbers which had been undertaken in the same laboratories and it made rapid progress in elucidating activity coefficients in more dilute solutions (up to 0·1 N) because, once the transport number has been found as a function of the concentration, the activity coefficient is given by measuring the potential of a comparatively simple cell. Thus in the case of sodium chloride one type of electrode only is needed, the silver–silver chloride electrode, and the difficult technique of the sodium amalgam electrode is not required. The method is limited, however, to salts towards at least one of whose ions there is known to be an electrode capable of nearly ideally reversible behaviour, and it is not surprising that the method has so far been applied almost exclusively to a series of chloride electrolytes. The cell:

$$\text{Ag, AgCl}|\text{NaCl}(m')|\text{NaCl}(m)|\text{AgCl, Ag}$$

is one in which, for each faraday of electricity passing, an equivalent of chloride ion is liberated at the left-hand electrode and formed at the right, whilst t_1 equivalents of sodium ion pass from left to right across the junction between the two solutions and t_2 equivalents of chloride pass in the opposite direction. The net result is the loss of t_1 moles of sodium chloride in the left-hand solution and a corresponding gain of t_1 moles at the right. Considering first the case when $m' = m + \mathrm{d}m$ the potential is:

$$\mathrm{d}E_t = -2kt_1\mathrm{d}\log\gamma m \qquad \ldots .(8.12)$$

or, should the transport number be dependent on the concentration, as in practice is the case, then for a finite difference:

$$E_t = -2k\int_{m'}^{m} t_1\mathrm{d}\log\gamma m$$

where the integration is to be carried out from the conditions prevailing in the left-hand solution up to those in the right-hand solution. The experimental side of the work is not difficult: using

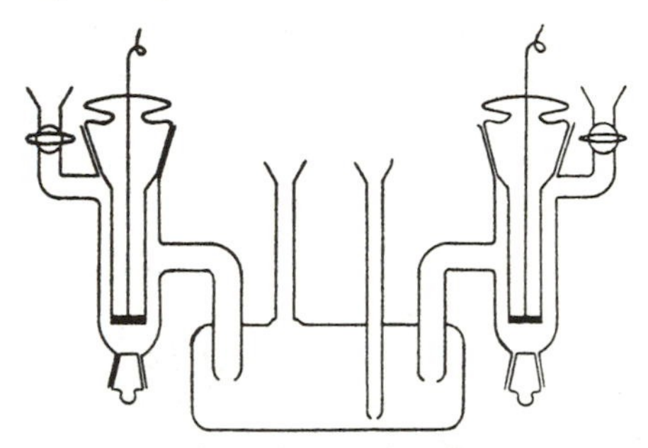

Figure 8.6. From HORNIBROOK, W. J., JANZ, G. J. and GORDON, A. R., *J. Amer. chem. Soc.*, 64 (1942) 513

silver–silver chloride electrodes based on the Carmody model but of much smaller dimensions[62] and forming the liquid junction in a manner similar to the 'sheared boundary' of the transport number experiment, very stable and reproducible potentials can be measured.

In subsequent work the electrodes were modified and the 'sheared boundary' method discarded because it introduced traces of grease into the solutions. *Figure 8.6* shows a simple design of cell used by GORDON[67] and very similar to the Rockefeller Institute cell. The platinum electrodes are much heavier and the boundary is formed by filling each electrode compartment and the side tubes with solution, after which the intervening compartment is filled with the heavier solution. The junction is, therefore, at one of the side tubes. Provided that no appreciable heat of mixing is involved at the junction, experience and theory agree that the potential is independent of the sharpness of the boundary region.

Some of the published methods of manipulating the experimental data from such cells in order to give the activity coefficients necessitate a series of approximations. The following method is more convenient. The experimental data consist of a series of values of the electromotive force, E_t, of a cell with transport, the molality on one side being kept fixed at some known value m'. The potential is related to the transport number and the activity coefficients by equation (8.12), *i.e.*, by:

$$-\,\mathrm{d}\log(\gamma m) = \frac{\mathrm{d}E_t}{2kt_1}$$

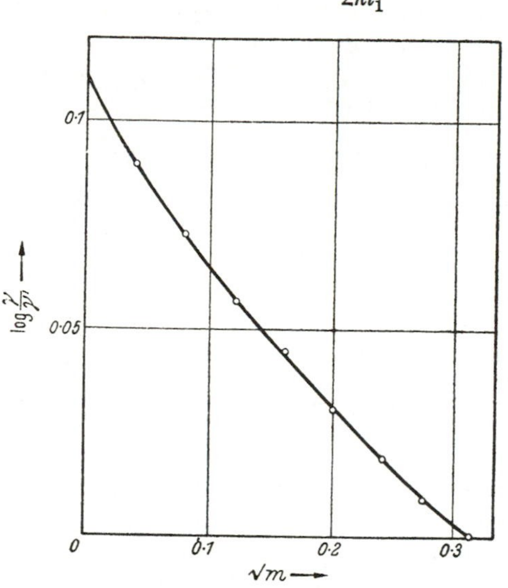

Figure 8.7. Values of log γ/γ' *against* $\sqrt{m}$ *from the data of* JANZ *and* GORDON *for sodium chloride.* $m' = 0{\cdot}1$

Now t_1 usually varies only slightly with m, so that if we define a quantity x by the equation:

$$\frac{1}{t_1} = \frac{1}{t_1'} + x$$

where t_1' is the transport number at m', x will be only a small fraction of $1/t_1'$. Hence:

$$-\,\mathrm{d}\log(\gamma m) = \frac{1}{2k}\left(\frac{1}{t_1'}\,\mathrm{d}E_t + x\mathrm{d}E_t\right)$$

and, t_1' being a constant, this can be integrated between m and m' to give, since $E_t = 0$ when $m = m'$:

$$-\log\frac{\gamma m}{\gamma' m'} = \frac{E_t}{2kt_1'} + \frac{1}{2k}\int_{m'}^{m} x\mathrm{d}E_t$$

or $$\log \gamma = \log \gamma' + \log \frac{m'}{m} - \frac{E_t}{2kt_1'} - \frac{1}{2k}\int_{m'}^{m} x\mathrm{d}E_t$$

(Care must be taken to give E_t the correct sign.)

The part of this expression involving the integral is now quite small and can easily be evaluated graphically or by tabulation without loss of accuracy. The activity coefficient γ', at the fixed concentration m', is now determined as follows, using the fact that as $m \to 0$, $\gamma \to 1$ by definition.

In *Figure 8.7* a plot of $\log \gamma/\gamma'$, against $\sqrt{m}$ taken from the data of JANZ and GORDON[68] for sodium chloride is extrapolated to zero

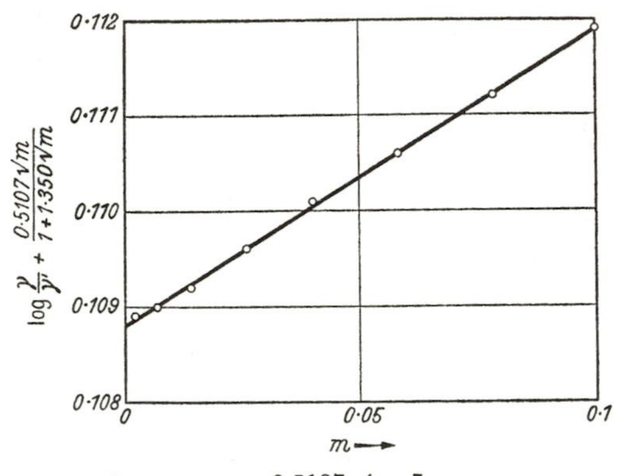

Figure 8.8. Plot of $\left[\log \gamma/\gamma' + \frac{0{\cdot}5107\sqrt{m}}{1 + 1{\cdot}350\sqrt{m}}\right]$ *against m for sodium chloride*

value of m and the intercept gives $-\log \gamma'$ as approximately 0·11. A more accurate extrapolation can be made by assuming the validity of the Debye-Hückel equation (8.4a).

We now plot the function $\left(\log \frac{\gamma}{\gamma'} + \frac{0{\cdot}5107\sqrt{m}}{1 + 1{\cdot}350\sqrt{m}}\right)$ against m as in *Figure 8.8* and find that the intercept is $-\log \gamma' = 0{\cdot}1088$ and values of $\log \gamma$ at other concentrations follow immediately.

In recent years the Rockefeller Institute workers[66, 69] have obtained data on hydrochloric acid, sodium chloride, potassium chloride, calcium chloride and lanthanum chloride. It should be noted that their results are expressed on the molarity concentration scale and their activity coefficient, f, is the mean *molar* activity coefficient. The Toronto school[67, 68, 70] have studied only three salts, sodium chloride, potassium chloride and calcium chloride, but they made measurements over the temperature range 15–45°. More recently an extensive study of the chlorides of lanthanum, cerium, praseodymium, neodymium, gadolinium, samarium, europium, erbium and ytterbium and the bromides of lanthanum, praseodymium, neodymium, gadolinium, holmium and erbium up to about 0·03 M has been published[71].

When measurements have been made on the same salt by independent workers, the agreement has usually been most encouraging; thus, for potassium chloride, Hornibrook, Janz and Gordon found $\gamma = 0{\cdot}8172$ and 0·7697 at 0·05 M and 0·1 M respectively compared with Shedlovsky and MacInnes' values of 0·8172 and 0·7701. For sodium chloride the Toronto school found $\gamma = 0{\cdot}7784$ at 0·1 M in exact agreement with the earlier measurements of Brown and MacInnes, although a recomputation by Shedlovsky, using a different modification of the Debye-Hückel equation, has given $\gamma = 0{\cdot}7744$. For calcium chloride, McLeod and Gordon found $\gamma = 0{\cdot}5769$ at 0·05 M compared with 0·5835 (since recomputed as 0·5826) from the Rockefeller Institute: the difference, however, arises from the transport numbers rather than the electromotive force measurements. We give in Appendix 8.9, the activity coefficients of some electrolytes at concentrations below 0·1 M, most of which have been determined in recent years by this method.

THE OSMOTIC PRESSURE

The osmotic pressure of a solution is determined by the condition that, for equilibrium across a semi-permeable membrane, the chemical potential of the pure solvent on one side of the membrane must be equal to the chemical potential of the solvent in the solution on the other side where it is subjected to a hydrostatic pressure equal to the osmotic pressure. Under this pressure the chemical potential of the solvent in the solution, $\bar{G}_A = \bar{G}^0_A + \boldsymbol{R}T \ln a_A$ becomes $\bar{G}^0_A + \bar{V}_A \Pi + \boldsymbol{R}T \ln a_A$, using equation (2.36) and neglecting the compressibility. Since this must equal the chemical potential of the pure solvent, we have $\bar{V}_A \Pi = -\boldsymbol{R}T \ln a_A$ and as the osmotic coefficient is defined by:

$$\ln a_A = -\frac{\nu m W_A}{1000} \phi \qquad \ldots .(2.16)$$

it follows that:

$$\Pi = \frac{\boldsymbol{R}T}{\bar{V}_A} \frac{\nu m W_A}{1000} \phi$$

A considerable amount of experimental skill was expended in the first fifteen years of this century in devising apparatus to measure osmotic pressures and to overcome the numerous experimental difficulties which seem to beset this subject. The effect is large: for example, a one molar solution of sucrose has an osmotic pressure of about 27 atm. at 25° and therefore the pressure should be measurable with accuracy at small concentrations of the solute; against this is to be put the difficulty of preparing truly semi-permeable membranes and the necessity, at least for data pertaining to concentrated solutions, of making allowance for the variation of $\bar{V}_A$ with concentration and with pressure. Very few accurate results have been obtained in spite of the spirited attack which was made on this problem early in the century. We can illustrate the accuracy by reference in *Table 8.3* to the osmotic coefficient of sucrose derived

Table 8.3

Osmotic Coefficient of Sucrose Solutions at 0°

m	Π (atm.)	$\phi^{(1)}$	$\phi^{(2)}$	$\phi^{(3)}$
1·651	43·84	1·182	1·179	1·185
2·373	67·68	1·259	1·262	1·273
3·273	100·43	1·354	1·351	1·369
4·120	134·71	1·437	1·433	1·459

$\phi^{(1)}$ Derived from vapour pressure measurements.
$\phi^{(2)}$ From osmotic pressure measurements, allowing for variation of $\bar{V}_A$ with concentration and pressure.
$\phi^{(3)}$ From osmotic pressure measurements, putting $\bar{V}_A = \bar{V}_A^0 = 18{\cdot}01$ ml/mole.

from direct vapour pressure measurements using the dynamic method and from osmotic pressure measurements allowing for the compressibility of the solution[72]. Accurate measurements of solutions of simple electrolytes (as distinct from polyelectrolytes) are not extensive. For calcium ferrocyanide both osmotic pressure and the vapour pressure have been measured at 0° (the latter by the dynamic method) and give the following values of the osmotic coefficient:

m	1·075	1·353	1·469	1·617	1·711
Π (atmospheres)	41·22	70·84	87·09	112·84	130·66
ϕ (from osmotic pressure)	0·557	0·756	0·853	0·995	1·086
ϕ (from vapour pressure)	0·562	0·759	0·854	1·004	1·100

THE POROUS-DISC OSMOMETER

The limitation of osmotic pressure measurements in the study of simple electrolytes arises from the difficulty of preparing membranes permeable to solvent molecules but impermeable to ions which may be little different in size from the solvent molecules. An elegant solution to this problem is provided in principle by the 'porous-disc osmometer', in which the 'membrane' consists of a path of solvent vapour, and is therefore perfectly impermeable to ions. In this method, which has been highly developed by WILLIAMSON[73], the drop in the chemical potential of the solvent brought about by the presence of the solute is matched by applying a *negative* pressure to

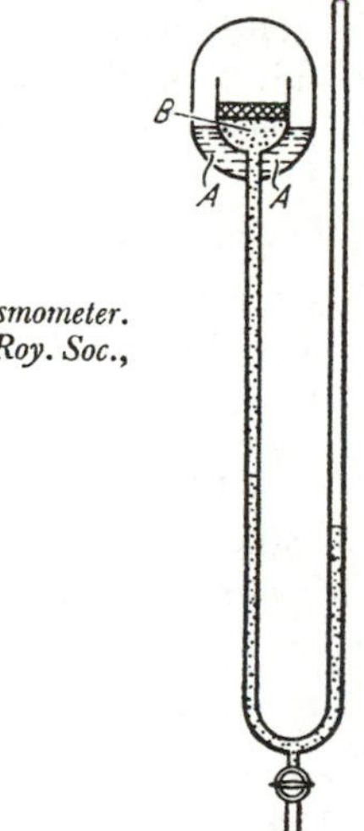

Figure 8.9. The porous-disc osmometer. From WILLIAMSON, A. T. *Proc. Roy. Soc.*, 195 A (1948) 98

the pure solvent. This is achieved by having a column of the solvent *under tension*, the intermolecular cohesive forces preventing the column from breaking. The principle of the method is indicated in *Figure 8.9.* The solution in the vessel *A* is equilibrated via the vapour phase with the pure solvent in the inner vessel *B*, which is held by capillary forces in the porous glass diaphragm against the tension due to the hanging column of solvent.

Owing to the enormous magnitude of osmotic pressures as compared with other colligative properties, and to the practical difficulty of establishing a column of liquid under a tension corresponding to more than a few decimetres in the height of the solvent column, the method is confined to solutions of very low molar concentration, and was in fact developed for the study of high polymers. It is, however, a method of great potential value for extremely dilute electrolyte solutions. A major experimental difficulty lies in the necssity for extreme uniformity of temperature in the equilibration vessel, which can be illustrated by the following figures: a one-thousandth molar solution of an ideal non-electrolyte solute in water at 25° would have an osmotic pressure equivalent to approximately 25 cm height of water. The vapour pressure lowering of such a solution would be approximately 0·0004 mm Hg; since the vapour pressure of water changes by approximately 1 mm/degree at 25°, a temperature difference of 0·0004° between the solvent and solution would wipe out the free energy difference which results in the osmotic pressure mentioned. In order to obtain quantitatively useful data for solutions of this concentration, the temperature must be uniform within 5×10^{-6} of a degree. Williamson has described the elaborate precautions necessary to ensure such constancy.

SOLUBILITY MEASUREMENTS

The condition for saturation of a solution is that the chemical potential of the solute is the same in the solid state and in the saturated solution:

$$\bar{G}_{\text{solid}} = \bar{G}^0_B + \nu \boldsymbol{R}T \ln (Qm\gamma_{\pm})$$

If there is another electrolyte present the solubility of the first electrolyte may be different but will still be determined by the condition:

$$\bar{G}_{\text{solid}} = \bar{G}^0_B + \nu \boldsymbol{R}T \ln (Qm'\gamma'_{\pm})$$

where Q is the factor tabulated in Appendix 2.1. Thus the ratio of the solubilities in the absence and in the presence of another electrolyte measures the influence of the added electrolyte on the activity coefficient of the first:

$$\frac{m}{m'} = \frac{\gamma'_{\pm}}{\gamma_{\pm}}$$

The method is a powerful one for studying the variation of the activity coefficient of a sparingly soluble salt in a mixed electrolyte solution; the accuracy of the method depends mainly on the analytical accuracy with which the solubility can be determined; hence the coordinated ammines of cobalt compounds have proved favourite electrolytes for such measurements because of the ease and accuracy with which the ammonia content can be measured. *Table 8.4* gives some results for the solubility of oxalotetramminecobaltic diamminodinitrooxalocobaltiate[74]

$$[Co(NH_3)_4C_2O_4]^+ [Co(NH_3)_2(NO_2)_2C_2O_4]^-$$

in sodium chloride solutions at 15°. The solubilities being expressed as molarities, it is convenient to express the activity coefficients on the molar scale. What results is a set of activity coefficients relative

Table 8.4

Activity Coefficient of $[Co(NH_3)_4C_2O_4]^+[Co(NH_3)_2(NO_2)_2C_2O_4]^-$ *in Sodium Chloride Solution at* 15°

Molarity NaCl	*Solubility* (m mole/l)	$\log y/y'$	$\log y/y' + 0{\cdot}0115$	y
0	0·4900	0	0·0115	0·974
0·0003	0·4935	0·0031	0·0146	0·967
0·001	0·5000	0·0087	0·0202	0·954
0·005	0·5220	0·0275	0·0390	0·914
0·01	0·5396	0·0419	0·0534	0·885
0·02	0·5646	0·0615	0·0730	0·845

to the value at a concentration corresponding to the solubility in the absence of sodium chloride, *i.e.*, at $4{\cdot}9 \times 10^{-4}$ mole/l in this case. By plotting $\log y/y'$ against the square root of the total ionic

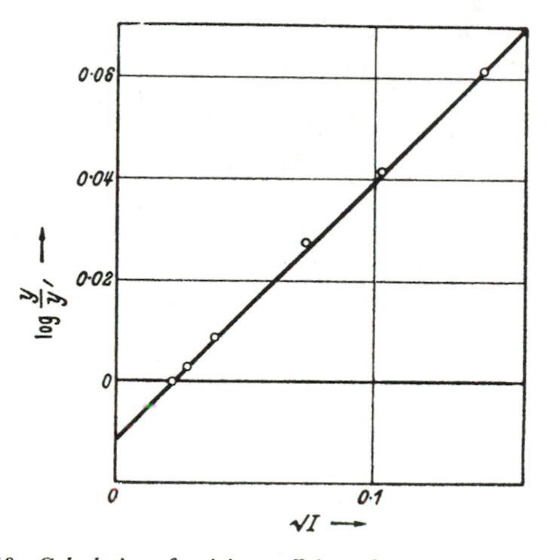

Figure 8.10. Calculation of activity coefficients from solubility measurements

strength (*Figure 8.10*) a straight line can be drawn extrapolating to $-$ 0·115 at $I = 0$. This is added to each value of $\log y/y'$ to give a set of activity coefficients relative to unity at infinite dilution.

MEASUREMENTS OF SOLUTE VAPOUR PRESSURE

Just as the solvent vapour pressure of a solution determined relative to its value for the pure solvent measures the solvent activity, so the vapour pressure of a solute measures the activity of the solute. Very few electrolytes have vapour pressures large enough to make this method feasible—the halide acids are well-known examples[75]. Even with these it is only in comparatively concentrated solutions that the solute vapour pressure is appreciable enough to be measured and hence the results have to be expressed relative to an arbitrarily assigned activity coefficient at one concentration, unless the value at this concentration can be obtained by some other method.

DETERMINATION OF ACTIVITY COEFFICIENTS BY THE 'SOLVENT-EXTRACTION' PROCESS

Although this method has not been much used, it holds promise for some special studies and can be described by reference to the work of GLUECKAUF, MCKAY and MATHIESON[76]. Through a series of six tubes, filled with aqueous solutions of uranyl nitrate and sodium nitrate in different proportions, a dibutyl-ether solution of uranyl nitrate was forced under slight pressure from a container. The ethereal solution entered at the bottom of the first tube, percolated through the first tube and passed through a side tube into the bottom of the second tube whence it percolated through the second solution and so on through all six solutions. Provided that sodium nitrate is insoluble in the ether and that water and ether are practically immiscible *even in the presence of uranyl nitrate*, the passage of the ethereal solution through the aqueous solutions will result in the addition to or removal from the aqueous solution of uranyl nitrate according as the chemical potential of this electrolyte is less than or greater than that of uranyl nitrate in the ether solution. No transfer of sodium nitrate or of water from one tube to another can occur if the solubility conditions already mentioned hold. If sufficient ethereal solution is percolated and equilibrium is reached, each of the six aqueous solutions is in a state where its uranyl nitrate activity is equal to the activity of this salt in the ether, that is to say, uranyl nitrate is present at the same activity in each of the six aqueous solutions; if m_B is the molality of uranyl nitrate and m_C that of sodium nitrate in any one tube, the activity of the uranyl nitrate is $m_B(2m_B + m_C)^2\gamma_{\pm}^3$ and this must have the same value in each aqueous solution. Just as in the isopiestic method it is the water activity which becomes equal in all the solutions because it is water which is the transportable component, so in this method it is the uranyl nitrate activity which becomes equal in all solutions because this is the component which can be moved by means of the ether solution.

ACTIVITY COEFFICIENTS BY SEDIMENTATION IN AN ULTRACENTRIFUGE

In discussing the measurement of transport numbers we noted that on subjecting a solution of uniform composition to a centrifugal field, an electromotive force is found between two electrodes at different points in the field. No concentration gradient is set up in these experiments (unless the centrifugal field acts for longer times than are usual in such transport number measurements). The system is, however, not one in equilibrium and if sufficient time is allowed, or better, if an ultracentrifuge is used, a concentration gradient is set up and the electromotive force falls to zero, *i.e.*, the centrifugal field is now compensated by a concentration gradient and not by an electrical potential gradient. The heavier particles are preferentially removed to the outer parts of the centrifuge tube but, in the case of an electrolyte solution, the oppositely charged ions cannot move independently, governed only by their individual masses, but must proceed as partners because no appreciable charge separation is permitted.

If we replace the EF term in equation (5.4) by $-\Delta\bar{G}$ we get:

$$\nu RT \ln \frac{\gamma' m'}{\gamma m} = 2\pi^2\omega^2(r_2^2 - r_1^2)(W_B - \rho\bar{V}_B)$$

where γ', m' refer to the point at a distance r_2 and γ, m to a point r_1.

The ultracentrifuge, however, introduces very high pressures in the tube and it is no longer permissible to take $\rho\bar{V}_B$ as independent of the position in the centrifugal field. Instead, we write:

$$\ln \gamma' = \ln \gamma + \ln \frac{m'}{m} + \frac{2\pi^2\omega^2}{\nu RT}(r_2^2 - r_1^2) W_B - \frac{4\pi^2\omega^2}{\nu RT}\int_{r_1}^{r_2} \rho\bar{V}_B r \, dr$$

If the point r_1 corresponds to the open end of the tube, this equation gives the activity coefficient γ' at atmospheric pressure and at a molality m' relative to γ at m; ρ and $\bar{V}_B$ are functions of r but $\bar{V}_B$ must be taken as the partial molal volume at the selected value of m'.

Whilst the general theory has been known for several years and some experimental work has been done[77], the method has recently been advanced in a way which suggests that it is going to be of widespread use. JOHNSON, KRAUS and YOUNG[78] used an ultracentrifuge at about 30,000 r.p.m. and measured the concentration gradient by following the change in the refractive index. The time required to attain equilibrium varied from three to ten days. For cadmium iodide they obtained results over the concentration range 0·2 to 0·8 M in remarkably good agreement with those already known from electromotive force measurements whilst, although the data for uranyl fluoride did not coincide with earlier freezing point measurements, the difference was reasonable in view of the 30° difference in the temperatures at which the two sets of measurements were made.

THE EFFECT OF TEMPERATURE ON THE ACTIVITY COEFFICIENT

Since

$$\frac{\partial \ln \gamma}{\partial T} = -\frac{\bar{L}_B}{\nu R T^2} \quad \ldots\ldots (2.30)$$

and $\bar{L}_B$ can be expressed as a function of temperature within the accuracy of experimental work by:

$$\bar{L}_B = \bar{L}_{B(T_s)} + \bar{J}_B(T - T_s)$$

where T_s is a selected reference temperature, then the activity coefficient should be represented as a function of temperature by an expression of the form:

$$\log \gamma = -\frac{A_1'}{T} + A_2' - A_3' \log T$$

where A_1', A_2' and A_3' are parameters characteristic of the electrolyte and its molality. There are few electrolytes for which measurements

Table 8.5

Activity Coefficient of Sodium Chloride at 1 M *calculated from:* $\log \gamma = 11{\cdot}4326 - 535{\cdot}45/T - 3{\cdot}9679 \log T$

Temperature	$\gamma_{obs.}$	$\gamma_{calc.}$
Freezing Point	0·639	0·634
0°	0·638	0·638
15°	0·654	0·653
25°	0·658	0·658
40°	0·655	0·660
60°	0·655	0·654
70°	0·648	0·648
80°	0·641	0·640
90°	0·632	0·631
100°	0·622	0·621

have been made over a sufficient temperature range to test this equation thoroughly; sodium chloride is one such electrolyte[79] and *Table 8.5* illustrates the concordance between the observed activity coefficients at 1 M and those calculated by this equation, putting $A_2' = 11{\cdot}4326$, $A_1' = 535{\cdot}45$ and $A_3' = 3{\cdot}9679$.

COMPARISON OF ACTIVITY COEFFICIENTS

The isopiestic vapour pressure method has the drawback of being only a comparative method; it measures the vapour pressure of a solution relative to that of another solution or the vapour pressure may be expressed as an osmotic coefficient, still, however, based on a set of values for some selected standard or reference electrolyte and, of course, any activity coefficients calculated from the data are still relative. On the other hand, a comparative method has some advantage in that it enables two sets of osmotic coefficients to be compared directly and it can examine consistencies between various sets of values. Experience has shown that four electrolytes are useful as reference solutes in the isopiestic method: potassium chloride, sodium chloride, sulphuric acid and calcium chloride; in addition sucrose is useful for work with non-electrolytes. Potassium chloride is obtainable in good quality, is easily recrystallized and is not appreciably hygroscopic; it is, however, saturated at about 4·8 M at 25° and hence its use as a standard is limited to solutions of a water activity of 0·85 or more. Sodium chloride is a somewhat more hygroscopic salt but, having a solubility in the region of 6 M, it can be used for water activities down to 0·76. The vapour pressures of the aqueous solutions of these salts are known with considerable accuracy. For solutions with a water activity below 0·76, the position is not so happy; sulphuric acid is one reference electrolyte with water activities as low as 0·07 at 20 M which can be made from pure material and analysed accurately by weight titration. Unfortunately, because of intermediate ion (HSO_4^-) formation, its solutions show complex behaviour and the isopiestic ratio of sulphuric acid solutions with respect to other electrolyte solutions is seldom one which can be plotted with ease. If a reference electrolyte can be found such that the isopiestic ratio can be plotted against the concentration to give a curve of simple form, measurements at an excessively large number of concentrations can be avoided. Calcium chloride can often be used to advantage in this way when other 2 : 1 salts are being measured. Although its solubility is 7·4 M at 25° it easily supersaturates and can be used to equilibrate with solutions down to a water activity of 0·18. It is advisable to prepare the stock calcium chloride solution from good grade calcium carbonate and hydrochloric acid and then, as a precaution, check its isopiestic ratio against a sodium chloride solution.

THE OSMOTIC AND ACTIVITY COEFFICIENTS OF SODIUM AND POTASSIUM CHLORIDE

To derive mean values for these coefficients we shall use the results of three different techniques, direct vapour pressure measurements, freezing point determinations and experiments on the potentials of concentration cells. The first of these is capable of precision but only at high concentrations, and it is doubtful if any results at concentrations below 1 M can stand comparison with those obtained by indirect methods. On the other hand, precise freezing point measurements have seldom explored the region of concentration above 1 M, whilst electromotive force measurements have to be treated with some caution in concentrated solutions because of troubles such as electrode solubility.

We may first consider the activity coefficient of sodium chloride at 0·1 M because it is a good illustration of the agreement that can be reached by different workers. BROWN and MACINNES[66] using a cell with transport and availing themselves of the transport number measurements of LONGSWORTH[80] found $-\log \gamma_{NaCl} = 0{\cdot}1088$. ALLGOOD and GORDON[81] made an independent determination of the transport number, although by essentially the same method, whilst JANZ and GORDON[68] repeated the cell measurements; the combination of these results gave exactly the same value, $-\log \gamma_{NaCl} = 0{\cdot}1088$. HARNED and COOK[82] studied cells without transport containing amalgam electrodes and by fitting their results to an extended Debye-Hückel equation a figure of 0·1085 ensued. A similar investigation[28, 67, 69] of potassium chloride at 0·1 M has given three values of $-\log \gamma_{KCl}$: 0·1134, 0·1137 and 0·1141. The isopiestic ratio of potassium chloride to sodium chloride is known even below 0·1 M and can be extrapolated back to zero concentration with some confidence, enabling us to calculate that at 0·1 M, $\log (\gamma_{NaCl}/\gamma_{KCl})$ should be 0·0048: the activity coefficients of potassium chloride can now be translated into values for sodium chloride to give $-\log \gamma_{NaCl} = 0{\cdot}1086$, 0·1089, 0·1093. We have used the results of four different laboratories and four different techniques to give six determinations of the activity coefficient of sodium chloride in 0·1 M solution at 25°: the average value of $-\log \gamma_{NaCl}$ is 0·1088 and the maximum deviation is only 0·0005.

At higher concentrations, up to 1 M, we rely more on amalgam cells; those containing sodium chloride give the activity coefficient directly; others containing potassium chloride[28], sodium bromide[29] or potassium bromide[24], require a knowledge of the isopiestic ratios between sodium chloride and these salts. Such ratios have been measured. In addition, we have accurate freezing point measurements on sodium chloride solutions and sufficient heat content and capacity data to calculate the temperature correction. As a result of this work, five separate determinations on sodium chloride are available and the agreement between them is satisfactory: the average deviation from the mean values is 0·0014 in γ.

Above 1 M it is better to assess the data in terms of the osmotic coefficient. Both Negus[83], using the technique of Lovelace, Frazer and Sease[84], and Olynyk and Gordon[85] have made direct vapour pressure measurements up to high concentrations: furthermore, Gibson and Adams[1] have measured the vapour pressure of the saturated solution at 20·28° and the correction to 25° is small. Similar determinations using potassium chloride[84], barium chloride[3] and sulphuric acid solutions[2, 86] have been

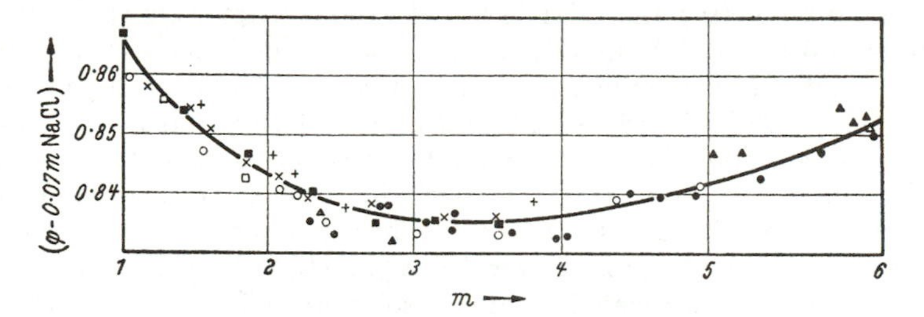

Figure 8.11. Deviation function for the osmotic coefficient of sodium chloride at 25°. *From* Robinson, R. A., *Proc. Roy. Soc., N.Z.*, 75 (1945) 203

○ Negus—sodium chloride
● Olynyk and Gordon—sodium chloride
× Lovelace, Frazer and Sease—potassium chloride
■ Harned and Cook—potassium chloride
□ Bechtold and Newton—barium chloride
△ Gibson and Adams—sodium chloride
+ Grollman and Frazer—sulphuric acid
▲ Shankman and Gordon—sulphuric acid

made and the isopiestic ratio of each with respect to sodium chloride has been measured carefully[5, 87] so that the vapour pressure measurements on these electrolytes can be used to give three sets of data for the vapour pressure (or osmotic coefficient) of sodium chloride. Actually two series of results are available for sulphuric acid, originating in different schools of chemistry, whilst the potassium chloride measurements were made at 20° and needed a special determination of the isopiestic ratio of sodium to potassium chloride at this temperature and a small correction of the calculated vapour pressures of sodium chloride solutions over a 5° interval. The amalgam cell work of Harned and Cook on potassium chloride gives the activity coefficient but from the potentials the solvent activity can be calculated by the method outlined on pp. 196–197.

Combining this with isopiestic data a new vapour pressure curve for sodium chloride solutions results. The collection of all these calculations is presented in *Figure 8.11* as a graph of (ϕ − 0·07m) against molality, from which the osmotic coefficient at round concentrations can be read. The activity coefficient follows as a matter of computation and further, since the isopiestic ratio of potassium chloride and sodium chloride has been the subject of much research, the osmotic and activity coefficients of potassium chloride are obtained, again as the result of fairly simple computation. In Appendix 8.3, we present a set of values of the water activity, osmotic coefficient, activity coefficient and relative molal vapour pressure lowering for these two salts which, we think, represent the best values. The data are supported by some recent measurements[88] by the dynamic method at 30° for 0·7 − 4 M potassium chloride and 3·7 − 5 M sodium chloride which, after a small correction is made to convert the data to 25°, show an average deviation from the values recorded in Appendix 8.3 of only 0·0008 in ϕ. The water activity of potassium chloride has also been measured[89] by a method in which the pure solvent vapour is isolated from the solution at 25° by a sensitive bellows pressure gauge and the temperature of the solvent lowered until the vapour pressures of solvent and solution are equal. The results for more concentrated solutions are particularly important; they agree with those in Appendix 8.3 within 0·0010 in ϕ.

THE WATER ACTIVITY OF SULPHURIC ACID SOLUTIONS

As we have already mentioned, sulphuric acid could be a most useful reference electrolyte for the isopiestic method because of its purity, ease of analysis and the wide range of water activity covered by its solutions with, however, the expensive disadvantage of needing platinum dishes. Unfortunately, the question of the vapour pressure of its solutions has not yet been finally settled. HARNED and HAMER[11] contrived two cells each of which gives the activity coefficient of the acid and each can be used to give the water activity by using some form of the Gibbs-Duhem equation. One cell, containing hydrogen and lead dioxide—lead sulphate electrodes, could be used to 7 M whilst the other cell, with hydrogen and mercury–mercurous sulphate electrodes, gave good results up to 17 M. Over the concentration range common to both there was excellent agreement between the water activities calculated from each cell; for example, at 7 M the figures of a_W = 0·5453 and 0·5458 show the widest difference in the two series and at other concentrations the agreement is even better. The direct vapour pressure measurements of SHANKMAN and GORDON[2] give values somewhat different; thus, for example, a_W was found to be 0·5497 at 7 M. The vapour pressure results agree with the electromotive force method at 2 M and 3 M, they are higher at concentrations up to about 8 M, but at higher concentrations still it is the vapour pressure figures which are the lower. Expressed in terms of a_W the differences may appear large but the differences should not be over-emphasized; put in terms of a cell potential, and this is what Harned and Hamer measured, a difference of one or two millivolts accounts for most of the discrepancy. Since there seemed little use in merely repeating what was clearly very careful work, STOKES[9] devised the method with the lengthy title of 'bithermal equilibration through the vapour phase' which has already been discussed. As sulphuric acid was too corrosive to be used in his apparatus, he measured the vapour pressure of sodium hydroxide solutions between 5 M and 14 M and also made a few measurements on sodium chloride and on calcium chloride solutions. The last were valuable in showing that the method was working properly, his value of $a_W = 0{\cdot}7464$ for 3·033 M calcium chloride comparing well with $a_W = 0{\cdot}7458$ by BECHTOLD and NEWTON[3] who used the dynamic method. Stokes' values for the vapour pressure of sodium hydroxide solutions must, therefore, be treated with considerable confidence and, the isopiestic ratios of this base to sulphuric acid having been measured with care, we arrive at a new determination for sulphuric acid. The concentration range 5–14 M for sodium hydroxide is equivalent, in the isopiestic sense, to the range 4–11·5 M for sulphuric acid and it is over this range that comparison with the work of Shankman and Gordon can be made. Whilst Stokes' results seem to be about 0·0008 in a_W above those of Shankman and Gordon, this difference is only about twice the reproducibility of either set of measurements and gives strong support to their data. Stokes concluded that the 'best' values for sulphuric acid were probably to be calculated from the sodium chloride–sulphuric acid isopiestic ratios up to 3 M acid; between 3 and 11·5 M the choice between his own results and those of Shankman and Gordon was difficult (though the difference was not significant in view of the likely experimental error of either method) but he preferred his own data because they gave a somewhat smoother vapour pressure curve. Above 11·5 M, of course, we rely entirely on the work of Shankman and Gordon with, however, the confidence inspired by the good agreement in the range where comparison is possible. Further measurements have been made[90] in solutions 24 M and over by a method similar to that of STOKES[9] except that the pure solvent at the lower temperature was replaced by a solution of sulphuric acid more dilute than the one at 25° with which it was equilibrated. The data for sulphuric acid are collected in Appendix 8.4; they have been substantiated recently[91] by direct vapour pressure measurements at 13·88, 18·51 and 27·74 M giving $a_w = 0{\cdot}2016$ (0·2016), 0·0993 (0·0996) and 0·0260 (0·0258) respectively, the figures in parentheses being interpolated from Appendix 8.4.

THE OSMOTIC AND ACTIVITY COEFFICIENTS OF CALCIUM CHLORIDE

These depend on isopiestic measurements against sodium chloride and sulphuric acid solutions[92]. They are anchored at 0·1 M by the osmotic and activity coefficients of McLeod and Gordon[70] and one check at 3·033 M referred to above. Appendix 8.5 contains data for this salt.

THE OSMOTIC AND ACTIVITY COEFFICIENTS OF SUCROSE

The isopiestic ratio of this solute to either sodium chloride or potassium chloride has been measured a number of times[5, 93] so that its osmotic and activity coefficients can be calculated with confidence (see Appendix 8.6).

GENERAL CONSIDERATION OF THE ACTIVITY COEFFICIENTS OF ELECTROLYTES

Appendix 8.10 contains extensive data for the osmotic and activity coefficients of electrolytes at 25° from 0·1 M upwards. *Figure 8.12* illustrates the variation with concentration of the activity coefficients of a few electrolytes.

We may now make a few remarks about the behaviour of the activity coefficients with changing concentration.

1. In dilute solutions, the activity coefficient decreases with increasing concentration; for many but not all electrolytes, the curve of the activity coefficient plotted against concentration shows a minimum and at high concentrations the activity coefficient may reach a very high value. An extreme example is found in uranyl perchlorate with $\gamma = 1457$ at 5·5 M. An extreme example in the opposite sense is found in cadmium iodide with $\gamma = 0{\cdot}0168$ at 2·5 M. In general we can recognize three kinds of behaviour; activity coefficients rising to very high values, which will be interpreted in the next chapter as evidence for extensive hydration of the ions; moderately low activity coefficients which are explained by Bjerrum ion-pair formation; very low activity coefficients resulting from complex ion formation.

2. Electrolytes with polyvalent cations usually have much higher activity coefficients than electrolytes of analogous valency type containing a polyvalent anion. Lanthanum chloride and potassium ferricyanide are a contrast in this respect. The explanation is thought to be found in the extensive hydration of the cations and the absence of hydration in large polyvalent anions.

3. The order of the activity coefficient curves is Li > Na > K > Rb > Cs for the chlorides, bromides, iodides, nitrates, chlorates

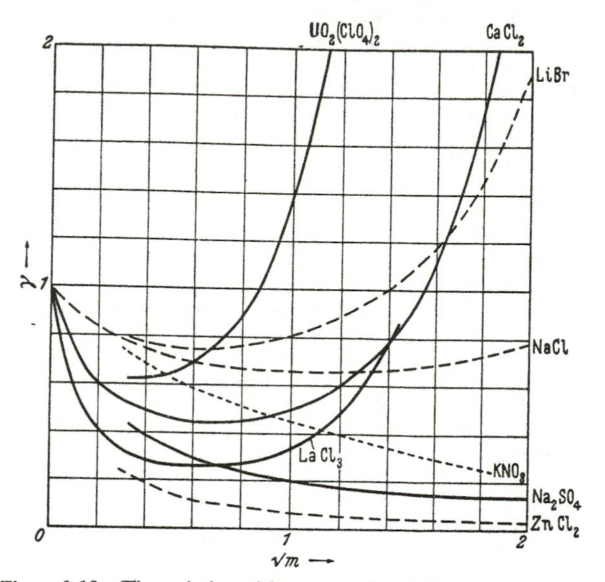

Figure 8.12. The variation with concentration of the activity coefficients of some electrolytes at 25°

and perchlorates. The order is reversed with the hydroxides, formates and acetates.

4. The order of curves is I > Br > Cl for lithium, sodium and potassium halides but is reversed for rubidium and caesium halides.

5. The potassium salts of the oxy-acids, like the nitrate, chlorate and perchlorate, have low activity coefficients and probably form ion-pairs. By contrast the perchlorates of bivalent metals have very high activity coefficients.

The last appendix of this chapter (8.11) contains values of the sulphuric acid, calcium chloride or sodium hydroxide concentrations (expressed as molalities and weight percentages) of solutions with round values of the water activity[94]. The appendix also gives the water activity of a number of saturated solutions which are useful in setting up controlled humidity chambers.

REFERENCES

1 Gibson, R. E. and Adams, L. H., *J. Amer. chem. Soc.*, 55 (1933) 2679
2 Shankman, S. and Gordon, A. R., *ibid.*, 61 (1939) 2370
3 Bechtold, M. F. and Newton, R. F., *ibid.*, 62 (1940) 1390
4 Bousfield, W. R., *Trans. Faraday Soc.*, 13 (1918) 401
5 Sinclair, D. A., *J. phys. Chem.*, 37 (1933) 495; Robinson, R. A. and Sinclair, D. A., *J. Amer. chem. Soc.*, 56 (1934) 1830; Scatchard, G., Hamer, W. J. and Wood, S. E., *ibid.*, 60 (1938) 3061
6 Stokes, R. H. and Robinson, R. A., *Trans. Faraday Soc.*, 37 (1941) 419
6a Morton, J. E., Campbell, A. D. and Ma, T. S., *Analyst*, 78 (1953) 722
7 Gordon, A. R., *J. Amer. chem. Soc.*, 65 (1943) 221
8 Randall, M. and White, A. M., *ibid.*, 48 (1926) 2514
8a Guggenheim, E. A. and Stokes, R. H., *Trans. Faraday Soc.*, 54 (1958) 1646
9 Stokes, R. H., *J. Amer. chem. Soc.*, 69 (1947) 1291
10 Stokes, R. H., *N. Z. J. Sci. Tech.*, 27 (1945) 75
11 Harned, H. S. and Hamer, W. J., *J. Amer. chem. Soc.*, 57 (1935) 27
11a Dorsey, N. E., 'Properties of ordinary water-substance', p. 562, Reinhold Publishing Corp., New York (1940); Giauque, W. F. and Stout, J. W., *J. Amer. chem. Soc.*, 58 (1936) 1144; Osborne, N. S., Stimson, H. F. and Ginnings, D. C., *J. Res. nat. Bur. Stand.*, 23 (1939) 197
11b Lewis, G. N. and Randall, M., 'Thermodynamics', McGraw-Hill Book Co. Inc., New York (1923)
12 Scatchard, G. and Prentiss, S. S., *J. Amer. chem. Soc.*, 55 (1933) 4355
13 Guggenheim, E. A. and Turgeon, J. C., *Trans. Faraday Soc.*, 51 (1955) 747
14 Harned, H. S. and Owen, B. B., 'The Physical Chemistry of Electrolytic Solutions,' p. 116, Reinhold Publishing Corp., New York (1950)
15 Smith, R. P., *J. Amer. chem. Soc.*, 61 (1939) 497
16 Hills, G. J. and Ives, D. J. G., *J. chem. Soc.*, (1951) 318
17 Hitchcock, D. I., *J. Amer. chem. Soc.*, 50 (1928) 2076
18 Harned, H. S. and Ehlers, R. W., *ibid.*, 54 (1932) 1350; 55 (1933) 2179; Bates, R. G. and Bower, V. E., *J. Res. nat. Bur. Stand.*, 53 (1954) 283
19 Åkerlöf, G. and Teare, J. W., *J. Amer. chem. Soc.*, 59 (1937) 1855
20 Keston, A. S., *ibid.*, 57 (1935) 1671
21 Harned, H. S., Keston, A. S. and Donelson, J. G., *ibid.*, 58 (1936) 989
22 Owen, B. B. and Foering, L., *ibid.*, 58 (1936) 1575
23 — *ibid.*, 57 (1935) 1526
24 Harned, H. S., *ibid.*, 51 (1929) 416
25 — and Douglas, S. M., *ibid.*, 48 (1926) 3095
26 — and Schupp, O. E., *ibid.*, 52 (1930) 3886
27 — and Nims, L. F., *ibid.*, 54 (1932) 423
28 — and Cook, M. A., *ibid.*, 59 (1937) 1290
29 — and Crawford, C. C., *ibid.*, 59 (1937) 1903
30 — and Swindells, F. E., *ibid.*, 48 (1926) 126

[31] HARNED, H. S., *ibid.*, 47 (1925) 676; — and HECKER, J. C., *ibid.*, 55 (1933) 4838; ÅKERLÖF, G. and KEGELES, G., *ibid.*, 62 (1940) 620
[32] — and COOK, M. A., *ibid.*, 59 (1937) 496
[33] ÅKERLÖF, G. and BENDER, P., *ibid.*, 70 (1948) 2366; see also STOKES, R. H., *ibid.*, 67 (1945) 1686
[34] HATTOX, E. M. and DE VRIES, T., *ibid.*, 58 (1936) 2126
[35] BRAY, U. B., *ibid.*, 49 (1927) 2372
[36] LAMER, V. K. and PARKS, W. G., *ibid.*, 53 (1931) 2040
[37] SCATCHARD, G. and TEFFT, R. F., *ibid.*, 52 (1930) 2272; ROBINSON, R. A and STOKES, R. H., *Trans. Faraday Soc.*, 36 (1940) 740; PARTON, H. N. and MITCHELL, J. W., *ibid.*, 35 (1939) 758; STOKES, R. H. and STOKES, J. M., *ibid.*, 41 (1945) 688; BATES, R. G. and VOSBURGH, W. C., *J. Amer. chem. Soc.*, 59 (1937) 1583; BATES, R. G., *ibid.*, 60 (1938) 2983; 61 (1939) 308; HARNED, H. S. and FITZGERALD, M. E., *ibid.*, 58 (1936) 2624
[38] TIPPETTS, E. A. and NEWTON, R. F., *ibid.*, 56 (1934) 1675
[39] HARNED, H. S. and MASON, C. M., *ibid.*, 54 (1932) 1439
[40] LUCASSE, W. W., *ibid.*, 47 (1925) 743
[41] HARNED, H. S. and HECKER, J. C., *ibid.*, 56 (1934) 650
[42] ÅKERLÖF, G., *ibid.*, 48 (1926) 1160
[43] HAMER, W. J., *ibid.*, 57 (1935) 9
[44] STOKES, R. H., *ibid.*, 67 (1945) 1686
[45] HARNED, H. S., MORRISON, J. O., WALKER, F., DONELSON, J. G. and CALMON, C., *J. Amer. chem. Soc.*, 61 (1939) 49. This paper is a summary and gives full references to earlier work.
[46] NONHEBEL, G. and HARTLEY, H., *Phil. Mag.*, 50 (1925) 729; KOSKIKALLIO, J., *Suomen Kem.*, 30B (1957) 38, 43, 111
[47] AUSTIN, J. M., HUNT, A. H., JOHNSON, F. A. and PARTON, H. N., private communication; OIWA, I. T., *J. phys. Chem.*, 60 (1956) 754
[48] WOOLCOCK, J. W. and HARTLEY, H., *Phil. Mag.*, 5 (1928) 1133; DANNER, P. S., *J. Amer. chem. Soc.*, 44 (1922) 2832; TANIGUCHI, H. and JANZ, G. J., *J. phys. Chem.*, 61 (1957) 688
[49] HARNED, H. S. and FLEYSHER, M. H., *ibid.*, 47 (1925) 82
[50] LUCASSE, W. W., *Z. phys. Chem.*, 121 (1926) 254
[51] MUKHERJEE, L. M., *J. Amer. chem. Soc.*, 79 (1957) 4040
[52] HARNED, H. S. and THOMAS, H. C., *ibid.*, 57 (1935) 1666; 58 (1936) 761
[53] HARNED, H. S. and CALMON, C., *ibid.*, 61 (1939) 1491; PATTERSON, A. and FELSING, W. A., *ibid.*, 64 (1942) 1478
[53a] HARNED, H. S. and ALLEN, D. S., *J. phys. Chem.*, 58 (1954) 191
[54] CLAUSSEN, B. H. and FRENCH, C. M., *Trans. Faraday Soc.*, 51 (1955) 708
[54a] MOORE, R. L. and FELSING, W. A., *J. Amer. chem. Soc.*, 69 (1947) 1076
[55] FEAKINS, D. and FRENCH, C. M., *J. chem. Soc.*, (1956) 3168
[55a] HARNED, H. S. and NESTLER, F. H. M., *J. Amer. chem. Soc.*, 68 (1946) 665; KNIGHT, S. B., CROCKFORD, H. D. and JAMES, F. W., *J. phys. Chem.*, 57 (1953) 463
[56] KNIGHT, S. B., MASI, J. F. and ROESEL, D., *J. Amer. chem. Soc.*, 68 (1946) 661
[56a] CLAUSSEN, B. H. and FRENCH, C. M., *Trans. Faraday Soc.*, 51 (1955) 1124
[57] WILLIAMS, J. P., KNIGHT, S. B. and CROCKFORD, H. D., *J. Amer. chem. Soc.*, 72 (1950) 1277
[57a] CROCKFORD, H. D. and SAKHNOVSKY, A. A., *ibid.*, 73 (1951) 4177
[58] SCATCHARD, G., *ibid.*, 48 (1926) 2026

[59] COVINGTON, A. K. and PRUE, J. E., *J. chem. Soc.*, (1955) 3696, 3701; (1957) 1567, 1930
[60] BACARELLA, A. L., GRUNWALD, E., MARSHALL, H. P. and PURLEE, E. L., *J. org. Chem.*, 20 (1955) 747; *J. phys. Chem.*, 62 (1958) 856
[61] CARMODY, W. R., *J. Amer. chem. Soc.*, 51 (1929) 2901; 54 (1932) 188. An extensive account of the silver–silver chloride electrode is given by JANZ, G. J. and TANIGUCHI, H., *Chem. Rev.*, 53 (1953) 397
[62] BROWN, A. S., *J. Amer. chem. Soc.*, 56 (1934) 646
[63] GÜNTELBERG, E., *Z. phys. Chem.*, 123 (1926) 199
[64] HARNED, H. S. and MORRISON, J. O., *Amer. J. Sci.*, 33 (1937) 161
[65] HILLS, G. J. and IVES, D. J. G., *J. chem. Soc.*, (1951) 311
[66] BROWN, A. S. and MACINNES, D. A., *J. Amer. chem. Soc.*, 57 (1935) 1356
[67] HORNIBROOK, W. J., JANZ, G. J. and GORDON, A. R., *ibid.*, 64 (1942) 513
[68] JANZ, G. J. and GORDON, A. R., *ibid.*, 65 (1943) 218
[69] SHEDLOVSKY, T. and MACINNES, D. A., *ibid.*, 58 (1936) 1970; 59 (1937) 503; 61 (1939) 200; SHEDLOVSKY, T., *ibid.*, 72 (1950) 3680
[70] MCLEOD, H. G. and GORDON, A. R., *ibid.*, 68 (1946) 58
[71] SPEDDING, F. H., PORTER, P. E. and WRIGHT, J. M., *ibid.*, 74 (1952) 2781; SPEDDING, F. H. and YAFFE, I. S., *ibid.*, 74 (1952) 4751
[72] EARL OF BERKELEY, HARTLEY, E. G. J. and BURTON, C. V., *Phil. Trans.* 209 (1909) 177; 218 (1919) 295
[73] WILLIAMSON, A. T., *Proc. Roy. Soc.*, 195 A (1948) 97
[74] BRONSTED, J. N. and LAMER, V. K., *J. Amer. chem. Soc.*, 46 (1924) 555
[75] BATES, S. J. and KIRSCHMAN, H. D., *ibid.*, 41 (1919) 1991
[76] GLUECKAUF, E., MCKAY, H. A. C. and MATHIESON, A. R., *J. chem. Soc.*, (1949) S 299
[77] PEDERSEN, K. O., *Z. phys. Chem.*, 170 A (1934) 41; SVEDBERG, T. and PEDERSEN, K. O., 'The Ultracentrifuge,' p. 53, Oxford University Press (1940); DRUCKER, C., *Z. phys. Chem.*, 180 A (1937) 359
[78] JOHNSON, J. S., KRAUS, K. A. and YOUNG, T. F., *J. Amer. chem. Soc.*, 76 (1954) 1436; *J. chem. Phys.*, 22 (1954) 878
[79] ROBINSON, R. A. and HARNED, H. S., *Chem. Rev.*, 28 (1941) 419
[80] LONGSWORTH, L. G., *J. Amer. chem. Soc.*, 54 (1932) 2741
[81] ALLGOOD, R. W. and GORDON, A. R., *J. chem. Phys.*, 10 (1942) 124
[82] HARNED, H. S. and COOK, M. A., *J. Amer. chem. Soc.*, 61 (1939) 495
[83] NEGUS, S. S., *Thesis*, Johns Hopkins University (1922)
[84] LOVELACE, B. F., FRAZER, J. C. W. and SEASE, V. B., *J. Amer. chem. Soc.*, 43 (1921) 102
[85] OLYNYK, P. and GORDON, A. R., *ibid.*, 65 (1943) 224
[86] GROLLMAN, A. and FRAZER, J. C. W., *ibid.*, 47 (1925) 712
[87] ROBINSON, R. A., *Proc. Roy. Soc.*, *N.Z.*, 75 (1945) 203
[88] SMITH, H. A., COMBS, R. L. and GOOGIN, J. M., *J. phys. Chem.*, 58 (1954) 997
[89] BROWN, O. L. I. and DELANEY, C. L., *ibid.*, 58 (1954) 255; ROBINSON, R. A., *ibid.*, 60 (1956) 501
[90] GLUECKAUF, E. and KITTS, G. P., *Trans. Faraday Soc.*, 52 (1956) 1074
[91] HORNUNG, E. W. and GIAUQUE, W. F., *J. Amer. chem. Soc.*, 77 (1955) 2744
[92] STOKES, R. H., *Trans. Faraday Soc.*, 41 (1945) 637
[93] ROBINSON, R. A., and STOKES, R. H., *J. phys. Chem.* 65 (1961) 1954
[94] STOKES, R. H. and ROBINSON, R. A., *Industr. Engng Chem.*, 41 (1949) 2013

9

THE THEORETICAL INTERPRETATION OF CHEMICAL POTENTIALS

THE problem of accounting for the thermodynamic properties of a solution is best regarded as that of finding a theoretical expression for the non-ideal part of the chemical potential (of either component) as a function of composition, temperature, dielectric constant and any other relevant variables. Once this is obtained, all the colligative and thermal properties of the solution are readily calculable. In practice, the activity coefficient of the solute is usually more convenient to handle than the chemical potential, and the problem accordingly becomes one of finding a theoretical expression for the activity coefficient.

The features peculiar to the activity coefficients of electrolytes are most readily grasped by comparison with those of non-electrolytes. In *Figure 9.1* the logarithm of the rational activity coefficient is plotted against the mole fraction for three simple non-electrolytes in aqueous solution. It will be seen that the activity coefficient may increase or decrease with rising concentration, but in both cases $\log f_B$ approaches zero in a linear manner, *i.e.*,

$$\frac{\partial \log f_B}{\partial N_B} \rightarrow \text{constant as } N_B \rightarrow 0$$

From the Gibbs-Duhem equation for a non-electrolyte:

$$N_A \frac{\partial \bar{G}_A}{\partial N_B} = -N_B \frac{\partial \bar{G}_B}{\partial N_B} \qquad \ldots .(9.1)$$

or, introducing the activity coefficients and remembering that $(N_A + N_B) = 1$:

$$\frac{\partial \ln f_A}{\partial N_B} \bigg/ \frac{\partial \ln f_B}{\partial N_B} = -\frac{N_B}{1 - N_B}$$

it follows that as $N_B \rightarrow 0$ either:

$$\frac{\partial \ln f_A}{\partial N_B} \rightarrow 0 \quad \text{or} \quad \frac{\partial \ln f_B}{\partial N_B} \rightarrow -\infty$$

As GUGGENHEIM[1] points out, statistical theory requires that long-range forces between the solute particles must operate if the second alternative is found to occur. But non-electrolytes are characterized by short-range forces between the solute particles, and consequently

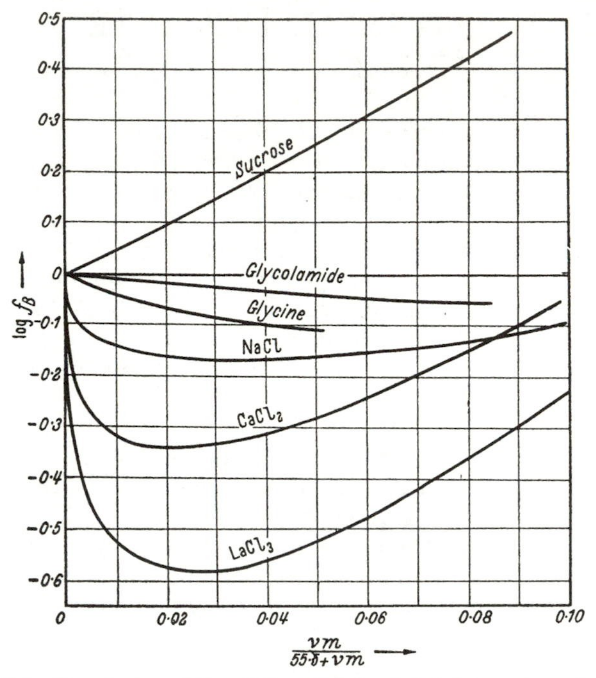

Figure 9.1. A comparison of the activity coefficients of electrolytes and non-electrolytes as a function of concentration

the first alternative applies so that, if $\ln f_A$ is expressed as a power series in N_B

$$\ln f_A = A_1 N_B^2 + A_2 N_B^3 + \ldots$$

there can be no term lower than the second power of N_B. In this case the logarithm of the activity coefficient f_B of the *solute* must, by equation (9.1), be represented by a power series commencing with the first power of N_B and consequently, in very dilute solution, a plot of $\log f_B$ against N_B will approximate closely to a straight line as is, indeed, the case for the three non-electrolytes, sucrose, glycine and glycolamide, illustrated in *Figure 9.1.* But should there be long-range forces acting, the behaviour must be different. In an electrolyte solution it would be expected that long-range electrostatic attractions and repulsions obeying the inverse square law would be found, in addition to short range Van der Waals forces, ion-dipole interactions, *etc.* But if $\ln f_B$ is represented by a series:

$$\ln f_B = aN_B^n + bN_B + cN_B^2 + \ldots$$

where n is any fraction between zero and unity, $(\partial \ln f_B/\partial N_B)$ must approach infinity as $N_B \to 0$. This is exactly what is found.

Figure 9.1 includes the curves of the activity coefficients of three electrolytes of different valency types (in this graph the abscissa is

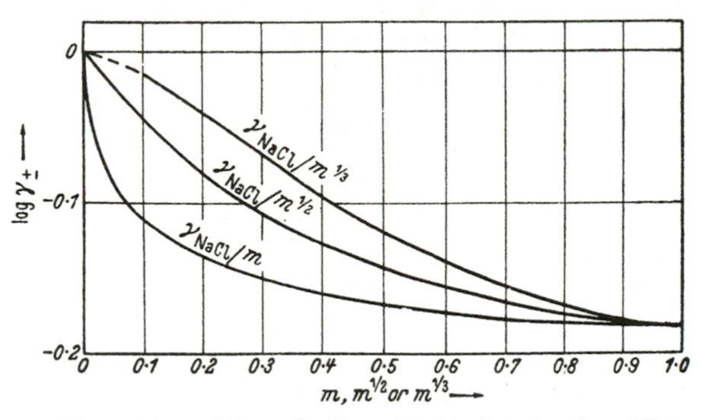

Figure 9.2. The activity coefficient of sodium chloride plotted against different powers of the molality

the 'mole fraction' of the electrolyte calculated on the basis of the total number of solute ions, *e.g.*, $N_B = \dfrac{3m}{55{\cdot}5 + 3m}$ for calcium chloride of molality m; this is a departure from our definition 2.21, but it seems the fairest basis for comparison with the non-electrolytes). The curves for the electrolytes show the infinite negative gradient as zero concentration is approached, which is a consequence of the long-range forces. At higher concentrations, the curves may flatten out and then rise more or less linearly, or may continue to fall. In this region the effects of short-range interactions become important and finally dominate the behaviour.

If the leading term of a power series expansion of $\log f_B$ involves a fractional power of the concentration, one would expect $\log f_B$ to approach linearity in that fractional power at low concentrations. In *Figure 9.2* $\log \gamma_\pm$ for sodium chloride is plotted against m, $m^{1/2}$ and $m^{1/3}$ respectively. It is evident that the slope approaches constancy in a very satisfactory manner when $m^{1/2}$ forms the abscissa, though in the experimental region the slope is also nearly constant when $m^{1/3}$ is used. It is easy to see why a linearity in the cube root might be expected: imagine the solute to form a regular ionic lattice in the solution. Its electrical potential energy can then be calculated by the methods used in the case of crystals, with the introduction of a dielectric constant (in the limit, that of the pure solvent) into the denominator of the expression for the coulomb energy of the crystal. This electrical energy will vary inversely with the distance between an ion and its nearest neighbours, and hence directly with $c^{1/3}$, or for dilute solutions with $m^{1/3}$ (where c is moles per litre and m is molality). In point of fact, if one identifies this electrical potential energy with $2\boldsymbol{R}T \ln \gamma_{\pm}$ and uses the Madelung constant of the sodium chloride lattice, one obtains the expression for water as solvent:

$$\log \gamma_{\pm} = -0{\cdot}29c^{1/3}$$

The slope of the uppermost curve in *Figure 9.2* is $-0{\cdot}26$ in the region 0·001–0·05 M. Similar calculations for calcium chloride likewise lead to a slope of $\log \gamma_{\pm}$ versus $c^{1/3}$ which is in fair agreement with that observed at moderate dilutions.

Such a lattice-model is obviously inadequate, since it takes no account of thermal disturbance of the lattice; it assumes, in fact, that the forces are sufficiently large to maintain a regular structure. If this were the case, it would be difficult to explain why the solute ions do not pull themselves together again to form a crystal. Clearly, at high enough dilutions, the interionic energy must become smaller than $\boldsymbol{k}T$, and in these circumstances the famous treatment by DEBYE and HÜCKEL[(2)] of the combined effects of Brownian motion and interionic forces becomes applicable. Furthermore, it leads, as will shortly be shown, to the result that the slope of $\log f_{\pm}$ against the *square root* of the concentration becomes constant at extreme dilution, and accounts quantitatively for the observed limiting slopes.

A complete theoretical account of the thermodynamic properties of electrolyte solutions must deal with both the long-range interionic forces and the short-range interactions between ions and solvent molecules, and this appears to be a formidable task. One can see intuitively, in a qualitative way, that the net effect of interionic attractions and repulsions will be to decrease the free energy of the solute as compared with uncharged particles and hence to decrease the activity coefficient, while the forces between ions and solvent dipoles will tend to hold the solvent in the solution, with a consequent decrease in the solvent vapour pressure from the ideal value, and a corresponding increase in the activity coefficient of the solute. The form of the curves in *Figure 9.1* suggests that these opposite effects are often of comparable magnitude at concentrations of the order of one molal. The short-range effects, however, depend approximately linearly on concentration, while the interionic effects approach linearity in the square root of concentration. There must, therefore, be a region where only the latter make any significant contribution to the non-ideality of the solution; we might tentatively expect that at one-thousandth molal, say, the long-range effects would exceed the short-range ones by a factor of the order of $\sqrt{1000}$, which would mean in practice that the short-range effects could be ignored, within the limits of experimental error, below this concentration. Theories dealing with the long-range interionic forces *only* can therefore be adequately tested by comparison with experiment for very dilute solutions. Though accurate thermodynamic data in this region are not easily obtained, there are some electrolytes for which reliable experimental studies have been made. We shall now consider the contribution which ion-ion interaction theory has to make to the thermodynamics of dilute electrolyte solutions.

THE CONTRIBUTION OF IONIC INTERACTIONS TO THE FREE ENERGY

The potential, ψ_j, at a distance r from a selected j-ion is:

$$\psi_j = \frac{z_j e}{\varepsilon} \cdot \frac{e^{\kappa a}}{1 + \kappa a} \cdot \frac{e^{-\kappa r}}{r} \quad \ldots.(4.13)$$

But an *isolated* ion of valency z_j in a medium of dielectric constant ε gives rise to a field of which the potential at distance r is given by:

$$\psi_j'' = \frac{z_j e}{\varepsilon r} \quad \ldots.(9.2)$$

By the principle of the linear superposition of fields the total potential at r, given by (4.13) may be treated as the sum of the potential, ψ_j'', due to the central ion and another potential, ψ_j', due to all the remaining ions:

$$\psi_j = \psi_j' + \psi_j''$$

Therefore by (9.2) and (4.13):

$$\psi_j' = \frac{z_j e}{\varepsilon r} \left[\frac{e^{\kappa a}}{1 + \kappa a} e^{-\kappa r} - 1 \right] \quad \ldots.(9.3)$$

This equation holds for all r down to $r = a$, *i.e.*, for the region in which equation (4.13) applies. Within the distance $r < a$, no other ions can penetrate and the potential due to the spherically symmetrical distribution of these other ions is therefore constant for all $r < a$ and equal to its value at $r = a$, which from (9.3) is given by:

$$\psi'_j = -\frac{z_j e}{\varepsilon} \frac{\kappa}{1 + \kappa a} \qquad \ldots.(9.4)$$

Thus the effect on the potential of the central ion of the resultant field of all other ions is the same as if the latter were distributed over a spherical surface at a distance $(a + 1/\kappa)$ from the centre. The net charge on this surface would of course be equal and opposite to the charge of the central ion. The quantity κ is sometimes described as the 'reciprocal thickness of the ionic atmosphere,' but it should be noted that this description is only accurate if the 'thickness' is measured from the distance $r = a$. In very dilute solutions $1/\kappa$ is large compared to a and the discrepancy is unimportant but if we consider a 1 M aqueous solution of a 1 : 1 electrolyte, $1/\kappa$ is approximately 3 Å, which is less than the normal distance of closest approach of two ions. The application of the Debye-Hückel treatment to concentrated solutions is sometimes criticized on the grounds that $1/\kappa$ becomes less than the ionic radius and therefore the model is inapplicable since the 'ionic atmosphere' is inside the 'ion.' Equation (9.4) shows that this is not the case, since the 'ionic atmosphere' is always outside the sphere $r = a$.

The electrical energy of the central ion itself is therefore reduced by the product of its charge $z_j e$ and this potential (9.4) due to its interactions with its neighbours. However, if we applied this argument to every ion in the solution, we should in effect be counting each ion twice: once as the central ion, and once as part of the surroundings of other ions. The change ΔG_j in the electrical energy of a j-ion due to ionic interactions is therefore:

$$\Delta G_j = -\frac{z_j^2 e^2}{2\varepsilon} \frac{\kappa}{1 + \kappa a} \qquad \ldots.(9.5)$$

The same result is obtained by an imaginary charging process in which the distribution of ions is kept fixed and their charges are all simultaneously built up gradually from zero to their actual values; or from the theorem of electrostatics that the mutual energy of a system of charges is one-half the sum of the products of the charges of each and the potentials due to the others.

If the linearized equation (4.8) is taken for ρ_j, as is done in deriving (4.13), the different hypothetical charging processes proposed by DEBYE and GÜNTELBERG[4] give the same result for ΔG_j, but this is no longer the case if the non-linear expression (4.6) is retained for ρ_j.

THE DEBYE-HÜCKEL FORMULA FOR THE ACTIVITY COEFFICIENT

The contribution of the electrical interactions with other ions to the free energy of a single j-ion being given by equation (9.5), it follows that the corresponding quantity for one mole of j-ions is:

$$\Delta\bar{G}_j\,(\text{el}) = -\frac{z_j^2 e^2 N}{2\varepsilon}\frac{\kappa}{1+\kappa a} \qquad \ldots.(9.6)$$

In obtaining this result the j-ion has been treated as a sphere of diameter a. If we now make the assumption that a solution of these entities would exhibit ideal behaviour in the absence of interionic forces, we may write the partial free energy of a mole of j-ions as:

$$\bar{G}_j = \bar{G}_j\,(\text{ideal}) + \Delta\bar{G}_j\,(\text{el})$$

or $\quad \bar{G}_j^0 + \boldsymbol{R}T\ln f + \boldsymbol{R}T\ln \mathcal{N}_j = \bar{G}_j^0 + \boldsymbol{R}T\ln \mathcal{N} + \Delta\bar{G}_j\,(\text{el})$

$\mathcal{N}_j$ being the mole fraction and f_j the rational activity coefficient of the j-ions and $\bar{G}_j^0$ referring to the hypothetical standard state. Hence

$$\ln f_j = \frac{\Delta\bar{G}_j\,(\text{el})}{\boldsymbol{R}T} = -\frac{z_j^2 e^2}{2\varepsilon \boldsymbol{k}T}\frac{\kappa}{1+\kappa a}$$

This gives the individual ionic activity coefficient of the j-ions, a quantity not separately determinable by experiment. The mean rational activity coefficient $f_\pm$ of an electrolyte dissociating into ν_1 cations of valency z_1 and ν_2 anions of valency z_2 is given by: (see p. 28)

$$\ln f_\pm = -\frac{e^2}{2\varepsilon \boldsymbol{k}T}\cdot\frac{\kappa}{1+\kappa a}\left(\frac{\nu_1 z_1^2 + \nu_2 z_2^2}{\nu_1+\nu_2}\right)$$

which, upon eliminating the ν's by means of the relation $\nu_1 z_1 = -\nu_2 z_2$, becomes:

$$\ln f_\pm = -\frac{|z_1 z_2| e^2}{2\varepsilon \boldsymbol{k}T}\frac{\kappa}{1+\kappa a}$$

Upon replacing κ by its definition $\kappa = \left(\dfrac{8\pi N e^2}{1000\varepsilon \boldsymbol{k}T}\right)^{1/2}\sqrt{I}$ this result takes the form:

$$\log f_\pm = -\frac{A|z_1 z_2|\sqrt{I}}{1+Ba\sqrt{I}} \qquad \ldots.(9.7)$$

where the constants A and B involve the absolute temperature and the dielectric constant of the solvent, as follows:

$$A = \sqrt{\frac{2\pi N}{1000}} \cdot \frac{e^3}{2{\cdot}303 k^{3/2}} \cdot \frac{1}{(\varepsilon T)^{3/2}} = \frac{1{\cdot}8246 \times 10^6}{(\varepsilon T)^{3/2}}$$

$$\text{mole}^{-1/2}\ \text{l}^{1/2}(\text{deg K})^{3/2} \qquad \ldots.(9.8)$$

$$B = \left(\frac{8\pi N e^2}{1000 k}\right)^{1/2} \frac{1}{(\varepsilon T)^{1/2}} = \frac{50{\cdot}29 \times 10^8}{(\varepsilon T)^{1/2}}$$

$$\text{cm}^{-1}\ \text{mole}^{-1/2}\ \text{l}^{1/2}\ (\text{deg K})^{1/2} \qquad \ldots.(9.9)$$

Values of A and B for water at various temperatures[3] are given in Appendix 7.1.

It is important to note that $(B\sqrt{I})$ is the fundamental quantity κ of the interionic attraction theory; Appendix 7.1 and equation (9.9) therefore also have applications in the theory of transport processes.

THE LIMITING LAW OF DEBYE AND HÜCKEL

Equation (9.7) contains, in addition to functions of temperature and concentration, the parameter a defined as the 'distance of closest approach' of the ions. Inasmuch as this is not known *a priori* (except as to order of magnitude), the formula for the activity coefficient is not expressible solely in terms of measurable quantities. However, it is clear that at very low values of $\sqrt{I}$, *i.e.*, in very dilute solutions, the term $(Ba\sqrt{I})$ will ultimately become negligible compared to unity, and (9.7) will approach the form:

$$\log f_{\pm} = -A|z_1 z_2|\sqrt{I} \qquad \ldots.(9.10)$$

This is the Debye-Hückel *limiting* law according to which $\log f_{\pm}$ approaches linearity in the square root of the concentration at high dilutions. It is not to be expected that it will be obeyed accurately at any usual experimental concentration, since the product (Ba) is in practice always of the order of unity. This means that even in a one-thousandth molar solution of a 1 : 1 electrolyte, the factor $(1 + \kappa a)$ or $(1 + Ba\sqrt{I})$ is about 1·03, and the value of $-\log f_{\pm}$ according to (9.7) is therefore 3 per cent different from the limiting law value. Nevertheless, the form (9.10) is an extremely useful guide to the behaviour of activity coefficients at high dilutions.

For many aqueous solutions the expression (9.7) is capable of representing the observed activity coefficients with very good accuracy by simply choosing a value of the parameter a, independent of concentration, and of a physically reasonable magnitude. This often holds up to an ionic strength of about $I = 0{\cdot}1$, when the ions are separated on the average by no more than about 20 Å. Here their mutual energy would be expected to be of the same order as $\boldsymbol{k}T$; it appears, therefore, that the simple distribution-function used in deriving the equation (9.7) is fairly adequate.

The derivation of equation (9.7) is such that the numerator of the right-hand side, $-A|z_1z_2|\sqrt{I}$, gives the effect of the long-range coulomb forces, while the denominator $(1 + Ba\sqrt{I})$ shows how these are modified by the short-range interactions between ions, which are represented by the crudest possible model, taking the ions to be non-deformable spheres of equal radii. In any actual solution, there will also be the short-range interactions between ions and solvent molecules to consider, as well as other types of short-range interactions between ions, which cannot be adequately represented by the rigid-spheres model. As was mentioned earlier, these are all likely to be of a type giving an approximately linear variation of $\log f_{\pm}$ with concentration. Consequently they can be included, in a highly empirical fashion, by adding to (9.7) a term linear in the concentration, thus:

$$\log f_{\pm} = -\frac{A|z_1z_2|\sqrt{I}}{1 + Ba\sqrt{I}} + bI \qquad \ldots.(9.11)$$

where now b as well as a is a constant adjustable to suit the experimental curve. Equations like (9.11) are widely used for the analytical representation of activity coefficients, especially for the non-associated 1 : 1 electrolytes, where they are usually capable of fitting the data within the experimental accuracy up to at least one molal.

A simpler form of equation (9.7) is due to GÜNTELBERG[4] who writes for aqueous solutions:

$$\log f_{\pm} = -\frac{A|z_1z_2|\sqrt{I}}{1 + \sqrt{I}} \qquad \ldots.(9.12)$$

that is to say he puts $a = 3{\cdot}04$ Å for all electrolytes at 25°. Although this equation has no adjustable parameters, it gives a fair representation of the behaviour of a number of electrolytes up to $I = 0{\cdot}1$; it is certainly superior to the limiting law as represented by equation (9.10). It can be greatly improved, however, by adding a term linear in the concentration:

$$\log f_{\pm} = -\frac{A|z_1z_2|\sqrt{I}}{1 + \sqrt{I}} + bI \qquad \ldots.(9.13)$$

a form which is due to GUGGENHEIM[5], b being an adjustable parameter. *Table 9.1* illustrates how equations (9.12) and (9.13) can be used to represent the data for sodium chloride.

Table 9.1

Activity Coefficient of Sodium Chloride at 25°

m	$-\log f$ (*obs.*)	$-\log f$ *Eq.* (9.10)	$-\log f$ *Eq.* (9.12)	$-\log f$ *Eq.* (9.13)
0·001	0·0155	0·0162	0·0157	0·0155
0·005	0·0327	0·0362	0·0338	0·0330
0·01	0·0446	0·0511	0·0465	0·0449
0·05	0·0859	0·1162	0·0933	0·0853
0·1	0·1072	0·1614	0·1227	0·1067

$b = 0{\cdot}16$ l.mole^{-1}

DAVIES[6] has modified equation (9.13) by putting $b = 0{\cdot}1|z_1 z_2|$; in this form it is useful as a guide to the behaviour of the activity coefficient of an electrolyte when no experimental measurements are available. *Table 9.2* shows how the activity coefficient of calcium chloride can be represented.

Table 9.2

Activity Coefficient of Calcium Chloride at 25°

$\sqrt{m}$	0·04	0·12	0·20	0·28
f (obs.)	0·864	0·694	0·596	0·535
f (Eq. 9.13)	0·862	0·682	0·579	0·519

$b = 0{\cdot}200$ l mole^{-1}

THE DEBYE-HÜCKEL EQUATIONS FOR SOLUTIONS CONTAINING MORE THAN ONE ELECTROLYTE

Equations (9.7) and (9.10) have been deduced for the special case of a single electrolyte in solution, *i.e.*, for a solute one mole of which dissociates into ν_1 moles of cations of valency z_1 and ν_2 moles of anions of valency z_2. The consideration of the case of an electrolyte solution containing more than one species of electrolyte (for example, a mixture of hydrochloric acid and calcium chloride) introduces only one difficulty. By following through the derivation of the limiting law (9.10) it is easy to show that this is equally valid for one electrolyte in a mixed electrolyte solution provided that κ is defined correctly in terms of $\sum_i n_i z_i^2$, *i.e.*, I is taken as $\frac{1}{2}\Sigma(c_i z_i^2)$. Thus, in a solution of hydrochloric acid and sodium chloride, each 0·005 N at 25°, the mean activity coefficients of hydrochloric acid and sodium chloride as given by the limiting law are 0·889.

But in a solution of hydrochloric acid and calcium chloride, the former at a concentration of 0·004 mole/l and the latter 0·002 mole/l, we still have $I = 0{\cdot}01$, but the activity coefficients of the two electrolytes are no longer equal because $f_{HCl} = 0{\cdot}889$ and $f_{CaCl_2} = 0{\cdot}790$. Moreover, it should be noted that if, as in the above example, the two electrolytes have an ion in common, the common ion is allowed for in both activity coefficients; thus f_{HCl} is an average activity coefficient for the hydrogen ions and all the chloride ions, those derived from the calcium chloride as well as those derived from hydrochloric acid.

Similarly, equation (9.7) is also appropriate for a mixed electrolyte solution provided attention is paid to the proper meaning of z_1, z_2 and I, although we may have some difficulty in giving a meaning to the quantity a.

A MORE EXACT TREATMENT OF THE FREE ENERGY DUE TO ELECTRICAL INTERACTIONS

The treatment of the previous sections, leading to equation (9.7) for the mean activity coefficient, regards the electrical free energy of the system of charged ions in the solvent as belonging exclusively to the partial molal free energy of the electrolyte. In reality, a small part of the electrical free energy belongs to the solvent; we may think of this part as being the free energy of the solvent in the electrical field of the ions. This originates in ion-dipole interactions, which appear in the Debye-Hückel treatment in the form of the dielectric constant of the solvent.

In a more detailed consideration of the thermodynamics of an imaginary charging process in which the ionic charges are simultaneously built up from zero to their actual values, FOWLER and GUGGENHEIM[7] show that the total electrical energy of the whole system is given by:

$$G^{el} = -\frac{\sum_i s_i z_i^2 e^2}{3\varepsilon} \kappa\tau\,(\kappa a) \qquad \ldots.(9.13a)$$

where s_i is the number of i-ions in the system of total volume V,

$$\kappa^2 = \frac{4\pi e^2}{\varepsilon kT}\frac{\Sigma s_i z_i^2}{V}$$

and the function $\tau(\kappa a)$ is defined by:

$$\tau(x) = \frac{3}{x^3}\,[\ln(1 + x) - x + x^2/2]$$

It is also convenient to introduce the function $\sigma(x)$ defined by:

$$\sigma(x) = \frac{3}{x^3}\left[1 + x - \frac{1}{1 + x} - 2\ln(1 + x)\right]$$

$$= \frac{3}{x^3}\int_0^x \left(\frac{x}{1 + x}\right)^2 \mathrm{dx}$$

which is tabulated in Appendix 2.2.

From (9.13a) one obtains by differentiating partially with respect to s_i, remembering that variations in s_i affect V, the result:

$$\bar{G}_j^{el} = -\frac{Nz^2_j e^2}{2\varepsilon}\frac{\kappa}{1 + \kappa a} + \frac{\bar{V}_j \,.\, kT}{24\pi \;\; a^3}(\kappa a)^3\,\sigma(\kappa a) \qquad \ldots.(9.13b)$$

where $\bar{V}_j$ is the molar volume of the j-ions. This differs from the previous value (9.6) by the term in $\bar{V}_j$ and leads to:

$$\ln f_j = -\frac{z_j^2 e^2}{2\varepsilon kT}\frac{\kappa}{1 + \kappa a} + \frac{\bar{V}_j}{24\pi N a^3}(\kappa a)^3\,\sigma(\kappa a) \qquad \ldots.(9.13c)$$

whence for a single electrolyte:

$$\ln f_\pm = -\frac{|z_1 z_2| e^2}{2\varepsilon kT}\frac{\kappa}{1 + \kappa a} + \frac{\bar{V}_B/\nu}{24\pi N a^3}(\kappa a)^3\sigma(\kappa a)$$

The second term is insignificant at values of κa small compared to unity, for then $\sigma(\kappa a)$ approximates to unity and $(\kappa a)^3$ is very small. At high concentrations, it is doubtful whether the theory is valid, but if we grant that it is, we may note that even at $\kappa a = 2$ (as in a 4 N solution of a 1 : 1 electrolyte) $(\kappa a)^3\sigma(\kappa a) \approx 1{\cdot}2$. Ignoring electrostriction, by putting $\bar{V}_B \approx \pi a^3 \nu N/6$, we find that the factor $\bar{V}_B/(24\pi N a^3 \nu) \approx 1/144$, so that the second term of (9.13c) alters $f_\pm$ by less than 1 per cent. It is therefore justifiably ignored in nearly all applications of the theory.

The solvent activity, a_A, may be obtained either by integration of the Gibbs-Duhem equation using equation (9.7) or directly from (9.13a) by partial differentiation of G^{el} with respect to the number of solvent molecules in the system. Here the differentiation leads to a term containing the partial molal volume, $\bar{V}_A$, of the solvent, and one finds:

$$\ln a_A = \ln \mathcal{N}_A + \frac{\bar{V}_A}{8\pi N a^3}\left[1 + x - \frac{1}{1+x} - 2\ln(1+x)\right]$$

$$= \ln \mathcal{N}_A + \frac{\bar{V}_A}{24\pi N a^3}(\kappa a)^3\,\sigma(\kappa a) \qquad \ldots.(9.13\text{d})$$

where $\mathcal{N}_A$ is the mole fraction of the solvent. Here the second term on the right is evidently of comparable magnitude to the corresponding term in (9.13c), but it can no longer be neglected since it represents the *whole* of the deviation of the solvent activity from ideality. The solvent activity is rather insensitive to non-ideality, since it is present in large excess relative to the solute. For this reason, the more sensitive osmotic coefficient, ϕ, is usually used to represent the solvent behaviour. We have for aqueous solutions of single electrolytes, using (9.13d):

$$\phi = -(55{\cdot}51/\nu m)\ln a_w$$

$$= -\frac{55{\cdot}51}{\nu m}\ln\frac{55{\cdot}51}{55{\cdot}51 + \nu m} - \frac{55{\cdot}51}{\nu m}\,\frac{\bar{V}_A}{24\pi N a^3}(\kappa a)^3\,\sigma(\kappa a)$$

$$\approx 1 - \frac{\boldsymbol{e}^2|z_1 z_2|}{6\varepsilon \boldsymbol{k}T}\,\kappa\sigma(\kappa a),$$

the last step requiring several approximations valid only for dilute solutions.

THE ION SIZE PARAMETER a

If the limiting law (9.10) is compared with experimental data, it is found that for fully dissociated strong electrolytes such as the alkali and alkaline-earth metal halides the observed values of $\log f$ lie *above* the straight line of slope $-A|z_1 z_2|$ when plotted against $\sqrt{I}$, the deviations increasing with concentration. The more complete form (9.7) shows the reason for this; the ion size parameter a must be positive, and this leads to values of $\log f$ greater than those given by the limiting law. Up to ionic strengths of about 0·1, it is often possible to fit the data very accurately using a values of the order of 4 Å in equation (9.7). The a value giving the best fit is, however, apt to vary somewhat with the concentration-range fitted, which suggests that at the higher concentrations equation (9.7) is inadequate, and the parameter a is being forced to take care of other short-range effects than those it was intended for. A change in a does produce, in dilute solutions, changes in $\log f$ which are approximately linear in concentration, and could thus compensate for small short-range effects. This can be seen by differentiating (9.7) partially with respect to a, giving

$$\delta \log f = \frac{AB|z_1 z_2|\delta a}{(1 + Ba\sqrt{I})^2} I \qquad \ldots.(9.14)$$

As long as the solution is dilute enough for the denominator of (9.14) to be constant within 10 per cent or 20 per cent, the effect is roughly linear in I.

Table 9.3

Activity Coefficients of Sodium Chloride at 25°

m	$I (= c)$	$-\log \gamma_{\pm}$ (Experimental Values)	$-\log f_{\pm}$ (Experimental Values)	$A\sqrt{I}$	$\frac{A\sqrt{I}}{1 + Ba\sqrt{I}}$ a = 4·8 Å	$\frac{A\sqrt{I}}{1 + Ba\sqrt{I}}$ a = 4·0 Å	$-\log f_{\pm}$ eq. (9.11)	10^4 $\delta \log f_{\pm}$
(1)	(2)	(3)	(4)	(5)	(6)	(7)	(8)	(9)
0·001	0·000997	0·0155	0·0155	0·0162	0·0154	0·0155	0·0154	0·1
0·002	0·001994	0·0214	0·0214	0·0229	0·0214	0·0216	0·0214	0
0·005	0·004985	0·0328	0·0327	0·0361	0·0325	0·0330	0·0327	0
0·01	0·009969	0·0447	0·0446	0·0510	0·0441	0·0451	0·0445	0·1
0·02	0·01993	0·0602	0·0599	0·0722	0·0590	0·0609	0·0598	0·1
0·05	0·04981	0·0866	0·0859	0·1142	0·0844	0·0882	0·0855	0·4
0·1	0·09953	0·1088	0·1072	0·1614	0·1077	0·1140	0·1085	1·3
0·2	0·1987	0·1339	0·1308	0·2280	0·1338	0·1437	0·1328	2·0
0·5	0·4940	0·1668	0·1593	0·3595	0·1703	0·1868	0·1596	0·3
1·0	0·9788	0·1825	0·1671	0·5060	0·1974	0·2198	0·1660	1·1
2·0	1·921	0·1755	0·1453	0·7089	0·2222	0·2510	0·1453	0
4·0	3·696	0·1061	0·0477	0·9831	0·2435	0·2786	0·0753	276
6·0	5·305	0·0060	− 0·0789	1·1780	0·2539	0·2922	0·0004	793

In calculating the values in column (8) equation (9.11) was used with the parameter $a = 4{\cdot}0$ Å, $b = 0{\cdot}055$ l.mole^{-1}, *i.e.*,

$$\log f_{\pm} = -\frac{0{\cdot}5115\sqrt{I}}{1 + 1{\cdot}316\sqrt{I}} + 0{\cdot}055I$$

Consequently it is not necessary to interpret changes in the a value giving the best fit as real changes in effective size of the ions with concentration. It is probably better in determining a to use equation (9.11) over a somewhat greater concentration range; this ensures that the a parameter will not have forced upon it some of the responsibility for a term linear in concentration. At the same time, it does demand accurate experimental data and careful curve-fitting. The concentration range should still not be unduly extended, for there are good reasons for supposing that the linearity in concentration of the short-range effects is itself limited to moderately dilute solutions. To illustrate these points, and to give some idea of the effectiveness of these formulae for the activity coefficient, the case of sodium chloride at 25° in water will be considered in some detail. The experimental activity coefficients are given in *Table 9.3* in the form of $\log f_{\pm}$; the data below 0·1 M are derived from very precise measurements of transport numbers and of the potentials of cells with transport by Brown and MacInnes recalculated with modern values of the constants by SHEDLOVSKY[7]. At and above 0·1 M, they are those computed by ROBINSON[8] as best values from a number of reliable sources. Columns (1) and (2) give the molality and the corresponding ionic strength in mole per litre units; columns (3) and (4) give the molal and rational mean activity coefficients (it will be noted that these are considerably different at the higher concentrations). Column (5) is the value predicted for $-\log f_{\pm}$ by the Debye-Hückel limiting expression 9.10.

As has been said, this expression is not accurate within the experimental error even at 0·001 M, and is some 300 per cent out at 1 M. Columns (6) and (7) are calculated from equation (9.7) using respectively $a = 4{\cdot}8$ Å and $a = 4{\cdot}0$ Å. The value 4·8 Å results in a moderately good reproduction of the experimental values up to 0·1 M, but the deviations then change sign, and are not proportional to the concentration at higher concentrations. The value 4·0 Å gives an almost perfect fit up to about 0·02 M, and thereafter the deviations are fairly well proportional to the concentration. As a result, equation (9.11) with $a = 4{\cdot}0$ Å and $b = 0{\cdot}055$, fits the data quite well up to 2 M, as shown in columns (8) and (9). Thereafter, the experimental activity coefficients lie increasingly above the calculated ones. It is clear that by altering the parameters a and b in equation (9.11) expressions could be obtained which would give a slightly better fit over a smaller range, or a slightly worse one over a greater range. In the absence of perfect experimental values, it is therefore not possible to decide the exact value of a, but it is clear that a value in the range 4·0–4·8 Å is appropriate.

Even better reproduction of the data can be obtained by adding to the expression for $\log f$ further arbitrary terms in higher powers of the concentration, or its logarithm, *etc.*; the parameters in these more complex equations are even more elastic, but they usually give a values close to those required for equation (9.11) over moderate ranges.

In the derivation of equation (9.7), a was defined as the distance from the centre of an ion, within which the centre of no other can penetrate. When two ions meet (especially if they are of opposite sign) one can well imagine that this distance will be somewhat variable, depending on the Brownian velocity of the ions along their line of centres. Sometimes this may be sufficient to cause penetration some distance into the hydration sheath. In general, therefore, one would expect that the average distance of closest approach would be greater than the sum of the crystallographic radii of the bare ions, though not necessarily by the thickness of a whole number of layers of water molecules. This is usually so: with sodium chloride, for instance, the radius-sum is $0{\cdot}95 + 1{\cdot}81 = 2{\cdot}8$ Å, as against 4·0 to 4·8 Å for a; with calcium chloride, 2·8 Å as against 5 Å; and with lanthanum chloride, 3·0 Å as against 6–7 Å. As might also be expected, the difference between the radius sum and a is greater for strongly hydrated than for weakly hydrated ions. This suggests the possibility of computing a from estimates of ionic hydration.

When the quantity a was introduced, it was with the assumption that it was the same for all ions. If the cation, say, were larger than the anion, there would be a region round each ion into which anions but not cations could penetrate; this would somewhat modify the formulae for the separate ionic activity coefficients and for the mean activity coefficients. Actually it is doubtful whether these calculations are justifiable; it is not easy to see whether they satisfy the requirements of the linear superposition of the ionic fields. In general it seems reasonable to say that since encounters between ions of opposite sign will be more frequent than those between ions of the same sign, the parameter a is likely to be nearly the same for anions and cations.

THE INFLUENCE OF ION-SOLVENT INTERACTIONS ON THE ACTIVITY COEFFICIENT

We have seen in Chapters 3 and 6 that there are good grounds for believing that the kinetic unit of the solute in many electrolyte solutions is an ion with several relatively firmly attached water molecules; further evidence for this state of affairs is provided by a comparison of the ionic size parameters, a, necessary in the Debye-Hückel equation (9.7), with the dimensions of the bare ions. This suggests that the activity coefficient predicted by the Debye-Hückel treatment is actually the mean rational ionic activity coefficient of the *hydrated* ions. It is, however, the invariable practice in computing activity coefficients, from whatever experimental data, to calculate the composition of the solution in terms of the number of moles of anhydrous solute in a fixed mass of solvent (molality), or in a given total number of moles of solute and solvent (mole fraction) or in a given volume of solution (molarity). It will be noted that if the molarity scale is used, the figure expressing the concentration of the solution is the same whether the solute is treated as solvated or not; this is an advantage of using the molarity scale, but one which is outweighed by the disadvantage that the molarity of a solution of given composition changes with temperature.

In the case of the molality and mole fraction scales, the figure expressing the composition will be different if the solute is considered as a solvated species. Furthermore, the chemical potential of the solute considered as a solvated species will be different from its value when considered as unsolvated. However, the total Gibbs free energy, G, of a fixed amount of solution is fixed, regardless of the method used for expressing its composition; and the chemical potential of the solvent, $\bar{G}_A$, being defined by $\bar{G}_A = \left(\frac{\partial G}{\partial n_A}\right)_{n_B, T, P}$ is therefore likewise unaffected by the method of expressing n_B—it is the free energy gain on adding one mole of solvent to an infinite amount of solution, regardless of whether part of the added solvent actually combines with the solute or not. These considerations provide a simple method of finding the relation between the rational activity coefficient of the *solvated* solute and the conventional activity coefficients computed with disregard of solvation. This method of deriving the required relation is more straightforward than that given by the authors in their original paper on the subject[10] and more fundamental than earlier treatments of the same subject by BJERRUM[11] and by HARNED[12]. Consider a quantity of solution containing one mole of anhydrous solute, B, dissociated into ν_1 moles of cations and ν_2 moles of anions, dissolved in S moles of solvent A. We now calculate the fixed total free energy of the system, G, in two ways: (*a*) considering the solute as unsolvated; (*b*) considering that a total of h moles of solvent are combined with the ν moles of ions (divided, if we wish, into h_1 moles of water combined with the ν_1 moles of cations, and h_2 moles of water combined with the ν_2 moles of anions). We denote the solvent by a subscript A, and distinguish chemical potentials and activity coefficients calculated on the basis of the solvated ions by primes: G', f', *etc.*

We have, in view of the arguments in the preceding paragraphs:

$$G = S\bar{G}_A + \nu_1\bar{G}_1 + \nu_2\bar{G}_2$$

and

$$G = (S - h)\bar{G}_A + \nu_1\bar{G}_1' + \nu_2\bar{G}_2'$$

whence, introducing for each chemical potential its expression in terms of the appropriate mole fraction and activity coefficient, and rearranging,

$$\nu_1(\bar{G}_1^0 - \bar{G}_1'^0)/RT + \nu_2(\bar{G}_2^0 - \bar{G}_2'^0)/RT + h\bar{G}_A^0/RT + h \ln a_A$$
$$+ \nu \ln \frac{S + \nu - h}{S + \nu} + \nu_1 \ln f_1 + \nu_2 \ln f_2 = \nu_1 \ln f_1' + \nu_2 \ln f_2' \quad \ldots.(9.15)$$

Now as $S \to \infty$ (*i.e.*, at infinite dilution) all the activity coefficients become unity and a_A becomes unity, so that all the logarithmic terms are zero and hence the sum of the first three terms on the left of (9.15), involving the chemical potentials in the standard states, is also zero. We therefore have, introducing the mean ionic activity coefficients instead of the sums of the separate ionic ones:

$$\ln f_{\pm}' = \ln f_{\pm} + \frac{h}{\nu} \ln a_A + \ln \frac{S + \nu - h}{S + \nu} \quad \ldots.(9.16)$$

(In arriving at this result we have implicitly assumed that the value of h in the actual solution is the same as at infinite dilution.) This result is in practice more useful in terms of the conventional mean molal activity coefficient $\gamma_{\pm}$ and the molality m; using the relations $S = 1000/(W_A m)$, where W_A is the molecular weight of the solvent, and $f_{\pm} = \gamma_{\pm}(1 + 0{\cdot}001\nu\, W_A m)$, (eq. 2.22) we obtain:

$$\ln f_{\pm}' = \ln \gamma_{\pm} + \frac{h}{\nu} \ln a_A + \ln[1 + 0{\cdot}001\, W_A(\nu - h)m] \quad \ldots.(9.17)$$

Or, putting $\ln a_A$ in terms of the osmotic coefficient

$$\phi = -\frac{1000}{\nu W_A m} \ln a_A$$

$$\ln f_{\pm}' = \ln \gamma_{\pm} - 0{\cdot}001\, W_A h m \phi + \ln[1 + 0{\cdot}001\, W_A(\nu - h)m] \quad \ldots.(9.18)$$

Since ϕ or a_A can be calculated if $\gamma_{\pm}$ is known over the range of composition up to that considered, or alternatively $\gamma_{\pm}$ calculated if ϕ or a_A is similarly known (see Chapter 2) we have in equations (9.17) and (9.18) a method of expressing the rational mean ionic activity coefficient of the solute, assumed solvated with h moles of solvent per mole of salt, in terms of the conventional activity coefficients.

The only extra-thermodynamic assumption used in deriving equations (9.17) and (9.18) has been that the value of h is unchanged on proceeding to infinite dilution; this means that in applying the equations to actual solutions, we are limited to cases where there are plenty of solvent molecules to go round among the solute particles, and where the forces between solute and solvent are at least approximately of a saturable nature. For example, if the aqueous copper ion consisted of a definite complex $Cu4H_2O^{++}$ of high stability, we would be able to use the equation with confidence up to molalities approaching 55·5/4, provided no other solvating ion were present. While the forces between ions and water molecules are at least mainly electrostatic, and therefore are not strictly saturable in the same way as 'chemical' binding forces, there is, as we have seen, strong reason to believe that water molecules in direct contact with the ion are subject to much greater forces of attraction than subsequent layers, and there is a geometrical limit to the number of such closest molecules. Consequently, we may reasonably expect the assumption of an h value independent of concentration to apply up to moderately high concentrations. In practice, as we shall see, the limit is often reached when about a quarter of the solvent molecules are combined with ions.

We propose to use equation (9.16) in combination with the Debye-Hückel equation (9.7), taking the latter to refer to the solvated ions. In effect we shall use equation (9.7) to deal with interionic forces, and equation (9.16) to deal with the ion-solvent forces; instead of assuming that the solution of the unsolvated ions would be ideal but for the interionic forces, we assume that the solution of the solvated ions would be ideal but for the interionic forces. Apart from the obvious advantage of using a model which certainly comes closer to the physical reality, this procedure will go a long way towards justifying another assumption implicit in the derivation of equation (9.7), *viz.*, that the dielectric constant ε is that of the pure solvent. The work of HASTED, RITSON and COLLIE[13] has shown that nearly all the observed lowering of the bulk dielectric constant by ionic solutes arises from effects in the first layer of water molecules round the ion. If this layer is reckoned as part of the solute particle and if other ions do not penetrate into it, the dielectric constant of the liquid outside may fairly be taken as that of the pure solvent.

The same conclusion is reached in a recent theoretical study by BUCKINGHAM[14], from a detailed calculation of the energies of the ion-dipole and quadrupole interactions of water molecules in the first layer with one another and with the ion; beyond this first shell, the effect of dielectric saturation is found to be negligible.

The question whether the ion-solvent forces are adequately dealt with by this model is hard to answer, the chief difficulty arising over the value to be given to the parameter h, the number of moles of 'bound' water per mole of solute. If this could be determined unequivocally by other means, without reference to the activity data, the position would be better; but we have seen that the determination of accurate hydration numbers is a very recalcitrant problem.

Some guidance can be obtained from the study of non-electrolyte solutions. For a detailed account of the theory of these we refer to the texts of HILDEBRAND and SCOTT[15] and of GUGGENHEIM[16]; here we merely summarize the main types of behaviour:

(*a*) *Ideal solutions*, in which $\log f_A = \log f_B = 1$, are rare; approximation to this behaviour occurs with chemically similar molecules such as benzene and cyclohexane. There is no volume or heat change on mixing the components and the entropy of mixing per mole of mixture is:

$$\Delta S^M(\text{ideal}) = -\boldsymbol{R}(N_A \ln N_A + N_B \ln N_B) \quad \ldots.(9.19)$$

(*b*) *Athermal solutions* resemble ideal solutions in having zero heat of mixing, but the entropy of mixing is not that given by equation (9.19). Their departure from ideality is ascribed to the differences of size and shape between the solvent and solute molecules. The entropy of mixing may be calculated on the assumption of a lattice-like structure for the solution, a solute molecule occupying several lattice-points whilst a solvent molecule occupies only one. In the simplest case the entropy of mixing is:

$$\Delta S^M = -\boldsymbol{R}\left[N_A \ln \frac{N_A}{N_A + rN_B} + N_B \ln \frac{rN_B}{N_A + rN_B}\right] \quad \ldots.(9.20)$$

where r is the ratio of the molal volume of the solute to that of the solvent. Another interpretation regards r as the ratio of 'free volumes' rather than molar volumes. Equation (9.20), along with $\Delta H^M = 0$, leads to the following equation for the molal activity coefficient:

$$\ln \gamma_B = \frac{0{\cdot}001\, W_A\, r(r-1)m}{1 + 0{\cdot}001\, W_A\, rm} - \ln(1 + 0{\cdot}001\, W_A\, rm) \quad \ldots.(9.21)$$

which, for small m, approximates to:

$$\ln \gamma_B \approx 0{\cdot}001\, W_A\, r(r-2)m$$

or

$$\ln f_B \approx 0{\cdot}001\, W_A\, (r-1)^2 m.$$

(*c*) In *regular solutions* the entropy of mixing is given by the ideal expression (9.19) but there is a non-zero heat of mixing. The molal and rational activity coefficients (with standard states chosen so that γ_B and $f_B \rightarrow 1$ at infinite dilution) are:

$$\ln \gamma_B = \frac{b}{\mathbf{R}T}\left[\frac{1}{(1 + 0{\cdot}001\,W_A m)^2} - 1\right] - \ln\,(1 + 0{\cdot}001\,W_A m)$$

$$\ln f_B = \frac{b}{\mathbf{R}T}\left[\frac{1}{(1 + 0{\cdot}001\,W_A m)^2} - 1\right]. \qquad \ldots .(9.22)$$

Approximation for small m leads to:

$$\ln f_B \approx -\,0{\cdot}001\,W_A\,(2\,bm)/(\mathbf{R}T) \qquad \ldots .(9.23)$$

In a mixture of two liquids, the quantity b (if independent of temperature) is the heat of mixing one mole of either component with a large amount of the other; heat evolved in the process corresponds to negative values of b, and in this case the activity coefficient will increase with concentration.

If the interaction between the components is fairly strong, it may be possible to treat it in terms of the formation of a definite complex·

$$B + h\,A \rightleftharpoons BA_h$$

with

$$K = a_{BA_h}/(a_B\,a_A^h)$$

If the concentration of B is small, the solvent activity, a_A, differs little from unity and the ratio of complex to free B is nearly constant. A good approximation to the actual behaviour may then be obtained by using equation (9.18) taking h as an average for the free and complexed solute and assuming the mixture of solute, solvent and complex to be ideal. Equation (9.18), with $f'_{\pm} = 1$, becomes:

$$\ln \gamma = -\,h \ln\,(1 - 0{\cdot}001\,W_A\,hm)$$
$$+\,(h - 1)\,\ln\,[1 + 0{\cdot}001\,W_A(1 - h)m] \qquad \ldots .(9.24)$$
$$\approx 0{\cdot}001\,W_A\,(2h - 1)m \quad \text{for small } m.$$

Comparing (9.23) and (9.24) we see that in dilute solutions a regular-solution behaviour with a heat of mixing ($h\mathbf{R}T$) affects the activity coefficient to the same extent as would the solvation of each molecule of solute with h molecules of solvent.

There are thus a variety of causes—solvation, heat of mixing and molecular size and shape—which have the effect of causing $\ln f_B$ to increase approximately linearly with concentration. Any or all of these may be present in electrolyte solutions and we need to consider their relevance to aqueous solutions. *Table 9.4* gives the activity coefficients of aqueous sucrose and glycerol solutions at 25°, along with the results calculated for (*a*) ideal solutions, (*b*) regular solutions with the b parameter indicated, (*c*) solutions in which the solute is present as a hydrated form carrying h water molecules, the mixture of hydrated solute and 'free' water behaving ideally and (*d*) athermal solutions, using either an arbitrary value of r in equation (9.21) selected to give agreement at 1 M or the ratio of the molal volumes of solute and water.

In the case of glycerol, which departs only slightly from ideality, all the theoretical expressions give a fair reproduction of γ, except equation (9.21) with r as the ratio of molar volumes of solute and

Table 9.4

Representation of Activity Coefficients of Aqueous Non-electrolytes at 25° by equations 9.21 *to* 9.24

Molality	0·1	0·5	1·0	2·0	3·0
γ_{ideal} ($f = 1$)	0·998	0·991	0·982	0·965	0·949
			Sucrose		
γ_{exptl}[17]	1·017	1·085	1·188	1·442	1·751
γ(eq. 9.22, $b = -5{\cdot}4\ RT$)	1·018	1·092	(1·188)	1·397	1·628
γ(eq. 9.24, $h = 5$)	1·017	1·087	(1·188)	1·449	1·822
γ(eq. 9.21, $r = 4{\cdot}38$)	1·019	1·094	(1·186)	1·368	1·545
γ(eq. 9.21, $r = \frac{\bar{V}_B}{\bar{V}_A} = 12$)	1·235	2·638	5·800	19·26	45·82
			Glycerol		
γ_{exptl}[17]	1·003	1·014	1·027	1·050	1·071
γ(eq. 9.22, $b = -1{\cdot}2\ RT$)	1·003	1·013	(1·025)	1·048	1·069
γ(eq. 9.24, $h = 1{\cdot}2$)	1·003	1·013	(1·026)	1·053	1·081
γ(eq. 9.21, $r = 2{\cdot}6$)	1·003	1·014	(1·026)	1·048	1·079
γ(eq. 9.21, $r = \frac{\bar{V}_B}{\bar{V}_A} = 4$)	1·015	1·072	1·141	1·276	1·403

solvent. For sucrose, the simple interpretation as an ideal solution of a pentahydrated solute by equation (9.24) gives representation considerably better than the other equations; equation (9.21) using the 'volume fraction statistics' is poor even when r is interpreted as a 'free volume ratio' chosen for best fit, and its failure when r is put equal to the ratio of molar volumes of solute and solvent is spectacular.

The statistical mechanical methods used in the derivation of equation (9.20) for the entropy of athermal mixing are applicable when the molecules are long chains. The equation does not, however, correctly describe the entropy of mixing of approximately spherical molecules and Hildebrand has recently presented increasingly strong experimental evidence that the ideal entropy of mixing given by equation (9.19) is more correct in this case. In particular, his studies[18] of solutions involving the very large non-polar and nearly spherical molecule octamethyl-*cyclo*tetrasiloxane, $(CH_3)_8Si_4O_4$, ($\bar{V}_B = 312$ ml/mole) are of great interest.

It would thus seem likely that the best course for dealing with moderately concentrated electrolyte solutions would be to use the Debye-Hückel expression (9.7) to give the electrical contribution to the free energy of the solvated ions, equations (9.21) and (9.22) to give the effects which would exist if the solvated ions were in the solution but uncharged, and equation (9.17) to relate the activity coefficient of the solvated ions to the conventional activity coefficient. This would involve the use of no less than four arbitrary parameters, *viz.*, a, the ion size parameter; h, the solvation number; r, the free-volume ratio; and b, the heat of mixing of the solvated ions with the solvent in the imagined absence of interionic effects. With four parameters at our disposal, it would be possible to fit almost any experimental data, and some attempt must clearly be made to determine some of them in terms of others, or in terms of other measurable properties of the solution, before the adequacy of such a treatment can be gauged.

Scatchard[19] in 1932 gave a theory for the activity coefficient of concentrated electrolytes, having some of the features just indicated. The non-electrolyte type of interactions, for instance, were dealt with by an expression similar to equation (9.22) above, and the treatment of ion–ion interactions was given by the Debye-Hückel expression, though with allowance for the variation of dielectric constant with composition. The ion-solvent interactions were, however, treated in a very different manner, in terms of an electrostatic salting-out effect. The solvated-ion model which we are proposing to use has some advantages in simplicity, for, as has been remarked, the dielectric constant may more reasonably be taken as constant, and the ion-solvent interactions are represented in a manner easier to grasp.

The simplest treatment of the effect of ionic solvation on the activity coefficient would, as suggested above, be simply to combine equation (9.17) with the Debye-Hückel expression, taking the latter to give the activity coefficient of the solvated ions, $\log f_{\pm}$. This course is consistent with the fact that the a values of the Debye-Hückel expression correspond to the dimensions of solvated ions. The resulting expression is:

$$\log \gamma_{\pm} = -\frac{A|z_1 z_2|\sqrt{I}}{1 + Ba\sqrt{I}} - \frac{h}{\nu}\log a_A - \log[1 + 0{\cdot}001\, W_A(\nu - h)m] \quad \ldots .(9.25)$$

This expression contains only two parameters, h and a, the 'non-electrolyte' contributions involving the parameters r and b (equations 9.21 and 9.22) being ignored.

Equation (9.25) has been extensively tested[10] for aqueous solutions and is remarkably successful with the non-associated electrolytes. The values of the parameters, a and h for thirty-six 1 : 1 and 2 : 1 electrolytes at 25° are given in *Table 9.5.*

Table 9.5

Constants of the Two-parameter Equation (9.25) *Giving Best Fits to the Experimental Activity Coefficients*

Salt	h	a (Ångstroms)	*Salt*	h	a (Ångstroms)
HCl	8·0	4·47	RbI	0·6	3·56
HBr	8·6	5·18	$MgCl_2$	13·7	5·02
HI	10·6	5·69	$MgBr_2$	17·0	5·46
$HClO_4$	7·4	5·09	MgI_2	19·0	6·18
LiCl	7·1	4·32	$CaCl_2$	12·0	4.73
LiBr	7·6	4·56	$CaBr_2$	14·6	5·02
LiI	9·0	5·60	CaI_2	17·0	5·69
$LiClO_4$	8·7	5·63	$SrCl_2$	10·7	4·61
NaCl	3·5	3·97	$SrBr_2$	12·7	4·89
NaBr	4·2	4·24	SrI_2	15·5	5·58
NaI	5·5	4·47	$BaCl_2$	7·7	4·45
$NaClO_4$	2·1	4·04	$BaBr_2$	10·7	4·68
KCl	1·9	3·63	BaI_2	15·0	5·44
KBr	2·1	3·85	$MnCl_2$	11·0	4·74
KI	2·5	4·16	$FeCl_2$	12·0	4·80
NH_4Cl	1·6	3·75	$CoCl_2$	13·0	4·81
RbCl	1·2	3·49	$NiCl_2$	13·0	4·86
RbBr	0·9	3·48	$Zn(ClO_4)_2$	20·0	6·18

From STOKES, R. H. and ROBINSON, R. A., *J. Amer. chem. Soc.*, 70 (1948) 1870.

Examination of this table reveals the following points: (*a*) The values of a lie in the range 3·5–6·2 Å, *i.e.*, they are much the same as those determined from simpler forms of equations such as equations (9.7) and (9.11). (*b*) The values of h for 1 : 1 chlorides lie in the order H > Li > Na > K > Rb which is just the reverse of the order of the radii of the bare ions. The same cation-order holds for the 1 : 1 bromides and iodides. The alkaline earth metal cations show a similar decrease in hydration number with increasing crystallographic radius. This behaviour is consistent with what is known of the hydration of these cations from other sources (except possibly in the case of hydrogen ion which will be taken up later). (*c*) For a given cation, the h values decrease in the order

$I^- > Br^- > Cl^-$. This implies that the largest anion is the most solvated, a conclusion at variance with both reasonable expectation and other experimental indications, and it is a serious weakness of the simple equation (9.25). (*d*) The solvation numbers h are not additive for the separate ions, *e.g.*, $h_{NaCl} - h_{KCl} = 1{\cdot}6$, but $h_{NaI} - h_{KI} = 3{\cdot}0$. It is thus not possible to ascribe a single value of h to each ion, but different values must be allowed depending on the nature of the companion ion. (*e*) The low h values for potassium, ammonium and rubidium salts indicate that the chloride, bromide and iodide ions can be little hydrated so that where high values of h are found most of the hydration must be attributed to the cation. This is quite reasonable since the anions considered are all large (~ 2 Å radius) and therefore have low surface charge. The resulting cation hydration numbers do, however, seem rather large. In particular, if one estimates the radius of spherical cations containing such amounts of bound water, one finds that the sum of this radius and the crystallographic radius of the anion exceeds the required *a* value by about 0·7 Å for the 1 : 1 salts and 1·3 Å for the 2 : 1 salts. This rather unsatisfactory situation was dealt with in our original paper[10] by assuming that when the anion and cation met, the anion could penetrate into the hydration sheath of the cation by these distances; the greater penetration when the cation is divalent (1·3 Å against 0·7 Å) arising from the greater attraction it has for the anion.

Using the concept of a limited penetration of the anion into the hydration sheath of the cation, it proved possible to obtain a relation between the parameters h and a, of sufficient accuracy to permit the calculation of the activity coefficients of the chlorides, bromides and iodides of the alkali and alkaline earth metals and hydrogen with quite good accuracy (better than 1 per cent) up to ionic strength often as high as $I = 4$. The relation between h and a is obtained as follows: The volume occupied by a water molecule in liquid water at 25° is 30 Å³. The volume of the hydrated cation (taking the anion as unhydrated) is therefore $(30h + V_1)$, where V_1 is the apparent molal volume of the ion in Å³. It was shown that V_1 could be estimated from the apparent molal volume of the salt, V_{app}, in solution by means of the formula:

$$V_1 = V_{app} - 6{\cdot}47 z_1 r_2^3 \qquad \ldots.(9.26)$$

where r_2 is the crystal radius of the anion (in Å). Since V_1 is usually only a small fraction of $30h$, it need not be calculated with great accuracy, and in practice it is sufficient to estimate the apparent molal volume at about 1 M and use this value at all concentrations. The radius of the hydrated cation, r_1', is then given by:

$$4\pi r_1'^3/3 = 30h + V_1 \qquad \ldots.(9.27)$$

and the value of the mean distance of closest approach of the ions is given by:

$$a = r_1' + r_2 - \Delta$$

where Δ is the 'penetration distance.' Since r_1' is a known function of h, a can be calculated from h and the known crystal radius of the anion. The resulting one-parameter equation for the activity coefficient (for an aqueous solution at 25°) is:

$$\log \gamma_\pm = -\frac{0{\cdot}5115|z_1z_2|\sqrt{I}}{1 + 0{\cdot}3291\sqrt{I}\left\{\left[\frac{3}{4\pi}(30h + V_1)\right]^{1/3} + r_2 - \Delta\right\}} - \frac{h}{\nu}\log a_w - \log[1 - 0{\cdot}018(h - \nu)m] \qquad \ldots.(9.28)$$

It is of course not strictly a one-parameter equation, for it involves Δ as well as h. However, Δ is constant for each class of salt (0·7 Å for 1 : 1 salts, 1·3 Å for 2 : 1 salts) so that only the parameter h has to be specified to give the activity coefficient of a salt of one of these valency types. The effectiveness of equation (9.28) is illustrated by the graphs in *Figure 9.3.* The limit of validity of this equation is generally reached when the product $hm \approx 12$, *i.e.*, when about one-fifth to one-quarter of the total water molecules are bound to ions as water of hydration. Above this limit, equation (9.28) usually predicts values which are higher than those observed, which suggests that the hydration number is beginning to fall owing to the effects of competition between neighbouring cations.

This treatment of solvation has been extended by GILLESPIE and OUBRIDGE[20] to solutions of metal sulphates in sulphuric acid as solvent.

Whilst there is no doubt of the success of both the two-parameter equation (9.25) and the one-parameter form (9.28), their theoretical basis is somewhat inadequate owing to the neglect of 'non-electrolyte' effects and there are, as shown above, some difficulties in accepting the h values as giving precise representations of the formulae of the hydrated ions.

Glueckauf's Treatment of Ionic Hydration

GLUECKAUF[21], recognizing the difficulty of interpreting the h values of equations (9.25) and (9.28) as actual sums of hydration

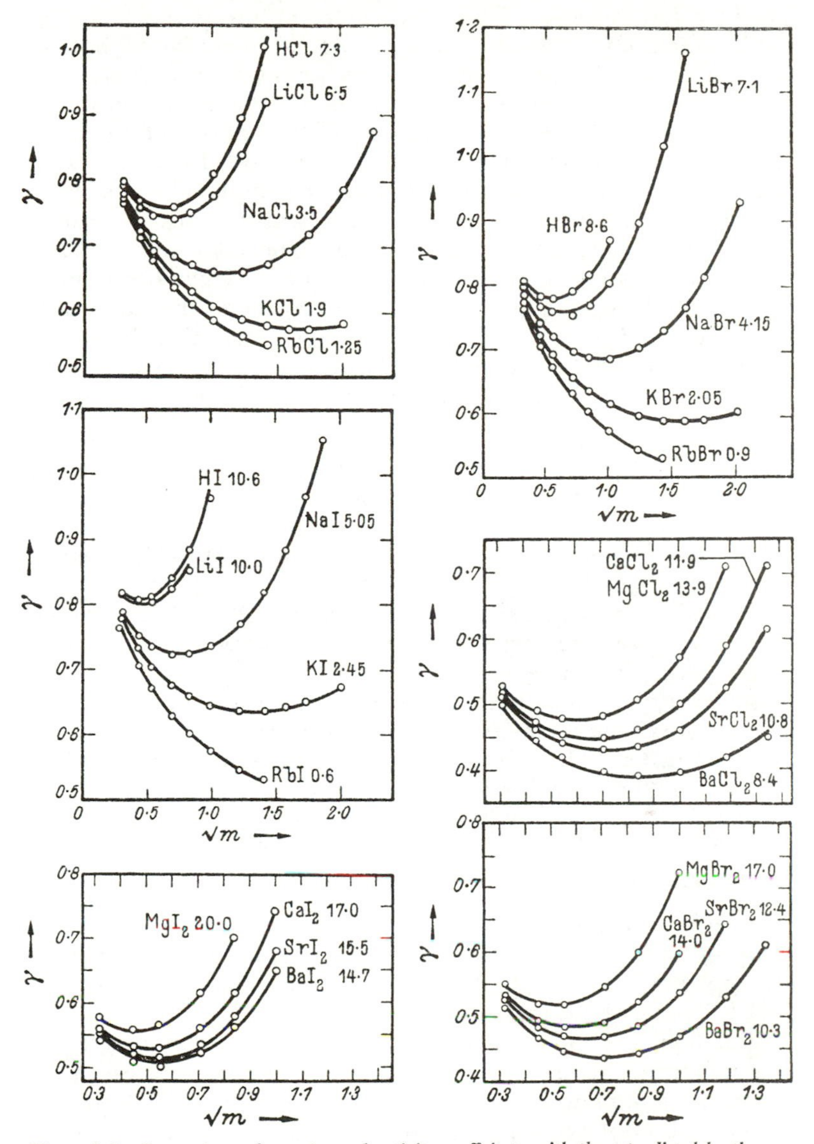

Figure 9.3. Comparison of experimental activity coefficients with those predicted by the one-parameter equation (9.28). *The full curves are calculated from equation* (9.28), *using the value for the 'hydration parameter' h following the formula of each salt. From* STOKES, R. H. *and* ROBINSON, R. A., *J. Amer. chem. Soc.*, 70 (1948) 1870

numbers for the ions of the electrolyte, proposed the following modification of the theory: If h_i is the actual hydration number of an ion i, the partial molal volume of the hydrated ion $\bar{V}_i'$ is $(h_i\bar{V}_A + \bar{V}_i)$ where $\bar{V}_i$ is the partial molal volume of the unhydrated ion as ordinarily defined. He then assumes that the entropy of mixing of the hydrated ions and the 'free' solvent is given by an equation analogous to (9.20), with $r_i = \bar{V}_i'/\bar{V}_A$, *i.e.*, he employs 'volume fraction statistics' instead of the 'mole fraction statistics' which we used in deriving equation 9.25. He also takes the total electrical free energy of the system as given by equation 9.13a. Upon differentiating the total free energy partially with respect to the number of moles of anhydrous solute in the system, and neglecting the second term on the right of equation 9.13c, he obtains for the molal activity coefficient $\gamma\pm$ (as ordinarily expressed on the basis of anhydrous solute);

$$\ln \gamma_{\pm} = \ln f_{\pm}' + \frac{0{\cdot}018\, mr\, (r + h - \nu)}{\nu(1 + 0{\cdot}018mr)} \qquad \ldots .(9.29)$$

$$+ \frac{h - \nu}{\nu} \ln (1 + 0{\cdot}018mr) - \frac{h}{\nu} \ln (1 - 0{\cdot}018mh)$$

This equation fits the experimental data as well as does equation (9.25), using as parameters h and a; the quantity r, now referring to the electrolyte as a whole, is $r = (\bar{V}_B + h\bar{V}_A)/\bar{V}_A$. We have shown[22] that a simple relation between actual volumes and effective volumes in solution enables one to dispense with a as an arbitrary parameter. The values of the hydration numbers required by equation 9.29 are considerably smaller than those of *Table 9.5*, *e.g.*

$$h_{HCl} = 4{\cdot}7,\ h_{NaCl} = 2{\cdot}7,\ h_{KCl} = 1{\cdot}7,\ h_{NH_4Cl} = 1{\cdot}1,$$

$$h_{CaBr_2} = 6{\cdot}2,\ h_{BaI_2} = 5{\cdot}5,\ h_{LaCl_3} = 10{\cdot}2.$$

Furthermore, the most serious anomaly of our earlier treatment now disappears: the new hydration numbers become nearly additive for the separate ions, and the hydration numbers of the ions Cl^-, Br^- and I^- are all about the same ($h \approx 0{\cdot}9$).

Equation 9.29 appears to differ from equation 9.25 not merely in the presence of terms in the molar volume ratio r, but also in the absence of a term in $\ln a_w$. This latter discrepancy is only apparent; we have shown[21] that it arises from the neglect of the very small second term of equation 9.13(b) in *both* treatments. If this is not neglected the two treatments differ only in being based on an entropy of mixing given by equation 9.19 in our case and on 9.20 in Glueckauf's. (In both cases the appropriate modification for an ionized solute is of course made.) In view of the convenience and effectiveness of equation 9.29, it is most unfortunate that the use of 'volume-fraction statistics' receives so little justification from tests on aqueous non-electrolytes (*Table 9.4*).

Summary of Theoretical Treatments of Chemical Potential in Electrolyte Solutions

The importance of ion-solvent interactions in modifying the activity coefficients of concentrated solutions was first recognized by BJERRUM[11], who used an equation similar to (9.25) except that the electrical term was represented less accurately by a term proportional to $c^{1/3}$. The appearance in 1923 of the DEBYE-HÜCKEL[2] treatment of electrical interactions shortly afterwards focussed attention on dilute solutions, and the ion-solvent interactions were usually treated quite empirically by adding terms linear in c to the expression for $\log f$. HÜCKEL[23], and later SCATCHARD[19], treated the ion-solvent interactions in terms of an electrostatic salting-out effect arising from the fact that the electrolyte lowers the dielectric constant; Scatchard also introduced a term corresponding to the thermal effect described by equation (9.22). The writers[10] combined Bjerrum's thermodynamic treatment of ion-solvent interactions with the Debye-Hückel treatment of ion-ion interactions, obtaining equation 9.25; and GLUECKAUF[21] modified this treatment by substituting 'volume-fraction statistics' for the conventional 'mole-fraction statistics.' EIGEN and WICKE[24] developed a treatment similar to that of Debye and Hückel, but employing a distribution function modified to allow for the co-volume of the ions (Chapter 4). MAYER[25] showed that the limiting Debye-Hückel square-root law could be established from a general statistical-mechanical treatment of the ion-ion interactions, which avoids the self-consistency difficulties inherent in the Poisson-Boltzmann equations, and developed the theory for finite ion sizes; POIRIER[26] applied the theory to actual solutions, obtaining fair agreement with experimental results. The calculations are laborious and have as yet been applied to very few salts.

There are many salts, normally regarded as strong electrolytes, for which the Debye-Hückel formula requires absurdly small or even negative values of the ion size parameter a. GRONWALL, LA MER and SANDVED[27] dealt with these cases by accepting the non-linear Poisson-Boltzmann equation (equation 4.2 with ρ given by equation 4.6) and solving for ψ by numerical integration. This treatment, though moderately successful, has been criticized on logical grounds[7]. BJERRUM's[28] theory of ion-association provides a more satisfying explanation of this type of behaviour; it is developed in detail, along with elaborations by FUOSS and KRAUS[29], in Chapter 14.

REFERENCES

1 GUGGENHEIM, E. A., 'Thermodynamics: An advanced treatment for chemists and physicists', p. 201, North Holland Publishing Co., Amsterdam (1949)
2 DEBYE, P. and HÜCKEL, E., *Phys. Z.*, 24 (1923) 185
3 MANOV, G. G., BATES, R. G., HAMER, W. J. and ACREE, S. F., *J. Amer. chem. Soc.*, 65 (1943) 1765; their values for A and B have been recalculated to conform with the dielectric constants recorded in Appendix 1.1
4 GUNTELBERG, E., *Z. phys. Chem.*, 123 (1926) 199
5 GUGGENHEIM, E. A., *Phil. Mag.*, 19 (1935) 588
6 DAVIES, C. W., *J. chem. Soc.*, (1938) 2093
7 FOWLER, R. H. and GUGGENHEIM, E. A., 'Statistical Thermodynamics' Ch. IX, Cambridge University Press (1949)
8 BROWN, A. S. and MACINNES, D. A., *J. Amer. chem. Soc.*, 57 (1935) 1356; SHEDLOVSKY, T., *ibid.*, 72 (1950) 3680
9 ROBINSON, R. A., *Proc. Roy. Soc., N.Z.*, 75 (1945) 203; STOKES, R. H. and LEVIEN, B. J., *J. Amer. chem. Soc.*, 68 (1946) 333
10 STOKES, R. H. and ROBINSON, R. A., *J. Amer. chem. Soc.*, 70 (1948) 1870
11 BJERRUM, N., *Medd. vetensk Akad. Nobelinst.*, 5 (1919) No. 16; *Z. anorg. Chem.*, 109 (1920) 275
12 HARNED, H. S. in TAYLOR, H. S., 'Treatise on Physical Chemistry', Vol. 2, p. 776, D. Van Nostrand & Co., New York, 1924
13 HASTED, J. B., RITSON, D. M. and COLLIE, C. H., *J. chem. Phys.*, 16 (1948) 1
14 BUCKINGHAM, A. D., *Disc. Faraday Soc.*, 24 (1957) 151
15 HILDEBRAND, J. H. and SCOTT, R. L., 'The Solubility of Non-electrolytes', Reinhold Publishing Co., New York, 1950
16 GUGGENHEIM, E. A., 'Mixtures', Clarendon Press, Oxford 1952; see also ref. 1
17 SCATCHARD, G., HAMER, W. J. and WOOD, S. E., *J. Amer. chem. Soc.*, 60 (1938) 3061
18 SHINODA, K. and HILDEBRAND, J. H., *J. Phys. Chem.*, 61 (1957) 789
19 SCATCHARD, G., *Phys. Z.*, 33 (1932) 22; *Chem. Rev.*, 19 (1936) 309
20 GILLESPIE, R. F. and OUBRIDGE, J. V., *J. chem. Soc.*, (1956) 80
21 GLUECKAUF, E., *Trans. Faraday Soc.*, 51 (1955) 1235
22 STOKES, R. H. and ROBINSON, R. A., *ibid.*, 53 (1957) 301
23 HÜCKEL, E., *Phys. Z.*, 26 (1925) 93
24 WICKE, E. and EIGEN, M., *Naturwissenschaften*, 38 (1951) 453; 39 (1952) 545; *Z. Elektrochem.*, 56 (1952) 551; 57 (1953) 319; *Z. Naturf.*, 8A (1953) 161
25 MAYER, J. E., *J. chem. Phys.*, 18 (1950) 1426
26 POIRIER, J. C., *ibid.*, 21 (1953) 965, 972
27 GRONWALL, T. H., LAMER, V. K. and SANDVED, K., *Phys. Z.*, 29 (1928) 358; LAMER, V. K., GRONWALL, T. H. and GREIFF, L. J., *J. phys. Chem.*, 35 (1931) 2245
28 BJERRUM, N., *K. danske vidensk. Selsk.*, 7 (1926) No. 9
29 FUOSS, R. M. and KRAUS, C. A., *J. Amer. chem. Soc.*, 55 (1933) 1019, 2387; 57 (1935) 1

10

THE MEASUREMENT OF DIFFUSION COEFFICIENTS

THE fundamental equations defining the diffusion coefficient have been introduced in Chapter 2. Before considering the theoretical interpretation of diffusion data, we shall discuss the various experimental methods which are available for obtaining them.

EXPERIMENTAL METHODS FOR THE STUDY OF DIFFUSION

The available methods may be grouped in several ways: perhaps the most obvious division is into steady-state methods based on the equation:

$$J = -D\frac{\partial c}{\partial x} \qquad \ldots .(2.53)$$

and other methods, based on the equation:

$$\frac{\partial c}{\partial t} = \frac{\partial}{\partial x}\left(D\frac{\partial c}{\partial x}\right) \qquad \ldots .(2.54)$$

Steady-state Methods

In true steady-state diffusion, a constant concentration is maintained at both ends of a column of liquid through which diffusion takes place; the flux of solute ultimately becomes independent both of time and of position in the column. When this steady state has been reached, the flux J and the concentration-gradient $\frac{\partial c}{\partial x}$ are measured, giving D by equation (2.53). Almost the only results obtained by this method are those of CLACK[1], who devoted many years to the development of suitable apparatus. The concentration at the lower end of his column was maintained at saturation by means of a reservoir of solid salt, while that at the upper end was maintained effectively at zero by means of a slow flow of water; the flux was determined analytically and the concentration gradient was measured at any desired level by an optical determination of the refractive index gradient. By integration of the concentration gradient, the concentration to which each value of D referred could be calculated; the method thus had the advantage that a single successful run provided values of D at all concentrations up to saturation. The experimental difficulties of establishing and maintaining the steady state were, however, too great to encourage widespread use of this method; chief among these were thermally and mechanically induced convection currents. Curiously enough, however, Clack's results for sodium and potassium chlorides appear to be several per cent *lower* than undoubtedly more reliable recent data obtained by the methods described below.

A much more important method based on equation (2.53) is the porous diaphragm technique introduced by NORTHROP and ANSON[2] and developed by McBAIN[3], HARTLEY[4] and STOKES[5]. The fundamental idea of this method is to eliminate the disturbing effects of vibration and of small temperature fluctuations by confining the diffusion process to the capillary pores of a sintered glass diaphragm; an idea excellent in itself, but one which introduces a number of new problems:

(*a*) *Calibration of the diaphragm*—Since it is not possible to measure the true length and cross-section of the diaphragm pores, the effective average values must be determined. This is done by performing diffusion experiments with a solute which has been studied by one of the absolute methods to be described later. At the time when the porous diaphragm technique was introduced, however, no absolute values reliable within less than about 2 per cent were available; this may have delayed the full development of the method.

(*b*) *Stagnant layers on the diaphragm*—It is essential that the diffusion process be confined entirely to the pores of the diaphragm. This means that the reservoirs of solution on either side must be maintained at a uniform concentration right up to the surface of the diaphragm. The originators of this method[2] approximated to this condition by placing the diaphragm horizontally with the denser solution in a closed reservoir above it, so that diffusion would lower the density on the upper side, and raise it on the lower side. This results in a gravity-induced streaming, which is easily seen by using a coloured salt solution; but later work has shown conclusively that a thin stagnant layer persists at the surface of the diaphragm. For a given solute and concentration this layer behaves very reproducibly, so that a precision of 0·1 per cent can readily be obtained, but the thickness of the layer varies with the solute and the concentration-gradient across the diaphragm so that systematic errors of several per cent can occur when different solutes are compared[5]. Mechanical stirring is therefore adopted to remove this layer; HARTLEY and RUNNICLES[4] used glass balls which rolled on the diaphragm as it rotated in a slanting position, while MOUQUIN and CATHCART[(6)] used balls which fell through the solutions as the cell was inverted end over end. Both these methods seem inadequate inasmuch as they do not ensure complete removal of the stagnant layers; furthermore, the departure of the diaphragm from the horizontal position encourages streaming of solution through the diaphragm when large density-differences are used, leading to high

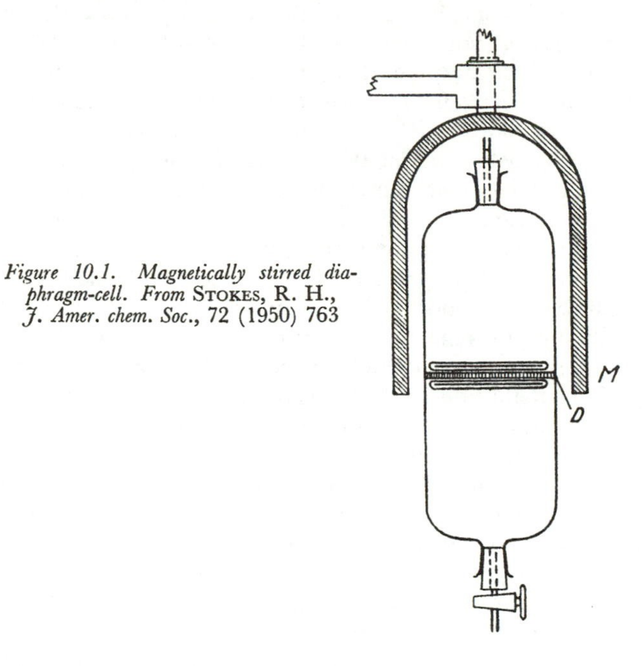

Figure 10.1. Magnetically stirred diaphragm-cell. From STOKES, R. H., *J. Amer. chem. Soc.*, 72 (1950) 763

results. The method finally adopted by Stokes is a logical development of that of Hartley and Runnicles. The cell is shown in *Figure 10.1*. The stirrers in this cell are sealed glass tubes slightly shorter than the diameter of the diaphragm; they enclose an iron wire and are caused to sweep over the diaphragm by a rotating permanent magnet mounted co-axially with the cell. The weights of the stirrers are so adjusted that the upper one sinks and the lower floats, both pressing lightly on the diaphragm. The stagnant layers are thus completely swept off.

(*c*) *Avoidance of streaming*—If the pores are too coarse, transport can occur by bulk streaming through the diaphragm as well as by diffusion; this is more likely to occur if the denser liquid is above the diaphragm. It can be reduced to negligible proportions by using a diaphragm of No. 4 porosity (average pore size $\sim 15\ \mu$) and by placing the denser solution *below* the diaphragm; this is permissible with the magnetic stirring system described above, as it is easily shown[5] that above a moderate threshold rate ($\sim$ 20 r.p.m.) the cell calibration is independent of the stirring-rate, *i.e.*, that the stirring is sufficient to ensure uniformity within each reservoir.

(*d*) *Surface transport effects*—A serious limitation on the use of diaphragm-cells for the study of diffusion in electrolyte solutions arises from adsorption effects on the large internal surface of the diaphragm, which may be of the order of a square metre in area. By comparing the results of diaphragm-cell measurements on dilute electrolyte solutions with absolute measurements it has been proved[5] that below about 0·05 M the former give substantially higher results, the error increasing as the solutions are made more dilute; it is of the order of 2 per cent at 0·01 M. The effect appears to arise through an enhancement of mobility in the electrical double layer on the pore walls; this has been confirmed by MYSELS and MCBAIN[7] by conductivity measurements in a cell in which a porous diaphragm is interposed between the electrodes. The double layer is most compact at high concentrations, and then makes a negligible contribution to the total transport. In more dilute solutions it occupies a larger proportion of the capillary cross-section, and its contribution to the transport is more marked. The result is that the diaphragm-cell method cannot safely be used at concentrations below 0·05 M; it is possible that with electrolytes of higher valency types than the 1 : 1 electrolytes this limit may be at even higher concentrations. The method is, however, quite reliable at higher concentrations, and with care an accuracy of 0·2 per cent in the diffusion coefficient may be expected.

In use, the cell is filled with an air-freed solution of approximately known concentration and one end is connected to a vacuum pump in order to remove air from the diaphragm. After eliminating any bubbles formed, the cell is thermostated, and the solution in the upper end is replaced by water or by a solution of lower concentration. The cell is then run for a few hours, in order to produce a steady state in the diaphragm; the upper solution is then replaced by water, or by a solution of accurately known concentration less than that in the lower end. The run is timed from this point, and proceeds for a matter of one to three days. The compartments are then sampled at a known time and the final solutions analyzed. The diffusion coefficient is calculated as follows: denote the concentrations at the beginning and end of the run by c_1, c_2, c_3, c_4 as shown in *Figure 10.2*, and the volumes of the compartments and diaphragm pores by V_1, V_2, V_3 respectively. Let the total effective cross-section of the diaphragm pores be A, and their effective average length along the diffusion path be l. It is now necessary to assume that the diaphragm is in a steady state during the experiment, to the extent that there is no tendency for solute to accumulate in or to be lost from the diaphragm. Thus, at any given time, the flux of solute across any plane in the diaphragm parallel to its surfaces is

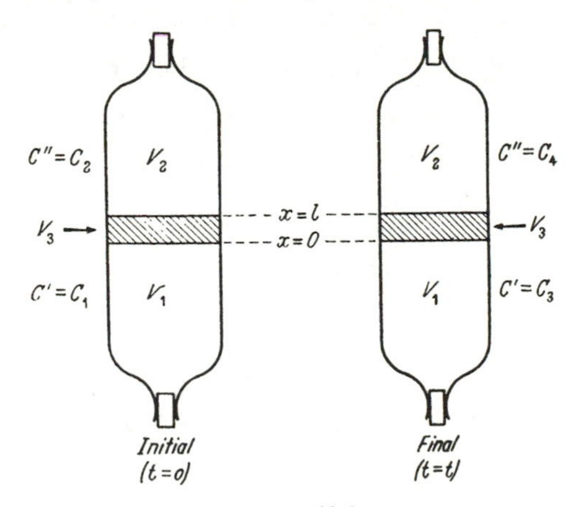

Figure 10.2

everywhere the same. This flux will, however, vary slowly with time, decreasing as the process of diffusion reduces the concentration-difference. To emphasize this we shall write it as $J(t)$.

Denoting the concentrations of the upper and lower compartments by c'' and c' respectively, the rates of change of these concentrations are related to the flux $J(t)$ by:

$$\frac{\mathrm{d}c'}{\mathrm{d}t} = -J(t)\frac{A}{V_1}$$

$$\frac{\mathrm{d}c''}{\mathrm{d}t} = J(t)\frac{A}{V_2}$$

Hence
$$\frac{\mathrm{d}(c' - c'')}{\mathrm{d}t} = -J(t)A\left(\frac{1}{V_1} + \frac{1}{V_2}\right) \qquad \ldots.(10.1)$$

We now introduce the average value of the diffusion coefficient D with respect to concentration over the concentration range c' to c'' prevailing at the time considered; this quantity is also a function of time, which we denote $\bar{D}(t)$:

Then

$$\bar{D}(t) = \frac{1}{c' - c''} \int_{c''}^{c'} D \, \mathrm{d}c = - \frac{1}{c' - c''} \int_{x=0}^{l} D \left(\frac{\partial c}{\partial x}\right) \mathrm{d}x$$

$$= \frac{lJ(t)}{c' - c''} \qquad \ldots . (10.2)$$

since $J(t) = - D \dfrac{\partial c}{\partial x}$ is constant for all points within the diaphragm at time t, x being the distance of the plane considered from the

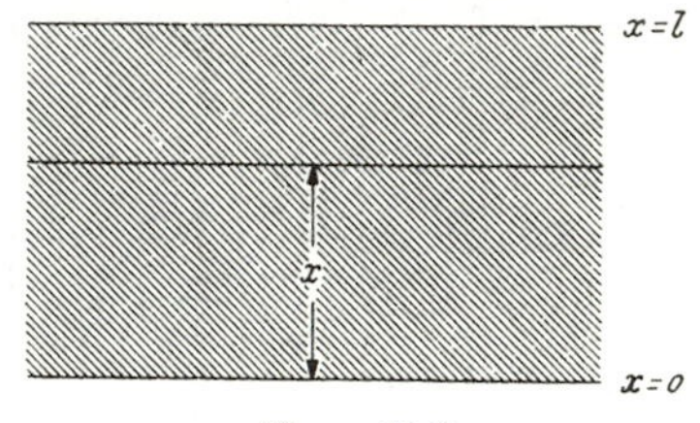

Figure 10.3

lower surface of the diaphragm (*Figure 10.3*). Combining (10.1) and (10.2) gives:

$$- \frac{\mathrm{d} \ln (c' - c'')}{\mathrm{d}t} = \frac{A}{l} \left(\frac{1}{V_1} + \frac{1}{V_2}\right) \bar{D}(t)$$

Hence integrating between the initial and final conditions shown in *Figure 10.2* we obtain:

$$\ln \frac{c_1 - c_2}{c_3 - c_4} = \frac{A}{l} \left(\frac{1}{V_1} + \frac{1}{V_2}\right) \int_{t=0}^{t=t} \bar{D}(t) \, \mathrm{d}t$$

We now denote by $\bar{D}$ the time-average of $\bar{D}(t)$ (which is itself already a concentration-average), *i.e.*, let $\bar{D} = \dfrac{1}{t} \displaystyle\int_0^t \bar{D}(t) \, \mathrm{d}t$ and also write β for the cell constant $(A/l) \left(\dfrac{1}{V_1} + \dfrac{1}{V_2}\right)$.

Then

$$\bar{D} = \frac{1}{\beta t} \ln \frac{c_1 - c_2}{c_3 - c_4} \qquad \ldots . (10.3)$$

The value $\bar{D}$ calculated from the initial and final concentrations and the time by means of (10.3) is therefore a rather complicated double average known as the diaphragm-cell integral coefficient, which it is not easy to convert immediately into the more fundamental differential diffusion coefficient D. Fortunately it has been demonstrated[8] that a negligible error is introduced in all ordinary cases if instead of using the exact relation:

$$\bar{D} = \frac{1}{t}\int_0^t \bar{D}(t)\,\mathrm{d}t$$

we treat the integrand as having a constant value equal to its value when the concentrations c' and c'' are half-way between their initial and final values; this constant value is then clearly equal to $\bar{D}$ as defined above and given by equation (10.3), and is related to the differential diffusion coefficient by:

$$\bar{D} = \frac{1}{c_{m'} - c_{m''}}\int_{c_{m''}}^{c_{m'}} D\,\mathrm{d}c \qquad \ldots.(10.4)$$

where $$c_{m'} = \frac{c_1 + c_3}{2} \quad \text{and} \quad c_{m''} = \frac{c_2 + c_4}{2}$$

The problem of computing D at various values of c from a set of $\bar{D}$ values obtained in experiments using various concentrations can be dealt with by a simple method of graphical approximation, provided that the Nernst limiting value is known (*i.e.*, that accurate limiting ionic conductivities are available); otherwise, some suitable analytical expression with arbitrary coefficients must be assumed for D as a function of c, and the coefficients determined so that equation (10.4) will fit the observed $\bar{D}$ values[9].

The cell calibration to determine β may be carried out using potassium chloride solutions, for which D is known as a function of c from absolute measurements.

The integral diffusion coefficient $\bar{D}$ corresponding to the initial and final concentrations is most readily computed as follows: a quantity $\bar{D}^0(c)$ is defined as the average D with respect to concentration over the range 0 to c,

$$\bar{D}^0(c) = \frac{1}{c}\int_0^c D\,\mathrm{d}c$$

This quantity has been computed[10] for potassium chloride at 25° from the D values of Harned and Nuttall[11] and of Gosting[12] and is given in *Table 10.1*. Then from equation (10.4) it is easily shown that:

$$\bar{D} = \left[\bar{D}^0(c_{m'}) - \frac{c_{m''}}{c_{m'}}\bar{D}^0(c_{m''})\right] \Big/ \left(1 - \frac{c_{m''}}{c_{m'}}\right)$$

Table 10.1

Integral Diffusion Coefficients of Potassium Chloride Solutions at 25°

c	$\bar{D}^0$	c	$\bar{D}^0$	c	$\bar{D}^0$
0	1·996	0·05	1·893	1·4	1·874
0·001	1·974	0·07	1·883	1·6	1·882
0·002	1·966	0·1	1·873	1·8	1·892
0·003	1·960	0·2	1·857	2·0	1·901
0·005	1·951	0·3	1·850	2·5	1·927
0·007	1·945	0·5	1·848	3·0	1·953
0·01	1·938	0·7	1·851	3·5	1·979
0·02	1·920	1·0	1·859	3·9	2·000
0·03	1·908	1·2	1·866		

From Stokes, R. H., *J. Amer. chem. Soc.*, 73 (1951) 3527.

It is usually simplest to start the experiment with pure solvent on the upper side of the diaphragm, *i.e.*, with $c_2 = 0$. The initial concentration c_1 on the lower side is not conveniently measurable in the usual design of cell, since it changes during the preliminary period of diffusion while the diaphragm is being brought into the steady-state. It can, however, readily be calculated from c_2, the final concentrations c_3 and c_4, and the volumes of the cell compartments and the diaphragm pores, using the fact that the total amount of solute in the system must be the same throughout. The volumes are measured by weighing the cell with the various parts filled in turn with water; the volume of the diaphragm being small compared to the reservoirs, the accuracy of about $\pm$ 0·02 ml. with which its volume can be determined in this way is sufficient. The small amount of solute in the diaphragm is assumed, for the purpose of calculating c_1, to be half at the concentration of the upper compartment and half at that of the lower. Thus c_1 is given by:

$$c_1 = c_3 + (c_4 - c_2) \frac{V_2 + \frac{1}{2}V_3}{V_1 + \frac{1}{2}V_3}$$

It is not, as a general rule, practicable to attempt to obtain a differential coefficient directly by working with only a small concentration-difference between the two sides of the cell, as in this case analytical errors are greatly magnified. However, if the measurements of concentration can be made by a method which permits the determination of concentration differences with high accuracy (*e.g.*, the Rayleigh interferometer) this course may be feasible and should be seriously considered where D is expected to vary rapidly with c; for in such cases the accurate evaluation of D from values of $\bar{D}$ referring to a wide concentration range is not easy.

It will be noted from equation (10.4) that the units in which c is expressed do not affect the value of $\bar{D}$; thus, for example, in self-diffusion work with radioactive tracers counting-rates may be substituted for the corresponding c's; or titration volumes may be similarly used where volumetric analyses are made.

METHODS INVOLVING SOLUTIONS OF THE EQUATION:

$$\frac{\partial c}{\partial t} = \frac{\partial}{\partial x}\left(D\frac{\partial c}{\partial x}\right)$$

In the more important absolute methods of measuring diffusion coefficients, the experimental methods are such as to require solution of the partial differential equation (2.54) for the appropriate boundary conditions. Though there are certain special cases in which equation (2.54) can be reduced by appropriate substitutions to an ordinary differential equation in a single independent variable, a general solution cannot always be found in this way; further, a general solution is possible only in the case where D is a constant. This means that the experimental conditions must usually be arranged so that the range of concentration in any one experiment is small enough to justify treating D as a constant.

MEASUREMENT OF SELF-DIFFUSION USING TRACER TECHNIQUE

Perhaps the most obvious example of a method in which this condition is fulfilled is the ANDERSON[13] capillary tube method for the study of self-diffusion. A uniform capillary tube of known length is filled with an isotopically 'tagged' solution, and immersed in a much larger vessel containing an isotopically normal solution of the same concentration, which may be gently stirred. At the mouth of the capillary, the concentration, c, of the tagged form is thus held at zero throughout the experiment. After a measured time the total amount of tagged material in the capillary is measured and compared with the initial amount.

The equation

$$\frac{\partial c}{\partial t} = D\frac{\partial^2 c}{\partial x^2} \quad (D = \text{constant}) \qquad \ldots .(10.5)$$

may be solved for this case as follows: assume that c can be expressed as a product of separate functions of x and t only,

$$c = F(x) \, . f(t)$$

Then
$$\frac{\partial c}{\partial t} = F(x)\,\frac{\mathrm{d}}{\mathrm{d}t}f(t)$$

and
$$\frac{\partial^2 c}{\partial x^2} = f(t)\,\frac{\mathrm{d}^2}{\mathrm{d}x^2}F(x)$$

Hence (10.5) becomes:

$$\frac{1}{Df(t)}\,\frac{\mathrm{d}}{\mathrm{d}t}f(t) = \frac{1}{F(x)}\,\frac{\mathrm{d}^2}{\mathrm{d}x^2}F(x) \qquad \ldots.(10.6)$$

Since the left and right sides of equation (10.6) are respectively functions of t only and of x only, the equation can be satisfied only if each side is separately equal to the same constant. Writing $-k^2$ for this constant we have the two equations:

$$\frac{\mathrm{d}}{\mathrm{d}t}f(t) = -k^2 Df(t)$$

and
$$\frac{\mathrm{d}^2}{\mathrm{d}x^2}F(x) = -k^2F(x) \qquad \ldots.(10.7)$$

The negative sign is necessary since if the constant were positive the solutions would lead to infinite values of the concentration as $t \to \infty$. Thus physically permissible solutions of the one-dimensional diffusion problem must be of the form:

$$c = b \exp\,(-k^2Dt)\,.\,F(x) \qquad \ldots.(10.8)$$

where $F(x)$ is a solution of (10.7) and b and k are constants. The most general solution is a linear combination of terms like the right-hand side of (10.8), with coefficients to be determined from the boundary-conditions. In the capillary-tube method, the boundary conditions for a tube closed at $x = 0$ and open at $x = a$ are:

$$\text{At } t = 0, \quad c = c_0 \quad \text{for} \quad 0 < x < a, \quad c = 0 \quad \text{for} \quad x > a$$

$$\text{At } t > 0, \quad c = 0 \quad \text{at} \quad x = a \quad \text{and} \quad \frac{\partial c}{\partial x} = 0 \quad \text{at} \quad x = 0$$

These conditions can be satisfied only if $k = \dfrac{2n+1}{2a}\pi$ where $n = 0, 1, 2,$ *etc.*, since $F(x)$ by equation (10.7) must clearly be a sine or cosine function. The solution is therefore:

$$c = \sum_{n=0}^{n=\infty} B_n \exp\,[-\pi^2(2n+1)^2Dt/(4a^2)]\cos\frac{\pi(2n+1)x}{2a}$$

By Fourier analysis, it is found that the coefficients B_n are given by:

$$B_n = (-1)^n \frac{4c_0}{\pi(2n+1)}$$

so that finally

$$\frac{c}{c_0} = \sum_{n=0}^{n=\infty} (-1)^n \frac{4}{\pi(2n+1)} \exp\left[-\pi^2(2n+1)^2 Dt/(4a^2)\right] \cos \frac{\pi(2n+1)x}{2a}$$

The *average* concentration in the tube at time t is:

$$c_{av} = \frac{1}{a}\int_0^a c \, \mathrm{d}x$$

whence

$$\frac{c_{av}}{c_0} = \sum_{n=0}^{n=\infty} \frac{8}{\pi^2(2n+1)^2} \exp\left[-\pi^2(2n+1)^2 \frac{Dt}{4a^2}\right] \quad \ldots.(10.9)$$

A graph of the right-hand side of (10.9) against Dt/a^2 can be prepared; interpolation on it at the experimentally determined value $\frac{c_{av}}{c_0}$ gives Dt/a^2 and hence D. It will be noted that provided the tube is uniform, its cross-section is not required, but only its length a. In computing the function (10.9) for the graph, very few terms need in practice be taken as the series converges very rapidly for reasonably large times. The ratio of the first term ($n = 0$) to the second ($n = 1$) is $9 \exp (2\pi^2 Dt/a^2)$. This ratio is greater than 1000 as soon as Dt/a^2 exceeds 0·24, and higher terms fall off even more rapidly. In a tube 5 cm in length, and with a diffusion coefficient of 10^{-5} cm^2 sec^{-1}, the first term of (10.9) is therefore amply sufficient after a week, though for the shorter times which are more practically convenient a few more terms must be taken. To illustrate the rate of change, it may be remarked that when $Dt/a^2 = 0{\cdot}24$, the average concentration in the tube has fallen to 45 per cent of its initial value.

This method has been extensively used for determining self- and tracer-diffusion coefficients of electrolytes, but agreement between different workers has often been poor, discrepancies of 10 per cent or more having been reported. In a critical study of the method, MILLS[27] has concluded that serious errors can arise from the mode of stirring of the large container into which the diffusion proceeds. Turbulent flow near the capillary mouth appears to lead to a 'scooping-out' of solution from the tube. On the other hand, if the solution is not stirred at all, a 'cloud' of the diffusing species may tend to accumulate at the mouth of the tube so that the boundary condition $c = 0$ for $x > a$ is not fulfilled. Mills has shown that correct results (*i.e.*, results in agreement with similar measurements using diaphragm-cells) can be obtained by arranging for slow controlled streamline flow past the capillary mouth. Another difficulty concerns the complete removal of all the active material from the tube at the end of the run, for radioactive counting; he overcomes this by not removing it. Instead, he surrounds the tube by a scintillation-counter crystal, making it possible to measure the decrease in activity continuously throughout the run. These improvements lead to a precision of a few tenths of one per cent in the measurement of tracer-diffusion coefficients.

MEASUREMENT OF DIFFUSION BY THE CONDUCTIMETRIC METHOD

It is characteristic of the capillary-tube method that it is permissible and indeed desirable to let diffusion proceed until the concentration-change even at the closed end of the tube is large: it may be termed a 'restricted diffusion' method in contrast to the free diffusion methods, in which an essential feature is that part of the diffusion column should be so remote from the region of the initial discontinuity that it undergoes no detectable concentration change. The optical methods to be described later are free-diffusion methods. Another important restricted-diffusion technique is the conductimetric method developed at Yale by HARNED and his collaborators[11, 14]. The diffusion channel of their cell is rectangular in cross-section (*A* in *Figure 10.4*) and its height *a* (about 5 cm) is accurately measured. It is closed permanently at the top, and at the bottom fits against a sliding plate containing two small reservoirs *B* and *C* which have the same cross-section as the channel *A*, so that by suitably sliding the plate either of them may be made to form a downward continuation of the channel. In an inverted position, the channel *A* is filled with conductivity water and the plate is placed in position with the reservoir *B* in line with *A*. On sliding the plate to the position shown, the excess water is carried off in *B*, leaving *A* completely filled. Reservoir *C* is filled with a salt solution of suitable concentration. The cell is then turned right way up and set up in an air-tight thermostated box with the most stringent precautions against mechanical vibration. After allowing a day for attainment of thermal equilibrium, the sliding plate is moved by a remote control so that the solution in reservoir *C* is in line with *A*, and salt diffuses into A. When a suitable amount has entered, the plate is moved back to the position shown and the main run begins. The concentration changes are followed by measuring the conductivity at two positions in the cell by means of pairs of very small

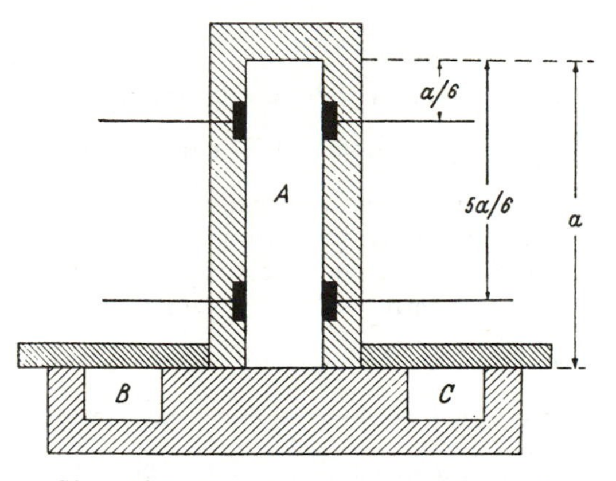

Figure 10.4. Harned's conductimetric diffusion cell (diagrammatic only)

electrodes set in opposite walls at heights $\frac{a}{6}$ and $\frac{5a}{6}$ above the sliding plate.

The boundary conditions are, since both ends of the cell are closed,

$$\frac{\partial c}{\partial x} = 0 \quad \text{at} \quad x = 0 \quad \text{and at} \quad x = a$$

and the appropriate Fourier-series solution[15] of equation (10.5) for the concentration c at a height x is:

$$c = c_0 + \sum_{n=1}^{n=\infty} B_n \exp\,(-\,n^2\pi^2 Dt/a^2)\cos\frac{n\pi x}{a}$$

where c_0 and the B_n's are constants. Hence the difference in concentration between the planes $x = \frac{a}{6}$ and $x = \frac{5a}{6}$ is

$$c_{a/6} - c_{5a/6} = \sum_{n=1}^{n=\infty} B_n \exp\left(-\,n^2\pi^2\frac{Dt}{a^2}\right)\left[\cos\frac{n\pi}{6} - \cos\frac{5n\pi}{6}\right] \quad \ldots.(10.10)$$

For even values of n, $\cos \frac{5n\pi}{6} = \cos \frac{n\pi}{6}$ and for odd n, $\cos \frac{5n\pi}{6} = -\cos \frac{n\pi}{6}$ so that all the terms for even n vanish since the factor in square brackets is zero; and for odd n the square bracket becomes $2 \cos \frac{n\pi}{6}$ which equals $\sqrt{3}$ for $n = 1$, 0 for $n = 3$, $-\sqrt{3}$ for $n = 5$ and 7, *etc.* Equation (10.10) therefore becomes:

$$c_{a/6} - c_{5a/6} = B_1' \exp(-\pi^2 Dt/a^2) + B_5' \exp(-25\pi^2 Dt/a^2) + \ldots$$

where $B_1' = B_1\sqrt{3}$, *etc.* Since the leading term of this expression exceeds the second term by the factor $\exp(24\pi^2 Dt/a^2)$ the series converges very rapidly even for small values of Dt/a^2, and after a few days only the first term need be considered at all. This rapid convergence is a result of the ingenious choice of the heights $\frac{a}{6}$ and $\frac{5a}{6}$ for the electrode pairs, which makes the term for $n = 3$ vanish at all times. The coefficient B_1' need not be determined, for by logarithmic differentiation one obtains:

$$\frac{\mathrm{d}}{\mathrm{d}t} \ln [c_{a/6} - c_{5a/6}] = -\frac{\pi^2 D}{a^2} \qquad \ldots .(10.11)$$

so that by plotting $\ln [c_{a/6} - c_{5a/6}]$ against the time t a straight line of slope $-\frac{\pi^2 D}{a^2}$ results.

In the early stages of the experiment the assumption of constant D may not be justified, but as the diffusion proceeds the concentration-differences become smaller, and D is more nearly constant throughout the solution. The remarkably constant values of D given by equation (10.11) after the first day are evidence for the validity of the theoretical treatment. The constant value attained can therefore be treated as the differential diffusion coefficient at the average concentration of the solution, which is found by allowing the cell solution to mix under the action of thermal convection after completing the run, and measuring its concentration conductimetrically.

This method demands great care and elaborate precautions to avoid trouble from vibration and thermal convection, owing to the long duration of the runs; but it is extremely important since it provides a means by which data accurate to 0·1 per cent can be obtained for electrolyte solutions more dilute than about 0·05 M, a concentration region of great theoretical interest. In this region the diaphragm-cell is untrustworthy owing to the surface transport effect mentioned above, and the optical methods discussed below (except that recently described by Bryngdahl—*see* p. 282) cannot give reliable results since the change of the refractive index between the ends of the diffusion column is too small. It is indeed remarkable that Harned and his collaborators have been successful in making measurements at concentrations as low as 0·001 M, for at these concentrations only minute density gradients exist and the stabilizing effect of gravity is therefore very slight; to preserve a stable column of diffusing liquid for a week or more under such conditions is a striking experimental achievement. Some idea of the precautions necessary to avoid thermal disturbances is given by the fact that, when it was desired to mix the cell solution thoroughly at the end of the run (in order to determine the average concentration), it was only necessary to place a heating lamp outside the sealed box containing the cell; absorption of radiation by the black platinum electrodes started convection currents which produced complete uniformity within a few hours.

OPTICAL METHODS

The various optical methods for determining diffusion coefficients employ some form of cell in which a sharp boundary can be established between two initially uniform columns of liquid of different concentration. There is thus a sharp discontinuity in refractive index at the beginning of the experiment; as diffusion proceeds, the discontinuity is replaced by an increasingly broad region of gradual change of refractive index, which is studied by suitable optical arrangements.

Whilst it is possible to obtain a Fourier series solution valid at any time for a cell of which the ends are closed at known distances from the initial boundary, this form of solution is not well adapted to the needs of the optical methods. A solution of much simpler form exists for the special case in which the two columns of liquid extend to a virtually infinite distance above and below the boundary, and is applicable to columns of finite length provided that the times considered are not long enough for detectable concentration changes to have reached the ends of the cell. This solution may be obtained from the general Fourier-series solution by a special method of summing the infinite series, but is more readily arrived at independently as follows:

In the cell shown in *Figure 10.5*, let one-dimensional diffusion in the x-direction proceed from an initially sharp boundary between two semi-infinite columns of liquid of initial concentrations c_1 and c_2. The boundary conditions are given by:

$$\begin{aligned} \text{at } t = 0 \quad & c = c_1 \quad \text{for} \quad 0 > x > -\infty \\ & c = c_2 \quad \text{for} \quad 0 < x < +\infty \\ \text{at } t > 0 \quad & c = c_1 \quad \text{at} \quad x = -\infty \\ & c = c_2 \quad \text{at} \quad x = +\infty \end{aligned} \qquad \ldots (10.12)$$

Define a new variable $y = x/(2\sqrt{Dt})$.

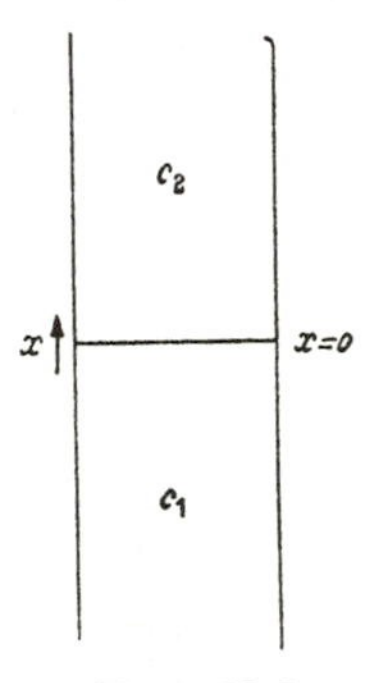

Figure 10.5

The boundary conditions (10.12) can then be stated more simply as:

$$\begin{aligned} c = c_1 \quad & \text{for} \quad y = -\infty \\ c = c_2 \quad & \text{for} \quad y = +\infty \end{aligned} \qquad \ldots .(10.13)$$

since at $t = 0, y \to -\infty$ and $+\infty$ for all finite x below and above the boundary respectively, and for positive $t, y = \pm\infty$ corresponds to $x = \pm\infty$. This reduction of the two sets of boundary conditions (10.12) to a single set (10.13) is clearly only possible because the columns are considered to extend to infinity; the reader who is not convinced of this may attempt a similar reduction for the case of finite columns.

In terms of the new independent variable y, equation (2.54) can be reduced from a partial to an ordinary differential equation. We have, assuming once more that D is constant,

$$\frac{\partial c}{\partial t} = \frac{dc}{dy}\frac{\partial y}{\partial t} = -\frac{x}{4\sqrt{D}}t^{-3/2}\frac{dc}{dy}$$

and $$\frac{\partial c}{\partial x} = \frac{\mathrm{d}c}{\mathrm{d}y}\frac{\partial y}{\partial x} = \frac{1}{2\sqrt{Dt}}\frac{\mathrm{d}c}{\mathrm{d}y}$$

therefore $$\frac{\partial}{\partial x}\left(D\frac{\partial c}{\partial x}\right) = \frac{1}{4t}\frac{\mathrm{d}^2c}{\mathrm{d}y^2}$$

Hence equation (2.54), *i.e.*,

$$\frac{\partial c}{\partial t} = \frac{\partial}{\partial x}\left(D\frac{\partial c}{\partial x}\right)$$

becomes:

$$\frac{\mathrm{d}^2c}{\mathrm{d}y^2} = -2y\frac{\mathrm{d}c}{\mathrm{d}y} \qquad \ldots.(10.14)$$

Equation (10.14) is readily solved $\left(\text{by the substitution } \frac{\mathrm{d}c}{\mathrm{d}y} = p\right)$ to give:

$$\frac{\mathrm{d}c}{\mathrm{d}y} = Ae^{-y^2}$$

where A is a constant. Hence

$$c = c_0 + A\int_0^y e^{-y^2}\,\mathrm{d}y$$

where c_0 is another constant. The constants c_0 and A can be evaluated from the boundary conditions (10.13), using the known definite integral:

$$\int_0^{\pm\infty} e^{-y^2}\,\mathrm{d}y = \pm\frac{\sqrt{\pi}}{2}$$

giving finally:

$$c = \frac{c_1 + c_2}{2} - \frac{c_1 - c_2}{\sqrt{\pi}}\int_0^y e^{-y^2}\,\mathrm{d}y$$

$$\frac{\mathrm{d}c}{\mathrm{d}y} = -\frac{c_1 - c_2}{\sqrt{\pi}}e^{-y^2}$$

At a given time t, therefore, the concentration c and its gradient at distance x are:

$$\left.\begin{aligned} \frac{\mathrm{d}c}{\mathrm{d}x} &= -\frac{c_1 - c_2}{2\sqrt{\pi Dt}}\exp\left(-\frac{x^2}{4Dt}\right) \\ c &= \frac{c_1 + c_2}{2} - \frac{c_1 - c_2}{2}\,\mathrm{erf}\left(\frac{x}{2\sqrt{Dt}}\right) \end{aligned}\right\} \qquad \ldots.(10.15)$$

where erf (a), the error-function of a, is defined by the definite integral

$$\operatorname{erf}(a) = \frac{2}{\sqrt{\pi}} \int_0^a e^{-z^2}\,dz$$

Values of the error-function are given in tables and books on probability theory; important special values are erf (0) = 0, erf (∞) = 1, erf ($-\infty$) = -1; and it has the property erf ($-a$) = $-$ erf (a), *i.e.*, it is an odd function.

In most of the optical methods the refractive index gradient $\frac{dn}{dx}$ is of more direct interest than the concentration and its gradient.

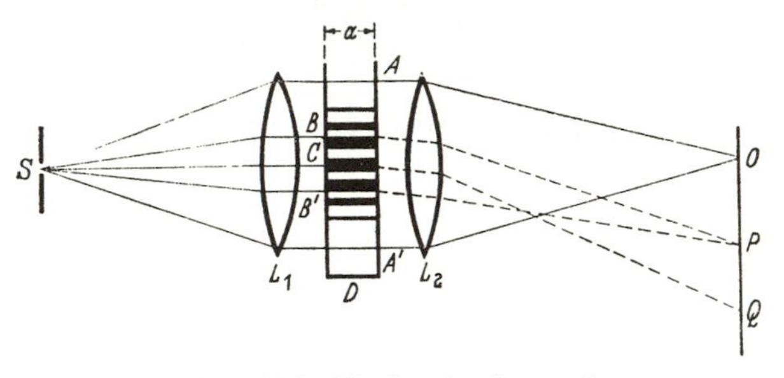

Figure 10.6. The Goüy interference effect

S—horizontal illuminated slit
L_1—collimating lens
L_2—focusing lens, focal length = b
D—diffusion cell, thickness = a
OPQ—interference pattern

Rays deviated by the diffusion-boundary are shown as broken lines.

The experimental conditions necessary to ensure a nearly constant value of D (*viz.*, that c_1 and c_2 should not be too far apart) are usually such that $\frac{dn}{dc}$ is constant also. Then (10.15) becomes:

$$\frac{dn}{dx} = -\frac{n_1 - n_2}{2\sqrt{\pi D t}} \exp\left(-\frac{x^2}{4Dt}\right) \qquad \ldots .(10.16)$$

THE GOÜY INTERFERENCE METHOD

In 1880 Goüy[16] reported a new interference phenomenon: when collimated light (*Figure 10.6*) from a horizontal slit was passed through a cell in which diffusion was occurring in a vertical direction, and the beam was brought to a focus by means of a lens, an interference pattern consisting of a finite number of horizontal bands was produced in the focal plane. The intensity of the bands is greatest near the optic axis (on which the initial boundary between the diffusing solutions is preferably situated); here they are closely spaced. Proceeding downwards from the optic axis, the interference bands become more widely spaced and less intense, ending with an especially wide one with an ill-defined lower boundary.

Qualitatively the origin of this interference pattern is easily seen: according to equation (10.16) the refractive index gradient in the cell is given by a Gaussian curve which is symmetrical about the

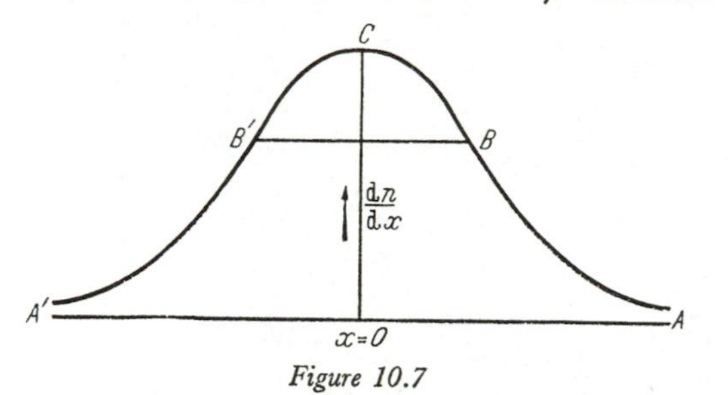

Figure 10.7

initial boundary at $x = 0$. At equal distances x above and below the boundary there are, therefore, conjugate regions of equal refractive index gradient. Parallel light passing through these regions will be bent down to the same extent in each, since each element of the solution can be considered to act as a prism tapering upwards. The refractive index gradient at various points in the cell is shown in *Figure 10.7*.

In *Figure 10.6*, the regions A and A' remote from the boundary have not yet experienced any concentration change; the refractive index at these positions is uniform, and light through A and A' is undeflected, coming to a focus on the optic axis O. At the centre of the cell, C, the refractive index gradient has always a maximum value $\left(\text{given by } \frac{dn}{dx} = -\frac{n_1 - n_2}{2\sqrt{\pi D t}}\right)$. Light passing through C therefore experiences the greatest deflection, forming the lowest band Q of the pattern. At intermediate points B and B', the light is deflected downward to a focus at P. The light-path lengths SBP and $SB'P$ are, however, not equal: when they differ by an integral number p of wavelengths, the two rays reinforce and form a bright band, and when the path lengths differ by $(p + \frac{1}{2})$ wavelengths they interfere and cancel giving a dark band. The path-length difference will obviously vary with the distance of B and B' from the optic axis C, giving rise to the alternating system of light and dark bands. This band-system is shown in *Figure 10.8*, from a photograph by J. R. Hall.

This qualitative explanation was all that was available until 1947, when a more complete theory of the phenomenon was given by KEGELES and GOSTING[17], and almost simultaneously by COULSON

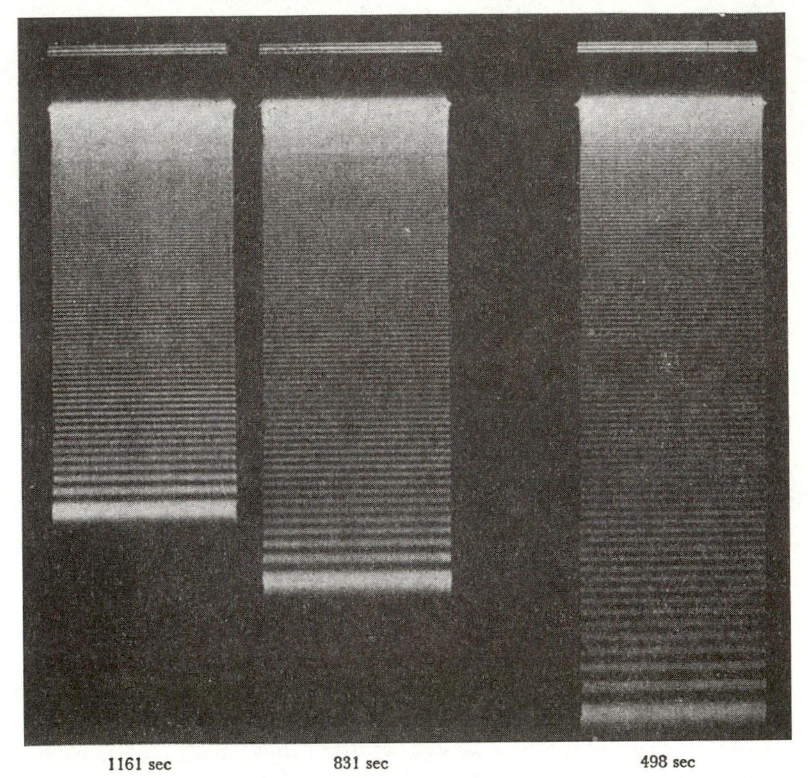

Figure 10.8. Goüy interference patterns. Produced by the diffusion of calcium chloride between concentrations of 3·48 *and* 3·58 *molar at* 25°. *The times are measured from the establishment of the sharp boundary. The three-line patterns at the top of each picture are the reference mark*

et al.[18]. An even more rigorous treatment has been given by GOSTING and ONSAGER[19]. These authors show that the reasoning based on geometric optics, in which pairs of rays through various parts of the cell are considered, is not quite adequate. Instead they use the methods of wave optics, according to which every portion of the wave front makes some contribution to the resultant amplitude at each point in the focal plane. This leads to a slight change in the conditions for reinforcement and cancellation. Instead of complete cancellation occurring when the points B and B' are so placed that the light-paths SBP and SBP' differ by $(p + \frac{1}{2})$ wavelengths, it is found to occur when they differ by very nearly $(p + \frac{3}{4})$ wavelengths. Similarly, maximum reinforcement occurs when the paths differ by very nearly $(p + \frac{1}{4})$ wavelengths instead of p wavelengths.

If the thickness of the cell along the optic axis is denoted by a, and the focal length of the second lens by b, it can be shown that, according to geometric optics, the lowest bright band of the pattern has its maximum intensity at a distance C_t below the centre of the undeviated image formed by the light passing through the unchanged solution (*i.e.*, $C_t = OQ$ in *Figure 10.6*) and that C_t is given by:

$$C_t = \text{ab}\,[(n_1 - n_2)/(2\sqrt{\pi Dt})] \qquad \ldots.(10.17)$$

Furthermore, the number of bands in the pattern bears a simple relation to the difference of refractive index, expressed as the number, j_m, of wavelengths, λ, retardation of the light passing through unchanged solution at A' compared with that passing through at A:

$$\frac{a(n_1 - n_2)}{\lambda} = m + \alpha = j_m \qquad \ldots.(10.18)$$

where α is a fraction, less than unity and m is an integer, one less than the number of bright bands in the pattern. Thus if the lowest bright band is numbered 0, the next lowest one 1, *etc.*, the mth band is the one next to the undeviated image. The main (integral) part of j_m can therefore be determined by simply counting the number of bands in the pattern; the fractional part can be found by a minor modification to the apparatus as described below, so that j_m is known for each experiment. Photographs of the interference pattern are taken at known times t after the start of the diffusion, and the distances from the undeviated image of various minima in the light-intensity are measured. The wave-optical theory[17] shows that if y_j is the displacement of the jth minimum from the undeviated position, counting the lowest minimum, *i.e.*, the one above the lowest bright band, as $j = 0$, then y_j and C_t are related by the following expression:

$$C_t = y_j e^{z^2} \qquad \ldots.(10.19)$$

where z is a dimensionless quantity given by the implicit expression:

$$f(z) \equiv \frac{2}{\sqrt{\pi}} \int_0^z e^{-z^2}\,\mathrm{d}z - \frac{2}{\sqrt{\pi}}\, z e^{-z^2} \equiv \frac{4}{\sqrt{\pi}} \int_0^z z^2 e^{-z^2}\,\mathrm{d}z = \frac{j + \frac{3}{4}}{j_m} \qquad \ldots.(10.20)$$

The rather clumsy-looking function $f(z)$ is, in fact, easily evaluated from tables of probability functions for a series of values of z, and e^{z^2} can then be tabulated against $f(z)$ (Appendix 10.1).

A further slight refinement[19] of the theory (involving the use of the zeros of the Airy integral) shows that the quantity ($\frac{3}{4}$) in the expression, $\frac{j + \frac{3}{4}}{j_m}$ in (10.20) should be replaced by slightly different values, the difference being appreciable only for the lowest fringes up to about $j = 5$. Appendix 10.2 gives these slightly modified values.

By the use of the tables in Appendixes 10.1 and 10.2, C_t may be calculated from the measured value of the displacement y_j of any chosen fringe (the jth) in the interference pattern. C_t should be constant for all the fringes in the pattern and the constancy provides a check on the correctness of the value of j_m. It is, in fact, possible to find j_m without using the Rayleigh interferometer modification of the apparatus, by counting fringes to find the integral part of j_m and then trying various fractional parts α [equation (10.18)] until a value is found which gives the best constancy for C_t for fringes corresponding to well separated j_m values. C_t can also be evaluated by measuring the maxima instead of the minima in the light-intensity though these are not so readily located as minima with the usual design of travelling microscope. For maxima, the only difference in the calculation is that $(j + \frac{3}{4})$ of equation (10.20) is replaced by $(j + \frac{1}{4})$; again the Airy integral refinement gives slightly different values from $\frac{1}{4}$ for the lowest fringes. From the constant value of C_t for a pattern photographed at time t, D is calculated by combining (10.17) and (10.18) to give:

$$D = \frac{j^2_m}{tC_t^2} \frac{b^2 \lambda^2}{4\pi} \qquad \ldots .(10.21)$$

The measurements are most conveniently made at the wavelength of the green mercury line, $\lambda = 5461$ Å. The distance b, the focal length of the second lens, must be very accurately known, and correct focus is critical. It is often more convenient to replace the two lenses of *Figure 10.6* by a single lens of focal length of the order of 20 cm which focuses an image of the slit on to the photographic plate. This is placed at least a metre from the lens so that the light through the diffusion cell is only slightly convergent; under these conditions the same theory holds, but b in equation (10.21) must now be taken as an 'optical distance' from the centre of the cell to the photographic plate, given by $b = \sum \frac{l}{n}$ where l is the distance through each medium (air, glass, thermostat-water, or solution) and *n* is its refractive index. In this method it is possible to fix the plate at a measured distance from the cell and then to focus by moving the lens or the source-slit through small known increments (*e.g.*, 0·001 in.) by means of a set of feeler gauges placed between the lens or slit mount and a fixed stop on the optical bench, taking a photograph at each setting; the setting which gives the sharpest image of the slit is correct.

Various methods are employed for forming the sharp boundary in the cell. A conventional Tiselius electrophoresis cell may be

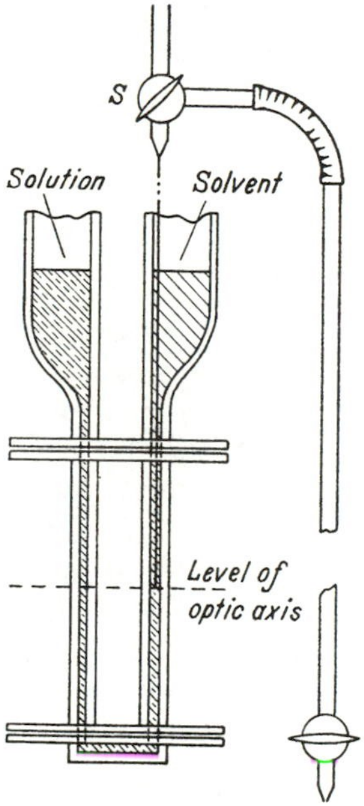

Figure 10.9. From GOSTING, L. J., HANSON, E. M., KEGELES, G. and MORRIS, M. S., *Rev. Sci. Instrum.*, 20 (1949) 209

used[20]; this gives a sheared boundary which has to be displaced so as to bring it into view, the initial position being obscured by the sliding faces (*Figure 10.9*). The boundary is disturbed by shifting and it is desirable to sharpen it again by drawing the disturbed solution out through a fine capillary tube. In another system[18, 21] the boundary is formed by allowing the two solutions to flow out through a horizontal slit in the wall of the cell at the level of the optic axis and then smoothly stopping the flow; this avoids the need for lubricated sliding surfaces with the consequent risks of leakage and grease contamination. A cell employing this system is shown in *Figure 10.10.*

Freedom from vibration, and adequate thermostating, are important requirements of the Goüy interference method, as of all methods in which the diffusion column is stabilized only by gravity.

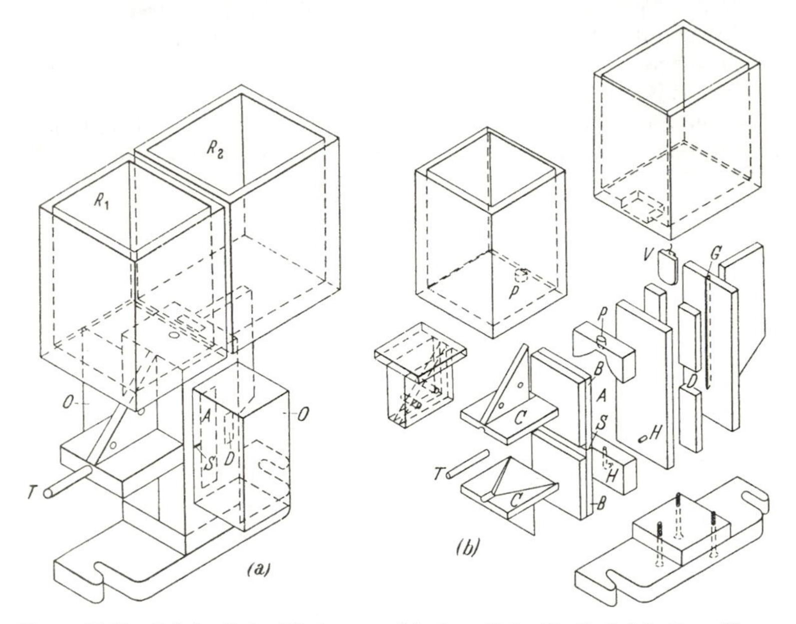

Figure 10.10. Cell for Goüy diffusiometer. (a) Assembled. (b) Exploded. From HALL, J. R., WISHAW, B. F. *and* STOKES, R. H., *J. Amer. chem. Soc.*, 75 (1953) 1556

R_1—reservoir for dilute solution
R_2—reservoir for concentrated solution
A—diffusion channel
S—sharpening slit
T—exit tube for solution from slits
D—reference-channel, filled with homogeneous concentrated solution for light forming reference marks
O—optical flats
C—collecting-channel for solution from S
H—hole admitting concentrated solution to bottom of channel A
P—hole admitting dilute solution to top of channel A
V—gate-valve for closing H during filling of cell.

In use, a thermostated water-jacket surrounds the whole of the cell except the outer faces of the two optical flats, to which it is fitted by rubber gaskets

For work within about 10° of room temperature it has been found sufficient[18, 21] to thermostat the cell by means of a jacket surrounding all of the cell except the outer faces of the optical flats, to which it is fitted by rubber gaskets; for more extreme temperatures, the cell must be completely immersed, which requires a thermostat with optically flat windows, or one in which the windows are formed by the lenses of the optical system.

In the interference pattern, the undeviated slit image is of very high intensity relative to the rest of the pattern, especially at short times when most of the light passes through uniform solution; it is consequently over-exposed as appears from *Figure 10.8*, and cannot be located accurately. Its intensity may be reduced by suitably placed filters at the plate or by specially shaped masks at the cell, but a more satisfactory way of locating it[18] is to form on the plate a reference mark lying a few millimetres above the undeviated image. The distance between this mark and the undeviated image can be accurately measured, and the reference mark is then used as a base line for the measurements of fringe positions. This reference mark is most conveniently made by placing at the cell a double stop consisting of two rectangular holes about one to two millimetres square and the same distance vertically apart. Light from the source-slit passes through this double stop, and through the thermostat-water or a channel in the cell which always contains uniform solution, and is then displaced upwards by passing through a tilted fixed optical flat. A Rayleigh interference pattern is thus formed at the plate, as shown in *Figure 10.8*, at a distance above the undeviated slit image which is determined by the angle and thickness of the tilted flat. While the main diffusion channel is filled with uniform solution, a similar double stop is placed over it also, and a similar Rayleigh pattern is formed at the plate, with its central fringe exactly where the undeviated slit image will be in the diffusion photographs. These two interference patterns are very well defined and can be located with high accuracy. During the diffusion exposures, the double stop is removed from the main channel, but the other one forming the reference mark remains in place.

The same system of double stops can be used[18] to determine the fractional part α of j_m (equation 10.18). For this purpose the double stops are placed over the diffusion channel while a sharp boundary is maintained between them, *e.g.*, by flowing the liquids out of the lateral sharpening-slit. Thus light from one of the rectangular holes passes through solution of refractive index n_1, and that from the other through solution of refractive index n_2. If the light-paths through the solutions differ by an integral number of wavelengths (*i.e.*, if j_m is integral) the pattern is identical in appearance and position with that formed when uniform solution fills the whole diffusion-channel; but if j_m is not integral, the pattern changes in appearance and position. The relative intensities of the lines alter, but the spacings between minima in the light intensity remain the same. The intensity-diagrams for various fractional values of α are

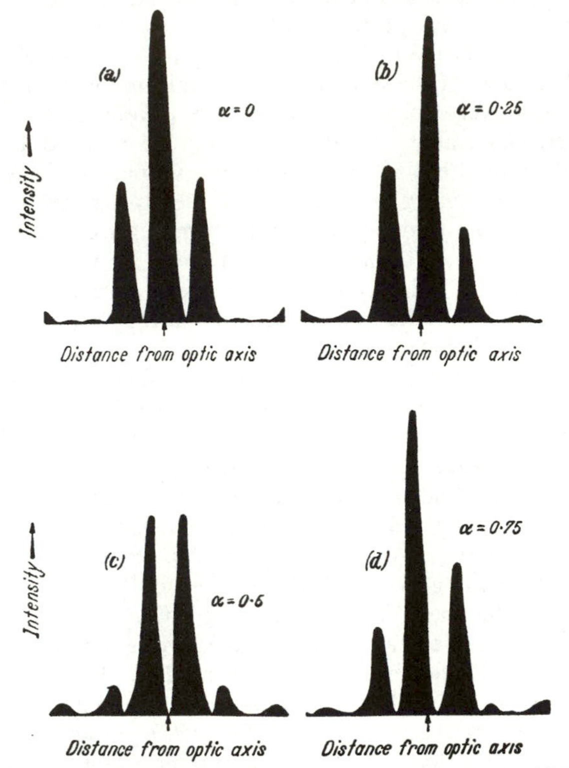

Figure 10.11. Theoretical light-intensity-distribution curves in the Rayleigh interference patterns formed by a double stop consisting of two slots of width d separated by a distance d. The quantity α *is the fractional part of the number of wavelengths difference in the light-path for the two slots (equation* 10.18). *The symmetrical pattern A is found when this light-path difference is zero or an integral number of wavelengths, as, for instance, in the upper (reference mark) patterns in Figure 10.8.*

The location of the optic axis in each pattern is shown by the small arrow-head, and points to the right of this are below the optic axis in the actual patterns photographed.

(Curves computed from formulae in reference 18)

shown in *Figure 10.11*; it will be seen that $\alpha = 0{\cdot}5$ gives a symmetrical four-line pattern centred on the optic axis, while $\alpha < 0{\cdot}5$ gives unsymmetrical patterns in which the nearest minimum to the optic axis is *above* it, and $\alpha > 0{\cdot}5$ gives similar patterns reflected in the optic axis. The theory shows[18] that if S be the spacing between minima in the Rayleigh pattern and s the distance from the nearest minimum to the optic axis, then:

$\alpha = 0{\cdot}5 - \frac{s}{S}$ if this minimum lies above the optic axis and

$\alpha = 0{\cdot}5 + \frac{s}{S}$ if it lies below the optic axis. The counting of fringes to obtain the integral part of j_m can usually be done satisfactorily under a travelling microscope, but occasionally it is in doubt within one integer due to over-exposure near the optic axis: this doubt is easily settled by the requirement that C_t should be constant for all fringes of the Goüy pattern.

No cell design can, in practice, give a mathematically sharp boundary at zero time; with the best designs, the degree of initial blurring of the boundary is equivalent to what would be reached in a time of five to fifteen seconds after the formation of an ideal boundary. This 'zero-time correction', Δt, depends on the density-gradient in the solution and on the diffusion coefficient, decreasing as these increase. It is allowed for by making a series of exposures at known times, *e.g.*, 5, 10, 20, 30 min. after the start of the experiment. The times may be recorded by photographing the dial of a stop-watch which is started when the boundary is formed[21], the exposure being controlled by the movement of the same shutter which exposes the fringe patterns; or by moving the plate mechanically at a measured rate[18]. The diffusion coefficient D given by equation (10.21) is then not constant for the several exposures, since the time t of equation (10.21) should be replaced by $t + \Delta t$. However, by plotting the D values for different times against $\frac{1}{t}$ a straight line of slope $D\Delta t$ is obtained, which extrapolates to the true value of D at $\frac{1}{t} = 0$. With a little experience it becomes possible to start the stop-watch approximately Δt seconds before the start of the diffusion, thus making the remaining correction very small and the extrapolation graph practically horizontal.

The Goüy method is probably the most exact of those at present available for measuring diffusion coefficients. It is, however, restricted to concentrations large enough to give a reasonable number of fringes in the interference pattern; at least 30 are desirable for 0·1 per cent accuracy. With special cell designs it is possible to obtain results accurate to 1 per cent with as few as 10 bands[22], but even this corresponds for electrolyte solutions to a concentration-difference between the upper and lower solutions of the order of

0·02 M in a cell of 2 cm length along the light-path. For the study of dilute electrolyte solutions the Goüy method must therefore give way to Harned's conductimetric method. Another limitation is that the solution must not absorb light of the wavelength used.

OTHER OPTICAL METHODS

While the Goüy method has in recent years become preferred among the optical methods, some earlier methods are of great interest, and retain considerable importance in the study of colloids.

In the Lamm scale method[23], a transparent scale ruled with horizontal lines (spaced, say, 0·2 mm apart) is illuminated by monochromatic light, and an image of it is focused on to a photographic plate by means of a long-focus lens. The diffusion cell is placed between the scale and the lens, the optical distance between the scale and the centre of the cell being b. The length of the cell along the light-path is denoted by a. When the cell is filled with *uniform* solution, the image of the scale formed on the plate is undistorted but magnified by a factor G depending on the relative positions of lens, scale and plate. When a diffusion-boundary is present in the cell, however, the light passing through regions of varying refractive index will be deflected downwards by amounts proportional to the refractive index gradient, and the image of the scale will be distorted, the scale-lines being displaced from their normal positions. Geometric optics show that the displacement of a scale-line is given by:

$$Z = Gab \frac{dn}{dx} \qquad \ldots .(10.22)$$

and the refractive index gradient $\frac{dn}{dx}$ at a distance x from the boundary by:

$$\frac{dn}{dx} = \frac{n_1 - n_2}{2\sqrt{\pi Dt}} \exp\left(-\frac{x^2}{4Dt}\right) \qquad \ldots .(10.23)$$

In these expressions, n_1, n_2, x, D, t have the same meanings as in the theory of the Goüy method. From the photographic measurements, values of $\frac{dc}{dx}$ at various values of x can be computed. It is sufficient to know $\frac{dc}{dx}$ in arbitrary units, since the constant of proportionality α between $\frac{dc}{dx}$ and $\frac{dn}{dx}$ is determined during the calculation, as follows:

Figure 10.12 represents the graph of $\alpha \frac{dc}{dx}$ versus x which can be plotted from the measurements at a time t after the start of the diffusion process. If D is independent of c, this graph is the Gaussian curve (10.23); if D varies with c it is more or less skewed. In the latter case equation (10.23) no longer holds, but equation (10.22) is still valid, so that the form of the curve of $\frac{dc}{dx}$ versus x can still be obtained. For this general case[23a], equation (2.54) reduces by the

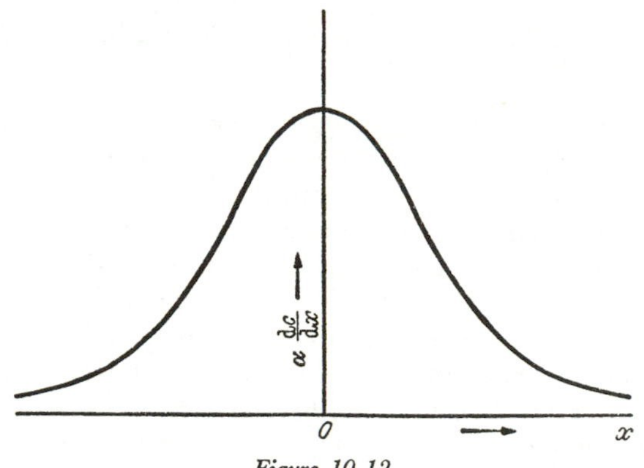

Figure 10.12

substitution $y = xt^{-1/2}$ (applicable under the boundary conditions for these experiments) to:

$$y \frac{dc}{dy} = -2 \frac{d}{dy}\left(D \frac{dc}{dy}\right) \qquad \ldots .(10.24)$$

$$\left.\begin{aligned} c = c_1 \quad \text{at} \quad y = -\infty \\ c = c_2 \quad \text{at} \quad y = +\infty \end{aligned}\right\} \qquad \ldots .(10.25)$$

It follows that at a fixed time t,

$$\alpha \frac{dc}{dx} = \alpha t^{-1/2} \frac{dc}{dy}$$

$$x = t^{1/2} y$$

The nth moment μ_n of the curve in *Figure 10.12* about the vertical axis through $x = 0$ is defined by:

$$\mu_n = \int_{-\infty}^{\infty} \alpha \frac{dc}{dx} x^n \, dx \qquad \ldots .(10.26)$$

From equations (10.24), (10.25) and (10.26) it is easily shown that:

$$\left.\begin{aligned}\mu_0 &= -\alpha(c_1 - c_2)\\ \mu_1 &= 0\\ \mu_2 &= -2\alpha t\int_{c_2}^{c_1} D\,\mathrm{d}c\end{aligned}\right\} \qquad \ldots.(10.27)$$

so that

$$\mu_2/\mu_0 = \frac{2t}{c_1 - c_2}\int_{c_2}^{c_1} D\,\mathrm{d}c = 2\bar{D}t \qquad \ldots.(10.28)$$

where $\bar{D}$ is the average value with respect to concentration of the diffusion coefficient over the concentration range of the experiment. In the ideal case where D is constant, the same results of course hold, but it may be more convenient to calculate D from the height and area of the curve, or from its points of inflection, because these characteristics are simply related to the moments for a true Gaussian curve. The calculation given implies that the position corresponding to $x = 0$, the initial boundary, is accurately known from the photograph; this is not so in practice, and the origin of x is actually located by means of the property $\mu_1 = 0$ (equation 10.27), *i.e.*, it is taken vertically below the centroid of the curve in *Figure 10.12*. The zero-th and second moments are then calculated with this origin.

In addition to the Lamm scale method, there are several other ways of obtaining the refractive index-gradient curve, of which the chief are the LONGSWORTH[24] 'schlieren scanning' method and PHILPOT[25] 'diagonal schlieren' method. These methods are important in the study of colloids, but have found little application in the field of simple electrolytes; for this reason and because of the complexity of the optical systems, they will not be described here. Other recent developments include the 'integral fringe' method[26], by which a photographic record of the *concentration* rather than its gradient is obtained; this is valuable in cases where the diffusion coefficient varies strongly with concentration, as with some high molecular weight solutes.

An extremely promising new optical method is described by BRYNGDAHL[28], whose apparatus incorporates a Savart plate which produces birefringent interferences. This has the effect of amplifying the refractive index differences in the diffusing solution, so that extremely small concentration differences between the upper and lower parts of the diffusion-cell may be used. For example, he reports the result $D = (5{\cdot}229 \pm 0{\cdot}011) \times 10^{-6}$ cm^2 sec^{-1} for sucrose at 25°, for diffusion into water from a solution containing only 0·0112 per cent of sucrose by weight. This is in excellent agreement with the result obtained by GOSTING and MORRIS[19] by extrapolation to zero concentration of their Goüy data, *viz.*, $5{\cdot}226 \times 10^{-6}$ $cm^2\,sec^{-1}$; the Goüy method however would be quite impracticable for direct use at this low concentration. The method will obviously be of great value in the study of diffusion in dilute electrolytes.

REFERENCES

1 CLACK, B. W., *Proc. phys. Soc.*, 36 (1924) 313

2 NORTHROP, J. H. and ANSON, M. L., *J. Gen. Physiol.*, 12 (1929) 543

3 MCBAIN, J. W. and DAWSON, C. R., *Proc. roy. Soc.*, 148 A (1935) 32

4 HARTLEY, G. S. and RUNNICLES, D. F., *ibid.*, 168 A (1938) 401

5 STOKES, R. H., *J. Amer. chem. Soc.*, 72 (1950) 763

6 MOUQUIN, H. and CATHCART, W. H., *ibid.*, 57 (1935) 1791

7 MYSELS, K. J. and MCBAIN, J. W., *J. Colloid Sci.*, 3 (1948) 45

8 GORDON, A. R., *Ann. N.Y. Acad. Sci.*, 46 (1945) 285

9 STOKES, R. H., *J. Amer. chem. Soc.*, 72 (1950) 2243; HAMMOND, B. R. and STOKES, R. H., *Trans. Faraday Soc.*, 49 (1953) 890

10 STOKES, R. H., *J. Amer. chem. Soc.*, 73 (1951) 3527

11 HARNED, H. S. and NUTTALL, R. L., *ibid.*, 69 (1947) 736; 71 (1949) 1460

12 GOSTING, L. J., *ibid.*, 72 (1950) 4418

13 ANDERSON, J. S. and SADDINGTON, K., *J. chem. Soc.*, (1949) S 381

14 HARNED, H. S. and FRENCH, D. M., *Ann. N.Y. Acad. Sci.*, 46 (1945) 267

15 CARSLAW, H. S. and JAEGER, J. C., 'Conduction of Heat in Solids,' Oxford University Press, 1947

16 GOÜY, G. L., *Compt. rend.*, 90 (1880) 307

17 KEGELES, G. and GOSTING, L. J., *J. Amer. chem. Soc.*, 69 (1947) 2516

18 COULSON, C. A., COX, J. T., OGSTON, A. G. and PHILPOT, J. St. L., *Proc. roy. Soc.*, 192 A (1948) 382

19 GOSTING, L. J., and ONSAGER, L. *J. Amer. chem. Soc.*, 74 (1952) 6066; GOSTING, L. J. and MORRIS, M. S., *ibid.*, 71 (1949) 1998

20 — HANSON, E. M., KEGELES, G. and MORRIS, M. S., *Rev. Sci. Instrum.*, 20 (1949) 209

21 HALL, J. R., WISHAW, B. F. and STOKES, R. H., *J. Amer. chem. Soc.*, 75 (1953) 1556

22 CREETH, J. M., *Biochem. J.*, 51 (1952) 10

23 LAMM, O., *Nova Acta Soc. Sci. Upsaliensis*, 10, No. 6, Series IV (1937) 15

23a STOKES, R. H., *Trans. Faraday Soc.*, 48 (1952) 887

24 LONGSWORTH, L. G., *Industr. Engng. Chem.* (Anal. Ed.), 18 (1946) 219

25 PHILPOT, J. St. L., *Nature*, 141 (1938) 283

26 — and COOK, G. H., *Research*, 1 (1948) 234; SVENSSON, H., *Acta chem. scand.*, 5 (1951) 72; LONGSWORTH, L. G., *J. Amer. chem. Soc.*, 74 (1952) 4155

27 MILLS, R., *J. Amer. chem. Soc.*, 77 (1955) 6116; MILLS, R. and GODBOLE, E. W., *Aust. J. Chem.* 11 (1958) 1

28 BRYNGDAHL, O., *Acta chem. scand.*, 11 (1957) 1017

11

THE THEORY OF DIFFUSION; CONDUCTANCE AND DIFFUSION IN RELATION TO VISCOSITY IN CONCENTRATED SOLUTIONS

TABLES OF DIFFUSION COEFFICIENTS OF ELECTROLYTE SOLUTIONS

APPENDICES 11.1 and 11.2 present values of the diffusion coefficients of a number of aqueous electrolytes as determined by the three most reliable of the modern methods, *viz.*, the conductimetric method of Harned, the Goüy interference method and the magnetically-stirred diaphragm-cell method. The first two of these are absolute methods; the conductimetric method has in most cases been restricted to solutions below 0·01 M, but for potassium chloride has been used up to 0·5 M. The Goüy method is at its best for solutions above 0·05 M, and in the case of potassium chloride in the range 0·1–0·5 M there is remarkably good agreement with the conductimetric method. The diaphragm-cell method is a relative one, and its calibration has been based on the absolute data for potassium chloride; results from it for potassium chloride agree with those from the absolute methods within about 0·2 per cent at other concentrations than the one used for calibration. However, since the diaphragm-cell gives an integral diffusion coefficient, there is some loss of accuracy in converting this to a differential diffusion coefficient: the differential diffusion coefficients derived from the original magnetically-stirred diaphragm-cell data for potassium chloride in the range 0·1 N–4 N showed an average deviation of approximately 0·5 per cent from later results by the Goüy method, most of the discrepancy arising from errors in two points near the minimum of the integral diffusion coefficient curve, which had affected the differential results over an appreciable range. In general, the accuracy of the differential diffusion coefficients listed may be taken as 0·2 per cent or better for the Goüy and conductimetric methods, and 0·3 per cent for the diaphragm-cell method.

THEORETICAL DISCUSSION OF DIFFUSION

Both diffusion and electrical conductance in electrolyte solutions involve the motion of ions and it is therefore to be expected that a relation will exist between the diffusion coefficient of an electrolyte and its equivalent conductivity. The most important differences between the two processes are (*a*) in conduction positive and negative ions move in opposite directions, whilst in diffusion they move in the same direction, and (*b*) in conduction, at the limit of extreme dilution, the various ions of an electrolyte move independently of one another, whereas in diffusion they are obliged to move at equal speeds, since otherwise a separation of electrical charge in the solution would result. Both processes can be regarded as deriving from small perturbations of the ordinary molecular motions; in conduction, the perturbing influence is the external electric field, and in diffusion, the concentration gradient. In Nernst's original derivation[1] of the relation between the two effects, the osmotic pressure was given the status of a driving force for diffusion, analogous to the electrical field in conduction. Though this approach leads to the correct result in the limiting case of infinite dilution, modern views on osmotic pressure do not favour regarding it as an actual pressure in the solution. Instead, the gradient of chemical potential in the solution, which has the dimensions of a force per unit quantity of solute, is treated as the virtual force producing diffusion; this course was first suggested by GIBBS[2] and later by GUGGENHEIM[3], HARTLEY[4] and by ONSAGER and FUOSS[5]. It is by no means easy to justify the use of the free energy, a quantity usually relevant to systems in equilibrium, in dealing with an irreversible process such as diffusion; thus it is well known that in general the rate of a chemical reaction is not directly related to the free energy change during the reaction. The step of equating the free energy change, occurring when the solution mixes by diffusion, with the work done by the diffusing particles against the resistance of the medium therefore requires careful scrutiny; this it has received at the hands of ONSAGER[6], DE GROOT[7] and others. Here we propose to accept its validity without further discussion, merely noting that diffusion is a slow process in which departures from equilibrium are small compared with chemical processes, and that in these circumstances the whole of the free energy change can be taken as energy dissipated by the viscous forces.

Each ion of the diffusing electrolyte can be regarded as moving under the influence of two forces, (*a*) the gradient of chemical potential for that ionic species, and (*b*) an electrical field produced by the motion of oppositely charged ions. The more mobile ions will tend to diffuse faster than the less mobile ones, but by doing so they will create on a microscopic scale a charge separation or gradient of electric potential in the solution. This will have the effect of increasing the speed of the slower ion and decreasing that of the faster; the resultant speeds of both must finally be equal, since it is an experimental fact that a macroscopic charge separation does not occur.

DIFFUSION OF A SINGLE ELECTROLYTE: THE NERNST-HARTLEY RELATION

The diffusion of a single electrolyte is especially amenable to exact theoretical treatment, since the condition of electrical neutrality requires that anions and cations must move at the same speed. Where more than two ionic species are present the situation becomes more complex, since there is an infinite number of ways of satisfying the electrical neutrality condition; general equations can be derived, but not necessarily solved for such cases.

For a single electrolyte, one 'molecule' of which gives ν_1 cations of algebraic valency z_1 and ν_2 anions of algebraic valency z_2, the following argument is used. The chemical potentials $\bar{G}_1$ and $\bar{G}_2$ of the cations and anions may be considered separately (although they are not separately measurable) provided that the final equations contain only the chemical potential of the solute as a whole,

$$\bar{G}_B = \nu_1\bar{G}_1 + \nu_2\bar{G}_2 \qquad \ldots.(2.7)$$

The forces on single ions due to the gradient of chemical potential are therefore:

$$-\frac{1}{N}\frac{\partial\bar{G}_1}{\partial x} \quad \text{and} \quad -\frac{1}{N}\frac{\partial\bar{G}_2}{\partial x}$$

respectively where N is the Avogadro number. The negative sign is used since the ionic motion is down the free energy gradient. The effect arising from the unequal mobilities of the ions may be represented as an electrical field of intensity E, which exerts on each ion an additional force given by z_1eE and z_2eE respectively. The total forces are therefore:

$$F_1 = -\frac{1}{N}\frac{\partial\bar{G}_1}{\partial x} + z_1eE$$

$$F_2 = -\frac{1}{N}\frac{\partial\bar{G}_2}{\partial x} + z_2eE$$

These forces, acting respectively on ions of absolute mobilities u_1 and u_2, are required to produce equal velocities v given by:

$$v = u_1\left(-\frac{1}{N}\frac{\partial \bar{G}_1}{\partial x} + z_1 eE\right) = u_2\left(-\frac{1}{N}\frac{\partial \bar{G}_2}{\partial x} + z_2 eE\right)$$

From these equations eE can be eliminated giving:

$$\frac{1}{z_1}\left(\frac{v}{u_1} + \frac{1}{N}\frac{\partial \bar{G}_1}{\partial x}\right) = eE = \frac{1}{z_2}\left(\frac{v}{u_2} + \frac{1}{N}\frac{\partial \bar{G}_2}{\partial x}\right)$$

whence, using the condition of electrical neutrality:

$$\nu_1 z_1 + \nu_2 z_2 = 0$$

one obtains:

$$v = -\frac{1}{N}\frac{u_1 u_2}{\nu_1 u_2 + \nu_2 u_1}\cdot\left(\nu_1\frac{\partial \bar{G}_1}{\partial x} + \nu_2\frac{\partial \bar{G}_2}{\partial x}\right) = \frac{-1}{N}\frac{u_1 u_2}{\nu_1 u_2 + \nu_2 u_1}\frac{\partial \bar{G}_B}{\partial x}$$

Now let c be the concentration of solute in moles per unit volume at the point considered. Then the flux of solute is:

$$J = cv = -\frac{u_1 u_2}{\nu_1 u_2 + \nu_2 u_1}\cdot\frac{c}{N}\cdot\frac{\partial \bar{G}_B}{\partial c}\frac{\partial c}{\partial x}$$

But the flux also defines the diffusion coefficient D in terms of the concentration gradient:

$$J = -D\frac{\partial c}{\partial x}$$

Therefore D is given by:

$$D = \frac{u_1 u_2}{\nu_1 u_2 + \nu_2 u_1}\frac{1}{N}\frac{\partial \bar{G}_B}{\partial \ln c} \qquad \ldots.(11.1)$$

Also, from the definition of the mean molar activity coefficient, the differential in equation (11.1) is:

$$\frac{\partial \bar{G}_B}{\partial \ln c} = RT(\nu_1 + \nu_2)\left(1 + \frac{\mathrm{d}\ln y_\pm}{\mathrm{d}\ln c}\right) \qquad \ldots.(11.2)$$

and finally the absolute ionic mobilities u may be expressed in terms of the limiting equivalent conductivities, λ^0, by equation (2.46),

giving: $$D = \frac{(\nu_1 + \nu_2)\lambda_1^0\lambda_2^0}{\nu_1|z_1|(\lambda_1^0 + \lambda_2^0)}\frac{RT}{F^2}\left(1 + \frac{\mathrm{d}\ln y_\pm}{\mathrm{d}\ln c}\right) \qquad \ldots.(11.3)$$

These formulae may be called the Nernst–Hartley relation. The limiting value of D at infinite dilution, where $\frac{\mathrm{d} \ln y_\pm}{\mathrm{d} \ln c} \rightarrow 0$, is given by:

$$D^0 = \frac{RT(\nu_1 + \nu_2)}{F^2 \nu_1 |z_1|} \frac{\lambda_1^0 \lambda_2^0}{(\lambda_1^0 + \lambda_2^0)} \qquad \ldots.(11.4)$$

an expression due to Nernst. Equivalent forms of equation (11.3) obtained by using the condition of electrical neutrality $\nu_1|z_1| = \nu_2|z_2|$ and the definition of the transport numbers $t_1^0 = \lambda_1^0/(\lambda_1^0 + \lambda_2^0) = \lambda_1^0/\Lambda^0$ are:

$$D = \frac{RT}{F^2} \frac{|z_1| + |z_2|}{|z_1 z_2|} \cdot \frac{\lambda_1^0 \lambda_2^0}{\lambda_1^0 + \lambda_2^0} \cdot \left(1 + \frac{\mathrm{d} \ln y_\pm}{\mathrm{d} \ln c}\right) \qquad \ldots.(11.5)$$

$$D = \frac{RT}{F^2} \frac{|z_1| + |z_2|}{|z_1 z_2|} \Lambda^0 t_1^0 t_2^0 \left(1 + \frac{\mathrm{d} \ln y_\pm}{\mathrm{d} \ln c}\right) \qquad \ldots.(11.6)$$

or

$$D = D^0(1 + \mathrm{d} \ln y_\pm / \mathrm{d} \ln c) \qquad \ldots.(11.7)$$

THE INTERPRETATION OF DIFFUSION COEFFICIENTS

Dilute Solutions

At high concentrations consideration must be given to the motion of the solvent molecules as well as those of the solute: even for non-electrolytes this involves some difficult concepts, and the situation for concentrated electrolytes is a very complex one. In very dilute solutions, however, the motion of the solvent can be disregarded, and the experimental diffusion coefficients can be regarded as describing the motion of the solute particles through a stationary solvent.

The activity factor, $\frac{\mathrm{d} \ln a_\pm}{\mathrm{d} \ln c} = \left(1 + c \frac{\mathrm{d} \ln y_\pm}{\mathrm{d}c}\right)$ is a separately available experimental quantity; interest therefore centres, for dilute solutions, on whether the mobility factor in equation (11.3):

$$\frac{RT}{F^2} \frac{\nu \lambda_1^0 \lambda_2^0}{\nu_1 |z_1| (\lambda_1^0 + \lambda_2^0)}$$

is applicable at finite concentrations, and if not, what corrections should be applied to it. This question can be examined experimentally by dividing the observed D values by the quantity $\left(1 + c \frac{\mathrm{d} \ln y}{\mathrm{d}c}\right)$, giving a quantity proportional to the actual mobility of the diffusing solute, which can be compared with the limiting value. In *Table 11.1* this comparison is made for several typical electrolytes in dilute solution at 25°.

Table 11.1

mole l^{-1}	KCl		LiCl		$CaCl_2$		$LaCl_3$	
	D	$D/f(y)$	D	$D/f(y)$	D	$D/f(y)$	D	$D/f(y)$
0*	1·993	1·993	1·366	1·366	1·335	1·335	1·293	1·293
0·001	1·964	1·998	1·342	1·366	1·249	1·320	1·175	1·307
0·002	1·954	2·001	1·335	1·366	1·225	1·319	1·145	1·316
0·003	1·945	2·001	1·330	1·367	1·201	1·310	1·126	1·325
0·005	1·934	2·004	1·323	1·368	1·179	1·310	1·105	1·331
0·007	1·925	2·005	1·317	1·368	—	—	1·084	1·327
0·010	1·917	2·009	1·313	1·369	—	—	—	—

$$f(y) = 1 + c\frac{\mathrm{d}\ln y_{\pm}}{\mathrm{d}c}$$

* The D values at $c = 0$ are Nernst limiting values calculated by equation (11.4), which at 25° becomes:

$$D^0/(\mathrm{cm^2\ sec^{-1}}) = 2{\cdot}661_2 \times 10^{-7}\frac{|z_1| + |z_2|}{|z_1 z_2|}\,\frac{\lambda_1^0\lambda_2^0}{\lambda_1^0 + \lambda_2^0}\,/(\mathrm{Int\ ohm^{-1}\ cm^2\ equiv^{-1}})$$

The table shows that the variation of the diffusion coefficient D with concentration is in each case many times greater than that of the quantity $D\Big/\left(1 + c\dfrac{\mathrm{d}\ln y}{\mathrm{d}c}\right)$, so that the greater part of the change in D may be attributed to the non-ideality in thermodynamic behaviour which is allowed for by the factor $\left(1 + c\dfrac{\mathrm{d}\ln y}{\mathrm{d}c}\right)$. It remains to consider whether the residual variation shown in the third column for each solute is experimentally significant. The accuracy of the diffusion coefficients themselves is about 0·2 per cent; and the factor $\left(1 + c\dfrac{\mathrm{d}\ln y}{\mathrm{d}c}\right)$ can, for the four electrolytes listed, be computed with similar accuracy from the activity coefficient data; we must conclude, therefore, that in general the actual mobility of the diffusing ions does vary slightly with concentration. In the case of potassium chloride, it increases by approximately 0·8% per cent between 0 and 0·01 molar; for lithium chloride it is constant within experimental error; for calcium chloride it decreases by approximately 2 per cent between 0 and 0·005 molar, and for lanthanum chloride it increases by approximately 2·5 per cent in the same concentration range. These small but real variations, occurring as they do at very low concentrations, suggest the possibility that they are due to interionic effects; and the manner in which they differ from salt to salt indicates that the theory which will adequately explain them will be of a complicated nature.

The form of the Nernst–Hartley expression (11.3) suggests the course of substituting, for finite concentrations, the actual ionic mobilities λ_1 and λ_2 for the limiting values λ_1^0 and λ_2^0. The Nernst factor may be rewritten as:

$$t_1^0 t_2^0 \Lambda^0/(\nu_1|z_1|)$$

where t_1^0 and t_2^0 are the limiting transport numbers of the anion and cation and Λ^0 is the limiting equivalent conductivity of the salt. The corresponding expression for finite concentrations is $\dfrac{t_1 t_2 \Lambda}{\nu_1|z_1|}$; the transport number product $t_1 t_2$ varies little with concentration, but Λ decreases by amounts of the order of 5 per cent to 20 per cent in the concentration range 0 to 0·01 molar for the salts of *Table 11.1.* It is clear, therefore, that the use of the actual ionic equivalent conductivities instead of the limiting values would severely overcorrect, as was pointed out by HARTLEY[(4)]; indeed, for potassium and lanthanum chlorides the effect would be in the wrong direction. The position is, then, that the *mobility* of the ions in diffusion varies much less with concentration than does their mobility in electrolytic conduction; and while the latter always decreases with increasing concentration, the former may increase, decrease or remain constant, depending on the salt considered. This difference between the two types of transport process is due to the fact that in diffusion the ions move in the same direction, while in conduction oppositely charged ions move in opposite directions. The mutual attraction of the ions in the latter case will clearly have the effect of retarding the motion of both species, whereas in diffusion the slower ions are accelerated and the faster ones retarded. The effects of the ionic interactions on conductance have been dealt with (Chapter 7) in terms of a relaxation process and an electrophoretic effect, the former arising from the disturbance of the symmetrical arrangement of the ions in the solution, and the latter from a transfer of force between the moving ions *via* the solvent. In the diffusion of a single electrolyte, it can be shown that the symmetry of the ionic distribution is not disturbed so that the relaxation-effect is absent; the primary result of ionic interactions in this case is the harmonic averaging of the ionic speeds as given by the Nernst expression, but there is also a small electrophoretic effect, which will now be discussed along the lines laid down by ONSAGER and FUOSS[5], but with some generalization[8].

THE ELECTROPHORETIC EFFECT IN DIFFUSION

Our discussion, as in the case of the conductivity problem, will be restricted to the case of a single electrolyte, the cation and anion being denoted by subscripts 1 and 2 respectively. In Chapter 7 a general equation (7.7) was derived for the electrophoretic contribution to the motion of an ion in such a solution, in terms of the unspecified forces k_1 and k_2 causing the motion.

In the case of diffusion, these forces can conveniently be evaluated in terms of the velocity of the ions and their absolute mobilities. Because the solution must remain electrically neutral at all points, both ions must diffuse with the same final velocity v. Therefore by the definition of the absolute mobility, we may write:

$$k_1 = (\boldsymbol{F}^2/\boldsymbol{N})|z_1|v/(t_1^0\Lambda^0) \qquad \ldots.(11.8)$$

$$k_2 = (\boldsymbol{F}^2/\boldsymbol{N})|z_2|v/(t_2^0\Lambda^0) \qquad \ldots.(11.9)$$

Equation (7.7) then gives:

$$\frac{\Delta v_1}{v} = (\boldsymbol{F}^2/\boldsymbol{N})\Sigma A_n \frac{\dfrac{z_1^{2n}}{t_1^0\Lambda^0} + \dfrac{z_1^n z_2^n}{t_2^0\Lambda^0}}{a^n(z_1 - z_2)} = \delta_1 \qquad \ldots.(11.10)$$

$$\frac{\Delta v_2}{v} = (\boldsymbol{F}^2/\boldsymbol{N})\Sigma A_n \frac{\dfrac{z_1^n z_2^n}{t_1^0\Lambda^0} + \dfrac{z_2^{2n}}{t_2^0\Lambda^0}}{a^n(z_1 - z_2)} = \delta_2 \qquad \ldots.(11.11)$$

(δ_1 and δ_2 are merely convenient abbreviations).

Now this means that a force which would produce a velocity v in the absence of the electrophoretic effect will actually produce velocities $v + \Delta v_1$ and $v + \Delta v_2$. Provided Δv_1 and Δv_2 are small compared to v, *i.e.*, that δ_1 and δ_2 are small compared to unity, we can therefore treat the electrophoretic effect in diffusion by increasing the mobilities of the ions by factors $(1 + \delta_1)$ and $(1 + \delta_2)$ respectively.

In the simple Nernst–Hartley treatment which leads to the formula $D = D^0\left(1 + c\dfrac{\mathrm{d}\ln y}{\mathrm{d}c}\right)$, the mobilities appear in the form of the factor $\dfrac{\lambda_1^0\lambda_2^0}{\lambda_1^0 + \lambda_2^0}$. We therefore replace this factor by:

$$\frac{\lambda_1'\lambda_2'}{\lambda_1' + \lambda_2'} = \frac{\lambda_1^0\lambda_2^0(1 + \delta_1)(1 + \delta_2)}{\lambda_1^0(1 + \delta_1) + \lambda_2^0(1 + \delta_2)} \qquad \ldots.(11.12)$$

which, on putting $\lambda_1^0 = t_1^0\Lambda^0$, $\lambda_2^0 = t_2^0\Lambda^0$ and expanding in series in δ_1 and δ_2 as far as the first powers, becomes:

$$\frac{\lambda_1'\lambda_2'}{\lambda_1' + \lambda_2'} = t_1^0 t_2^0\Lambda^0 + t_1^0 t_2^0\Lambda^0(t_1^0\delta_2 + t_2^0\delta_1) \qquad \ldots.(11.13)$$

On inserting the values of δ_1 and δ_2 according to equation (11.11) and simplifying, this becomes:

$$\frac{\lambda_1'\lambda_2'}{\lambda_1' + \lambda_2'} = t_1^0 t_2^0\Lambda^0 + \frac{F^2}{N}\Sigma A_n \frac{(z_1^n t_2^0 + z_2^n t_1^0)^2}{a^n(z_1 - z_2)} \quad \ldots.(11.14)$$

We now replace the Nernst–Hartley expression (11.5) by:

$$D = \frac{RT}{F^2}\frac{|z_1| + |z_2|}{|z_1 z_2|}\frac{\lambda_1'\lambda_2'}{\lambda_1' + \lambda_2'}\left(1 + c\frac{\mathrm{d}\ln y}{\mathrm{d}c}\right) \ldots.(11.15)$$

from which we obtain, since $z_1 - z_2 = |z_1| + |z_2|$:

$$D = \left(1 + c\frac{\mathrm{d}\ln y}{\mathrm{d}c}\right)(D^0 + \Sigma\Delta_n) \qquad \ldots.(11.16)$$

where D^0 is the Nernst limiting value of the diffusion coefficient given by:

$$D^0 = \frac{RT}{F^2}\frac{\nu}{\nu_1|z_1|}\frac{\lambda_1^0\lambda_2^0}{\lambda_1^0 + \lambda_2^0} = \frac{RT}{F^2}\frac{|z_1| + |z_2|}{|z_1 z_2|}\frac{\lambda_1^0\lambda_2^0}{\lambda_1^0 + \lambda_2^0}$$

and the electrophoretic terms Δ_n are given by:

$$\Delta_n = kT A_n \frac{(z_1^n t_2^0 + z_2^n t_0)^2}{a^n|z_1 z_2|} \qquad \ldots.(11.17)$$

The coefficients A_n are functions of the dielectric constant and viscosity of the solvent, the temperature and the dimensionless concentration-dependent quantity κa and are defined in equation (7.8).

It will be recalled from Chapter 7 that this expression for the electrophoretic effect has been derived on the basis of a Boltzmann distribution function for the ions, but has employed the Debye–Hückel expression for the potential (Eq. 4.13). The latter, however, is actually based on the distribution function obtained by taking only the first power term of the expansion of the exponential Boltzmann function, except for symmetrical electrolytes where the square term, in $\left(\frac{e\psi}{kT}\right)^2$, can also justifiably be included. This means that to be consistent we should accept in applying equations (11.16) and (11.17) only the first-order electrophoretic term ($n = 1$), except in the case of symmetrical electrolytes where the second order term ($n = 2$) is acceptable. It is therefore convenient to consider the cases of symmetrical and unsymmetrical valency types separately. Here we depart from the treatment given by ONSAGER and FUOSS[5], who proposed that in *all* cases equation (11.16) should be taken as far as the second-order term Δ_2.

Symmetrical Types

Here we put $|z_1| = |z_2| = z$. The factor $(z_1^n t_2^0 + z_2^n t_1^0)^2$ appearing in Δ_n then reduces to $|z|^{2n}(t_2^0 - t_1^0)^2$ for odd n, and $|z|^{2n}$ for even n. Thus, in contrast to the case of conduction, no terms vanish identically; but in the special case where the ions have nearly equal mobilities ($t_2^0 \approx t_1^0$), all the odd-order terms will be negligible. The condition is realized for aqueous potassium chloride, bromide or iodide solutions. The second-order term can never vanish, and is always positive since A_n is positive for even n. Upon substituting $n = 1$ and $n = 2$ in the general formula for Δ_n we have,

$$\Delta_1 = -\frac{\boldsymbol{k}T}{6\pi\eta}(t_2^0 - t_1^0)^2 \frac{\kappa}{1 + \kappa a}$$

$$\Delta_2 = \frac{\boldsymbol{k}T}{12\pi\eta}\frac{\boldsymbol{e}^2}{\varepsilon \boldsymbol{k}T}(\kappa a)^2 \left(\frac{e^{\kappa a}}{1 + \kappa a}\right)^2 Ei(2\kappa a)\left(\frac{|z|}{a}\right)^2$$

The function of (κa) appearing in Δ_2:

$$\phi_2(\kappa a) = (\kappa a)^2 \left(\frac{e^{\kappa a}}{1 + \kappa a}\right)^2 Ei(2\kappa a)$$

is given in *Table 7.1* where higher-order functions of the same type are also tabulated. In practice we shall be interested in the application of these formulae to 1 : 1 electrolytes at 25° in water, when they reduce to:

$$\left.\begin{aligned} \Delta_1 &= -\,8{\cdot}07 \times 10^{-6}(t_2^0 - t_1^0)^2\sqrt{c}/(1 + \kappa a) \\ \Delta_2 &= +\,8{\cdot}77 \times 10^{-21}\phi_2(\kappa a)/a^2 \end{aligned}\right\} \qquad \ldots.(11.18)$$

where c is expressed in mole/l and a in cm.

Unsymmetrical Types

Here the requirements of self-consistency limit us to taking only the first-order electrophoretic term ($n = 1$) which gives:

$$\Delta_1 = -\frac{\boldsymbol{k}T}{6\pi\eta}\frac{(z_1 t_2^0 + z_2 t_1^0)^2}{|z_1 z_2|}\frac{\kappa}{1 + \kappa a} \qquad \ldots.(11.19)$$

reducing for aqueous solutions at 25° to:

$$\Delta_1 = -8{\cdot}07 \times 10^{-6} \frac{(z_1 t_2^0 + z_2 t_1^0)^2}{|z_1 z_2|} \frac{\sqrt{I}}{1 + 0{\cdot}3291 \times 10^8 a\sqrt{I}} \quad \ldots\ldots(11.20)$$

the ionic strength I being computed on the mole/l scale and a being measured in cm. The higher-order terms may of course be evaluated; STOKES[8] and ADAMSON[9] found that, in contrast to the case of 1 : 1 electrolytes, they do not converge satisfactorily, terms as high as the fifth order being comparable to the first-order term. DYE and SPEDDING[10] have in effect evaluated the series $\Sigma\Delta_n$ to infinity by numerical integration; though this overcomes the difficulty of the slow convergence, it is doubtful if the results are applicable since the theory is self-consistent only as far as the first-order terms for unsymmetrical electrolytes. We therefore proceed to examine the experimental data on the basis of the equations:

$$D = (D^0 + \Delta_1 + \Delta_2)\left(1 + c\frac{\mathrm{d}\ln y}{\mathrm{d}c}\right) \quad \ldots\ldots(11.21)$$

for symmetrical electrolytes, and

$$D = (D^0 + \Delta_1)\left(1 + c\frac{\mathrm{d}\ln y}{\mathrm{d}c}\right) \quad \ldots\ldots(11.22)$$

for unsymmetrical ones. ONSAGER and FUOSS[5] proposed the use of equation (11.21) for all cases, and (11.21) is known as the Onsager–Fuoss expression.

TESTS OF THE THEORY OF THE ELECTROPHORETIC EFFECT IN DIFFUSION

It has been shown earlier that one test of the theory of the electrophoretic effect is provided by the variation of transport numbers with concentration; in that case the relaxation effect does not enter the final equations because it affects the velocity of both ions in the same proportion. In the diffusion of a single electrolyte, the relaxation effect is also absent, for the more physical reason that there is no mean motion of ions relative to one another, owing to the necessity of preserving electrical neutrality at all points. The testing of the theoretical equations for the electrophoretic effect in diffusion is, however, less straightforward, since the diffusion coefficient depends also on the concentration gradient of the free energy. It would be possible to compute this gradient separately from the theory, combine this with the equation for the electrophoretic effect and test the resulting expression for D directly

However, one would then not know whether deviations from the theory were attributable to failure of the electrophoretic or of the free energy portion of the calculation. It is therefore more convenient to use experimentally determined values of the free energy gradient, obtainable from activity coefficient data; but unless these are of high precision, the small electrophoretic effect may be masked. Furthermore, the tests should be confined to fairly dilute solutions, since diffusion in more concentrated solutions involves further considerations of viscosity, hydration and volume-effects. The most suitable diffusion data are thus those obtained by Harned's conductimetric method, and given in Appendix 11.1; but unfortunately accurate activity coefficient data are not available for all of the salts at the concentrations in question ($< 0{\cdot}01$ M).

DILUTE 1 : 1 ELECTROLYTES

For potassium and sodium chlorides, reliable activity coefficients below 0·01 M have been obtained from cells with transport and moving-boundary transport numbers, with good agreement between independent workers. For lithium chloride, the activity coefficients at 0° obtained by the freezing-point method[11] may be used, since the corrections to 25° at these low concentrations are certain to be small; the resulting values of the factor $\left(1 + c\dfrac{\mathrm{d}\ln y}{\mathrm{d}c}\right)$ differ by less than 0·1 per cent from those obtained by extrapolating the 25° data available from vapour pressures above 0·1 M, by means of a Debye-Hückel equation of the type given in equation (9.11). For these three electrolytes, therefore, the data are adequate to test the theory at low concentrations. The values of $D\Big/\left(1 + c\dfrac{\mathrm{d}\ln y}{\mathrm{d}c}\right)$ which contain the *experimental* quantities, have been given in *Table 11.1*. We now require to compute the electrophoretic corrections Δ_1 and Δ_2 of equations (11.21) and (11.18). These are functions of both concentration and ion size, and the first-order term Δ_1 also involves the factor $(t_1^0 - t_2^0)^2$. *Figure 11.1* shows the form of the terms for 1 : 1 electrolytes at 25° in water, for ion sizes of 3·6 and 5 Å, which are about the upper and lower limits of ionic diameters encountered with simple non-associated electrolytes. It will be seen that Δ_2 is much more sensitive to ion size than Δ_1 and that both change only slowly with concentration above one molar. The factor $(t_2^0 - t_1^0)^2$ has the following values at 25°: HCl, 0·4115; HBr, 0·4032; LiCl, 0·1072; LiBr, 0·1141; NaCl, 0·0430; NaBr. 0·0479; NaI, 0·0441; KCl, KBr, KI $< 0{\cdot}001$.

Since the diffusion coefficients are of the order of 2×10^{-5} cm^2 sec^{-1}, this means that Δ_1 is negligible for the potassium halides but affects the third or fourth significant figure in the other cases. Below 0·01 molar, the value of a used will not be very critical, but at higher concentrations it will make a substantial difference

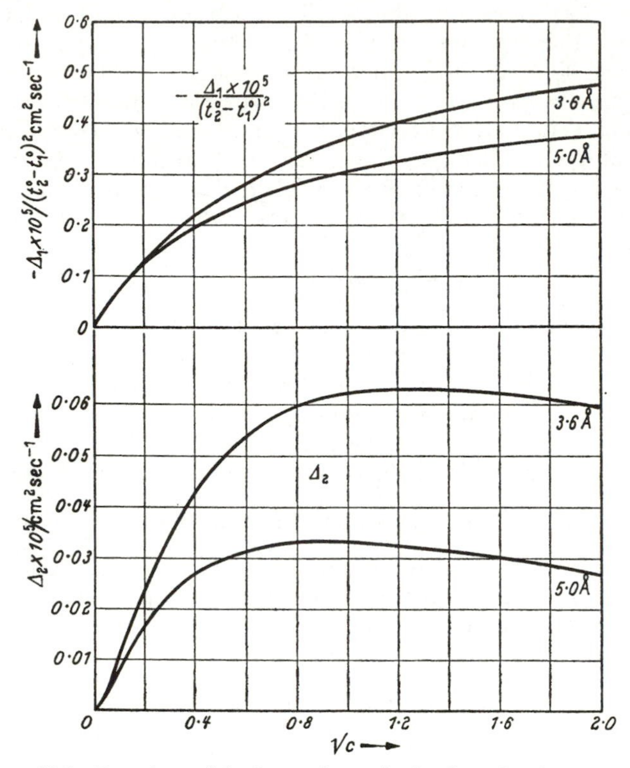

Figure 11.1. Dependence of the first- and second-order electrophoretic corrections Δ_1 *and* Δ_2 *on concentration and ion-size. (Aqueous* 1 : 1 *electrolytes at* 25°)

especially in Δ_2. We use here the a values which have been found to give a satisfactory account of the activity coefficient data (see *Table 9.5*). The quantity $(D^0 + \Delta_1 + \Delta_2)$ is given at a few concentrations below 0·01 molar in *Table 11.2*, together with the a value used in its computation; the value found experimentally, $D\Big/\left(1 + c\,\frac{\mathrm{d}\ln y}{\mathrm{d}c}\right)$ from *Table 11.1* is included for comparison.

Though the electrophoretic effects in this concentration range are small, the theory gives fair agreement with observation. The virtual constancy of the calculated mobility for lithium chloride arises from the cancellation of Δ_1 and Δ_2; for potassium chloride the small increase with concentration arises from Δ_2 alone, Δ_1 being negligible; in sodium chloride Δ_2 is somewhat larger numerically than Δ_1. GUGGENHEIM[12] has pointed out in a more detailed analysis of the data for sodium chloride that the electrophoretic effects are not as great as the random experimental errors, a fact which is masked in *Table 11.2* by the presentation of smoothed values in the columns headed 'obs.' Nevertheless the extent of agreement provides some

Table 11.2

Values of $10^5(D^0 + \Delta_1 + \Delta_2)/(\text{cm}^2\,\text{sec}^{-1})$ *at* 25°

c mole/l	LiCl (4·32 Å)		NaCl (3·97 Å)		KCl (3·63 Å)	
	calc.	obs.	calc.	obs.	calc.	obs.
0	1·366	(1·366)	1·610	(1·610)	1·993	(1·993)
0·001	1·366	1·366	1·611	1·611	1·995	1·998
0·002	1·366	1·366	1·612	1·613	1·996	2·001
0·005	1·366	1·368	1·614	1·617	1·999	2·004
0·01	1·368	1·369	1·616	1·618	2·003	2·009

evidence in favour of the theory of the electrophoretic effect. Better support is found in the theory of conductance, where the electrophoretic effect is much larger since the ions move in opposite directions. In the particular case of potassium chloride, the Onsager-Fuoss expression 11.21 holds even at half-molar concentration, but this is a fortunate coincidence, attributable mainly to the fact that the ions are little hydrated and the viscosity little different from that of water. In other cases the calculated mobilities fall increasingly above the observed values as the concentration is increased. This cannot be attributed to the failure of the theory of the electrophoretic effect; it is due rather to the neglect of other effects, which are discussed later in this chapter.

HIGHER SYMMETRICAL VALENCY TYPES

With salts of higher valency type than uni-univalent the theory encounters a number of difficulties which are at present only partly solved. With symmetrical valency types of double or higher charge, an appreciable fraction of the ions are present as closely associated pairs. Zinc and magnesium sulphates have been studied up to 0·005 molar by HARNED's school[17], who find that the observed values in this region are up to 10 per cent higher than the predictions of the equation:

$$D = (D^0 + \Delta_1 + \Delta_2)\left(1 + c\frac{\mathrm{d}\ln y}{\mathrm{d}c}\right) \quad \ldots.(11.21)$$

when a value of $a = 3{\cdot}64$ Å is used in the computation of the electrophoretic terms. They have also shown that if the degree of ion-pair formation computed from the conductivity data is taken into account, and the ion-pair is assumed to have a constant mobility equal to that of an ion for which $\lambda^0 = 44$ (for $ZnSO_4$) or 46 (for $MgSO_4$), the observed and calculated values can be brought into satisfactory agreement. The ratio D obs./D calc. (where D calc. is the value obtained by equation (11.21) for a fully ionized 2 : 2 electrolyte) is shown to be given by:

$$D \text{ obs.}/D \text{ calc.} = 1 + (1 - \alpha)\left[\frac{\lambda_{12}^0(\lambda_1^0 + \lambda_2^0)}{\lambda_1^0\lambda_2^0} - 1\right] \quad \ldots.(11.23)$$

where α is the fraction of non-paired ions, and λ_{12}^0 is the mobility of the ion-pair in equivalent conductance units. The result indicates that the ion-pair has a higher mobility than the dissociated part of the electrolyte, which is explained on the ground that its formation results in the loss of water of hydration by the zinc ion. The success of this treatment is, however, somewhat reduced by the fact that it employs an ion size $a = 3{\cdot}64$ Å for bi-bivalent electrolytes; other evidence about the size of the zinc and magnesium ions indicates a minimum acceptable value of 6 Å for the unpaired ions. However, the value 3·64 is consistent with that needed to interpret the transport numbers of cadmium sulphate (see *Table 7.8*).

UNSYMMETRICAL VALENCY TYPES

A few salts of unsymmetrical valency type have been studied, *e.g.*, the alkaline earth chlorides[13], lanthanum chloride[14], some alkali sulphates[15], and potassium ferrocyanide[16]. Of these, calcium and strontium chlorides are the only ones in which we are reasonably certain that ion-pair formation is negligible, though it may well be only slight in the alkali sulphates at the concentrations in question (below 0·005 M).

For calcium chloride, the original Onsager-Fuoss equation (11.21) definitely breaks down, the observed diffusion coefficient at 0·005 M being some 5 per cent lower than that predicted by equation (11.21). However, the 'self-consistent' equation:

$$D = (D^0 + \Delta_1)\left(1 + c\frac{\mathrm{d}\ln y}{\mathrm{d}c}\right) \quad \ldots.(11.22)$$

gives a satisfactory account of the data[8]. The term Δ_2 is rejected on the grounds that its inclusion implies a distribution function inconsistent with the Poisson equation; this term is, however, twice as large as Δ_1 and the higher terms (Δ_3, Δ_4 . . .) do not converge satisfactorily so that the theory cannot be regarded as adequate in

Table 11.3

Diffusion Coefficients of Dilute Aqueous 2 : 1 *and* 1 : 2 *Electrolytes at* 25°

c mole/l.	$CaCl_2$ (a = 4·73Å)			Li_2SO_4			Na_2SO_4		
	$f(y)$	$D_{calc.}$	$D_{obs.}$	$f(y)$	$D_{calc.}$	$D_{obs.}$	$f(y)$	$D_{calc.}$	$D_{obs.}$
0	1·000	1·336	—	1·000	1·041	—	1·000	1·230	—
0·001	0·947	1·257	1·249	0·950	0·989	0·990	0·945	1·162	1·175
0·002	0·924	1·223	1·225	0·939	0·978	0·974	0·927	1·141	1·160
0·005	0·900	1·185	1·179	0·917	0·955	0·950	0·892	1·097	1·123

Notes: (*a*) $D_{obs.}$ by conductimetric method of Harned (see Appendix 11.1). Later values of D for $CaCl_2$ are 1·263, 1·243 and 1·213 at c = 0·001, 0·002, 0·005 respectively.

(*b*) $D_{calc.}$ from the 'self-consistent' equation:

$$D_{calc.} = (D^0 + \Delta_1)\left(1 + c\frac{d \ln y_\pm}{dc}\right)$$

(*c*) $y_\pm$ from freezing-point data for Li_2SO_4 and Na_2SO_4, and from e.m.f. measurements for $CaCl_2$. For the sulphates, Δ_1 is negligible, hence no ion-size parameter is needed.

(*d*) D in $cm^2 \, sec^{-1} \times 10^{-5}$.

(*e*) $f(y) = \left(1 + c\frac{d \ln y_\pm}{dc}\right)$

(*f*) Values at $c = 0$ from equation (11.4).

its present form. For the strontium chloride results[13] on the other hand, the Onsager-Fuoss equation (11.21) seems to be more satisfactory than the 'self-consistent' equation (11.22). This anomaly has not been resolved. In the case of sodium and lithium sulphates, calculation shows that Δ_1 is very small, since the factor $(z_1t_2^0 + z_2t_1^0)^2$ is only 0·0247 for the former and 0·00042 for the latter, in contrast to 0·4706 for calcium chloride. Therefore, if only the first-order electrophoretic term is relevant for these electrolytes, there should be no detectable electrophoretic effect at all. The experimental results do not settle the matter definitely, owing to uncertainties about the activity coefficients at these low concentrations, which make it difficult to compute the factor $\left(1 + c\frac{\mathrm{d}\ln y}{\mathrm{d}c}\right)$ with sufficient accuracy; one must choose between using freezing-point data which refer to 0° and extrapolation from the region above 0·1 M at 25° where vapour pressure and e.m.f. measurements are available. The freezing-point activity data are probably to be preferred, as the corrections from 0° to 25° will be very small at the concentrations below 0·005 M to which the diffusion coefficients refer.

Table 11.3 shows moderately good agreement between the observed values and those calculated by equation (11.22) for calcium chloride and lithium sulphate, but for sodium sulphate the diffusion coefficients are definitely higher than theory predicts. This may possibly be due to ion-pair formation, which would be more pronounced in sodium sulphate than in lithium sulphate; but even in the other cases the theory is by no means as successful as for 1 : 1 electrolytes. Lanthanum chloride conforms only poorly with the theoretical equation[8]; it has been shown that the series $\Sigma\Delta_n$ of equation (11.16) is initially divergent with alternating signs for this case, so that agreement with the present form of the theory cannot be expected.

DIFFUSION OF AN INCOMPLETELY DISSOCIATED ELECTROLYTE

In associated electrolytes it is necessary to recognize that an appreciable fraction of the transport of solute may occur as a result of the motion of ion-pairs (or larger aggregates); in the extreme case of a weak electrolyte, the covalent molecular form is the predominating diffusing entity. Ion-association affects the diffusion coefficient in two ways: first, it reduces the activity of the solute as compared with a fully dissociated electrolyte, and hence leads to lower values of the gradient of free energy with concentration; and secondly, when two particles merge into one they offer less resistance to motion through the liquid; this has the effect of increasing the diffusion coefficient. The effect on the free energy gradient need not be considered, since we use experimental values of the factor $\left(1 + m\frac{\mathrm{d}\ln \gamma}{\mathrm{d}m}\right)$ in comparing observed and calculated diffusion coefficients. The chemical potentials of the associated and dissociated forms of the solute are the same, since they exist in equilibrium, and the free energy gradient of the solute is therefore the same for both forms.

Denoting the absolute mobilities of the ions by u_1 and u_2 and that of the ion-pair or molecule by u_{12}, and with a degree of dissociation α, we thus obtain for dilute solutions of associated symmetrical electrolytes:

$$D = 2kT\left(1 + c\frac{\mathrm{d}\ln y_{\pm}}{\mathrm{d}c}\right)\left[\alpha\frac{u_1 u_2}{u_1 + u_2} + (1 - \alpha)u_{12}\right] \quad \ldots.(11.24)$$

In actual solutions the mobility terms (in the square brackets) will mutually influence each other. The electrophoretic effect already discussed for non-associated electrolytes will of course be operative though at an *ionic* concentration of αc rather than c. An essentially similar effect will operate between the neutral diffusing particles and their neighbours, whether molecules or ions; this is not easy to evaluate without arbitrary assumptions about the distribution of the associated particles. Neglecting it, one would obtain the relation:

$$D = [\alpha(D^0 + \Delta_1 + \Delta_2) + 2(1 - \alpha)D^0_{12}]\left(1 + c\frac{\mathrm{d}\ln y_{\pm}}{\mathrm{d}c}\right) \quad \ldots.(11.25)$$

where D^0_{12} represents the (hypothetical) diffusion coefficient of an isolated ion-pair or molecule at infinite dilution and is defined by $D^0_{12} = kTu_{12}$. HARNED and HUDSON[17] first derived and tested an equation equivalent to (11.25) for zinc sulphate at 25°. Their values for α were obtained from conductivity estimates, and their diffusion-coefficients measured in the range 0·001–0·005 M indicated a reasonably constant value for the diffusion coefficient D^0_{12} of the ion-pairs.

For 1 : 1 electrolytes, the proportion of the associated form is small at low concentrations, so that it is more difficult to estimate D^0_{12} with any accuracy. However, some recent measurements on concentrated ammonium nitrate solutions[18] have been interpreted with fair quantitative success on this basis. In this case an equation similar to equation (11.66) was used in order to allow for the effect of volume restraints and of the diffusion of the solvent; hydration of the ions of ammonium nitrate was assumed to be negligible. The final equation was:

$$D_{\text{obs.}} = [\alpha(D^0 + \Delta_1 + \Delta_2) + 2(1 - \alpha)D_{12}]\left(1 + m\frac{\mathrm{d}\ln \gamma_{\pm}}{\mathrm{d}m}\right)\left(1 + 0{\cdot}036\, m\frac{D^*_{\mathrm{H_2O}}}{D^0}\right)\frac{\eta^0}{\eta} \quad \ldots.(11.26)$$

As in the work of Harned and Hudson, the degree of dissociation, α, was estimated from the conductivities, though at the high concentrations (0·1 to 8 M) of these measurements the theory used in calculating the degree of dissociation is necessarily very approximate. Nevertheless, the diffusion coefficients of ammonium nitrate calculated from equation (11.26) with $D_{12}^0 = 1{\cdot}5 \times 10^{-5}$ agreed within 2 per cent with the experimental values up to 6 mole/litre. The value $D_{12}^0 = 1{\cdot}5 \times 10^{-5}$ for the ion-pair was not arbitrarily assumed, but was calculated from the mobilities of the separate ions by considering them to merge into an ellipsoidal body.

Diffusion of a weak electrolyte. This is essentially the same problem as the diffusion of an incompletely dissociated electrolyte. Goüy measurements have been made on citric acid[19] and acetic acid[20] in water. The degree of ionization of these substances being known from conductance data, equation (11.25) can be rearranged to yield a convenient extrapolation for D_{12}^0 if the activity data are available. An alternative method which was used in the case of citric acid, leads to the conclusion that the function:

$$D_{12}' = [(1 - \alpha/2)\, D\eta/\eta^0 - (\alpha/2)\, D_i^0]/(1 - \alpha) \qquad \ldots.(11.27)$$

when plotted against concentration extrapolates linearly to D_{12}^0 at zero concentration. Here D is the measured diffusion coefficient and D_i^0 is the limiting Nernst value for the completely dissociated ionic form. The limiting diffusion coefficients D_{12}^0 of *molecular* citric acid and acetic acid are found to be $0{\cdot}657 \times 10^{-5}$ and $1{\cdot}201 \times 10^{-5}$ cm^2 sec^{-1} respectively. It is interesting to compare these with the limiting Nernst values ($\boldsymbol{RT\lambda^0/F^2}$) for the monocitrate ion ($0{\cdot}81 \times 10^{-5}$ cm^2 sec^{-1}) and the acetate ion ($1{\cdot}088 \times 10^{-5}$ cm^2 sec^{-1}). The citric acid molecule has a considerably lower mobility than its anion, whilst the acetic acid molecule has a higher one. This, taken in conjunction with the facts that the acetate ion has a low mobility for its size and that the activity coefficients of metal acetates are high, suggests that the acetate ion interacts fairly strongly with water molecules. The effect of the charge of the monocitrate ion, on the other hand, appears to be predominantly a structure-breaking one.

VISCOSITY AND IONIC MOTION IN CONCENTRATED SOLUTIONS

In Chapter 7 and in pp. 286–302 we have discussed the way in which the motion of ions is affected by electrical interactions with other ions. The mathematical treatment of these effects is at present strictly valid only for low concentrations, owing to approximations which must be made in order to give manageable results. In regard to concentrated solutions, many workers adopt a counsel of despair, confining their interest to concentrations below about 0·02 M, while others maintain that an adequate theory of the behaviour of pure fused salts is an essential pre-requisite to the understanding of transport processes in concentrated electrolytes. Nevertheless we believe that useful information can be gained from the study of transport processes in concentrated solutions, and their practical importance justifies the attempt.

The treatment of interionic effects on both chemical potential and ion mobility leads (apart from minor corrections) to results which may be summarized as

$$\Upsilon - \Upsilon_0 \propto \kappa/(1 + \kappa a) \qquad \ldots.(11.28)$$

where Υ denotes diffusion coefficient, conductance, or the logarithm of an activity coefficient and the subscript zero refers to infinite dilution. Now the quantity $\kappa/(1 + \kappa a)$, though varying as $\sqrt{c}$ at low concentrations, changes only slowly at high concentrations, and variations in the ion size parameter a are insufficient to cause *large* variations in behaviour between one salt and another of the same valency type. Yet such large variations do occur, and become very marked at concentrations of a few molar. In the case of thermodynamic properties, they can be explained by the effects of ion-solvent interactions, which have an important influence at high concentrations. In the case of the transport properties, another effect, negligible in dilute solutions, becomes important at high concentrations. This effect is connected with the changed viscosity of the solution; we shall not say it is caused by the changed viscosity, but for brevity we shall refer to it as the viscosity-effect.

The Viscosity of Electrolyte Solutions

Viscosity, the force required to produce unit rate of shear between two layers separated by unit distance, is an important property of liquids. For methods of measurement, the reader is referred to standard text-books on practical physical chemistry(21); the papers of G. Jones and collaborators(22) and recent publications from the U.S. Bureau of Standards(23) should then be studied to dispel any idea that really accurate measurements are easily made. It is usual to calibrate viscometers by means of pure water, for which careful absolute viscosity measurements have been made. The most recent of these has resulted in an appreciable change from the values(24) which had been accepted since 1919, and a close approach to those obtained in the classical work of Thorpe and Rodger(25) in 1894. Calibration at two or more points is highly desirable, and is most conveniently obtained by using water at several temperatures (*see* Appendix 1.1).

For solutions the relative viscosity, $\eta_{rel} = \eta/\eta^0$ is often used, η^0 denoting the viscosity of the pure solvent at the same temperature. The 'specific viscosity' $\eta^\star$, defined by

$$\eta^\star = (\eta_{rel} - 1)/c \qquad \ldots .(11.29)$$

is also useful, particularly since it often changes only slowly with concentration and temperature.

It is clear that the electrical forces between ions in adjacent layers of an electrolyte solution will increase the viscosity. The mathematical treatment of this effect was given by FALKENHAGEN and collaborators[26] who showed that the limiting law is of the form

$$\eta_{rel} = 1 + A_1\sqrt{c} \qquad \ldots .(11.30)$$

the constant A_1 being as usual a function of solvent properties, ionic charges and mobilities, and temperature. Numerically, A_1 is fairly small, *e.g.*, for KI and Li_2SO_4 at 25° in water, the calculated values[27] are $A_1 = 0{\cdot}0050$ and $0{\cdot}0167$ mole$^{-\frac{1}{2}}$ $l^{\frac{1}{2}}$ respectively. The corresponding experimental values are 0·0047 and 0·0167. This agreement, however, does not mean that the theory is of practical use for calculating viscosities, since the small square-root term is quickly swamped by a much larger linear term, as expressed by the equation of JONES and DOLE[27]:

$$\eta_{rel} = 1 + A_1\sqrt{c} + A_2c \qquad \ldots .(11.31)$$

(The coefficient A_2 is usually referred to in the literature as the 'viscosity B-coefficient,' a terminology we are obliged to depart from here to avoid confusion with our other B symbols.) This coefficient is highly specific for the electrolyte and temperature, *e.g.*, $-$ 0·014 mole^{-1} l for KCl and + 0·567 mole^{-1} l for $LaCl_3$, at 25°. Equation (11.31) is usually valid up to a few tenths molar. The A_2 coefficients are found to be fairly accurately additive properties of the constituent ions, and several independent workers[28, 29, 30] have agreed that individual ionic A_2 values can be based on $A_2(K^+, 25°) = -0{\cdot}007$ mole^{-1} l. The A_2-values are strongly correlated with the entropy of solution of the ions (see p. 16). Negative values are found with those ions which exert a 'structure-breaking' effect on water, *e.g.*, Rb^+, Cs^+, I^-, ClO_3^-, NO_3^-, and such values become less negative or even change to positive as the temperature is raised. The reason is clearly that at the higher temperatures the water structure is already so broken by thermal agitation that the ion can scarcely make matters worse. These negative values of A_2 appear to be confined to aqueous solutions, and even here they seldom cause a decrease of more than 10 per cent in the viscosity. More typical are fairly large positive values of A_2, found with ions which are strongly hydrated, *e.g.*, at 25° the A_2 values for Na^+, Li^+, Mg^{++}, La^{+++} are 0·0863, 0·1495, 0·3852 and 0·5888 mole^{-1} l, respectively. A recent tabulation by KAMINSKY[28] is reproduced in Appendix 11.3. Similar behaviour is found with large non-electrolyte molecules such as glycerol and sucrose.

The viscosity increase found with large solute particles was explained by EINSTEIN[31] as due to interference of the particles with the stream-lines in the liquid; by classical hydrodynamic methods, treating the liquid as a viscous continuum with rigid spherical obstructions at the surface of which the liquid is at rest, he obtained the result

$$\eta_{\text{rel}} = 1 + 2{\cdot}5\phi \qquad \ldots.(11.32)$$

valid at low concentrations, with ϕ denoting the volume-fraction occupied by the obstructions. Later work[32] has extended this limiting theory to higher concentrations, giving

$$\ln \eta_{\text{rel}} = \frac{2{\cdot}5\phi}{1 - Q\phi} \qquad \ldots.(11.33)$$

where Q is an interaction parameter dealing with mutual interference between the spheres, and with their Brownian motion; various authors agree only that it does not differ greatly from unity. Since ϕ is a volume fraction, we can replace it by $c\bar{V}$ where c is the molar concentration, and $\bar{V}$ is an 'effective rigid molar volume' expressed in litres per mole. This gives:

$$\log \eta_{\text{rel}} = \frac{A_3 c}{1 - Q'c} \qquad \ldots.(11.34)$$

where $A_3 = 2{\cdot}5\bar{V}/2{\cdot}303$, and $Q' = Q\bar{V}$ is an arbitrary constant. Equation (11.34) gives an excellent representation of the viscosities of strongly hydrated electrolyte solutions and of solutions of large non-electrolyte molecules, often up to the point where the viscosity is five or ten times that of water, though it is necessarily less exact than equation (11.31) in dilute electrolyte solutions owing to the omission of the small term in $\sqrt{c}$. The connection between equations (11.31) and (11.34) is apparent on expanding the logarithm in (11.34) when we obtain for small c:

$$\eta_{\text{rel}} - 1 = 2{\cdot}303 A_3 c \qquad \ldots.(11.35)$$

$$\approx A_2 c \text{ (ignoring } A_1\sqrt{c} \text{ in (11.31))}$$

i.e., $A_2 = 2{\cdot}303 A_3$. *Table 11.4* gives the values of A_3 and Q' required by equation (11.34) at 25° for a few substances, along with the value of $A_2/2{\cdot}303$ from KAMINSKY's[28] table (Appendix 11.3).

If the theory underlying equation (11.34) is taken literally, the 'rigid volumes' $\bar{V}$ of this table should represent the molar volumes of the solutes including any water of hydration which is held too firmly to participate in the viscous shearing process. The molar volume of sucrose in solution, on the unhydrated basis, is 0·212 l mole^{-1}, and that of glycerol is 0·071 l mole^{-1}. The $\bar{V}$ values therefore indicate that the sucrose molecule acts, on Einstein's model, as if it includes (0·350 − 0·212)/0·018 = 7·7 molecules of water, and

Table 11.4

Constants of Equation (11.34) *for Viscosities of Concentrated Aqueous Solutions at* 25°, *and the 'Rigid Volume'* $\bar{V}$

Solute	A_3 / l mole^{-1}	Q' / l mole^{-1}	$\bar{V}$ / l mole^{-1}	$A_2/2{\cdot}303$ / l mole^{-1}
Sucrose	0·380	0·231	0·350	(0·3816)[a]
Glycerol	0·0959	0·0363	0·0883	—
NaCl	0·0379	0·0589	0·0349	0·0344[(b)]
LiCl	0·0586	0·0079	0·0540	0·0619[(b)]
$MgCl_2$	0·147	0·078	0·135	0·161[(b)]

(a)—from measurements on dilute solutions (< 0·02 M): JONES G. and TALLEY, S. K., *J. Amer. Chem. Soc.* 55 (1933) 624.
(b)—from Appendix 11.3.

the glycerol molecule, one molecule of water. In Chapter 9 we showed that the activity coefficients are consistent with ideal-solution behaviour with 5 molecules of water of hydration for sucrose and 1·2 for glycerol. Another estimate of the 'solvation' of sucrose may be made from its limiting diffusion coefficient[33] (0·5226 × 10^{-5} cm sec^{-1} at 25°) which gives, using equation (2·51), a Stokes'—law radius of 4·69 Å. Applying to this the 'correction-factor' 1·05 indicated by *Table 6.2*, we calculate a 'hydrodynamic volume' of 0·301 l/mole, corresponding to 4.7 molecules of attached water. This agrees remarkably well with the 'thermodynamic' value of 5, but is less than is indicated by the viscosity result. Bearing in mind the possible effects of departure from spherical shape, there is no ground for dissatisfaction, however; rather, one may be astonished that the theory gives such reasonable results. Since the chloried ion causes a small *decrease* in viscosity, the 'rigid volumes' of *Table 11.4* must be regarded as essentially those of the cations; after subtracting the estimated molar volume of the cation (which is actually negative) they suggest 'hydrodynamic' hydration numbers of 2–3 for Na^+, 3–4 for Li^+, and 9–10 for Mg^{++}. These viscosity calculations thus provide yet another set of hydration numbers, of reasonable magnitude, but differing materially from estimates by other means (*cf. Tables 6.3, 9.5*, and p. 331). *Table 11.4* encourages the view that where large increases in viscosity occur in electrolyte solutions, they are mainly a direct hydrodynamic result of the distortion of stream-lines by particles considerably larger than water molecules. It should be emphasized that the above considerations refer only to approximately spherical particles; long-chain ions and molecules, in particular, need special treatment which will be found in text-books of colloid chemistry.

'Microscopic Viscosity' and the Mobility of Dissolved Particles

In order to understand the motion of ions in concentrated electrolytes, we need an answer to the question : how is the mobility of ions related to the change in viscosity of the solution, bearing in mind that this changed viscosity is itself produced by the ions of interest?

A direct answer could be given if we had an exact treatment of the interionic effects which also alter the mobilities. In concentrated solutions, however, we can do no more than estimate the magnitude of interionic effects; it is therefore more profitable to seek other information which may bear on the viscosity relationship.

In Chapter 6 we have seen that comparison of the temperature-dependence of viscosity and ionic mobility is of some value. A viscosity change produced in this way is, however, of a different nature from the isothermal change produced by the presence of dissolved particles; the latter is concerned with the distortion of stream-lines, while the former is due to changes in the relative magnitudes of thermal agitation and intermolecular forces. The relation between ionic mobility and viscosity can scarcely be expected to be the same in these two cases. We shall now summarize some relevant experimental results.

The Influence of Large Neutral Molecules on the Limiting Mobilities of Ions

Limiting conductances[34] and transport numbers[35] for a number of simple electrolytes have been measured at 25° in aqueous 10 per cent mannitol, 10 per cent and 20 per cent glycerol, and 10 per cent and 20 per cent sucrose. Limiting ion mobilities in the non-electrolyte solutions and in water, are summarized in *Table 11.5*.

The ratio of the viscosity of water to that of the mixed solvent is also shown in *Table 11.5.* The following generalizations can be made about the results:

(1) All the ions studied suffer some reduction in mobility through the presence of the added non-electrolyte.

(2) Different ions are affected differently by a given non-electrolyte, though for simple small ions the R values cluster about an

Table 11.5

Relative Ion Mobilities R in Aqueous Non-electrolyte Solutions at 25°

Ion	Sucrose		Glycerol		Mannitol
	10%	20%	10%	20%	10%
H^+	0·841	0·684	—	—	0·837
K^+	0·812	0·627	0·817	0·648	0·797
Na^+	0·810	0·621	0·815	0·647	0·790
Li^+	0·802	0·610	—	—	0·778
Ag^+	0·800	0·607	0·801	0·632	0·780
Ca^{++}	0·787	0·585	—	—	—
Mg^{++}	0·788	0·582	—	—	—
La^{+++}	0·778	0·567	—	—	—
$N(n\ Am)_4^+$	0·761	0·550	—	—	—
Cl^-	0·815	0·631	0·813	0·644	0·800
NO_3^-	0·810	0·624	0·817	0·644	0·803
Br^-	0·807	0·619	0·806	0·632	0·797
ClO_4^-	0·803	0·612	—	—	—
I^-	0·796	0·604	0·799	0·617	0·792
η^0/η	0·756	0·525	0·775	0·579	0·747

The quantity R is the ratio of the limiting mobility of the ion in the mixed solvent to its value in water, as given in Appendix 6.1.

Solvent compositions are in percent non-electrolyte to total solution, by weight.

From Steel, B. J., Stokes, J. M. and Stokes, R. H., *J. phys. Chem.* 62 (1958) 1514.

average, *e.g.*, $R \approx 0{\cdot}80$ for many monovalent ions in 10 per cent sucrose. Hydrogen ion is less retarded than any other, and there is a fair degree of correlation between the size of the ion (allowing for probable hydration of many cations) and the extent of retardation.

(3) Different non-electrolytes have slightly different effects, in the sense that the relation between the viscosity and the mobility of a given ion is not quite the same for different non-electrolytes.

(4) In no case is the mobility reduced to the full extent that the increase of viscosity would demand, *i.e.*, the behaviour is not consistent with Stokes' law (or Walden's rule), but approximates to that described by

$$\lambda \eta p = \text{const} \qquad \ldots.(11.36)$$

where p is less than unity. This relation holds fairly accurately for a given ion in a given non-electrolyte solution, but the index p varies with both the ion and the non-electrolyte. For a given ion, p is approximately a linear function of the molar volume $\bar{V}$ of the non-electrolyte, decreasing as $\bar{V}$ increases, while for a given non-electrolyte p increases with the size of the ion, approaching unity for very large ions. The same relation, but in general with a

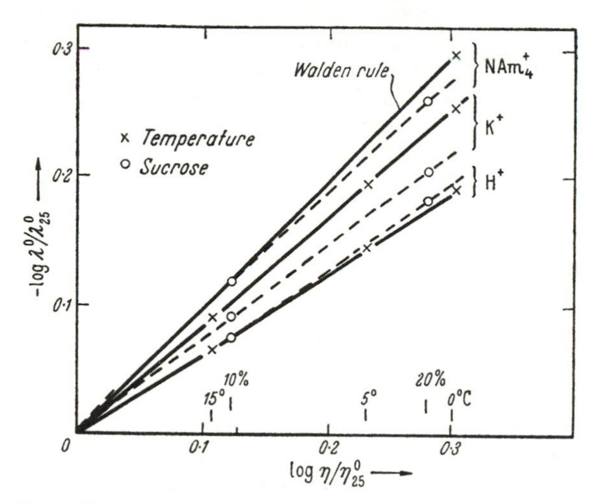

Figure 11.2. Effects on various ions of viscosity increases caused by:
(a) addition of sucrose (circles),
(b) lowering of temperatures (crosses).
Viscosities and equivalent conductivities are expressed relative to water at 25°.

different value of p, describes fairly accurately the variation of λ^0 in pure water as solvent when the viscosity is increased by lowering the temperature. *Figure 11.2* compares the effects of these two kinds of viscosity-increase for a few ions. The large tetra-amyl ammonium ion has $p \approx 1$ for both kinds of viscosity-increase, *i.e.*, it approximates to Stokes' law or Walden's rule as would be expected. It is remarkable, however, that hydrogen ion, with the much lower value $p = 0{\cdot}63$, does not seem to discriminate between addition of sucrose or mannitol and lowering of temperature as causes of viscosity-increase. This is doubtless connected in some way with its abnormal transport mechanism, (p. 121) which is believed to be limited mainly by the ease of *rotation* of water molecules. Most other ions, of which K^+ is typical, show considerable differences in their responses to the two kinds of viscosity-increase. It is thus not feasible to treat the effect of added non-electrolytes in terms of a change in 'structural temperature' of the water.

The 'Obstruction-effect'

The experimental work described above shows that when the viscosity of water is increased by molecular solutes, the resistance encountered by an ion moving through the solution is also increased, but not in direct proportion to the viscosity as would be predicted by simple hydrodynamic considerations. Presumably this direct proportionality would be observed for really large moving particles, but they would have to be large not merely in relation to the water molecules, but even in relation to the molecules or ions causing the viscosity increase. (It should be noted that in the conductance and transport number experiments discussed above, the concentration of the moving particles themselves is made effectively zero by suitable extrapolations, so that we do not have to consider their own effect on the viscosity.)

This situation is sometimes described by saying that the 'microscopic viscosity' of the solution is lower than the measured viscosity, but this statement does not constitute an explanation of the effect. An alternative possibility is that the increased viscosity and the increased resistance experienced by the moving particle are not related as cause and effect, but are two parallel effects of a common cause, that cause being the obstructive action of the added solute. In viscous flow, the solute molecules or ions distort the stream-lines, introducing a rotational quality to the previously irrotational flow; in conductance and diffusion, they lengthen the effective paths of the moving particles. This suggestion was first advanced by WANG[36] in connection with the self-diffusion of water molecules in protein solutions. We shall not make direct use of his treatment, since the result can be obtained more conveniently by the following argument, which also brings to notice a point not dealt with by Wang.

We shall idealize the actual situation to the following model: the moving particles and the solvent molecules are both of negligible size compared to the added solute molecules causing the obstruction; the latter are regarded as rigid spheres in a continuous medium. We shall discuss the motion of ions in terms of the passage of electric current through such a system. The added non-electrolyte must now be regarded as a set of insulating spheres dispersed in a random manner through a conducting continuum, and we wish to compare the conductances of, for example, a unit cube of this material and a unit cube of the same conducting medium in the absence of the insulating spheres. When a single insulating sphere is introduced into a uniform infinite conductor, the standard methods of electrical theory show that the current-lines are distorted as shown in *Figure 11.3*; this will clearly cause an increase in resistance. The actual problem deals with a large array of such spheres comparatively close together, and has been discussed by FRICKE[37] in connection with the conductance of blood, where the blood corpuscles form 'obstructions' in the plasma. (His treatment, like Wang's, deals with the more general case of ellipsoidal obstructions.) However, it is instructive to approach this problem in another way:

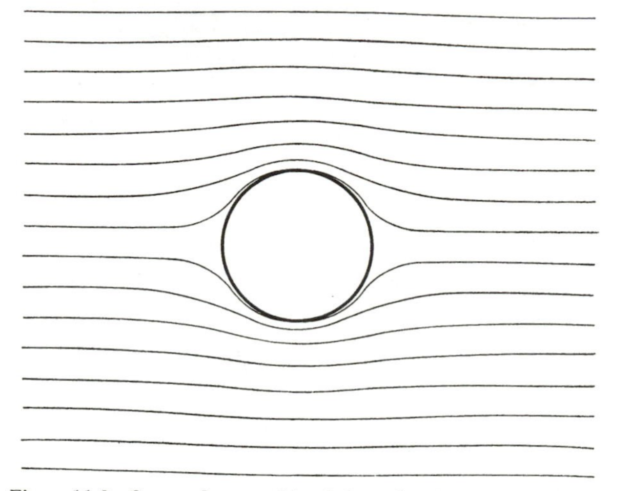

Figure 11.3. Current flow round insulating sphere in conducting medium

it is a well-known principle of electricity that problems of steady current flow in conductors and of lines of force in insulators are formally identical, the only change necessary in the mathematics being the substitution of specific conductances for dielectric constants. Now the corresponding problem in dielectric theory is: what is the effective dielectric constant of a medium of dielectric constant ε_0 in which are suspended spherical particles of a different dielectric constant ε_1? This problem has received attention at intervals since it was first discussed by Rayleigh in 1892; all investigators have concluded that the size of the spheres is irrelevant, only the fraction of the total volume which they occupy appearing in the equations. In *Table 11.6* the results of the main investigations are summarized, along with the corresponding results for the present case of electrical conductance. (It is notable that though the formal mathematics is the same, the electrical conductance of the spheres can take the value zero, whereas the dielectric constant can never be less than unity.)

Though all these formulae reduce for low volume fractions ϕ of the insulating spheres to the same result, *viz.*

$$K_{sp}/K_{sp(0)} = 1 - 3\phi/2 \qquad \ldots.(11.37)$$

they differ somewhat at higher volume fractions. (The upper limit for ϕ, when the spheres are in mutual contact in close packing, is

Table 11.6

The 'Obstruction-effect' in Conductance from Analogy with the Corresponding Dielectric Problem

Dielectric	Conductance
Medium: ε_0	$K_{sp(0)}$
Spherical obstructions: ε_1	$K_{sp(1)} = 0$
Mixture: ε	K_{sp}
Volume fraction of obstructions: ϕ	ϕ
RAYLEIGH, (1892) $\dfrac{\varepsilon - \varepsilon_0}{\varepsilon + 2\varepsilon_0} = \dfrac{\varepsilon_1 - \varepsilon_0}{\varepsilon_1 + 2\varepsilon_0}\phi$	$\dfrac{K_{sp}}{K_{sp(0)}} = \dfrac{1-\phi}{1+\phi/2} \approx 1 - 1{\cdot}5\phi$
BRUGGEMAN (1935) $\dfrac{\varepsilon_1 - \varepsilon}{\varepsilon_1 - \varepsilon_0} = (1-\phi)\left(\dfrac{\varepsilon}{\varepsilon_0}\right)^{1/3}$	$\dfrac{K_{sp}}{K_{sp(0)}} = (1-\phi)^{3/2} \approx 1 - 1{\cdot}5\phi$
BÖTTCHER (1945) $\dfrac{\varepsilon - \varepsilon_0}{3\varepsilon} = \dfrac{\varepsilon_1 - \varepsilon_0}{\varepsilon_1 + 2\varepsilon}\phi$	$\dfrac{K_{sp}}{K_{sp(0)}} = 1 - 1{\cdot}5\phi$

RAYLEIGH, J. W. *Phil. Mag.* (5) 34 (1892) 481.
BRUGGEMAN, D., *Ann. Physik. Lpz.* (5) 24 (1935) 636.
BÖTTCHER, C. J. F., *Rec. Trav. Chim., Pays-Bas.* 64 (1945) 47.
See also EL SABEH, S. H. and HASTED, J. B., *Proc. Phys. Soc.* 66B (1953) 611.

$\phi = 0{\cdot}7405$ which would give a negative conductance on several of the formulae; they are not intended to apply under such extreme conditions.) Böttcher's formula has been tested by measuring the dielectric constant of suspensions of salts in organic liquids, and is satisfactory at volume fractions as high as 0·5.

Equation 11.37 would not be immediately applicable to the *equivalent* conductances of *Table 11.5*, even if the model were valid for the solutions considered. Allowance must first be made for the fact that when calculating equivalent conductances of the electrolytes in the non-electrolyte solutions, the whole volume of the solution is taken as the basis for the concentration calculation. In formula 11.37 on the other hand, $K_{sp(0)}$ represents the specific conductance of the electrolyte solution *outside* the insulating spheres. Since the latter occupy a volume-fraction ϕ, we obtain:

$$R = \frac{\Lambda^0}{\Lambda^0_{(\phi=0)}} = \frac{1 - 3\phi/2}{1 - \phi} \approx 1 - \phi/2 \text{ for small } \phi \qquad \ldots.(11.38)$$

This may be compared with the corresponding result for the relative viscosity of the non-electrolyte solution, which by equation (11.32) gives for small ϕ:

$$\eta^0/\eta \approx 1 - 2{\cdot}5\phi \qquad \ldots.(11.39)$$

Thus the addition of sufficient large non-electrolyte molecules to lower the fluidity of water by 5 per cent should lower the mobility of ions in the solution by only 1 per cent, if this 'obstruction' model of the situation is valid. The other extreme model, in which the non-electrolyte solution is treated as a viscous continuum in which the ions move under Stokes'-Law conditions, would lead one to expect a 5 per cent decrease in mobility for a 5 per cent decrease in fluidity. *Table 11.5* shows that the effects actually found are intermediate between these two extremes, *i.e.*, there is for small ions about 3 to 4 per cent decrease in mobility for 5 per cent decrease in fluidity. This seems reasonable, since the ions and the non-electrolyte molecules are in fact of comparable sizes, whereas the obstruction model considers the ions to be much smaller than the non-electrolyte molecules, and the Stokes'-Law model considers them to be much larger.

Furthermore, the largest ion studied—the tetra-*n*-amylammonium ion—shows the nearest approach to the predictions of the Stokes'-Law model, while the chloride ion is the least affected by increased viscosity (though even here the effect is much larger than the 'obstruction' model predicts).

BROERSMA[38] has developed the hydrodynamic theory of a liquid containing suspended or dissolved particles which cause a change in the local viscosity, falling off with an inverse power of distance from the particle. With suitable choice of the parameters describing this change, the theory promises considerable success in the treatment of problems of conductance, viscosity and diffusion.

In the meantime, we can conclude only that there is no universal quantitative relation between the mobility of an ion and the viscosity of the medium, at least when the viscosity-change is produced by adding non-electrolytes. The position may be expected to be even more complicated when the viscosity-change is produced by ions.

SELF-DIFFUSION AND TRACER-DIFFUSION IN ELECTROLYTE SOLUTIONS

In a uniform liquid, the molecules are continually moving about in a random manner, and a given molecule at one point has a definite probability of arriving at some other point within a given time. This motion constitutes true self-diffusion, but is a process which can never be detected because of the indistinguishability of the molecules. A close approach to it is realized in the case of the inter-diffusion of two isotopically different species which differ only enough to be distinguishable, but have nearly identical dimensions and force fields. Since mixtures of isotopic species show practically ideal thermodynamic behaviour, the gradient of chemical potential for each isotope is also ideal, *i.e.*, it is simply the gradient of ($\boldsymbol{R}T \ln c$). The 'driving force' for the inter-diffusion of isotopic species may be considered to arise solely from the term contributed to the free energy of mixing by the entropy of mixing, which obeys the ideal relation:

$$\Delta S_{(\text{mixing})} = -\boldsymbol{R}(\mathcal{N}_1 \ln \mathcal{N}_1 + \mathcal{N}_2 \ln \mathcal{N}_2) \qquad \ldots .(11.40)$$

$\mathcal{N}_1$ and $\mathcal{N}_2$ being the mole fractions of the two species.

A closely-related process occurs when an ion of one kind in very small amount diffuses in a large excess of other electrolyte; the name 'tracer-diffusion' has been given to this process. Examples are the diffusion of radioactive sodium ion present in tracer amounts in an otherwise uniform solution of (*a*) potassium chloride, or (*b*) of sodium chloride. In case (*b*) the diffusion coefficient of the tracer species is assumed to be identical with the true self-diffusion coefficient of sodium ion in the sodium chloride solution. In case (*a*), since the ionic environment of the tracer ion is effectively unchanged during diffusion, its activity coefficient remains practically constant, so that the 'driving force' is once more the gradient of ($\boldsymbol{R}T \ln c$). In both cases the 'diffusion potential' is negligibly small, so that the movement of the tracer ions is not tied to that of ions of opposite sign. The electrophoretic effect, which involves the concentration of the diffusing ions, may also be neglected, since the concentration of the diffusing radioactive species is extremely low. Rather unexpectedly, however, the relaxation effect now becomes important, though in ordinary diffusion, owing to the preservation of the symmetry of the 'ionic atmosphere', it is negligible. The reason for this is that in self- or tracer-diffusion the tracer ion is moving relative to a background of non-diffusing ions, whereas in the ordinary diffusion of a single electrolyte all the ions are moving with the same velocity.

These tracer-diffusion coefficients, incidentally, are the quantities which appear in the formulae of polarographic theory for the limiting diffusion-current at the dropping mercury electrode.

But for the complications introduced by the interionic relaxation-effect (*cf.* p. 136) tracer- and self-diffusion studies in electrolytes would provide valuable information on viscosity-effects. The tracer-diffusion of iodide ion in alkali chloride solutions[39] has been studied with especial care both by conventional radioactive tracer techniques and by chemical analysis as means of following the iodide ion, with excellent agreement ($\sim 0{\cdot}5$ per cent) between the two methods. Fairly comprehensive data are also available for sodium ion, chloride ion, and hydrogen ion; the data are given in *Table 11.7.*

THEORETICAL EXPRESSIONS FOR THE RELAXATION-EFFECT IN SELF-DIFFUSION

Onsager[6] has discussed the problem of the diffusion of an ion present in vanishingly small amounts in a solution of another electrolyte, as a special case of diffusion in multi-component systems and Gosting and Harned[46] have shown that his formulae can be applied to the case of self-diffusion. In our notation, Onsager's equation for the diffusion coefficient D_j^* of an ion j present in vanishingly small amounts in an otherwise uniform electrolyte solution becomes:

$$D_j^* = u_j \left[\boldsymbol{k}T - \frac{\kappa z_j^2 \boldsymbol{e}^2}{3\varepsilon} (1 - \sqrt{\mathrm{d}(u_j)}) \right] \quad \ldots .(11.41)$$

The function $\mathrm{d}(u_j)$ depends on the mobilities and valencies of the various ions present, and is discussed below; all the other symbols have already been introduced. Now (11.41) may be rewritten as:

$$D_j^* = \boldsymbol{k}Tu_j \left[1 - \frac{\kappa z_j^2 \boldsymbol{e}^2}{3\varepsilon \boldsymbol{k}T} (1 - \sqrt{\mathrm{d}(u_j)}) \right]$$

$$= D_j^{*0} \left[1 - \frac{\kappa z_j^2 \boldsymbol{e}^2}{3\varepsilon \boldsymbol{k}T} (1 - \sqrt{\mathrm{d}(u_j)}) \right] \quad \ldots .(11.42)$$

It is instructive to compare equation (11.42) with equation (7.9): both deal with a relaxation effect, the former in tracer- or self-diffusion, and the latter in electrical conduction. In conduction the relaxation effect changes the applied field by the factor:

$$1 + \frac{\Delta \mathrm{X}}{\mathrm{X}} = 1 - \frac{\kappa |z_1 z_2| \boldsymbol{e}^2}{3\varepsilon \boldsymbol{k}T} \cdot \frac{q}{1 + \sqrt{q}} \quad \ldots (11.43)$$

while in tracer- or self-diffusion it changes the virtual force acting on the tracer-ion j by the factor:

$$1 - \frac{\kappa z_j^2 e^2}{3\varepsilon \boldsymbol{k} T}(1 - \sqrt{\mathrm{d}(u_j)}) \qquad \ldots .(11.44)$$

The only difference between expressions (11.43) and (11.44) lies in the valency factor and the mobility functions, z_j^2 replacing $|z_1 z_2|$ and $(1 - \sqrt{\mathrm{d}(u_j)})$ replacing $q/(1 + \sqrt{q})$. These two mobility functions are both dimensionless quantities, q being defined by equation (7.10). The definition of $\mathrm{d}(u_j)$ is rather more complicated, especially in the general case covered by Onsager's treatment. However, when the only kinds of ions present are the ions 2 and 3 of the main electrolyte and the tracer-species 1, the definition becomes:

$$\mathrm{d}(u_1) = \frac{|z_1|}{|z_2| + |z_3|}\left(\frac{|z_2|\lambda_2^0}{|z_1|\lambda_2^0 + |z_2|\lambda_1^0} + \frac{|z_3|\lambda_3^0}{|z_1|\lambda_3^0 + |z_3|\lambda_1^0}\right) \qquad \ldots .(11.45)$$

and for the case of most immediate interest, where all the ions are univalent and species 1 is an isotopic form of species 2 so that $\lambda_1^0 \approx \lambda_2^0$, (11.45) simplifies further to:

$$\mathrm{d}(u_1) = \frac{\lambda_2^0 + 3\lambda_3^0}{4(\lambda_2^0 + \lambda_3^0)} = \frac{1 + 2t_3^0}{4} \qquad \ldots .(11.46)$$

where t_3^0 is the limiting transport number of the ion 3, *i.e.*, of the ion of opposite sign to the tracer-ion 1. (It may be noted that in the special case where the anion and cation have equal mobilities, as is nearly the case in aqueous potassium chloride, for example, the mobility functions $q/(1 + \sqrt{q})$ and $(1 - \sqrt{\mathrm{d}(u_1)})$ become identical, both taking the value $1 - \sqrt{0{\cdot}5} = 0{\cdot}2929$. In this case expressions (11.43) and (11.44) are not merely similar but identical.)

Because of the similarity of form of equations (11.42) and (7.9), we can simply take over the numerical evaluations of the quantities in 7.9 for use in equation (11.42). Thus, referring to equations (7.29) and (7.31), we see that equation (11.42) becomes:

$$D_j^* = D_j^{*0}\left[1 - \frac{2{\cdot}801 \times 10^6}{(\varepsilon T)^{3/2}}(1 - \sqrt{\mathrm{d}(u_j)})z_j^2\sqrt{I}\right] \qquad \ldots .(11.47)$$

and for aqueous solutions of 1 : 1 electrolytes at 25°:

$$D_j^* = D_j^{*0}[1 - 0{\cdot}7816(1 - \sqrt{\mathrm{d}(u_j)})\sqrt{c}] \qquad \ldots .(11.48)$$

Equations (11.47) and (11.48) thus represent the Onsager limiting law for tracer- or self-diffusion *at low total ionic strengths.* The limiting value D_j^{*0} is given by the Nernst expression:

$$D_j^{*0} = \frac{RT\lambda_j^0}{|z_j|F^2}$$

$$= 2{\cdot}661 \times 10^{-7} \frac{\lambda_j^0}{|z_j|} \text{ at } 25° \qquad \ldots.(11.49)$$

Table 11.7

Tracer and Self-diffusion Coefficients of Ions in Alkali Chloride Solutions at 25°

Supporting Electrolyte	Tracer Ion	Concentration of Supporting Electrolyte, mole/l					
		0·1	0·5	1	2	3	4
		$D/D°$	$D/D°$	$D/D°$	$D/D°$	$D/D°$	$D/D°$
KCl	Na^+(*a*)	$0{\cdot}99_3$	$0{\cdot}98_7$	$0{\cdot}98_3$	$0{\cdot}96_4$	$0{\cdot}94_3$	$0{\cdot}92_8$
NaCl	Na^+(*b*)	$0{\cdot}97_4$	$0{\cdot}95_9$	$0{\cdot}92_5$	$0{\cdot}84_9$	$0{\cdot}77_2$	$0{\cdot}69_7$
LiCl	Na^+(*b*)	$0{\cdot}96_1$	$0{\cdot}93_1$	$0{\cdot}88_4$	$0{\cdot}79_4$	$0{\cdot}72_0$	—
KCl	Cl^-(*c*)	$0{\cdot}96_7$	$0{\cdot}96_3$	$0{\cdot}96_2$	$0{\cdot}93_8$	$0{\cdot}91_0$	$0{\cdot}87_3$
NaCl	Cl^-(*c*)	$0{\cdot}96_1$	$0{\cdot}91_2$	$0{\cdot}87_2$	$0{\cdot}79_4$	$0{\cdot}71_3$	$0{\cdot}62_1$
LiCl	Cl^-(*c*)	$0{\cdot}95_2$	$0{\cdot}89_4$	$0{\cdot}82_8$	$0{\cdot}73_5$	$0{\cdot}63_8$	—
HCl	Cl^-(*b*)	$0{\cdot}97_6$	$0{\cdot}95_0$	$0{\cdot}92_4$	$0{\cdot}86_6$	$0{\cdot}80_7$	$0{\cdot}76_5$
KCl	I^-(*d*)	$0{\cdot}96_2$	$0{\cdot}95_1$	$0{\cdot}93_8$	$0{\cdot}90_6$	$0{\cdot}86_0$	—
NaCl	I^-(*d*)	$0{\cdot}95_7$	$0{\cdot}90_7$	$0{\cdot}86_3$	$0{\cdot}77_5$	$0{\cdot}68_1$	$0{\cdot}59_7$
LiCl	I^-(*d*)	$0{\cdot}95_1$	$0{\cdot}89_3$	$0{\cdot}83_3$	$0{\cdot}72_5$	$0{\cdot}63_0$	$0{\cdot}54_4$
KCl	H^+(*e*)	$0{\cdot}87_0$	$0{\cdot}85_0$	$0{\cdot}82_7$	$0{\cdot}75_7$	$0{\cdot}67_8$	—
NaCl	H^+(*e*)	$0{\cdot}86_2$	$0{\cdot}80_6$	$0{\cdot}74_8$	$0{\cdot}64_2$	$0{\cdot}52_5$	$0{\cdot}42_5$
LiCl	H^+(*e*)	$0{\cdot}83_3$	$0{\cdot}76_4$	$0{\cdot}67_5$	$0{\cdot}51_0$	$0{\cdot}38_5$	$0{\cdot}28_0$

Data from the following sources, interpolated to round concentrations where necessary:

(*a*) MILLS, R., *J. Phys. Chem.*, 61 (1957) 1258.
(*b*) MILLS, R., private communication, (1958); *J. Amer. chem. Soc.*, 77 (1955) 6116.
(*c*) MILLS, R., *J. phys. Chem.*, 61 (1957) 1631.
(*d*) STOKES, R. H., WOOLF, L. A. and MILLS, R., *ibid.*, 61 (1957) 1634.
(*e*) WOOLF, L. A., *Thesis*, University of New England, 1958.

$D° = \frac{RT}{F^2}\lambda°$; $D°$ for Na^+, Cl^-, I^- and H^+ has respectively the values 1·333 2·032, 2·045 and $9{\cdot}308 \times 10^{-5}$ cm² sec⁻¹.

No reliable studies of tracer-diffusion have yet been made at concentrations of supporting electrolyte low enough to test the limiting law (11.42). The data in concentrated supporting electrolytes lie well above the predictions of the limiting law, a situation similar to that arising in studies of conductance, ordinary diffusion, and activity. There is little doubt that the introduction of a finite ion size into the theory would raise the predicted values but it seems unlikely that it will explain the large differences in the tracer diffusion coefficients at high concentrations. It must be expected that the viscosity of the supporting electrolyte will play an important

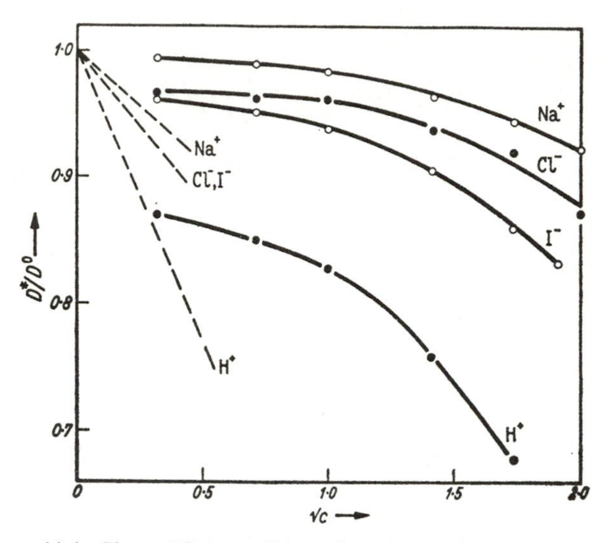

Figure 11.4. Tracer diffusion coefficients of ions in potassium chloride solutions

role, and this is confirmed by a detailed consideration of the result in *Table 11.7*.

Potassium chloride causes only a small change of viscosity with concentration, and the tracer diffusion coefficients of ions in potassium chloride solutions are presumably governed by interionic effects. In *Figure 11.4* D/D^0 is plotted against $\sqrt{c_{KCl}}$ for four ions; the limiting slope given by equation 11.42 is also shown. Evidently not even approximate reliance may be placed upon the limiting theory in the range of supporting electrolyte concentration above 0·1 M. The sodium ion suffers only a slight retardation amounting to 2 per cent at 1 molar KCl, while the hydrogen ion suffers 17 per cent retardation at the same concentration.

If we consider the same ion in different supporting electrolytes, the curves for sodium chloride and lithium chloride fall progressively below that for potassium chloride, suggesting that the viscosity is important. Indeed, the results for iodide ion fall on a single smooth curve for both LiCl and NaCl as supporting electrolytes, when plotted against viscosity instead of concentration; the same is true for sodium ion and for chloride ion, but not for hydrogen ion. The results for potassium chloride as supporting electrolyte do not lend themselves to this form of plotting, as the viscosity is nearly constant.

It also appears that in lithium and sodium chloride solutions, sodium ion is less affected by the viscosity of the supporting electrolyte than are iodide or chloride ions. A very tentative explanation is that in a concentrated solution the sodium ion will tend to have as its immediate neighbours chloride ions which have little effect on the local viscosity; the chloride and iodide ions, on the other hand, will be predominantly surrounded by sodium ions or lithium ions which cause a marked increase in the local viscosity. This suggestion is consistent with the ideas of BROERSMA[38].

CONDUCTANCE AND VISCOSITY IN CONCENTRATED SOLUTIONS

In conductance, the interionic effects are even more complicated than in tracer-diffusion, for one is measuring the motion of all the ions, cations in one direction and anions in the other. Surprisingly, however, the equations developed for dilute solutions continue to give a reasonable account of the conductance up to quite high concentrations, though of course without the nearly perfect quantitative fit which can be obtained at low concentrations. An equation which proves fairly successful is that proposed by WISHAW and STOKES[18]; this is equation 7.27, with the relaxation-factor $(1 + \Delta X/X)$ given by Falkenhagen's earlier expression 7.13, and with the introduction of the relative viscosity of the solution:

$$\Lambda\eta/\eta^0 = \left(\Lambda^0 - \frac{B_2\sqrt{c}}{1 + \kappa a}\right)\left(1 + \frac{\Delta X}{X}\right) \quad \ldots.(11.50)$$

Though the equation has little theoretical justification at high concentrations it is most effective, requiring only the single arbitrary parameter a to reproduce the conductances of fully dissociated 1 : 1 electrolytes up to concentrations of many moles per litre with an accuracy of a few per cent, as shown in *Figure 11.5*. There is a tendency for the measured conductances of viscous salt solutions to be slightly higher than the equation predicts, if the a value is selected to give reasonable fit at the lower concentrations. A better fit can be obtained by using a fractional power of the relative viscosity, but this amounts to introducing a second arbitrary parameter.

A bibliography of recent conductance and viscosity data for a number of concentrated solutions appears in Appendix 6.3.

MUTUAL DIFFUSION IN CONCENTRATED ELECTROLYTES

In the ordinary diffusion of a salt in a concentration gradient, both ions must move in the same direction at the same speed, to maintain electrical neutrality. The main interionic effect is therefore the

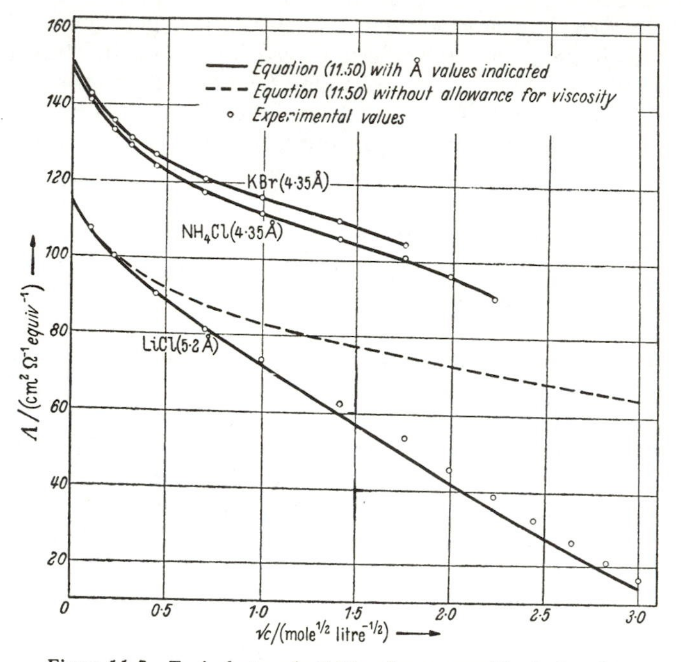

Figure 11.5. Equivalent conductivities of concentrated 1 : 1 *electrolytes*

harmonic averaging of the ion mobilities in accordance with equation 11.4; the electrophoretic effect is relatively small, and levels off to a nearly constant value at high concentrations as shown in *Figure 11.1.* The relaxation-effect, which presents the greatest theoretical difficulties in conductance and tracer-diffusion, is fortunately absent since the ionic distribution remains symmetrical. From the success which the theory of the electrophoretic effect has in representing transport numbers, we may reasonably argue that it is equally valid in diffusion, even at several moles per litre. We therefore make use of equation 11.21 in the ensuing discussion of diffusion in concentrated solutions.

For concentrated solutions a number of effects, negligible for the dilute solutions so far considered, become of great importance. These are:

1. That solvent molecules will in general move in the opposite direction to the solute.
2. That some of the ions may carry with them a permanently attached layer of solvent molecules, which acts as a part of the diffusing solute entity.
3. That the viscous forces may be considerably modified by the presence of large numbers of ions.

Our theoretical treatment of these phenomena will be based on that due to HARTLEY and CRANK[40]. We consider first a non-electrolyte solution containing only two types of diffusing entity, the molecules A and B and also restrict ourselves to the case where the partial volumes, $\bar{V}_A$ and $\bar{V}_B$, of both components are constant; this means in practice that the coefficients discussed will be 'differential' values referring to diffusion between two solutions differing only slightly in concentration. In this case the diffusion coefficient measured experimentally will be in terms of the flux across a plane P so fixed that the total volumes on each side of it remain constant; that is, across a plane fixed with respect to the apparatus. This measured diffusion coefficient is denoted by D_A^V for component A, and by D_B^V for component B; and we have, denoting the fluxes of moles of A and B across unit area of the plane P by J_A^V and J_B^V:

$$J_A^V = -D_A^V \frac{\partial C_A}{\partial x}, \quad J_B^V = -D_B^V \frac{\partial C_B}{\partial x}$$

C_A and C_B being the concentrations of A and B in moles per unit volume. The fluxes of *volumes* of A and B through the plane P are therefore:

$$-D_A^V \bar{V}_A \frac{\partial C_A}{\partial x} \quad \text{and} \quad -D_B^V \bar{V}_B \frac{\partial C_B}{\partial x}$$

But since there is no net transfer of volume across the plane P, it follows that the sum of these two quantities is zero:

$$D_A^V \bar{V}_A \frac{\partial C_A}{\partial x} + D_B^V \bar{V}_B \frac{\partial C_B}{\partial x} = 0 \qquad \ldots .(11.51)$$

Also, since C_A and C_B are the numbers of moles of A and B in unit volume of solution:

$$\bar{V}_A C_A + \bar{V}_B C_B = 1$$

Differentiating this with respect to x gives:

$$\bar{V}_A \frac{\partial C_A}{\partial x} + \bar{V}_B \frac{\partial C_B}{\partial x} = 0 \qquad \ldots .(11.52)$$

and on comparing (11.51) and (11.52) it is evident that for both to hold it is necessary that:

$$D_A^V \equiv D_B^V$$

except for the trivial cases $\bar{V}_A = 0$ or $\bar{V}_B = 0$. Thus the diffusion of a binary system where the partial volumes are constant can be described by a single diffusion coefficient, which may be called the

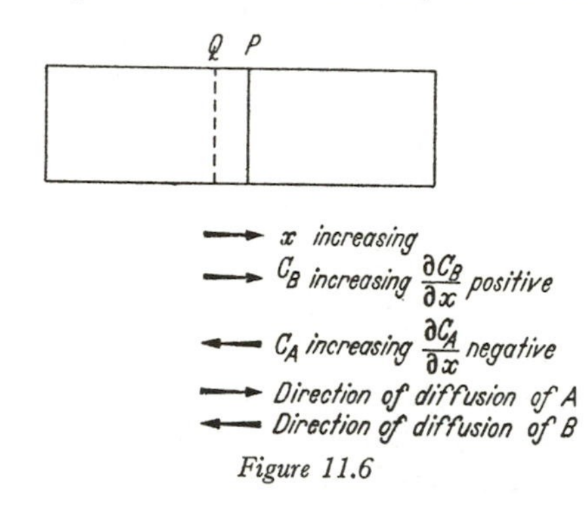

Figure 11.6

mutual diffusion coefficient, and denoted simply by D^V; the same value of D^V will be found whether one measures and calculates from the concentration of component A or component B.

Next, Hartley and Crank introduce the idea of an 'intrinsic diffusion coefficient' of each component, here denoted by D'_A and D'_B. The passage of component A through the volume-fixed plane P just discussed must necessitate the passage of an *equal volume* of B in the opposite direction, in order to preserve the fixed volumes on each side of the plane. The total flow of each component is regarded as made up partly of a true diffusion-flux and partly of a 'bulk flow' which originates in the volume-difference between the two components. A plane Q may be imagined (though not as a rule constructed) so that no 'bulk flow' occurs through it; and the 'intrinsic diffusion coefficients' D'_A and D'_B are defined in terms of the flux across unit area of such a plane. (A better physical idea of the meaning of these intrinsic diffusion coefficients may be obtained as follows: Imagine two solutions of slightly differing compositions to be placed in a porous-diaphragm diffusion cell such as that shown in *Figure 10.1*. Now imagine that the action of gravity is abolished and that both ends of the cell are opened. The liquids will stay in place, having no reason to do otherwise, but there will be no artificial restraints upon the motion of the liquids, such as are normally imposed by the closure of the ends. Under these conditions, the rates of transfer of the two components across the diaphragm will be governed by the intrinsic diffusion coefficients D'_A and D'_B. In the normal use of the cell with the ends closed, the rate of transfer of both components is governed by the mutual diffusion coefficient D^V.)

Since the partial volumes are constant, the concentration-gradients $\frac{\partial C_A}{\partial x}$ and $\frac{\partial C_B}{\partial x}$ must be opposite in sign; suppose that C_A increases to the left (see *Figure 11.6*), and C_B to the right, and that the distance x is measured from left to right. Then on the right of the plane Q there will be a rate of *increase* of volume, due to the entry of A, given by $-\bar{V}_A D'_A \frac{\partial C_A}{\partial x}$ and a rate of *decrease* of volume, due to the outward passage of B, given by $+\bar{V}_B D'_B \frac{\partial C_B}{\partial x}$. The net rate of increase of the volume V' on the right of Q is therefore given by:

$$\frac{\partial V'}{\partial t} = -\left(\bar{V}_A D'_A \frac{\partial C_A}{\partial x} + \bar{V}_B D'_B \frac{\partial C_B}{\partial x}\right) \quad \ldots.(11.53)$$

This expression, since we are considering unit cross-section of the plane Q, also gives the rate at which the plane Q moves away from the fixed plane P. Since no bulk flow occurs through Q, the motion of this plane with respect to P must be due to a bulk flow through P; so that expression (11.53) also represents the bulk flow through P from right to left. The bulk flow therefore involves a transport of component A from the left to the right of P (*i.e.*, in the direction of diffusion of A), given by:

$$J_A \text{ (bulk flow)} = -C_A \frac{\partial V'}{\partial t}$$

This transport of A across P by bulk flow is superimposed on the transport of A across P by 'pure' diffusion, which is given by:

$$J_A \text{ (pure diffusion)} = -D'_A \frac{\partial C_A}{\partial x}$$

The total flux of A across P is therefore:

$$J_A \text{ (total)} = -D'_A \frac{\partial C_A}{\partial x} + C_A \left(\bar{V}_A D'_A \frac{\partial C_A}{\partial x} + \bar{V}_B D'_B \frac{\partial C_B}{\partial x}\right) \quad \ldots.(11.54)$$

But the total flux J_A across the fixed plane $\boldsymbol{P}$ also defines the experimentally measured mutual diffusion coefficient D^V:

$$J_A\ (\text{total}) = -D^V \frac{\partial C_A}{\partial x} \qquad \ldots.(11.55)$$

Combining equations (11.54), (11.55) and (11.52) now yields:

$$D^V = D'_A + \bar{V}_A C_A(D'_B - D'_A) \qquad \ldots.(11.56)$$

Now the intrinsic diffusion coefficient D'_A of A at a finite concentration C_A is related to its value D^0_A at infinite dilution ($C_A = 0$) by the factor ($\text{d}\ln a_A/\text{d}\ln C_A$) which expresses the effect of the deviation of the solution from ideal behaviour. It is also probable, but by no means certain, that the bulk viscosity of the solution compared with that of the pure liquid B should also be introduced; we denote this relative viscosity by η/η^0_B. The activity a_A in the thermodynamic factor may be expressed on any scale of concentration we like, since the logarithmic differentiation will eliminate any constant conversion factors; we choose the mole fraction scale (with mole fractions $\mathcal{N}_A$ and $\mathcal{N}_B$) for later convenience. This gives:

$$\left.\begin{aligned} D'_A &= D^0_{AB} \frac{\text{d}\ln \mathcal{N}_A f_A}{\text{d}\ln C_A} \frac{\eta^0_B}{\eta} \\ D'_B &= D^0_{BB} \frac{\text{d}\ln \mathcal{N}_B f_B}{\text{d}\ln C_B} \frac{\eta^0_B}{\eta} \end{aligned}\right\} \qquad \ldots.(11.57)$$

where D^0_{AB} is the diffusion coefficient of A at infinite dilution in B, and D^0_{BB} is the (self) diffusion coefficient of B in pure B.

Since $C_A = \dfrac{\mathcal{N}_A}{\mathcal{N}_A \bar{V}_A + \mathcal{N}_B \bar{V}_B}$, one finds on logarithmic differentiation, and using $\mathcal{N}_A + \mathcal{N}_B = 1$,

$$\frac{\text{d}\ln C_A}{\text{d}\ln \mathcal{N}_A} = 1 - \frac{\mathcal{N}_A(\bar{V}_A - \bar{V}_B)}{\mathcal{N}_A \bar{V}_A + \mathcal{N}_B \bar{V}_B} = \frac{\bar{V}_B C_A}{\mathcal{N}_A}$$

and similarly

$$\frac{\text{d}\ln C_B}{\text{d}\ln \mathcal{N}_B} = \frac{\bar{V}_A C_B}{\mathcal{N}_B}$$

Hence equation (11.57) becomes:

$$\left.\begin{aligned} D'_A &= D^0_{AB} \frac{\mathcal{N}_A}{\bar{V}_B C_A} \frac{\text{d}\ln \mathcal{N}_A f_A}{\text{d}\ln \mathcal{N}_A} \frac{\eta^0_B}{\eta} \\ D'_B &= D^0_{BB} \frac{\mathcal{N}_B}{\bar{V}_A C_B} \frac{\text{d}\ln \mathcal{N}_B f_B}{\text{d}\ln \mathcal{N}_B} \frac{\eta^0_B}{\eta} \end{aligned}\right\} \qquad \ldots.(11.58)$$

Furthermore, the Gibbs-Duhem equation gives:

$$\frac{\mathrm{d}\ln \mathcal{N}_A f_A}{\mathrm{d}\ln \mathcal{N}_A} = \frac{\mathrm{d}\ln \mathcal{N}_B f_B}{\mathrm{d}\ln \mathcal{N}_B}$$

Using this along with equations (11.58) and (11.56) gives for the mutual diffusion coefficient D^V:

$$D^V = \frac{\mathrm{d}\ln \mathcal{N}_A f_A}{\mathrm{d}\ln \mathcal{N}_A} \cdot \frac{\eta_B^0}{\eta}\left[D_{AB}^0 \mathcal{N}_A\left(\frac{1}{\bar{V}_B C_A} - \frac{\bar{V}_A}{\bar{V}_B}\right) + D_{BB}^0 \mathcal{N}_B \frac{C_A}{C_B}\right]$$

Since $\dfrac{\mathcal{N}_A}{\mathcal{N}_B} = \dfrac{C_A}{C_B}$ and $C_A\bar{V}_A + C_B\bar{V}_B = 1$, the square bracket simplifies giving finally:

$$D^V = \frac{\mathrm{d}\ln \mathcal{N}_A f_A}{\mathrm{d}\ln \mathcal{N}_A} \frac{\eta_B^0}{\eta} [\mathcal{N}_B D_{AB}^0 + \mathcal{N}_A D_{BB}^0] \quad \ldots.(11.59)$$

which, by symmetry, can also be written:

$$D^V = \frac{\mathrm{d}\ln \mathcal{N}_B f_B}{\mathrm{d}\ln \mathcal{N}_B} \frac{\eta_A^0}{\eta} [\mathcal{N}_A D_{BA}^0 + \mathcal{N}_B D_{AA}^0] \quad \ldots.(11.60)$$

This is Hartley and Crank's expression (in a slightly modified form) for the mutual diffusion coefficient at any concentration. If volume effects and the counter-diffusion of the solvent are ignored, as was done in the derivation of the expressions for dilute electrolytes, one obtains an expression of the form:

$$D = D_{BA}^0 \frac{\mathrm{d}\ln \mathcal{N}_B f_B}{\mathrm{d}\ln C_B} \quad \ldots.(11.61)$$

for comparison with (11.60), when we see that the more complete expression differs from (11.61) in three respects: first, the activity factor in (11.60) is a differential with respect to mole fraction instead of concentration; secondly, the diffusion coefficient of the solvent also appears; and thirdly, the relative viscosity of the solution has been introduced.

The theory is capable of straightforward extension to the case of the diffusion of a single electrolyte in solution, making allowance for the possible hydration of the ions. We let B denote the solvated electrolyte, 1 mole of which is associated with h moles of bound water, and let A denote the *free* water. These we shall treat as the diffusing entities. In accordance with the notation used in the discussion of the chemical potential of solvated ions (Chapter 9), primed symbols will be used to denote quantities in which account is taken of solvation. Equation (11.60) now becomes:

$$D^V = \frac{\mathrm{d}\ln \mathcal{N}'_B f'_B}{\mathrm{d}\ln \mathcal{N}'_B} \cdot \frac{\eta^0_A}{\eta} \cdot [\mathcal{N}'_A D^0_{BA} + \mathcal{N}'_B D^0_{AA}]$$

The mole fraction $\mathcal{N}'_B$, like the term $\mathcal{N}_B$ which we have used in the derivation of formula (11.60) means simply the ratio of the number of the diffusing entity B to the total number of diffusing entities of both species. Since the diffusion of the ions of the electrolyte is restricted by the condition of electrical neutrality, it is permissible to treat the partial volumes, concentrations, *etc.*, as those of the hydrated electrolyte as a whole, without considering the separate ionic quantities. The only place in (11.60) at which consideration must be given to the fact of ionization is in the expression $\mathrm{d}\ln \mathcal{N}'_B f'_B$; this may be written $\mathrm{d}\ln \mathcal{N}'_B f'_B = \mathrm{d}\ln a'_B$, since we have remarked that any scale of activity can legitimately be used. Now the solute is hydrated and ionized; let it produce ν ions per 'molecule' so that $a_B = (a_\pm)^\nu$ denotes the conventional activity as computed for the unhydrated solute. Then because of the hydration, and for an aqueous solution, $\mathrm{d}\ln a'_B = \mathrm{d}\ln a_B + h\,\mathrm{d}\ln a_w$ (where a_w = water activity). Also by the Gibbs-Duhem relation:

$$\mathrm{d}\ln a_w = -\frac{m}{55{\cdot}51}\,\mathrm{d}\ln a_B$$

whence

$$\mathrm{d}\ln a'_B = (1 - 0{\cdot}018hm)\,\mathrm{d}\ln a_B = (1 - 0{\cdot}018hm)\nu\,\mathrm{d}\ln a_\pm \qquad \ldots.(11.62)$$

where $a_\pm$ is the mean activity of the unhydrated solute.

In formula (11.60), we also have to consider the meaning to be given to the limiting intrinsic diffusion coefficients D^0_{BA} and D^0_{AA}. In the case of mixtures of liquids which are non-electrolytes their meaning is clear. For an electrolyte solution, however, it is necessary that formula (11.60) shall reduce to the Nernst limiting value as $\mathcal{N}'_B \rightarrow 0$. This means that because of the factor ν in (11.62) we must put for electrolytes $D^0_{BA} = D^0/\nu$ where D^0 is the Nernst limiting value; or more completely, if we include the electrophoretic corrections, $D^0_{BA} = (D^0 + \Delta_1 + \Delta_2)/\nu$. D^0_{AA} is the diffusion coefficient of water in the infinitely dilute solution, *i.e.*, in the absence of any interfering non-ideal effects and volume restraints: we therefore put it equal to the self-diffusion coefficient of pure water, $D^*_{H_2O}$. Now, with the special meaning which applies here to the concept of 'mole fraction', we can put $\mathcal{N}'_A$ and $\mathcal{N}'_B$ in terms of the ordinary molality and the hydration number h:

$$\mathcal{N}'_B = \frac{m}{55{\cdot}51 - hm + m}, \quad \mathcal{N}'_A = \frac{55{\cdot}51 - hm}{55{\cdot}51 - hm + m} \quad \ldots.(11.63)$$

Hence

$$\frac{\mathrm{d}\ln \mathcal{N}'_B}{\mathrm{d}\ln m} = \frac{1}{1 + 0{\cdot}018(1-h)m} \quad \ldots.(11.64)$$

we therefore rearrange (11.60) to read:

$$D^V = D^0_{BA} \cdot \frac{\mathrm{d}\ln a'_B}{\mathrm{d}\ln m} \cdot \frac{\mathrm{d}\ln m}{\mathrm{d}\ln \mathcal{N}'_B}\left[\mathcal{N}'_A + \mathcal{N}'_B \frac{D^0_{AA}}{D^0_{BA}}\right]\frac{\eta^0_A}{\eta} \quad \ldots.(11.65)$$

and put $D^0_{BA} = D^0/\nu$, $D^0_{AA} = D^*_{H_2O}$, $\mathcal{N}'_A$ and $\mathcal{N}'_B$ as given by (11.63), $\dfrac{\mathrm{d}\ln m}{\mathrm{d}\ln \mathcal{N}'_B}$ as given by (11.64) and $\mathrm{d}\ln a'_B$ as given by (11.62), obtaining:

$$D^V = D^0\left(\frac{\mathrm{d}\ln a_\pm}{\mathrm{d}\ln m}\right)(1 - 0{\cdot}018hm)\left[1 + 0{\cdot}018m\left(\nu\frac{D^*_{H_2O}}{D^0} - h\right)\right]\frac{\eta^0_A}{\eta}$$

$$\ldots.(11.66)$$

Equation (11.66) was first derived from equation (11.60) by AGAR[44]. In (11.66), electrophoresis is neglected; if it is included, we merely write $(D^0 + \Delta_1 + \Delta_2 + \ldots)$ for D^0. The activity factor $\dfrac{\mathrm{d}\ln a_\pm}{\mathrm{d}\ln m}$ may of course also be written in the alternative form

$$(1 + m\,\mathrm{d}\ln\gamma/\mathrm{d}\,m)$$

The diffusion coefficient D^V given by (11.66) represents that which would be obtained if the volume-concentration and flux in the diffusion experiment were computed on the basis of the hydrated solute; however, since the volume-concentration of the electrolyte is unaffected by any considerations of hydration of the ions, it is the same as the diffusion coefficient D obtained by the ordinary computation with the concentration in moles of anhydrous solute per c.c. and the flux also in moles of anhydrous solute per cm^2 per sec.

For a uni-univalent electrolyte at m values small enough to justify neglecting the square of $(0{\cdot}018hm)$, and including the electrophoresis corrections in the main D^0 factor, but neglecting them in the small correction term $\nu\dfrac{D^*_{H_2O}}{D^0}$, (11.66) becomes:

$$D = (D^0 + \Delta_1 + \Delta_2)\left(1 + m\frac{\mathrm{d}\ln\gamma}{\mathrm{d}m}\right)\left[1 + 0{\cdot}036m\left(\frac{D^*_{H_2O}}{D^0} - h\right)\right]\frac{\eta^0}{\eta}$$

$$\ldots.(11.67)$$

We shall now use equation (11.67) in interpreting the D values of 1 : 1 electrolytes at concentrations up to a few moles per litre. The value of the self-diffusion coefficient of water has been the subject of much recent research. PARTINGTON, HUDSON and BAGNALL[41] obtained $2{\cdot}43 \times 10^{-5}$ cm^2 sec^{-1} at 25°, with an estimated accuracy of $\pm$ 0·5 per cent, by using the magnetically-stirred porous diaphragm-cell method of Stokes with heavy water as tracer; WANG[42] obtained by the Anderson capillary-tube method the values (at 25°): $(2{\cdot}34 \pm 0{\cdot}08) \times 10^{-5}$ using heavy water, $(2{\cdot}44 \pm 0{\cdot}07) \times 10^{-5}$ using tritiated water, and $(2{\cdot}66 \pm 0{\cdot}12) \times 10^{-5}$ using H_2O^{18}. The mean of Wang's three values is $2{\cdot}45 \times 10^{-5}$, in excellent agreement with that of Bagnall and Partington; we shall therefore adopt the value of $2{\cdot}4_4 \times 10^{-5}$ for $D^*_{H_2O}$ at 25° in equation (11.67).

In the application of this equation, we note that all the quantities except the hydration number h are capable of calculation from experimental limiting mobilities and thermodynamic data; the terms Δ_1 and Δ_2 also involve assuming some value for the ion size parameter a, but since this appears only in small correction terms, the exact value chosen is not highly critical. We shall use the a values given in *Table 9.5.*

The factor $\left(1 + m\dfrac{\mathrm{d}\ln\gamma}{\mathrm{d}m}\right)$ can be computed from the tabulated activity coefficients, or from the osmotic coefficients, ϕ, since by a simple application of the Gibbs-Duhem relation we obtain:

$$\left(1 + m\frac{\mathrm{d}\ln\gamma}{\mathrm{d}m}\right) \equiv 1 + \frac{\sqrt{m}}{2}\frac{\mathrm{d}\ln\gamma}{\mathrm{d}\sqrt{m}} = \phi + m\frac{\mathrm{d}\phi}{\mathrm{d}m} \equiv \phi + \frac{\sqrt{m}}{2}\frac{\mathrm{d}\phi}{\mathrm{d}\sqrt{m}} \qquad \ldots.(11.68)$$

Any of these equivalent forms can be used, different ones being best adapted to different concentration regions. The slopes involved can be obtained graphically or by the numerical method of RUTLEDGE[43]. Values of D, D^0, Δ_1, Δ_2, $\left(1 + m\dfrac{\mathrm{d}\ln\gamma}{\mathrm{d}m}\right)$, $\dfrac{\eta}{\eta^0}$ and the ratio:

$$f(D) = D_{\text{obs.}}\Big/\left[(D^0 + \Delta_1 + \Delta_2)\left(1 + m\frac{\mathrm{d}\ln\gamma}{\mathrm{d}m}\right)\right] \qquad \ldots.(11.69)$$

are given in *Table 11.8* at a number of concentrations, m, for sodium chloride at 25°. It will be seen from equation (11.67) that the plot of $\dfrac{\eta}{\eta^0}f(D)$ versus m should be a straight line of slope $0{\cdot}036\,(D^*_{H_2O}/D^0 - h)$. It should be noted that the D values, *etc.*, of this table are given at round molalities rather than at round volume concentrations as in Appendix 11.2; the quantities Δ_1 and Δ_2 are calculated at round values of (κa) which are converted to concentrations and thence to molalities, then graphically interpolated to the round molalities.

A comparison of the last two columns of *Table 11.8* shows that the relative viscosity factor can play an extremely important part.

Table 11.8

Application of Equation (11.67) *to Sodium Chloride Solutions at* 25°

m	D obs.	Δ_1	Δ_2	$1+m\dfrac{\mathrm{d}\ln\gamma_\pm}{\mathrm{d}m}$	$\dfrac{\eta}{\eta^0}$	$f(D)$ eq. (11.69)	$f(D)\dfrac{\eta}{\eta^0}$
0	1·610*	0	0	1·000	1·000	(1·000)	(1·000)
0·01	1·547	− 0·003	+ 0·009	0·955	1·001	1·001	1·002
0·05	1·506	− 0·006	+ 0·024	0·927	1·004	0·997	1·001
0·1	1·484	− 0·008	+ 0·032	0·917	1·009	0·989	0·998
0·2	1·478	− 0·010	+ 0·040	0·914	1·018	0·985	1·002
0·3	1·477	− 0·011	+ 0·043	0·915	1·027	0·982	1·009
0·5	1·474	− 0·013	+ 0·049	0·927	1·046	0·965	1·009
0·7	1·475	− 0·014	+ 0·050	0·946	1·065	0·946	1·007
1·0	1·482	− 0·015	+ 0·052	0·970	1·094	0·927	1·014
1·5	1·494	− 0·016	+ 0·052	1·031	1·147	0·879	1·008
2·0	1·511	− 0·017	+ 0·051	1·096	1·205	0·838	1·010
3·0	1·538	− 0·018	+ 0·049	1·245	1·341	0·752	1·008
4·0	1·567	− 0·019	+ 0·047	1·410	1·509	0·678	1·023

* Nernst limiting value = D^0.

$$f(D) = D_{\mathrm{obs.}} \Big/ \left[(D^0 + \Delta_1 + \Delta_2)\left(1 + m\frac{\mathrm{d}\ln\gamma_\pm}{\mathrm{d}m}\right)\right]$$

For calculating Δ_1 and Δ_2: a = 3·97Å.

We have taken over the bulk viscosity of the solution from the theory for non-electrolytes, but it is by no means certain that this step is justified: the change of the bulk viscosity brought about by adding ions is not necessarily a fair measure of the change in frictional resistance experienced by the ions. For this reason we shall give two parallel sets of results, one in which we write:

$$\frac{\eta}{\eta^0} f(D) = 1 + 0{\cdot}036m\left(\frac{D^*_{\mathrm{H_2O}}}{D^0} - h\right) \qquad \ldots.(11.70)$$

to obtain h, and another in which we write:

$$f(D) = 1 + 0{\cdot}036m\left(\frac{D^*_{\mathrm{H_2O}}}{D^0} - h'\right) \qquad \ldots.(11.71)$$

to obtain an alternative hydration number h'.

In view of our earlier discussions of viscosity effects, we might expect the actual hydration numbers to be between h and h', but nearer to the former.

In the case of sodium chloride we see that both $f(D)$ and $\frac{\eta}{\eta^0} f(D)$ approach unity at $m = 0$ in a satisfactory manner; it must be remembered that the experimental error in D is about 0·2–0·3 per cent and that in the factor $\left(1 + m \frac{\mathrm{d} \ln \gamma}{\mathrm{d}m}\right)$ is probably 0·2 per cent, at least for the more concentrated solutions. Consequently we should be satisfied if the functions $f(D)$ or $\frac{\eta}{\eta^0} f(D)$ are linear in m with a scatter of about 0·5 per cent, and extrapolate to unity with a similar tolerance. It should also be remembered that the calculation of the electrophoretic corrections is of doubtful validity for the higher concentrations where also Δ_2 is rather sensitive to the value of the ion-size parameter, and that our neglect of the squares of terms in $0{\cdot}018m$, in obtaining equation (11.67) from equation (11.66) may well involve some departure from linearity at the higher molalities. For all these reasons, it seems advisable to use only the data up to 1 M for the evaluation of h and h'. We find for sodium chloride, from a graph of $\frac{\eta}{\eta^0} f(D)$ versus m, $\frac{\eta}{\eta^0} f(D) = 1 + 0{\cdot}014m$ (average deviation $\pm$ 0·2 per cent up to 1 M) and $f(D) = 1 - 0{\cdot}072m$ (average deviation $\pm$ 0·2 per cent up to 1 M), whence putting $D^*/D^0 = 2{\cdot}44/1{\cdot}610 = 1{\cdot}51$ we obtain:

$$h_{\mathrm{NaCl}} = 1{\cdot}1_2 \quad \text{or} \quad h'_{\mathrm{NaCl}} = 3{\cdot}5_1$$

It follows that either of these values, inserted in equations (11.70) and (11.69) in the case of h, or in equations (11.71) and (11.69) in the case of h' will reproduce the observed diffusion coefficients of sodium chloride with an average accuracy of 0·2 per cent, *i.e.*, within experimental error. The results for the ten 1 : 1 halides in Appendix 11.2 are also capable of representation by these equations, in the concentration-range up to 1 M, the values needed for the two alternative parameters h and h' being given in *Table 11.9*. The 'deviations' listed indicate the percentage accuracy with which the equations are capable of reproducing the observed diffusion coefficients, using the given values of h or h'. It is clear that there is little to choose between the two equations on the score of accuracy. Equations (11.69) and (11.71), in which the viscosity factor is omitted, give at first sight a more reasonable set of hydration numbers, in that the values of h' increase in the order: iodide < bromide < chloride as would be expected from the order of the sizes of the bare ions. On the other hand, the reverse order, as found for the h values, is the same as that found in the treatment of activity coefficients in terms of hydration (Chapter 9). Furthermore,

Table 11.9

'Hydration Numbers' from Diffusion

Solute	KCl	KBr	KI	NaCl	NaBr	NaI	LiCl	LiBr	NH_4Cl	HCl	HBr
h	0·8	1·2	1·5	1·1	1·2	2·2	2·9	2·9	0·5	2·1	2·3
Average deviation	0·2%	0·4%	0·6%	0·2%	0·3%	0·5%	0·5%	0·6%	0·2%	0·2%	0·2%
Maximum deviation	0·3%	0·5%	1·2%	0·5%	0·6%	0·8%	1·0%	1·3%	0·5%	0·4%	0·5%
h'	0·6	0·3	− 0·3	3·5	2·8	3·0	6·3	5·6	0·2	3·7	3·2
Average deviation	0·2%	0·4%	0·6%	0·2%	0·3%	0·5%	0·3%	0·5%	0·2%	0·2%	0·2%
Maximum deviation	0·4%	0·5%	1·0%	0·4%	0·8%	0·6%	0·4%	0·9%	0·5%	0·4%	0·5%

the h values are all positive, whereas h' for potassium iodide is negative and so cannot be physically interpreted as a hydration number; and the h values are more nearly additive for the constituent ions than are the h' values. The following sets of ionic hydration numbers are capable of giving the h value for any salt in *Table 11.9* within 0·1, except for sodium iodide and bromide where the additive values differ by 0·3 and 0·2 respectively from the observed.

Ionic Hydration Numbers from Diffusion

Ion	NH_4^+	K^+	Na^+	Li^+	Cl^-	Br^-	I^-
(*a*)	0·5	0·9	1·2	2·8	0·0	0·2	0·7
(*b*)	0·0	0·4	0·7	2·3	0·5	0·7	1·2

The two sets (*a*) and (*b*) both lead to the same sums for positive and negative pairs; the latter set based on $h(NH_4^+) = 0$, is perhaps the more reasonable.

The above treatment of diffusion in concentrated solutions, while serving to indicate the more important effects which have to be considered, is in no sense final or completely satisfactory. The parameter h is useful in that it enables one to represent the observed diffusion coefficients within about one half of one per cent up to one molar concentration by means of equations (11.66) or (11.67).

It does not follow that h therefore necessarily represents only the hydration of the solute; there may be other short-range effects which would result in the appearance of a multiplier of the form $(1 - am)$ in the expression for the diffusion coefficient. Comprehensive investigations of the diffusion of non-electrolyte solutions in the light of equation (11.59) will no doubt help to provide information on such effects.

In the case of the salts, the interpretation of h as the number of water molecules moving with the ion as part of the diffusing unit may be tentatively accepted. No surprise need be felt that these h values from diffusion are smaller than those obtained from the treatment of activity data in terms of hydration, since in the latter treatment h was introduced as the effective number of molecules bound by the ion-solvent forces, and would therefore include contributions from water molecules beyond the first layer, which would not be firmly enough bound to move as a unit with the ion. These 'hydration numbers' h, are, however, somewhat lower than the majority of estimates of ionic hydration by other methods.

For the halogen acids, it is no longer legitimate to interpret h as the number of water molecules moving with the diffusing ions, for one would then be obliged to claim that at least 1·6 mole of water were moving with the hydrogen ion. But the hydrogen ion moves, in the main, by a series of proton-jumps from one water molecule to another, and the volume transfer in this process is negligible. It is of course possible that in addition to this proton-jump mechanism some ordinary motion of clusters of water molecules with a proton at their centre occurs; but one would have to assume such clusters to be rather numerous and large to account for an apparent hydration number of 1·6. A more likely explanation, for which we are indebted to Dr. J. N. Agar[44], is that water molecules close to a proton are not available as arrival points for another proton after one of its jumps; this would lead to the ease of such jumps diminishing more or less linearly with concentration and so producing a decrease in the diffusion rate, similar in magnitude to that occurring with salts because of the motion of water of hydration.

CONCENTRATED SOLUTIONS OF POLYVALENT ELECTROLYTES

The theory of diffusion for higher valency types in concentrated solution is even more tentative than for the 1 : 1 electrolytes. One reason for this is that the theory of the electrophoretic corrections is less satisfactory, even for very dilute solutions; also experimental data of any worth-while accuracy are very sparse. Only for three salts [18, 20, 45] at 25° are adequate experimental data at present available. The observed diffusion coefficients are plotted as curve I in *Figure 11.7* which also shows some theoretical curves. Curve IV represents the function:

$$D_{\text{calc.}} = D^0 \left(1 + m \frac{\mathrm{d} \ln \gamma}{\mathrm{d}m}\right) (1 - 0{\cdot}018hm) \left[1 + 0{\cdot}018m \left(\frac{3D^*_{\mathrm{H_2O}}}{D^0} - h\right)\right] \quad \ldots .(11.72)$$

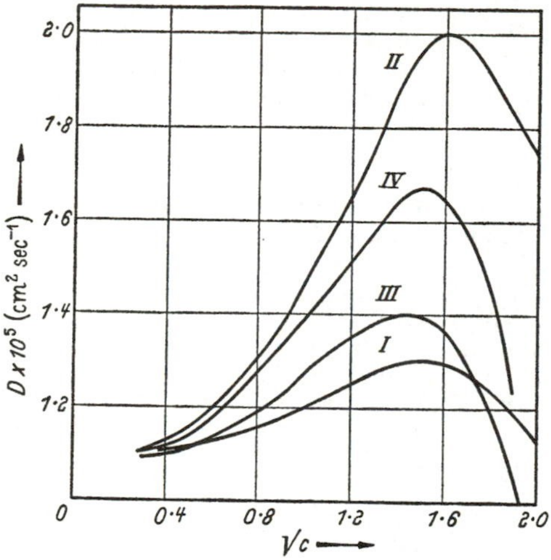

Figure 11.7. Observed and calculated diffusion coefficients of calcium chloride at 25°: *curve I, experimental; curve II, equation* (11.73) *with* $h = 0$; *curve III, equation* (11.73) *with* $h = 4$; *curve IV, equation* (11.72) *with* $h = 9$

without a viscosity factor, for $n = 9$; curves II and III show the function:

$$D_{\text{calc.}} = D^0 \left(1 + m \frac{\mathrm{d} \ln \gamma}{\mathrm{d}m}\right) (1 - 0{\cdot}018hm) \left[1 + 0{\cdot}018m \left(\frac{3D^*}{D^0} - h\right)\right] \frac{\eta^0}{\eta} \quad \ldots .(11.73)$$

with $h = 0$ and $h = 4$ respectively. Electrophoretic corrections have been omitted. It will be seen that all these functions approximately reproduce the form of the experimental curve, giving a rise followed by a maximum. Quantitative agreement is, however, far from satisfactory.

REFERENCES

[1] NERNST, W., *Z. phys. Chem.*, 2 (1888) 613

[2] GIBBS, J. W., 'Collected Works' Vol. 1, p. 429 Longmans, Green & Co., New York, (1928)

[3] GUGGENHEIM, E. A., *J. phys. Chem.*, 33 (1929) 842

[4] HARTLEY, G. S., *Phil. Mag.*, 12 (1931) 473

[5] ONSAGER, L., and FUOSS, R. M., *J. phys. Chem.* 26 (1932) 2689

[6] — *Phys. Rev.*, 37 (1931) 405; 38 (1931) 2265; *Ann. N.Y. Acad. Sci.*, 46 (1945) 241

[7] DE GROOT, S. R., 'Thermodynamics of Irreversible Processes', North-Holland Publishing Co., Amsterdam (1951)

[8] STOKES, R. H., *J. Amer. chem. Soc.*, 75 (1953) 4563

[9] ADAMSON, A. W., *J. phys. Chem.*, 58 (1954) 514

[10] DYE, J. L. and SPEDDING, F. H., *J. Amer. chem. Soc.*, 76 (1954) 888

[11] SCATCHARD, G., and PRENTISS S. S., *ibid.*, 55 (1933) 4355

[12] GUGGENHEIM, E. A., *Trans. Faraday Soc.*, 50 (1954) 1048

[13] HARNED, H. S. *et al.*, *J. Amer. chem. Soc.*, 71 (1949) 2781; 75 (1953) 4168; 76 (1954) 2064; 77 (1955) 265

[14] HARNED, H. S. and BLAKE, C. A., *ibid.*, 73 (1951) 4255

[15] HARNED, H. S. and BLAKE, C. A., *ibid.*, 73 (1951) 2448, 5882

[16] HARNED, H. S. and HUDSON, R. M., *ibid.*, 73 (1951) 5083

[17] HARNED, H. S. and HUDSON, R. M., *ibid.*, 73 (1951) 3781, 5880

[18] WISHAW, B. F. and STOKES, R. H., *ibid.*, 76 (1954) 2065

[19] MÜLLER, G. T. A. and STOKES, R. H., *Trans. Faraday Soc.*, 53 (1957) 642

[20] VITAGLIANO, V. and LYONS, P. A., *J. Amer. chem. Soc.*, 78 (1956) 4538

[21] WEISSBERGER, A., 'Physical Methods of Organic Chemistry' Vol. 1. Interscience Publishers, Inc., New York, (1945)

[22] JONES, G. *et al.*, *J. Amer. chem. Soc.*, 55 (1933) 624; 4124; 57 (1935) 2041; 58 (1936) 619, 2558; 59 (1937) 484; 62 (1940) 335, 338

[23] SWINDELLS, J. F., COE, J. R. and GODFREY, T. B., *J. Res. nat. Bur. Stand.*, 48 (1952) 1

[24] BINGHAM, E. C. and JACKSON, R. F., *Bull. U.S. Bur. Stand.*, 14 (1919) 59

[25] THORPE, T. E. and RODGER, J. W., *Phil Trans.*, *A*185 (1894) 397

[26] FALKENHAGEN, H. and DOLE, M., *Phys. Z.*, 30 (1929) 611; FALKENHAGEN, H. and VERNON, E. L., *ibid.*, 33 (1932) 140; *Phil. Mag.*, (7) 14 (1932) 537; FALKENHAGEN, H. and KELBG, G., *Z. Elektrochem.*, 56 (1952) 834. See also PITTS, E., *Proc. Roy. Soc.*, 217A (1953) 43

[27] JONES, G. and DOLE, M., *J. Amer. chem. Soc.*, 51 (1929) 2950

[28] KAMINSKY, M., *Z. phys. Chem. Frankfurt*, 8 (1956) 173

[29] COX, W. M. and WOLFENDEN, J. H., *Proc. Roy. Soc.*, 145A (1934) 475

[30] GURNEY, R. W., 'Ionic Processes in Solution', McGraw-Hill, New York (1954), p. 162

[31] EINSTEIN, A., *Ann. Phys.*, 19 (1906) 289

[32] VAND, V., *J. phys. Chem.*, 52 (1948) 277

[33] GOSTING, L. J., and MORRIS M. S., *J. Amer. chem. Soc.*, 71 (1949) 1998

[34] STOKES, R. H. and STOKES, J. M., *J. phys. Chem.*, 60 (1956) 217; 62 (1958) 497

[35] STEEL, B. J. and STOKES, R. H., *ibid.*, 62 (1958) 450; STEEL, B. J., STOKES, R. H. and STOKES, J. M., *ibid.*, 62 (1958) 1514

[36] WANG, J. H., *J. Amer. chem. Soc.*, 76 (1954) 4755

[37] FRICKE, H., *Phys. Rev.*, 24 (1924) 575; *J. phys. Chem.*, 57 (1953) 934

[38] BROERSMA, S., *J. chem. Phys.*, 28 (1958) 1158
[39] STOKES, R. H., WOOLF, L. A. and MILLS, R., *J. phys. Chem.*, 61 (1957) 1634
[40] HARTLEY, G. S. and CRANK, J., *Trans. Faraday Soc.*, 45 (1949) 801
[41] PARTINGTON, J. R., HUDSON, R. F. and BAGNALL, K. W., *Nature*, 169 (1952) 583
[42] WANG, J. H., ROBINSON, C. V. and EDELMANN, I. S., *J. Amer. chem. Soc.* 75 (1953) 466
[43] MARGENAU, H. and MURPHY, G. M., 'The Mathematics of Physics and Chemistry,' p. 456; D van Nostrand & Co., Inc., New York 1943
[44] AGAR, J. N., Private communication (1950)
[45] HALL, J. R., WISHAW, B. F. and STOKES, R. H. *J. Amer. chem. Soc.* 75 (1953) 1556; LYONS, P. A. and RILEY, J. F., *ibid.*, 76 (1954) 5216
[46] GOSTING, L. and HARNED, H. S., *ibid.*, 73 (1951) 159

12

WEAK ELECTROLYTES

THE idea of a weak acid developed in the early days of physical chemistry when it was noticed that a very large number of acids, most of them organic, obeyed a rule to which Ostwald was led by applying the law of mass action, $\alpha^2 c/(1 - \alpha) = K$, to his extensive measurements of the conductivity of acids. For the so-called strong acids, the ionic concentrations, judged by conductivity values and using the relation $\alpha = \Lambda/\Lambda^0$, led to values of the ionization 'constant' which varied markedly with the dilution. This was one of the 'anomalies of strong electrolytes' and many attempts were made to circumvent the law of mass action and preserve the ionic theory before it was realized that interionic forces were so important in solutions of strong electrolytes and were, indeed, sufficient to resolve this anomaly.

Interionic effects are, however, not negligible even in the case of weak acids. The effects enter in two ways. Taking one example from a series of measurements which later we will discuss in more detail, the specific conductivity of 0·02 N acetic acid at 25° is 0·00023132 in marked contrast to a solution of hydrochloric acid of the same concentration for which $K_{sp} = 0{\cdot}0091448$. The conventional equivalent conductivity of acetic acid is obtained by multiplying its specific conductivity by the factor 1000/0·02 to give 11·566. It is known from other measurements that the equivalent conductivity of acetic acid at infinite dilution is $\Lambda^0 = 390{\cdot}71$ so that *if the sum of the equivalent conductivities of hydrogen and acetate ions is the same at all concentrations*, the concentration of each of these ions is:

$$c_{\mathrm{H}^+} = c_{\mathrm{A}^-} = \frac{1000 K_{sp}}{\Lambda^0} = 0{\cdot}000592 \text{ mole l}^{-1}$$

and only a fraction, $\alpha = 0{\cdot}0296$, of the acetic acid molecules are dissociated.

Introducing the law of mass action, we get:

$$K = \frac{\alpha^2 c}{1 - \alpha} = 1{\cdot}806 \times 10^{-5}$$

Two errors occur in this derivation. First of all, the equivalent conductivity at an *ionic* concentration αc (*approximately* 0·000592 N) is not 390·71; it is slightly less because the interionic forces reduce it from its limiting value at infinite dilution; it will be shown later that it is 387·16 so that $\alpha = 0{\cdot}02987$ and $K = 1{\cdot}840 \times 10^{-5}$.

Another interionic effect is allowed for by introducing the activity coefficient product into the ionization constant:

$$K = \frac{y_{\mathrm{H}^+} y_{\mathrm{A}^-} \alpha^2 c}{y_{\mathrm{HA}}(1 - \alpha)}$$

Later a value of approximately 0·946 will be obtained for $y_{\mathrm{H}^+} y_{\mathrm{A}^-}/y_{\mathrm{HA}}$: this makes $K = 1{\cdot}740 \times 10^{-5}$.

These two corrections act in opposite directions; whilst they are not large in magnitude, they introduce changes into the ionization constant well beyond the very small errors of experiment found in the best measurements.

We see, therefore, that the Ostwald method was saved by the fact that the ionic strength of a solution of a weak acid is very small and the interionic forces are therefore small: it is essentially the appreciable magnitude of the interionic forces in a solution of a strong acid which leads to the failure of the law of mass action, the so-called 'anomaly of strong electrolytes'.

The situation is even less happy if we try to measure an ionization constant by the usual combination of a hydrogen (or quinhydrone) electrode in a buffered solution of the partially neutralized acid together with a calomel electrode connected by a 'salt bridge'. We cannot be sure to what extent the liquid junction potential between the salt bridge and the solution has been eliminated and as, moreover, it is usual to use moderately large concentrations of the buffered acid, the activity coefficient term is appreciable. The trend to-day is to discard cells with liquid junctions and to devise suitable combinations which are in principle concentration cells without transport. This trend can be over-emphasized and one should not lose sight of the use which can be made of cells with liquid junctions if results of only moderate accuracy are required. Such cells are easy to set up, results are obtained rapidly, and for many purposes give a good approximation to the ionization constant. But if results of the highest accuracy are wanted, the cell without liquid junction should be selected.

The perfection of technique in relation to ionization constants can be appreciated by reading two papers on this subject, one[1] using electromotive force measurements and the other[2] relying on conductivity measurements. Each paper is a classic.

IONIZATION CONSTANTS FROM CONDUCTIVITY MEASUREMENTS

MacInnes and Shedlovsky had the advantage of great experience in conductivity measurements on non-associated electrolytes. They then measured the conductivity of acetic acid at 25° in the range $c = 0{\cdot}00003$ to $c = 0{\cdot}2$. Because of the incomplete ionization of this electrolyte, the limiting conductivity, Λ^0_{HAc}, could not be determined by direct extrapolation of these data. Instead, it was determined as $\lambda^0_{\mathrm{H}^+} + \lambda^0_{\mathrm{Ac}^-}$, these figures being obtained by applying Kohlrausch's law to the known limiting conductivities of the strong electrolytes, hydrochloric acid, sodium chloride and sodium acetate. A first approximation to the degree of ionization of acetic acid was then obtained by the formula: $\alpha = \Lambda/\Lambda^0$. The approximate nature of this relation arises from the fact that the actual solution contains a concentration of ions, αc, which, though small, is not zero; the exact relation is $\alpha = \Lambda/\Lambda_i$ where Λ_i is the equivalent conductivity of a hypothetical fully ionized solution of acetic acid at a concentration αc. MacInnes and Shedlovsky estimated Λ_i by combining empirical equations for the conductivity of the strong electrolytes, hydrochloric acid, sodium chloride and sodium acetate, thus obtaining an improved approximation to α and a new value of αc; these successive approximations to α converge rapidly. They then computed K_a, the ionization constant, from the formula $K_a = \dfrac{\alpha^2 y_\pm^2 c}{1-\alpha}$, using for the activity coefficient the value predicted by the Debye-Hückel limiting law at the concentration αc. The more complete theory of the conductivity available today somewhat simplifies the calculation of Λ_i, and we shall use the experimental results of MacInnes and Shedlovsky to illustrate this simpler procedure.

For Λ_i we shall use equation (7.36) in the form:

$$\Lambda_i = \Lambda^0 - (B_1\Lambda^0 + B_2)\sqrt{\alpha c}/(1 + Ba\sqrt{\alpha c})$$

Since the actual ionic concentrations are very low ($\alpha c < 0{\cdot}002$), the value of Λ_i will not be very sensitive to the exact value of a, and an estimated value of 4 Å may be employed. The activity coefficient, $y_\pm$, at the ionic concentration αc may similarly be computed from the Debye-Hückel expression:

$$\log y_\pm \approx \log f_\pm = -A\sqrt{\alpha c}/(1 + Ba\sqrt{\alpha c})$$

again with $a = 4$ Å.

In *Table 12.1* the major stages of the calculation are shown for some of the concentrations studied by MacInnes and Shedlovsky. It will be seen from this table that measurements at concentrations less than $c = 0{\cdot}006$ lead to K_a values constant within the experimental error, and we can put $K = 1{\cdot}752 \times 10^{-5}$. However, the results at higher concentrations show a small downward trend with increasing concentration. This is probably due to neglect of the activity coefficient of the undissociated molecule, the use of $f_\pm$ instead of $y_\pm$, the possible effect of changing viscosity on the conductivity of the solution and even dimerization of the acid[(3)]. All these effects are likely to be approximately linear in the concentration of the acid, so that by plotting the value of log K_a in the last column of *Table 12.1* against the concentration, an extrapolation

Table 12.1

Calculation of Ionization Constant of Acetic Acid at 25°

c	$\Lambda_{obs.}$	$\Lambda/\Lambda^0 \approx \alpha$	Λ_i	$\Lambda/\Lambda_i = \alpha$	$-2 \log f_\pm$	$K_a \times 10^5$
0·00002801	210·38	0·5384	390·13	0·5393	0·0039	1·753
0·00011135	127·75	0·3270	389·81	0·3277	0·0061	1·754
0·0002184	96·493	0·2470	389·62	0·2477	0·0074	1·752
0·0010283	48·146	0·1232	389·05	0·12375	0·0113	1·751
0·002414	32·217	0·0825	388·63	0·08290	0·0141	1·752
0·005912	20·962	0·0537	388·10	0·05401	0·0178	1·750
0·02	11·566	0·0296	387·16	0·02987	0·0241	1·740
0·05	7·358	0·0188	386·27	0·01905	0·0302	1·726
0·1	5·201	0·0133	385·46	0·013493	0·0357	1·700
0·2	3·651	0·0093	384·54	0·009494	0·0420	1·653

should eliminate these effects. Such treatment of the results leads to $K_a = 1{\cdot}753 \times 10^{-5}$; since we have worked with the molarity as the concentration unit, this ionization constant is on the molarity scale. On the molality scale it would, by equation (2.42), be $1{\cdot}758 \times 10^{-5}$.

IONIZATION CONSTANTS FROM ELECTROMOTIVE FORCE MEASUREMENTS

The electromotive force method has the advantage of greater experimental ease. It depends essentially on the construction of a cell:

$$H_2|HA, NaA, XY|X$$

where HA is a weak acid and X is an electrode reversible to one ion of the electrolyte XY whose ionic concentration is known, *i.e.*, usually XY and HY must be strong electrolytes. Since the cell:

$$H_2|HCl|AgCl, Ag$$

has been studied so extensively, it is natural that the study of weak acids should commence with the cell:

$$H_2|HA(m), NaA(m'), NaCl(m'')|AgCl, Ag$$

This is essentially a cell containing hydrogen ions derived from the weak acid and chloride ions from sodium chloride, together with two electrodes reversible to these ions. The potential of the cell is therefore:

$$E = E^0 - k \log \gamma_{H^+}\gamma_{Cl^-}m_{H^+}m_{Cl^-} \qquad \text{....(12.1)}$$

We now introduce the law of mass action:

$$K_a = \frac{\gamma_{H^+}\gamma_{A^-}m_{H^+}m_{A^-}}{\gamma_{HA}m_{HA}} \qquad \text{....(12.2)}$$

where γ_{HA} is the activity coefficient of the undissociated part of the weak acid and *not* the ionic activity coefficient product, $\gamma_{H^+}\gamma_{A^-}$.

Equations (12.1) and (12.2) give:

$$E - E^0 + k \log \frac{m_{Cl^-}m_{HA}}{m_{A^-}} = -k \log \frac{\gamma_{H^+}\gamma_{Cl^-}}{\gamma_{H^+}\gamma_{A^-}} \gamma_{HA}K_a$$

Since $m_{Cl^-} = m''$, $m_{A^-} = m' + m_{H^+}$, $m_{HA} = m - m_{H^+}$ then, unless the acid is moderately strong, $m_{A^-} \approx m'$ and $m_{HA} \approx m$, and:

$$E - E^0 + k \log \frac{mm''}{m'} = -k \log K_a - k \log \frac{\gamma_{H^+}\gamma_{Cl^-}}{\gamma_{H^+}\gamma_{A^-}} \gamma_{HA} \qquad \text{....(12.3)}$$

To take one example from the paper of Harned and Ehlers, at $m = 0{\cdot}04922$, $m' = 0{\cdot}04737$, $m'' = 0{\cdot}05042$, $E = 0{\cdot}57977$ at 25° and $E^0 = 0{\cdot}22239$ whence the left-hand side of the above equation is 0·28164 V and, to a first approximation, $K_a = 1{\cdot}729 \times 10^{-5}$. From cell measurements with different concentrations of the components, values of $\log K_a$, not corrected for the activity coefficient term, are plotted against the total ionic strength and the curve extrapolated to zero concentration to give the limiting value of $\log K_a$. For acetic acid at 25°, Harned and Ehlers found $K_a = 1{\cdot}754 \times 10^{-5}$ which agrees remarkably well with the result of MacInnes and Shedlovsky.

The method has the advantage of simplicity and speed; it is not difficult to extend the experiment over a temperature range by measuring the potential at 5° intervals from 25° to 60°, then coming down at 5° intervals to 0° and back again to 25°, with a triple check at 25°. There is some reason to believe that the electrodes work best over the 0°–40° temperature range. If very accurate results are needed, the same cell can be measured with hydrochloric acid as electrolyte and a new determination of E^0 made, thus obviating any minor difference due to the method of preparing the electrodes. The quinhydrone electrode can be used[(4)] with acids like chloroacetic acid which are reduced by hydrogen.

Instead of equating m_{A^-} to m' and m_{HA} to m, it is more correct to put $m_{A^-} = (m' + m_{H^+})$ and $m_{HA} = (m - m_{H^+})$, m_{H^+} being calculated either from $m_{H^+} \approx K_a \dfrac{m}{m'}$ or from $E \approx E^0 - k \log m_{H^+}m''$. For stronger acids, like formic acid, a series of successive approximations may be necessary. For acids of even lower pK, the difficulty of calculating m_{H^+} becomes more formidable as has been recognised by BATES[5] and by KING and KING[6]. For example, sulphamic acid has $pK = 0{\cdot}988$ at 25° and an approximation to m_{H^+} is not good enough; it could be evaluated from the e.m.f. data if $\gamma_{H^+}\gamma_{Cl^-}$ could be calculated by a Debye-Hückel expression with a finite a term in the denominator. Unfortunately, the extrapolated value of pK is not independent of the a value selected, *e.g.* $pK = 0{\cdot}988$ if $a = 3{\cdot}85$Å and $pK = 1{\cdot}084$ if $a = 6{\cdot}00$Å and there is no way of finding which is the correct a value from the e.m.f. data alone. Conductivity measurements have, however, been made on sulphamic acid[7] and, as we have already seen, an activity coefficient term is also needed in the calculation of the ionization constant. The resulting pK for sulphamic acid is again dependent on the a value selected but in this method an increase in a decreases the apparent pK whereas the converse is true in the e.m.f. method and King and King noted that the two methods led to the same extrapolated pK value ($0{\cdot}988$) if $a = 3{\cdot}85$Å in both calculations. This they took as the most probable pK value although they noted that the agreement might be fortuitous in that the a value referred to sulphamic acid ions in one method and to these ions and those of sodium chloride in the other method.

If the acid is very weak, boric acid[8] for example, allowance must be made for hydrolysis:

$$A^- + H_2O \rightleftharpoons HA + OH^-$$

Now $m_{HA} = m + m_{OH^-}$ and $m_A = m' - m_{OH^-}$ and m_{OH^-} comes from the approximation: $m_{OH^-} = \dfrac{K_w m'}{K_a m}$. If the acid is polybasic, for example phosphoric acid in its second stage of dissociation[(9)], the activity coefficient term in equation (12.3) is no longer small: we now get a term $k \log \dfrac{\gamma_{H^+}\gamma_{Cl^-}\gamma_{H_2PO_4^-}}{\gamma_{H^+}\gamma_{HPO_4^{--}}}$, which is dealt with by using a

Debye-Hückel approximation: $-\log \gamma \approx Az^2\sqrt{I}$, or even the extended Debye-Hückel equation (9.11) for each ionic activity coefficient. But these are details which affect the computations; the principle of the method is remarkably simple and straightforward. It should be added that special experimental technique is needed for the first ionization constant of carbonic acid[10], when it is necessary to maintain a constant H_2–CO_2 ratio in the gas around the 'hydrogen' electrode. The amino-acids are dealt[11] with simply, by considering them as dibasic acids derived from $\overset{+}{N}H_3{\cdot}R{\cdot}COOH$. An interesting problem occurs when a polybasic acid has ionization constants not far different from one another, so that the acid molecule and its several derived ions may all be present in significant amounts during the neutralization. This effect is very marked with citric acid, and has been treated by BATES[12].

The electromotive force method has been applied to a number of weak acids, often over a temperature range and the field of mixed solvents has been explored. The extent of this work is illustrated in Appendix 12.1 in which are collected some of the more recent data pertaining to a temperature of 25° as well as the numerical values of the parameters of an equation which gives the temperature variation of the dissociation constant over a range from 0° to 50° or 60°. With few exceptions, the values recorded in this appendix were obtained by one of the two methods outlined above—conductimetric or potentiometric with proper allowance for the interionic forces.

The application of the method to a weak base involves nothing fundamentally different for a base such as ammonia has the ammonium ion as its conjugate acid and:

$$NH_4^+ + H_2O \rightleftharpoons NH_3 + H_3O^+$$

with an ionization constant:

$$K_a = \frac{\gamma_{H^+}\, \gamma_{NH_3}\, m_{H^+}\, m_{NH_3}}{\gamma_{NH_4^+}\, m_{NH_4^+}}$$

whereas if the ionization is treated as that of a base:

$$K_b = K_w/K_a$$

The cell:

$$H_2|NH_4OH\ (m),\ NH_4Cl\ (m')|AgCl,\ Ag$$

in which the hydrogen ions come from the solvent and are in equilibrium with the hydroxyl ions of the base, works well if adequate presaturators are provided for the stream of hydrogen gas to prevent loss of ammonia from the cell and if a correction is made for the solubility of silver chloride in the ammoniacal solution, for which purpose the instability constant of the ammine $Ag(NH_3)_2^+$ must be known: the hydrolysis of the ammonium ion requires a further correction and the appreciable vapour pressure of ammonia has to be studied if the electromotive force is to be corrected to unit hydrogen pressure. With attention to these matters, a reliable set of potentials can be measured over a temperature range, 0 — 50°. BATES and PINCHING[13] found $K_b = 1{\cdot}77 \times 10^{-5}$ at 25° To reduce the correction due to solubility of the electrode material, the silver-silver iodide electrode has been used by OWEN[14] who found $K_b = 1{\cdot}75 \times 10^{-5}$. Another method, which avoids difficulties due to volatility of the base and dissolution of the electrode-material, is that of 'partial hydrolysis',[15] In the cell:

$$H_2|NH_4A(m),\ NaCl|AgCl,\ Ag$$

where NH_4A is the ammonium salt of a weak acid with ionization constant K_A, because of the equilibria between the various species in this solution, four equations can be deduced:

$$K_a = \frac{a_{H^+}\, m_{NH_3}}{\gamma_{NH_4^+}\, m_{NH_4^+}} = \frac{a_{H^+}}{\gamma_{NH_4^+}} \cdot \frac{\alpha_1}{1-\alpha_1}$$

$$K_A = \frac{a_{H^+}\, \gamma_{A^-}\, m_{A^-}}{m_{HA}} = a_{H^+}\, \gamma_{A^-} \cdot \frac{1-\alpha_2}{\alpha_2} \qquad \ldots.(12.3a)$$

$$m = m_{NH3} + m_{NH_4^+} = m_{HA} + m_{A^-}$$

and the condition of electrical neutrality:

$$m_{H^+} + m_{NH_4^+} = m_{OH^-} + m_{A^-}$$

From the last equation it can be seen that $m_{NH_4^+} = m_{A^-}$ only if, the solution is neutral; if, as in this case, the solution is alkaline, $m_{NH_4^+}$ is slightly greater than m_{A^-}:

$$(\alpha_2 - \alpha_1)m = m_{OH^-} \qquad \ldots.(12.3b)$$

From these four equations, it follows to a very good approximation that:

$$K_a K_A = a_{H^+}^2$$

or, more accurately:

$$K_a K_A = a_{H^+}^2 \cdot \frac{\alpha_1}{\alpha_2} \cdot \frac{1-\alpha_2}{1-\alpha_1}$$

whence

$$\frac{1}{2}(pK_a + pK_A) = (E - E^0)/k + \log m_{Cl^-} + \log \gamma_{Cl^-} - \frac{1}{2}\log\frac{\alpha_1}{\alpha_2}\cdot\frac{1-\alpha_2}{1-\alpha_1}$$

As the last term is of the order 0·001, it suffices to calculate α_2 approximately from equation (12.3a) and then α_1 from (12.3b). It is desirable that α_2 be greater than 0·1 to ensure adequate buffer capacity but if the degrees of hydrolysis are too large, the electrode material becomes soluble: hence pK_a and pK_A should differ by less than two units. The first experiments were made with the base tris(hydroxymethyl)aminomethane, $(CH_2OH)_3C.NH_2$, with the addition of equimolar amounts of potassium *p*-phenolsulphonate. Measurements were also made on the more conventional cell:

$$H_2|MOH, MCl|AgCl, Ag$$

The ionization constants found by the two methods were the same, confirming that the method of 'partial hydrolysis' was satisfactory. Subsequent measurements gave $K_b = 1{\cdot}77 \times 10^{-5}$ for ammonia at 25°.

THE SPECTROPHOTOMETRIC METHOD

Ionization constants can also be measured by colorimetric methods or by ultraviolet spectroscopy. *Figure 12.1* shows the ultraviolet absorption spectrum of *p*-nitrophenol[16] in a number of buffer solutions: it can be seen that as the pH decreases absorption in the region of 3170Å becomes more pronounced whilst that at 4070Å, which is marked in alkaline solution, diminishes to zero in acid solutions. This suggests that absorption at 3170Å is due to the uncharged *p*-nitrophenol molecule and that at 4070Å to the negatively charged anion, there being an isosbestic point at 3500Å where the extinction coefficients of the two species are equal and the two can be mixed in any proportion (at constant total molality) without change in absorption; thus the solution has the same optical density at any pH. At a wave length not that of the isosbestic point, however, the optical density does depend on the pH and:

$$D = \varepsilon_{HR}\, c_{HR} l + \varepsilon_{R^-}\, c_{R^-} l$$

where D is the observed optical density, ε_{HR} and ε_{R^-} are extinction coefficients, c_{HR} and c_{R^-} concentrations and l is the cell length. Or:

$$D = D_1(1 - \alpha) + D_2\alpha$$

where D_1, D_2 and D are the optical densities of three solutions of the same total concentration of acid, measured in cells of the same length, D_1 referring to a solution of low pH, D_2 to a solution of

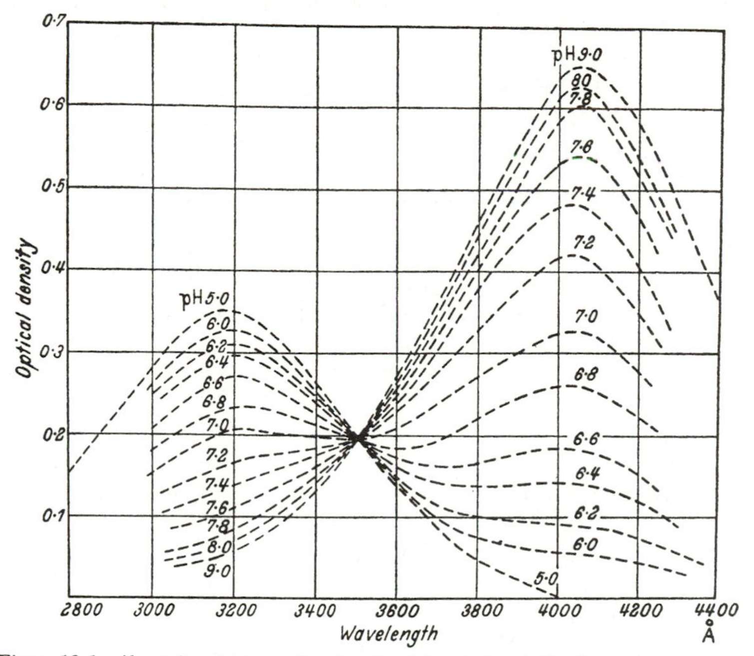

Figure 12.1. Absorption spectrum of p-nitrophenol in solutions buffered at various pH values. Concentration of p-nitrophenol: 0·000036 N. (*From* BIGGS, A. I., *Trans. Faraday Soc.*, 50 (1954) 800)

high pH and D to one of intermediate pH in which a fraction α of the acid is present in the ionized form. The ionization constant of the acid is then given by:

$$pK = \text{pH} - \log \frac{\alpha}{1 - \alpha} - \log \gamma_{\text{R}^-}$$

where $\text{pH} \equiv -\log \gamma_{\text{H}^+} m_{\text{H}^+}$ refers to a standard buffer in which the acid is dissolved for the measurement of D: the buffer should be so selected that its pH is about equal to the pK of the acid. The activity coefficient is calculated by Davies' equation (9.13) although it has been found better to use 0·2 as the coefficient of the linear term. This may well be due to the fact that the method has so far been applied to organic acids with anions larger than those of the simpler electrolytes, requiring a larger a term: in Davies' equation

$a = 3$Å, and the difference can be compensated by raising the coefficient of the linear term. The pK value derived by this method should be independent of the nature of the buffer mixture, at least in the concentration range in which the assumptions about the

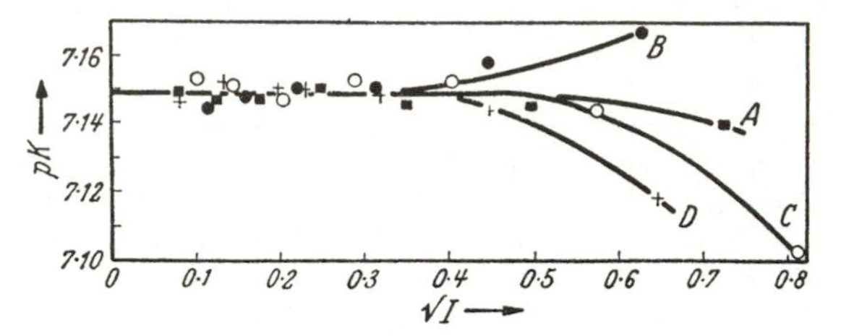

Figure 12.2. Ionization constant of p-nitrophenol from measurements in four buffer mixtures

A NaH_2PO_4 : Na_2HPO_4 : $NaCl = 1 : 0{\cdot}9819 : 1$
B NaH_2PO_4 : Na_2HPO_4 : $NaCl = 1 : 0{\cdot}6376 : 1$
C NaH_2PO_4 : Na_2HPO_4 : $NaCl = 1 : 1{\cdot}529 : 1$
D NaH_2PO_4 : $Na_2HPO_4 = 1 : 1$

(*From* ROBINSON, R. A. *and* BIGGS, A. I., *Trans. Faraday Soc.*, 51 (1955) 901)

activity coefficient term can be expected to hold. *Figure 12.2* shows that this is true for p-nitrophenol[17] in four buffer solutions up to a total ionic strength of 0·1 and gives us with confidence $pK = 7{\cdot}14_9$ at 25°.

POLYBASIC ACIDS

Ionization constants can also be measured by potentiometric titration: the method need not be described in detail, for it has been dealt with elsewhere[18], but we will give a general formula for the hydrogen ion activity of a solution of a polybasic acid when titrated with alkali. Suppose an $n-$ basic acid ionizes in n stages: $H_nA \rightarrow H_{n-1}A \rightarrow H_{n-2}A \rightarrow \ldots\ldots \rightarrow HA \rightarrow A$, losing a hydrogen ion at each stage. Let the $(n + 1)$ species carry $p, q, r \ldots\ldots z$ negative charges. Thus if H_nA were citric acid, $n = 3$, $p = 0$, $q = 1$, $r = 2$, $s = 3$. H_nA need not be a neutral molecule; if it were the $NH_3^+{\cdot}NH_3^+$ cation, then $n = 2, p = -2, q = -1, r = 0$ or if it were the $NH_3^+{\cdot}CH_2{\cdot}COOH$ ion, then $n = 2$, $p = -1$, $q = 0$, $r = 1$. In any case, there will be n equations of the form:

$$[H_{n-1}A] = \frac{K_1}{a_{H^+}} \frac{\gamma_p}{\gamma_q} [H_nA]$$

$$[H_{n-2}A] = \frac{K_2}{a_{H^+}} \frac{\gamma_q}{\gamma_r} [H_{n-1}A] = \frac{K_1 K_2}{a_{H^+}^2} \frac{\gamma_p}{\gamma_r} [H_nA]$$

— — — — — — —

$$[A] = \frac{K_n}{a_{H^+}} \frac{\gamma_{z-1}}{\gamma_z} [HA] = \frac{K_1 K_2 \ldots K_n}{a_{H^+}^n} \frac{\gamma_p}{\gamma_z} [H_nA]$$

where square brackets denote concentrations and activity coefficients are introduced with proper reference to the charge on each species as indicated by the subscripts. The total concentration of acid is given by:

$$c = [H_nA] + [H_{n-1}A] + \dots + [A]$$

and the condition of electrical neutrality is:

$$[H^+] + xc = p[H_nA] + q[H_{n-1}A] + \dots + z[A] + [OH^-] - pc$$

where xc is the concentration of alkali-metal cation resulting from the addition of alkali during titration. If H_nA is a neutral molecule, $p = 0$; if it is a positively charged acid, such as the $NH_3^+ \cdot NH_3^+$ ion, p has a negative value and the last pc term refers to an anion such as chloride which must accompany the positive ion. By elimination of the concentration terms from these equations, there results:

$$Z(K_1K_2 \dots K_n) + \dots + Ra_{H^+}^{n-2} K_1K_2 + Qa_{H^+}^{n-1} K_1 = Pa_{H^+}^n$$

where

$$P = \{xc + [H^+] - [OH^-]\}/\gamma_p$$
$$Q = \{(1 - x)c - [H^+] + [OH^-]\}/\gamma_q$$
$$R = \{(2 - x)c - [H^+] + [OH^-]\}/\gamma_r$$
$$— \quad — \quad — \quad — \quad — \quad — \quad —$$
$$Z = \{(n - x)c - [H^+] + [OH^-]\}/\gamma_z$$

If $[H^+]$ and $[OH^-]$ are negligible compared to the terms in c (and this will usually hold for $4 < \text{pH} < 9$) this equation reduces, for a monobasic acid such as acetic acid, to:

$$(1 - x)\ K_1/\gamma_1 = xa_{H^+}/\gamma_o$$

For a dibasic acid, it becomes:

$$(2 - x)\ K_1K_2/\gamma_2 + (1 - x)\ K_1a_{H^+}/\gamma_1 = xa_{H^+}^2/\gamma_o$$

or:

$$K_1K_2 + \frac{1-x}{2-x}\frac{\gamma_2}{\gamma_1} a_{H^+} K_1 = \frac{x}{2-x}\frac{\gamma_2}{\gamma_o} a_{H^+}^2$$

again assuming that $[H^+]$ and $[OH^-]$ are negligible. Thus $(1 - x)\gamma_2\, a_{H^+}/\{(2 - x)\gamma_1\}$ can be plotted against $x\gamma_2 a_{H^+}^2/\{(2 - x)\gamma_o\}$ and K_1 and K_1K_2 evaluated from the slope and the intercept of the plot, a method devised by Speakman.[19] Alternatively, we can write:

$$K_1 = xa_{H^+}{}^2/\{\gamma_o[(2 - x)\ K_2/\gamma_2 + (1 - x)a_{H^+}/\gamma_1]\}$$
$$K_2 = [xa_H{}^2/\gamma_o - (1 - x)K_1\, a_{H^+}/\gamma_1]/[(2 - x)\ K_1/\gamma_2]$$

whence K_1 and K_2 can be evaluated by successive approximations. The corresponding equation for $NH_3^+ \cdot NH_3^+$ would be:

$$(2 - x)\, K_1K_2/\gamma_o + (1 - x)\, K_1a_{H^+}/\gamma_1 = xa^2_{H^+}/\gamma_2$$

and for the $NH_3^+ \cdot CH_2 \cdot COOH$ ion:

$$(2 - x)K_1K_2/\gamma_1 + (1 - x)K_1a_{H^+}/\gamma_o = xa^2_{H^+}/\gamma_1$$

The ionization of a dibasic acid occurs by two paths:

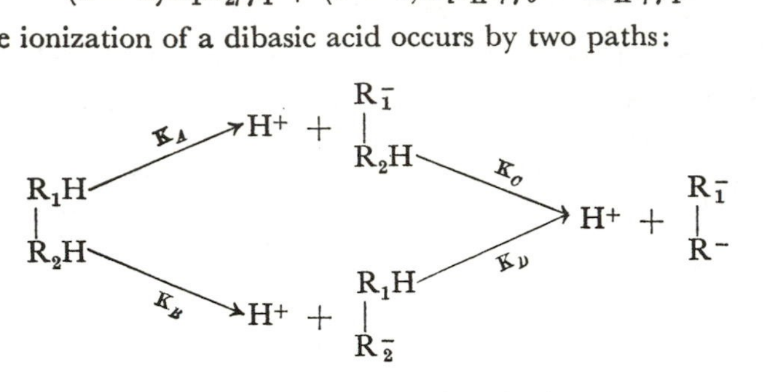

and it must be emphasized that the four ionization constants are not measured by direct experimental methods. Experiment does give the two ionization constants:

$$K_1 = \frac{[H^+]\{[R_1^- \cdot R_2H] + [R_1H \cdot R_2^-]\}}{[R_1H \cdot R_2H]}$$

and
$$K_2 = \frac{[H^+][R_1^- \cdot R_2^-]}{[R_1^- \cdot R_2H] + [R_1H \cdot R_2^-]}$$

where the square brackets denote concentrations and the activity coefficients have been omitted for the sake of brevity. Clearly, $K_1 = K_A + K_B$ and $1/K_2 = 1/K_C + 1/K_D$ and, since the free energy difference between $(2H^+ + R_1^- \cdot R_2^-)$ and $R_1H \cdot R_2H$ must be independent of the nature of the intermediate ion, $K_A \cdot K_C = K_B \cdot K_D$.

If the acid is symmetrical, $R_1 = R_2$, as in oxalic acid and its homologues, then $K_A = K_B$ and $K_C = K_D$, so that $K_1 = 2K_A$ and $K_2 = 1/2 \cdot K_C$. If the negative charge on the $R_1^- \cdot R_2H$ ion is very far distant from the remaining hydrogen atom, so that it is without effect on the second ionization, we would expect K_A to equal K_C and K_B to equal K_D. In this extreme case, we would then have $K_1/K_2 = 4$*. The effect of the negative charge would, however, be to make it more difficult for the second hydrogen to ionize, so that

* The general formula for an n-basic acid is:

$$K_1 = \frac{2n}{n-1} K_2 = \frac{3n}{n-2} K_3 = \frac{4n}{n-3} K_4 = \ldots$$

$K_A > K_C$ and $K_B > K_D$, whence $K_1/K_2 > 4$. This is what is found: for azelaic acid, $COOH(CH_2)_7COOH$, the first constant is six times as large as the second whereas for oxalic acid it is one thousand times larger.

The acid, $R_1H{\cdot}R_2H$, need not be symmetrical nor, indeed, need it be an uncharged molecule, for in the case of glycine hydrochloride we regard the cation, $NH_3^+{\cdot}CH_2{\cdot}COOH$, as a dibasic acid and formulate the ionization as:

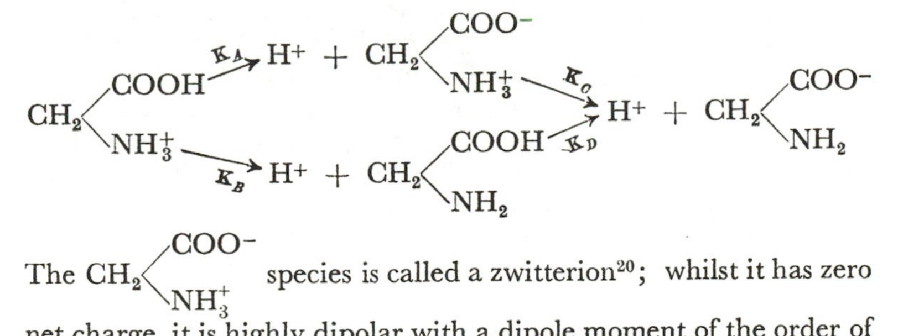

The $CH_2\langle^{COO^-}_{NH_3^+}$ species is called a zwitterion[20]; whilst it has zero net charge, it is highly dipolar with a dipole moment of the order of 13 Debye units and cannot be regarded as a particle with no long range forces, a treatment which may be valid for the neutral molecule, $CH_2\langle^{COOH}_{NH_2}$. As before, $K_AK_C = K_BK_D$, but it is no longer valid to equate K_A to K_B: in practice these are usually of different orders of magnitude. The relative amounts of zwitterion and neutral molecule coexisting in the solution are given by:

$$K_Z = \frac{[NH_3^+CH_2COO^-]}{[NH_2CH_2COOH]} = \frac{K_A}{K_B} = \frac{K_D}{K_C}$$

and it will be observed that the ratio is independent of the hydrogen ion concentration. The inequality of K_A and K_B and of K_C and K_D makes the problem more difficult to treat in that one further assumption has to be introduced. This usually takes the form that the effect of the carboxyl group on the ionization of the NH_3^+ group is not altered by esterifying the carboxyl group: for example, that K_E for the ethyl ester of glycine hydrochloride, a quantity which can be measured directly, is the same as K_B for glycine hydrochloride itself, a quantity which cannot be measured directly. Another method[21] is to extrapolate from the ionization constants of the ethyl, propyl and butyl esters: it should be noted that, with *p*-aminobenzoic acid, the methyl ester does not fit into the sequence of the other esters. With some assumption of this nature, ionization constants of an aminoacid can be determined and thence the fraction of the zwitterion present.[22] For the aminoacids which go to form the proteins, the first ionization produces almost entirely zwitterions to the practical exclusion of the neutral molecules: to be exact, in the case of glycine the ratio of zwitterion to neutral molecule is $2{\cdot}6 \times 10^5$. It is different for the aminobenzoic acids, for the *o*-, *m*- and *p*-isomers the fraction of zwitterion is 0·17, 0·70 and 0·12 respectively.

Returning now to a consideration of the symmetrical dibasic acid, BJERRUM[23] recognised that the failure of the ratio K_1/K_2 to equal the theoretical value of 4, can be explained by introducing into the free energy term an allowance for the electrical work to be done in dissociating the hydrogen ion under the influence of the charged carboxyl group distant R away: this work will be $\boldsymbol{e}^2/(\varepsilon R)$, the probability of finding a hydrogen ion at the second carboxyl group will be increased by the factor $\exp\{\boldsymbol{e}^2/(\varepsilon \boldsymbol{k} TR)\}$ and the second ionization constant will be decreased by the same factor: hence we expect:

$$\frac{K_1}{K_2} = 4 \exp \left\{\frac{\boldsymbol{e}^2}{\varepsilon \boldsymbol{k} TR}\right\}$$

The effect of a dipolar substituent on the ionization constant of a monobasic acid has been dealt with similarly by EUCKEN[24]; if μ is the dipole moment and ζ the angle of inclination, Eucken derives the equation:

$$\frac{K_1}{K_2} = \exp \left\{\frac{\boldsymbol{e}\mu \cos \zeta}{\boldsymbol{k} TR^2}\right\}$$

GANE and INGOLD[25] have measured the ionization constants of the series of dibasic acids from malonic to azelaic acid. For glutaric acid and the higher homologues, Bjerrum's equation gives reasonable values of R but for malonic and succinic acid the R values are much too low. Similarly Eucken's equation, applied to acetic and chloroacetic acids, gives too small a distance between the dipole and the carboxyl group. The theory would therefore appear to be sound in its application to long, thin molecules but not to shorter, more spherical molecules. That is to say, the theory applies when the electrical forces operate mainly through the solvent and we can use the macroscopic dielectric constant of the solvent in Bjerrum's equation. This is not justifiable for a more or less spherical molecule and KIRKWOOD and WESTHEIMER[26] have elaborated Bjerrum's work. They consider a model in which the acid occupies a spherical or ellipsoidal cavity in the solvent, the cavity having a dielectric constant, $\varepsilon = 2$, the value of liquid paraffins. The equations they deduce represent observed ionization constants with reasonable assumptions about the size and configuration of the molecules.

THE EFFECT OF THE SOLVENT ON THE IONIZATION CONSTANT

The addition of another liquid to water usually reduces the dielectric constant; for example, a water-dioxan mixture containing 82 per cent dioxan has a dielectric constant of only 9·5. If this mixture is used instead of water as a solvent for a weak acid, the electrostatic forces between the cations and anions are increased and more opportunities are provided for the formation of covalent bonds. A decrease in the dielectric constant of the solvent should, therefore, be accompanied by a decrease in the ionization constant of a weak acid dissolved in it. This prediction has been amply confirmed by experiment. To quote only one example of the very large changes which are observed, the ionization constant of acetic acid in water at 25° is $1{\cdot}754 \times 10^{-5}$; in 82 per cent dioxan it is $3{\cdot}1 \times 10^{-11}$. It is natural therefore to seek some relation between the dielectric constant and the ionization constant, but before considering this it would be well to study first the energy changes which accompany the transfer of a strong acid from one solvent medium to another.

Much attention has been given recently to the properties of hydrochloric acid in different solvent media (see Appendix 8.2), by studying cells such as:

$$H_2|\text{HCl in 20 per cent methanol}|\text{AgCl, Ag.}$$

Not only do such cells give information about the energy changes occurring during the transfer of hydrochloric acid from one solvent to another, but they are a pre-requisite if the Harned-Ehlers cell is to be used for studies of weak acids in mixed solvents. Moreover, we shall see later that the problems associated with this cell are closely related to those of the unbuffered cell:

$$H_2|\text{acetic acid in sodium chloride solution}|\text{AgCl, Ag.}$$

Previously the potential of the cell:

$$H_2|\text{HCl in water as solvent}|\text{AgCl, Ag}$$

has been written:

$$E = E^0_m - 2k \log \gamma m$$

except that now we have introduced a subscript m to emphasize that the concentration and the activity coefficient are measured on the molality scale. But there is no reason why we should not use the mole fraction scale and indeed it may have theoretical advantages, so that we could have:

$$E = E^0_N - 2k \log fN_B$$

The third possibility is:

$$E = E^0_c - 2k \log yc$$

We have therefore three standard potentials:

$$E^0_N = \underset{N_B \to 0}{\text{Lt}} [E + 2k \log N_B]$$

$$E^0_m = \underset{m \to 0}{\text{Lt}} [E + 2k \log m]$$

$$E^0_c = \underset{c \to 0}{\text{Lt}} [E + 2k \log c]$$

From the definitions of N_B, m and c it follows that:

$$E^0_m = E^0_N + 2k \log 1000/W_A$$

$$E^0_c = E^0_m + 2k \log d_0$$

Similar measurements could be made for hydrochloric acid in another *pure* solvent such as methanol; the three E^0 values would of course not be the same as in water; moreover, all the activity coefficients would be measured relative to unity at infinite dilution in pure methanol.

Now suppose we had the hydrochloric acid cell with a 20 per cent methanol–80 per cent water mixture as solvent. We have a choice of two methods of defining any one of the three standard cell potentials. We could ignore the composite nature of the solvent and treat it just as a medium in which to dissolve the acid. By measuring the potentials at a series of acid concentrations we could obtain the standard potential as:

$${}^sE = {}^sE^0_N - 2k \log {}^s_sf\, N_B$$

and

$${}^sE^0_N = \underset{N_B \to 0}{\text{Lt}} [{}^sE + 2k \log N_B] \qquad \ldots.(12.4)$$

This ungainly notation is used to indicate by the superscripts that the measurements are made in a mixed solvent medium and the subscripts mean that the activity coefficient is measured relative to unity at infinite dilution in this particular solvent medium. The corresponding equation for pure water as solvent would be:

$${}^wE = {}^wE^0_N - 2k \log {}^w_wf\, N_B$$

We might, however, prefer to consider the 20 per cent methanol cell in another way. We might say that it is nothing more than the cell with water as solvent to which has been added a certain proportion of methanol. As the 'water cell' has been studied so thoroughly, why not retain ${}^{w}E_N^0$? We are entitled to do so, writing:

$$ {}^{s}E = {}^{w}E_N^0 - 2k \log {}_{\mathrm{W}}^{\mathrm{S}}f N_B $$

but it is most important to recognize that the activity coefficient of the acid in the mixed solvent is now measured relative to unity at infinite dilution in water and not relative to unity at infinite dilution in the mixed solvent. The standard potential is now given by;

$$ {}^{w}E_N^0 = \mathop{\mathrm{Lt}}_{N_B \to 0} [{}^{s}E + 2k \log N_B + 2k \log {}_{\mathrm{W}}^{\mathrm{S}}f] \quad \ldots.(12.5) $$

where the last term does not disappear at infinite dilution in the mixed solvent. Instead, we see from equations (12.4) and (12.5) that:

$$ {}^{w}E_N^0 - {}^{s}E_N^0 = \mathop{\mathrm{Lt}}_{N_B \to 0} 2k \log {}_{\mathrm{W}}^{\mathrm{S}}f $$

Since the standard potentials both in water and in the mixed solvent are calculable from experimental measurements, then $\mathop{\mathrm{Lt}}_{N_B \to 0} \log {}_{\mathrm{W}}^{\mathrm{S}}f$ is also calculable. Is it a quantity of any significance? It represents the activity coefficient of hydrochloric acid at infinite dilution in 20 per cent methanol solution relative to unity at infinite dilution in pure water. At infinite dilution in either medium the interionic effects are absent. We are therefore measuring the effect of transferring a pair of ions from one solvent to another under conditions where the only effects are ion-solvent interactions. Owen[27] calls this 'primary medium effect'. We naturally suspect that there ought to be a relation between the primary medium effect and the dielectric constants of the solvents. This is an important matter to which we shall return, but first let us consider if there are any other kinds of medium effect. Consider the cell:

Ag, AgCl|HCl in water|H_2|HCl in 20 per cent methanol|AgCl, Ag

with the acid at the same mole fraction in each solvent. The cell reaction consists of the transfer of hydrochloric acid from the aqueous to the methanolic solution and the potential of the cell is:

$$ {}^{s}E - {}^{w}E = {}^{s}E_N^0 - {}^{w}E_N^0 - 2k (\log {}_{\mathrm{S}}^{\mathrm{S}}f - \log {}_{\mathrm{W}}^{\mathrm{W}}f) \quad \ldots.(12.6) $$

By making the mole fraction of hydrochloric acid the same in each half-cell, we have eliminated any energy change due to concentration changes, that is to say the energy change is zero except for changes departures on departures from the laws of ideal solutions.

The mole fraction scale is particularly suited for considering such changes. But even if we can avoid energy changes resulting directly from a difference in concentration in the two half cells, the last term of equation (12.6) still represents a complicated operation—the transfer of hydrochloric from a finite concentration in water to infinite dilution in water, its transfer from infinite dilution in water to infinite dilution in 20 per cent methanol and finally a transfer from infinite dilution in 20 per cent methanol to a finite concentration in this solvent. But we can simplify this by noting that the potential of the cell could equally well be written:

$$^{s}E - {}^{w}E = -2k\,(\log {}^{\mathrm{S}}_{\mathrm{W}}f - \log {}^{\mathrm{W}}_{\mathrm{W}}f)$$

and we have already found that:

$$^{w}E^{0}_{N} - {}^{s}E^{0}_{N} = \mathop{\mathrm{Lt}}_{N_B \to 0} 2k \log {}^{\mathrm{S}}_{\mathrm{W}}f$$

so that:

$$\log \frac{{}^{\mathrm{S}}_{\mathrm{W}}f}{{}^{\mathrm{W}}_{\mathrm{W}}f} = \mathop{\mathrm{Lt}}_{N_B \to 0} \log {}^{\mathrm{S}}_{\mathrm{W}}f + \log \frac{{}^{\mathrm{S}}_{\mathrm{S}}f}{{}^{\mathrm{W}}_{\mathrm{W}}f} \qquad \ldots .(12.7)$$

The term on the left Owen calls the 'total medium effect', total in the sense that it measures the total change in chemical potential attending the movement of hydrochloric acid at finite but equal concentrations in two solvents. Equation (12.7) shows that it is composed of two effects, the primary medium effect given by the first term on the right of the equation, determined by the difference of the ion-solvent interactions at infinite dilution in each solvent and, in addition, a further effect given by the last term of equation (12.7). This Owen terms the 'secondary medium effect'. Its significance is this: ${}^{\mathrm{S}}_{\mathrm{S}}f$ measures the difference in the 'non-ideal' part of the chemical potential of hydrochloric acid at a finite concentration and at infinite dilution in 20 per cent methanol. It will be given by some form of the Debye-Hückel equation and one factor which will be important will be the dielectric constant of the medium. But ${}^{\mathrm{W}}_{\mathrm{W}}f$ measures a difference in the 'non-ideal' part of the chemical potential for the same concentration change in pure water and again the dielectric constant is important. In fact, the secondary medium effect should be given to a first approximation by

$$\log \frac{{}^{\mathrm{S}}_{\mathrm{S}}f}{{}^{\mathrm{W}}_{\mathrm{W}}f} = \frac{1{\cdot}825 \times 10^{6}}{T^{3/2}} \left(\frac{\sqrt{c_w}}{\varepsilon_w^{3/2}} - \frac{\sqrt{c_s}}{\varepsilon_s^{3/2}} \right)$$

when the subscripts, s and w, designate the solvent. Thus if the Debye-Hückel equation accounts for the activity coefficients in the separate solvents, it will give an equally good measure of the secondary medium effect.

These considerations are summarized as follows: the total medium effect on the transfer of an electrolyte from a finite concentration in one solvent to a similar concentration in another solvent is a composite one. The secondary medium effect results mainly from a difference in ion–ion interactions in the two solvents and is determined to a large degree by the dielectric constant of each medium. The primary medium effect is independent of concentration and results from a difference of the ion-solvent interactions; it also should be largely dependent on the dielectric constants.

The simplest explanation of the primary medium effect is given by the Born equation for the energy of transfer of an ion of radius r from one solvent medium to another:

$$\frac{e^2}{2}\left(\frac{1}{\varepsilon_w} - \frac{1}{\varepsilon_s}\right)\frac{1}{r}$$

or for a mole of a 1 : 1 electrolyte:

$${}^{w}E^0_N - {}^{s}E^0_N = \frac{Ne^2}{2F}\left(\frac{1}{\varepsilon_s} - \frac{1}{\varepsilon_w}\right)\left(\frac{1}{r_1} + \frac{1}{r_2}\right)$$

Thus, if we could assume that the radius term does not change with the nature of the solvent, the standard cell potential should be a linear function of the reciprocal of the dielectric constant. Measurements have been made in recent years of ${}^{s}E^0_N$ for a number of solvent mixtures. The interpretation of the data is not easy. At one time it was thought[(28)] that, by postulating that the hydrogen ion of hydrochloric acid was associated with one water molecule and assuming that the water activity could be equated to the mole fraction of water in the solvent mixture, the function $({}^{s}E^0_N - k \log \mathcal{N}_W)$ would be the correct one to use. With the data available in 1941 this seemed true: the data for a number of mixed solvents, plotted in this way, fell on a single curve. More recent measurements on more solvent mixtures suggests that the problem is not so simple[(29)]; indeed, FEAKINS and FRENCH[30] discard the Born term and find a relation between ${}^{S}E^0_c$, the standard potential on the *molar* scale, and ϕ_w, the *volume* fraction of water in the mixed solvent:

$${}^{S}E^0_c = {}^{W}E^0_c - 2{\cdot}5\, k \log \phi_w$$

the coefficient, 2·5, denoting that 2·5 molecules of water accompany the transfer of a hydrogen ion from one solvent to another. This relation holds for eleven solvent mixtures down to $\phi_w = 0{\cdot}7$: exceptions are glucose, glycol and dioxan-water mixtures.

We can now consider the effect of the solvent on the ionization constant of a weak acid. The free energy change on the dissociation of such an acid is $-RT \ln K$; this is the energy change when a mole of undissociated acid in its standard state is replaced by an equivalent amount of its ions each in the hypothetical standard state. Then $RT \ln \frac{{}^{s}K}{{}^{w}K}$ (${}^{w}K$ and ${}^{s}K$ being the ionization constants in water and in a mixed solvent respectively) measures the change in free energy when a mole of undissociated acid is transferred from

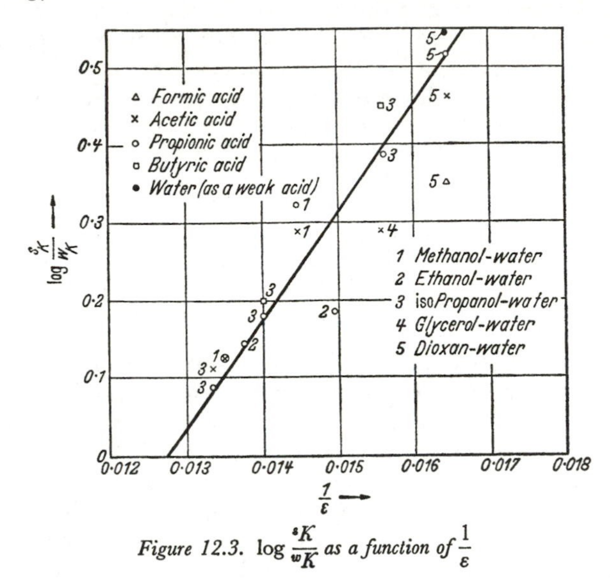

Figure 12.3. $\log \frac{{}^{s}K}{{}^{w}K}$ *as a function of* $\frac{1}{\varepsilon}$

the mixed solvent to pure water and the ions are transferred in the opposite direction. Furthermore, if we work on the mole fraction scale for the ionization constant, these transfers occur between states of the same mole fraction and there is no energy term corresponding to 'ideal gas expansion'. Moreover, the transfers occur between states of unit activity coefficient; there is, therefore, no term to be introduced for interionic effects. The term $RT \log \frac{{}^{s}K}{{}^{w}K}$ should measure the effect of the solvent on the ions and the undissociated molecules. Finally, we lose no generality if, in a comparison of acids in different solvents, we put $K = 1$ for each acid in water.

Figure 12.3 is a graph on which are plotted against $\frac{1}{\varepsilon}$ the ionization constants of a number of weak acids each relative to unity for the ionization constant of the acid in water. The points do cluster round a line although there are considerable departures from it: the straight line in the figure would, by Born's equation, assuming $r_1 = 3{\cdot}73$ Å for the hydrogen ion, lead to 1·2 Å for the carboxylic anions, a small but not impossible value. Just as with the behaviour of hydrochloric acid in different solvent media, Born's equation gives a first approximation to the properties of weak acids in different solvents, but it is evident that some highly specific effects must be allowed for if we are to give a complete account of weak acids.

THE EFFECT OF TEMPERATURE ON THE IONIZATION CONSTANT

Potential measurements of the Harned-Ehlers cell (see p. 340) have been made over a temperature range (usually 0–60°) for a number of weak acids, and the ionization constant can be calculated at each temperature. Since:

$$\boldsymbol{R}\frac{\partial \ln K}{\partial T} = -\frac{\partial\,(\Delta\bar{G}^0/T)}{\partial T} = \frac{\Delta\bar{H}^0}{T^2}$$

the information is not limited merely to a series of ionization constants at different temperatures but can be expanded to embrace the heat content change on ionization (at infinite dilution) and (if the analysis is sufficiently detailed), the temperature coefficient of the heat content, *i.e.*, the difference in heat capacity between the ions and the undissociated molecule. Many equations have been proposed to represent the temperature variation of the ionization constant, but it has not always been appreciated that the very method by which these constants are reported in the literature imposes limitations on the equations we can use. The experimental results are a set of potentials at regularly spaced temperature intervals. It has been asserted that these can be represented within the experimental error by a quadratic in the temperature; indeed, in some cases the reported potentials may have been smoothed by means of this quadratic. The potentials being proportional to a free energy change, then this also must be quadratic in the temperature within the experimental error. Hence we can write[31]:

$$\Delta\bar{G}^0 = -\boldsymbol{R}T\ln K = (A - CT + DT^2)$$

and by ordinary thermodynamic methods it follows that:

$$\Delta\bar{S}^0 = (C - 2DT)$$
$$\Delta\bar{H}^0 = (A - DT^2)$$
$$\Delta\bar{C}_p^0 = (-2DT)$$
$$2{\cdot}303\boldsymbol{R}\log K = -\frac{A}{T} + C - DT \qquad \ldots.(12{\cdot}8)$$

A number of equations have been proposed to express the ionization constant as a function of temperature and many of them do represent the observed data faithfully: there may be theoretical grounds for preferring them, but equation (12.8) is more closely related to the experimental results and is therefore adequate as a compact method of recording a set of results. Only one example is known, cyanoacetic acid[32], where additional terms are needed to represent the data. Equation (12.8) predicts that there will be a maximum value of the ionization constant at a temperature $T_{max} = \sqrt{A/D}$ at which it will be given by:

$$2{\cdot}303\boldsymbol{R} \log K = C - 2\sqrt{AD}$$

At this temperature $\Delta H^0 = 0$. For many weak acids this maximum is found about room temperature: for example, it is 22·5°C for acetic acid. In Appendix 12.1 we give the values of the parameters necessary for calculating the ionization constants of some weak acids and bases.

This equation is usually valid over the temperature range 0° to about 60° but it has been tested over the more extended range 0° to 90° for acetic acid in 50 per cent glycerol-water solution[(33)]. *Table 12.2* shows how well equation (12.8) represents the experimental results.

Table 12.2

Ionization Constant of Acetic Acid in 50 per cent Glycerol-water Solution

$$\log K_a = -\frac{1321{\cdot}43}{T} + 3{\cdot}4148 - 0{\cdot}014268T$$

Temp.	$K_a \times 10^6$	
	obs.	*calc.*
0°C	4·778	4·784
10	5·097	5·105
20	5·316	5·303
30	5·378	5·375
40	5·330	5·333
50	5·184	5·187
60	4·951	4·953
70	4·654	4·653
80	4·315	4·307
90	3·935	3·931

THE UNBUFFERED CELL CONTAINING A WEAK ACID

The cell:

$$H_2|HA(m), NaCl(m)|AgCl, Ag \qquad I$$

where HA is a weak acid, seems on first inspection to involve only the simplest considerations, for we write the potential:

$$E = E^0 - k \log \gamma_{H^+}\gamma_{Cl^-}m_{H^+}m$$

We have made the salt molality equal to the total acid molality to economize on symbols: the cell works equally well if the concentrations are different. $\gamma_{H^+}\gamma_{Cl^-}$ has some special properties: it is the activity coefficient product of hydrochloric acid in a solution in which the hydrogen ion concentration is very small since the only hydrogen ions are those formed by the dissociation of the weak acid. In Chapter 15 we shall see that the activity coefficient of hydrochloric acid in a solution of sodium chloride in which the molalities of both components are allowed to vary but the total molality is kept constant, is subject to a very simple empirical law:

$$\tfrac{1}{2} \log \gamma_{H^+}\gamma_{Cl^-} = -0{\cdot}1393 + 0{\cdot}037m_{HCl}$$

The numerical quantities are specific for 25° and the particular total molality considered; 0·5 M in this case. It is worth while calculating the activity coefficient of hydrochloric acid in some acid–sodium chloride mixtures at a total molality of 0·5 M:

m_{HCl}	m_{NaCl}	γ_{HCl}
0·5	0	0·757
0·25	0·25	0·741
0·10	0·4	0·732
0·05	0·45	0·729
0·01	0·49	0·726
0·001	0·499	0·726
0	0·5	0·726

If HA is acetic acid and $m = 0{\cdot}5$, the hydrogen ion concentration is about 0·003 M so that clearly we can equate γ_{HCl} to 0·726. The point to be emphasized is that γ_{HCl} is practically independent of acid concentration as long as this is small: moreover γ_{HCl} can be obtained from experiments on hydrochloric acid–sodium chloride mixtures. A measurement of the potential of Cell I therefore gives information on three matters[34]. First of all it gives m_{H^+}, the hydrogen ion concentration in a solution of a weak acid and the salt of a strong acid, as:

$$-k \log m_{H^+} = E - E^0 + k \log \gamma_{H^+}\gamma_{Cl^-} + k \log m$$

Secondly, by introducing the law of mass action in the form:

$$\gamma_{\mathrm{A}}^{2} \frac{m_{\mathrm{H}^+}^{2}}{(m - m_{\mathrm{H}^+})} = K_a$$

γ_{A}^{2} being an abbreviation for $\gamma_{\mathrm{H}^+}\gamma_{\mathrm{A}^-}/\gamma_{\mathrm{HA}}$, and introducing a Debye-Hückel approximation for γ_{A}, we can extrapolate to $I = 0$ to get K_a. *Figure 12.4* shows this extrapolation for 0·1 M acetic acid in sodium chloride solution, the limiting value of $\log K_a$ being $-4{\cdot}75$ and $K_a = 1{\cdot}72 \times 10^{-5}$. Thirdly, using this value of the ionization constant, we can calculate γ_{A}. It has been shown that this activity coefficient term behaves in some ways like the activity

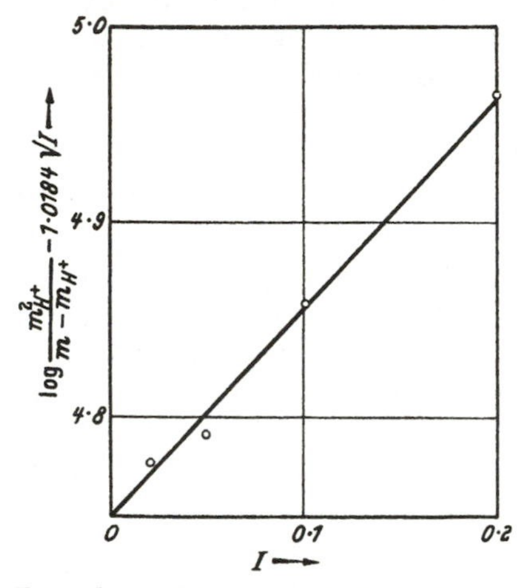

Figure 12.4. Extrapolation of data from unbuffered cell to give the ionization constant of acetic acid

coefficient of hydrochloric acid in a salt solution; with increasing salt concentrations it first diminishes in value, passes through a minimum at about 0·5 M and then increases to values which may exceed unity if the salt concentration is very large. The term differs from the activity coefficient of hydrochloric acid, however, in one important respect; at any given value of the total ionic molality, the activity coefficient of hydrochloric acid in different salt solutions is in the order:

$$\gamma_{\mathrm{HCl(LiCl)}} > \gamma_{\mathrm{HCl(NaCl)}} > \gamma_{\mathrm{HCl(KCl)}}$$

whereas the reverse is true of the γ_{A} term. Cell I is therefore an important source of information about the behaviour of weak acids in salt solutions but unfortunately the exact treatment of the problem is not as simple as we have suggested. So far we have assumed that the term $\gamma_{H^+}\gamma_{Cl^-}$ introduced early in the theory, is not influenced by the presence of the undissociated acetic acid molecules, *i.e.*, we have ignored the medium effect on both $\gamma_{H^+}\gamma_{Cl^-}$ and on γ_A. These complications have been considered by OWEN[27, 34] and are worth while describing in some detail because they illustrate the importance of the medium effect. Moreover, whilst this is not the best method of finding an ionization constant, it does give information not obtainable from the Harned-Ehlers cell. In the following discussion we follow the treatment given by OWEN[27]. The correct equation for Cell I is:

$$E = {}^wE^0 - k \log {}^s_w\gamma_{H^+} {}^s_w\gamma_{Cl^-} m_{H^+} m$$

where the activity coefficient term differs from that used before (and which should have been written ${}^w_w\gamma_{H^+} {}^w_w\gamma_{Cl^-}$). We now put this equation in the form:

$$-k \log m_{H^+} - 2k \log \frac{{}^s_w\gamma_{HCl}}{{}^w_w\gamma_{HCl}} = E - {}^wE^0 + 2k \log {}^w_w\gamma_{HCl} + k \log m$$

The second term is the total medium effect on hydrochloric acid. The right-hand side of the equation is identical with our first estimate of $-k \log m_{H^+}$ which we now see to be erroneous by a factor dependent on the medium effect. The right-hand side, however, contains quantities all of which are known or are measurable; it will be convenient to call it $-k \log m'_{H^+}$.

Next we write the equilibrium equation as

$$2 \log m_{H^+} - \log (m - m_{H^+}) + 2 \log {}^s_w\gamma_A = \log {}^wK$$

or

$$2 \log m'_{H^+} - \log (m - m_{H^+}) + 2 \log {}^w_w\gamma_A = \log {}^wK - 2 \log \frac{{}^w_w\gamma^2_{HCl}\, {}^s_w\gamma_A}{{}^s_w\gamma^2_{HCl}\, {}^w_w\gamma_A}$$

We ignore the medium effect on m_{H^+} in so far as it concerns the $(m - m_{H^+})$ term. Using a Debye-Hückel approximation for ${}^w_w\gamma_A$ the left-hand side can be plotted against the total molality and extrapolated to $I = 0$: we have already done that in *Figure 12.4* for 0·1 M acetic acid and the extrapolation can be repeated for each molality of acetic acid at which measurements have been made. (Four such extrapolations are shown in *Figure 12.5.*) The limiting value at $I = 0$ is:

$$\log {}^wK - 2 \underset{I \to 0}{\text{Lt}} \log \frac{{}^w_w\gamma^2_{HCl}\, {}^s_w\gamma_A}{{}^s_w\gamma^2_{HCl}\, {}^w_w\gamma_A} \quad \ldots .(12.9)$$

i.e., we have got the correct ionization constant except for a term which contains the primary medium effect. An extrapolation like that shown in *Figure 12.4* has eliminated the secondary medium

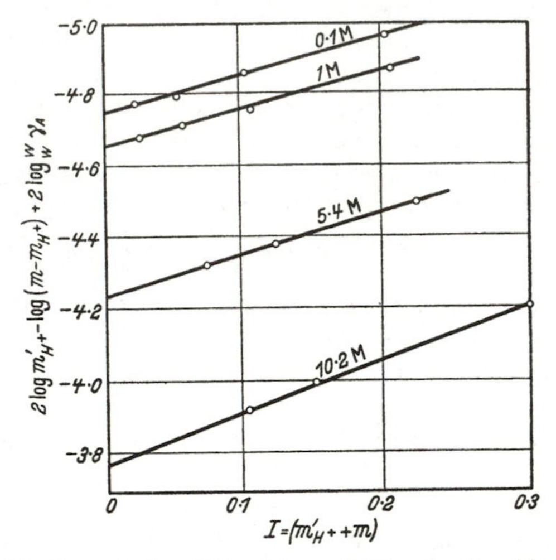

Figure 12.5. Extrapolation of data at four molalities of acetic acid to eliminate the secondary medium effect

effect. If we have a set of such extrapolated values at a series of acetic acid molalities, we now make a second extrapolation by plotting the quantity in (12.9) against the acetic acid molality (*Figure 12.6*). The result of this second extrapolation at $m = 0$ is $\log {}^wK$.

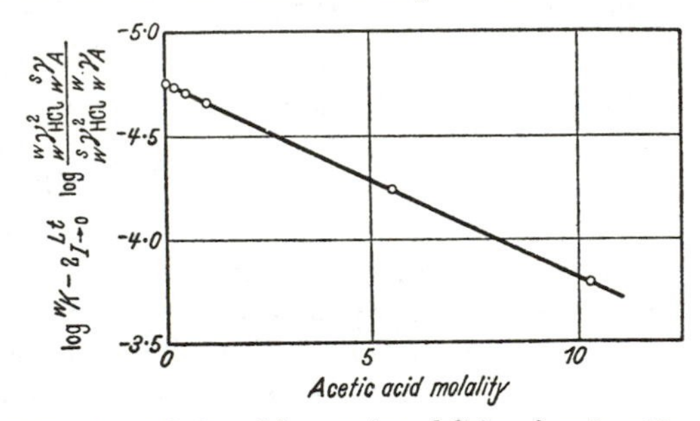

Figure 12.6. Extrapolation of data at six molalities of acetic acid to eliminate the primary medium effect

In order to perform the second extrapolation in *Figure 12.6*, values at $I = 0$ were read from *Figure 12.5*. Values could, however, be read for a given non-zero value of I, plotted like *Figure 12.6* and extrapolated to zero acetic acid molality. What would be the significance of these extrapolated values? They represent values of $\left(\log \frac{m_{H^+}^2}{m - m_{H^+}} + \log {}_w^w\gamma_A^2\right)$ at zero acid concentration but finite salt concentration. But $\log {}_w^w\gamma_A$ was replaced, for the purpose of extrapolation, by a Debye-Hückel approximation which can now be taken out again to give $\left(\frac{m_{H^+}^2}{m - m_{H^+}}\right)$. This, divided into ${}^wK$, gives the true value of $\gamma_A^2 = \gamma_{H^+}\gamma_{A^-}/\gamma_{HA}$ in water as solvent at a given sodium chloride concentration.

THE IONIZATION CONSTANT OF WATER

Water is a very weak acid and the determination of its ionization constant requires special methods. The equation for the potential of the cell:

$$H_2|NaOH(m), NaCl(m')|AgCl, Ag:$$

combined with

$$K_w = \frac{\gamma_{H^+}\gamma_{OH^-}m_{H^+}m_{OH^-}}{a_{H_2O}}$$

gives:

$$E - E^0 + k \log \frac{m'}{m} = -k \log K_w - k \log \frac{\gamma_{H^+}\gamma_{Cl^-}a_{H_2O}}{\gamma_{H^+}\gamma_{OH^-}} \quad \ldots\ldots(12.10)$$

and extrapolation of the left-hand side of this equation against the total ionic strength gives $-k \log K_w$ as the limiting value when $I = 0$. *Figure 12.7* shows two such extrapolations for cells containing lithium hydroxide–lithium chloride and potassium hydroxide–potassium chloride. The ionization constant of water has been deduced from measurements on a number of such cells[35] with good agreement (see Appendix 12.2), and some measurements have been made with mixed solvents, in particular dioxan[36]. In water itself at 25° the ionization constant is $1{\cdot}008 \times 10^{-14}$; in 20 per cent, 45 per cent and 70 per cent dioxan it is $23{\cdot}99 \times 10^{-16}$, $18{\cdot}09 \times 10^{-17}$ and $13{\cdot}95 \times 10^{-19}$ respectively.

The ionization constant increases with temperature and can be represented[31] by an equation of the form of (12.8)

$$-\log K_w = \frac{4471{\cdot}33}{T} - 6{\cdot}0846 + 0{\cdot}017053T$$

The heat content change at 25° is $\Delta\bar{H}^0 = 13522$ cal mole^{-1} and the heat capacity change is $\Delta\bar{C}_p^0 = -46{\cdot}53$ cal degree^{-1} mole^{-1}. This equation predicts a maximum in K_w at 239°C, a temperature well outside the usual range: however, from experiments on the hydrolysis of ammonium acetate[37], evidence has been found that the ionization constant of water has a maximum value about 220°C.

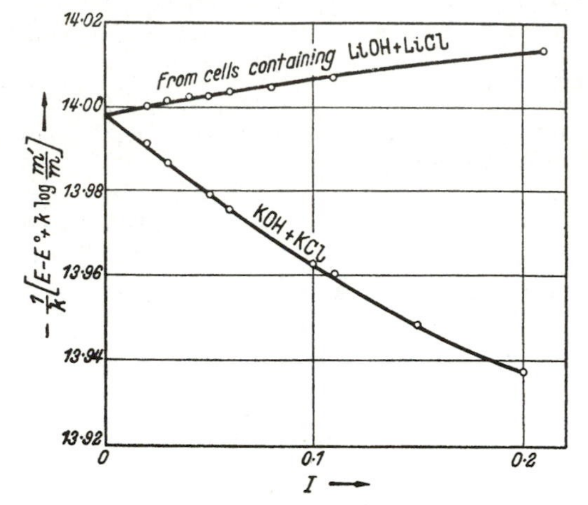

Figure 12.7. Extrapolation of electromotive force data to give the ionization constant of water at 25°

IONIC ACTIVITY COEFFICIENT PRODUCT OF WATER IN SALT SOLUTIONS

Equation (12.10) can be used to get $\gamma_{H^+} \gamma_{OH^-}/a_{H_2O}$. The left-hand side of the equation contains only experimentally measurable quantities and K_w has been calculated. $\gamma_{H^+}\gamma_{Cl^-}$ can be found by a method similar to that described when discussing the unbuffered weak acid cell. This $\gamma_{H^+}\gamma_{Cl^-}$ is the activity coefficient product of hydrochloric acid at the very low hydrogen ion concentration of these alkaline solutions in the presence of a considerable amount of chloride and since this is separately measurable, the ionic activity coefficient product of water can be calculated. It is a quantity which varies with the total ionic molality very much as does $\gamma_{H^+}\gamma_{Cl^-}$, but at any selected total molality it has the highest value in caesium chloride solutions and the smallest in lithium chloride solutions. In this respect it resembles $\gamma_{H^+} \gamma_{A^-}/\gamma_{HA}$ for a weak acid.

THE HYDROGEN ION ACTIVITY OF SOME SOLUTIONS

Cells such as: H_2|Aqueous buffer solution|AgCl, Ag I

give $\log(a_{H^+}\gamma_{Cl^-}) \equiv \log(m_{H^+}\gamma_{H^+}\gamma_{Cl^-})$ without ambiguity and with an accuracy depending only on the accuracy with which the e.m.f. and the concentration of chloride ion can be determined, for:

$$(E - {}^wE^0)/k + \log m_{Cl^-} = -\log (a_{H^+}\gamma_{Cl^-}) \quad \ldots .(12.11)$$

Sometimes $-\log (a_{H^+}\gamma_{Cl^-})$ is written $p(a_{H^+}\gamma_{Cl^-})$ or $p_w(a_{H^+}\gamma_{Cl^-})$ to emphasize that the value of E^0 is for water as solvent, and activity coefficients are given relative to the standard state in water.

pa_{H^+} differs from $p(a_{H^+}\gamma_{Cl^-})$ by $\log \gamma_{Cl^-}$, which cannot be found without some extrathermodynamic assumption. For aqueous solutions at 25° BATES and GUGGENHEIM[38] proposed the convention:

$$-\log \gamma_{Cl^-} = A\sqrt{I}/(1 + 1{\cdot}5\sqrt{I}) \quad \ldots .(12.12)$$

This is equation (9.7) with $a = 4{\cdot}56$ Å, and represents the activity coefficients of sodium chloride with remarkable fidelity up to $I = 0{\cdot}1$. Thus the Bates–Guggenheim convention is equivalent to assuming that γ_{Cl^-} in any solution is equal to γ_{NaCl} in a solution of sodium chloride of the same total ionic concentration ($I \leqslant 0{\cdot}1$).

Both $p(a_{H^+}\gamma_{Cl^-})$ and pa_{H^+} are therefore known if the e.m.f. of a cell of type I has been measured. Many such measurements have been made; moreover, by making such measurements at several chloride ion concentrations, a simple extrapolation can be made to give pa_{H^+} at its limiting value at zero chloride ion concentration. Appendix 12.3, *Table 1* gives pa_{H^+} values for seven buffer solutions made from well-defined, readily available substances.

Cells with liquid-junction are in more frequent use. Consider:

$$H_2|\text{Solution } X|\text{saturated KCl}|\text{calomel}, \quad E_X \quad \ldots .\text{II}$$
$$H_2|\text{Solution } S|\text{saturated KCl}|\text{calomel}, \quad E_S \quad \ldots .\text{III}$$

a combination which is essentially the cell:

$$H_2|\text{Solution } X|\text{saturated KCl}|\text{Solution } S|H_2 \quad \ldots .\text{IV}$$

The hydrogen electrode is often replaced by a glass electrode. The *operational* definition of pH is:

$$\text{pH}(X) - \text{pH}(S) = (E_X - E_S)/k \quad \ldots .(12.13)$$

where pH(X) is the value of pH to be determined in the solution of X and pH(S) is the value assigned to some standard, S. The British practice[39] is to adopt a single standard, 0·05 M potassium hydrogen phthalate, and to assume that for it:

$$\text{pH}(S) = pa_{H^+} = 4{\cdot}000 + 5 \times 10^{-5}(t - 15)^2 \qquad 0 \leqslant t \leqslant 55$$
$$= 4{\cdot}000 + 5 \times 10^{-5}(t - 15)^2 - 0{\cdot}002(t - 55) \qquad 55 \leqslant t \leqslant 95$$

BATES[40], however, relies more on three standards: the phthalate solution, the equimolal phosphate mixture and the borax solution. For each of these, it is assumed that $pH(S) = pa_{H^+}$, using values of $pa_{H}{}^{+}$ given in Appendix 12.3, *Table 1*.

The question now arises as to how closely $pH(X)$, defined operationally by equation (12.13), can be identified with pa_{H^+}. A partial answer can be obtained by making pH measurements with S the standard phosphate mixture and X either the phthalate or borax standard. The results seem to depend, to some extent, on the way the liquid-junction is made, but, with care in making the junction, agreement within less than 0·01 between pH and pa_{H^+} is possible in the pH range 4–9. Above pH 9, pa_{H^+} seems to be higher than pH and the equation:

$$pa_{H^+} = \text{pH} + 0{\cdot}014\ (\text{pH} - 9{\cdot}18) \qquad \ldots.(12.14)$$

has been proposed[41] to correct pH meter readings.

Whether $\text{pH} = pa_{H^+}$ at $\text{pH} < 4$ is a question more difficult to answer. It seems certain that, for 0·05 M potassium tetraoxalate, pa_{H^+} is higher than pH by 0·02. It is not certain whether this is a peculiarity of the tetraoxalate buffer solution or a difference inherent in all solutions of low pH. For calibration of electrodes for use at low pH, solutions of 0·01 M hydrochloric acid, $p(a_{H^+}\gamma_{Cl^-}) = 2{\cdot}087$, $pa_{H^+} = 2{\cdot}043$ and 0·1 M hydrochloric acid, $p(a_{H^+}\gamma_{Cl^-}) = 1{\cdot}197$, $pa_{H^+} = 1{\cdot}087$, are useful. Appendix 12.3, *Tables 2* and *3* contain some other useful reference values.

If the solvent is non-aqueous or only partially aqueous (the term 'non-aqueous' can conveniently cover both categories), $p(a_{H^+}\gamma_{Cl^-})$ can be obtained without ambiguity from cells like type I but with a non-aqueous solvent. But there are now two values of $p(a_{H^+}\gamma_{Cl^-})$; if we use ${}^{w}E^{0}$ as in equation (12.11) then we get $p_w(a_{H^+}\gamma_{Cl^-})$, where activity coefficients are relative to the standard state in water; if we use ${}^{s}E^{0}$, we get $p_s(a_{H^+}\gamma_{Cl^-})$, where activity coefficients are relative to the standard state in the non-aqueous solvent.

pa_{H^+} (strictly speaking, $p_s a_{H^+}$ sometimes written pa^{*}_{H}) is also a well-defined quantity, equal to $p_s(a_{H^+}\gamma_{Cl^-}) + \log {}_{s}\gamma_{Cl^-}$, provided that we use a conventional value of ${}_{s}\gamma_{Cl^-}$. This can be done by using equation (9.7) for $\log {}_{s}\gamma_{Cl^-}$ with $a = 4{\cdot}56$ Å as before and values of the A and B parameters suitable to the temperature and the dielectric constant of the medium. Some pa_{H^+} values in 50 per cent water–methanol[42] have been determined for acetate, succinate and phosphate buffer solutions over both a temperature and a concentration range. At 25°, pa_{H^+} values were as follows: equimolal (0·025 M) acetic acid, sodium acetate and sodium chloride, 5·529; equimolal (0·025 M) sodium hydrogen succinate and sodium chloride, 5·734; equimolal (0·01 M) potassium dihydrogen phosphate, disodium hydrogen phosphate and sodium chloride, 8·021.

If deuterium oxide is the solvent, the basic measurement is the e.m.f. of a cell of type I but with a deuterium gas electrode and a solution of deuterium chloride in deuterium oxide. The standard potential of the cell is determined; at 25° it is 0·21266 V compared with 0·22234 V for ordinary water. Cell I is then used with a deuterium gas electrode and a buffer solution in deuterium oxide to give values of $p(a_{D^+}\gamma_{Cl^-})$, whence values of pa_{D^+} are derived using a modified Bates–Guggenheim convention. The acetate and phosphate buffer solutions were studied(43) from 5°–50°. At 25° we have: 0·05 M CH_3COOD + 0·05 M CH_3COONa, $pa_{D^+} = 5{\cdot}230$ and 0·025 M KD_2PO_4 + 0·025 M Na_2DPO_4, $pa_{D^+} = 7{\cdot}428$

An operational pH scale for non-aqueous solvents could be defined by cells such as II, III and IV, using the same solvent in all cell compartments. It is not known to what extent pa_{H^+} and pH in non-aqueous solvents would be self-consistent.

More is known about the cell(44):

$H_2|X$ in non-aqueous solvent|saturated aqueous KCl|
aqueous calomel electrode V

The operational pH is still: $\text{pH}(X) - \text{pH}(S) = (E_X - E_S)/k$, where E_X is the e.m.f. of cell V and E_S that of cell III containing a standard buffer solution such as the phosphate mixture in *aqueous* solution. We suppose that the non-aqueous solution contains at least a small amount of chloride ion so that measurements of $p(a_{H^+}\gamma_{Cl^-})$ can be made in cells without liquid-junctions. If we use ${}^wE^0$, we get $p_w(a_{H^+}\gamma_{Cl^-})$.

Cell V must have a liquid-junction potential E_j at the interface between the aqueous and non-aqueous solutions. The best that we can hope for is that ($\text{pH} - E_j/k$) will measure hydrogen ion activity:

$$\text{pH} - E_j/k = p_w(a_{H^+}\gamma_{Cl^-}) + \log {}_w\gamma_{Cl^-}$$

But $\log {}_w\gamma_{Cl^-}/{}_S\gamma_{Cl} = \log {}_m\gamma_{Cl^-}$ is the medium effect—the 'primary' medium effect of equation (12.7). Hence,

$$\text{pH} - p_w(a_{H^+}\gamma_{Cl^-}) - \log {}_s\gamma_{Cl^-} = E_j/k + \log {}_m\gamma_{Cl^-} \quad \ldots.(12.15)$$

The first two terms on the left can be measured, the third is determined by convention. $\log {}_m\gamma_{Cl^-}$ depends only on the nature and composition of the solvent and it is not unreasonable to hope that E_j will likewise be independent of the nature and amount of the solutes. Experiment shows that this is so for water–methanol, water–ethanol and water–dioxan solvents. This, however, only proves that E_j is, at least to a good approximation, a constant for any one solvent medium. Equation (12.15) can be rearranged thus:

$$p_s(a_{H^+}) \text{ or } pa^*_H = pH - \delta \qquad \ldots\ldots(12.16)$$

where $$\delta = E_j/k - \log {}_m\gamma_{H^+}$$

Again, δ should depend only on the nature and the composition of the solvent and its value can be determined by measuring pH and pa^*_H values for a few solutions, using cells with and without liquid-junctions, respectively. Thus a correction factor can be provided to convert pH values into pa^*_H for each solvent composition. Some δ values for water–methanol solvents are as follows:

Wt. per cent methanol	10	20	30	40	50	60	70
δ	$0{\cdot}00_2$	$0{\cdot}01_2$	$0{\cdot}03_9$	$0{\cdot}08_1$	$0{\cdot}12_3$	$0{\cdot}14_9$	$0{\cdot}11_5$

As an example, for 52·1 per cent methanol, $\delta = 0{\cdot}130$. A solution of 0·004996 M borax and 0·009992 M potassium chloride in this solvent gave $p_s(a_{H^+}\gamma_{Cl^-}) = 9{\cdot}525$, $pa^*_H = 9{\cdot}432$. The pH value as measured by a combination of cell III (aqueous phosphate, pH 6·865) and cell V gave pH = 9·565. Applying the correction $\delta = 0{\cdot}130$, gives $pa^*_H = 9{\cdot}435$, which compares very well with that obtained from a cell without liquid-junction.

REFERENCES

HARNED, H. S. and EHLERS, R. W., *J. Amer. chem. Soc.*, 54 (1932) 1350; HARNED, H. S. and OWEN, B. B., *Chem. Rev.*, 25 (1939) 31
2 MACINNES, D. A. and SHEDLOVSKY, T., *J. Amer. chem. Soc.*, 54 (1932) 1429
3 KATCHALSKY, A., EISENBERG, H. and LIFSON, S., *ibid.*, 73 (1951) 5889
4 WRIGHT, D. D., *ibid.*, 56 (1934) 314
5 BATES, R. G., *J. Res. nat. Bur. Stand.*, 47 (1951) 127
6 KING, E. J. and KING, G. W., *J. Amer. chem. Soc.*, 74 (1952) 1212
7 TAYLOR, E. G., DESCH, R. P. and CATOTTI, A. J., *ibid.*, 73 (1951) 74
8 OWEN, B. B., *J. Amer. chem. Soc.*, 56 (1934) 1695
9 NIMS, L. F., *ibid.*, 55 (1933) 1946; BATES, R. G. and ACREE, S. F., *J. Res. nat. Bur. Stand.*, 30 (1943) 129
10 HARNED, H. S. and DAVIS, R., *J. Amer. chem. Soc.*, 65 (1943) 2030
11 — and OWEN, B. B., *ibid.*, 52 (1930) 5091
12 BATES, R. G. and PINCHING, G. D., *ibid.*, 71 (1949) 1274
13 BATES, R. G. and PINCHING, G. D., *J. Res. nat. Bur. Stand.*, 42 (1949) 419
14 OWEN, B. B., *J. Amer. chem. Soc.*, 56 (1934) 2785
15 BATES, R. G. and PINCHING, G. D., *J. Res. nat. Bur. Stand.*, 43 (1949) 519; *J. Amer. chem. Soc.*, 72 (1950) 1393
16 BIGGS, A. I., *Trans. Faraday Soc.*, 50 (1954) 800
17 ROBINSON, R. A. and BIGGS, A. I., *ibid.*, 51 (1955) 901
18 KOLTHOFF, I. M. and FURMAN, N. H., 'Potentiometric Titrations', John Wiley and Sons, Inc., New York 2nd ed., 1931
19 SPEAKMAN, J. C., *J. chem. Soc.*, (1940) 855; see also ANG, K. P., *J. phys. Chem.*, 62 (1958) 1109

[20] ADAMS, E. Q., *J. Amer. chem. Soc.*, 38 (1916) 1503; BJERRUM N., *Z. phys. Chem.*, 104 (1923) 147
[21] ROBINSON, R. A. and BIGGS, A. I., *Aust. J. Chem.*, 10 (1957) 128
[22] COHN, E. J. and EDSALL, J. T., 'Proteins, Aminoacids and Peptides,' Reinhold Publishing Corp., New York (1943)
[23] BJERRUM, N., *Z. phys. Chem.*, 106 (1923) 219
[24] EUCKEN, A., *Angew. Chem.*, 45 (1932) 203
[25] GANE, R. and INGOLD, C. K., *J. chem. Soc.*, (1928) 1594
[26] KIRKWOOD, J. G. and WESTHEIMER, F. H., *J. chem. Phys.*, 6 (1938) 506, 513
[27] OWEN, B. B., *J. Amer. chem. Soc.*, 54 (1932) 1758
[28] HARNED, H. S. and CALMON, C., *ibid.*, 61 (1939) 1491; ROBINSON, R. A. and HARNED, H. S., *Chem. Rev.*, 28 (1941) 419
[29] CROCKFORD, H. D., 'Symposium on Electrochemical Constants,' p. 153, Washington (1951)
[30] FEAKINS, D. and FRENCH, C. M., *J. chem. Soc.*, (1957) 2581
[31] HARNED, H. S. and ROBINSON, R. A., *Trans. Faraday Soc.*, 36 (1940) 973
[32] FEATES, F. S. and IVES, D. J. G., *J. chem. Soc.*, (1956) 2798
[33] HARNED, H. S. and NESTLER, F. M. H., *J. Amer. chem. Soc.*, 68 (1946) 966
[34] — and ROBINSON, R. A., *ibid.*, 50 (1928) 3157; HARNED, H. S. and OWEN, B. B., *ibid.*, 52 (1930) 5079; HARNED, H. S. and MURPHY, G. M., *ibid.*, 53 (1931) 8; HARNED, H. S. and HICKEY, F. C., *ibid.*, 59 (1937) 1284
[35] — and SCHUPP, O. E., *ibid.*, 52 (1930) 3892 [CsOH + CsCl]; HARNED, H. S. and HAMER, W. J., *ibid.*, 55 (1933) 2194 [KOH + KCl]; HARNED, H. S. and COPSON, H. R., *ibid.*, 55 (1933) 2206 [LiOH + LiCl]; HARNED, H. S. and HAMER, W. J., *ibid.*, 55 (1933) 4496 [NaOH + NaBr and KOH + KBr]; HARNED, H. S. and MANNWEILER, G. E., *ibid.*, 57 (1935) 1873 [NaOH + NaCl]; HARNED, H. S. and DONELSON, J. G., *ibid.*, 59 (1937) 1280 [LiOH + LiBr]; HARNED, H. S. and GEARY, C. G., *ibid.*, 59 (1937) 2032 [$Ba(OH)_2$ + $BaCl_2$]; HARNED, H. S. and PAXTON, T. R., *J. phys. Chem.*, 57 (1953) 531
[36] HARNED, H. S. and FALLON, L. D., *J. Amer. chem. Soc.*, 61 (1939) 2374
[37] NOYES, A. A., KATO, Y. and SOSMAN, R. B., *ibid.*, 32 (1910) 159
[38] BATES, R. G. and GUGGENHEIM, E. A., *Pure appl. Chem.*, 1 (1960) 163
[39] 'pH Scale', British Standard 1961: 1641, British Standards Institution, London
[40] BATES, R. G., 'Determination of pH', John Wiley and Sons, Inc., New York (1964)
[41] BOWER, V. E. and BATES, R. G., *J. Res. nat. Bur. Stand.*, 55 (1955) 197; BATES, R. G. and BOWER, V. E., *Anal. Chem.*, 28 (1956) 1322
[42] PAABO, M., ROBINSON, R. A. and BATES, R. G., *J. Amer. chem. Soc.*, 87 (1965) 415
[43] GARY, R., BATES, R. G. and ROBINSON, R. A., *J. phys. Chem.*, 68 (1964) 1186, 3806; 69(1965)2750
[44] VAN UITERT, L. G. and HAAS, C. G., *J. Amer. chem. Soc.*, 75 (1953) 451; GUTBEZAHL, B. and GRUNWALD, E., *ibid.*, 75 (1953) 565; DE LIGNY, C. L. and REHBACH, M., *Rec. Trav. chim. Pays-Bas.* 79 (1960) 727; BATES, R. G., PAABO, M. and ROBINSON, R. A., *J. phys. Chem.*, 67 (1963) 1833; ONG, K. C., ROBINSON, R. A. and BATES, R. G., *Anal. Chem.*, 36 (1964) 1971

13

THE 'STRONG' ACIDS

THE common acids, hydrochloric, nitric, perchloric and sulphuric, have many properties in common with other electrolytes but their dissociation into hydrogen (or H_3O^+) ions and their ability to act as solvents themselves, endow them with some characteristics which are described separately in this chapter.

AQUEOUS HYDROCHLORIC ACID

The thermodynamic properties of aqueous hydrochloric acid show a striking resemblance to those of lithium chloride. (See Appendix 8.10.) The osmotic and activity coefficients of hydrochloric acid, the chlorides of the alkali-metals and ammonium chloride form a very regular group of non-intersecting curves, the coefficients at any given concentration decreasing in the order:

$$H^+ > Li^+ > Na^+ > K^+ > NH_4^+$$

The activity coefficients can be quantitatively accounted for by the combination of the Debye-Hückel theory with the concept of ionic hydration which was discussed in Chapter 9. The values of the 'hydration parameter' (h) required in equation (9.25) at 25° are:

HCl, 8·0; LiCl, 7·1; NaCl, 3·5; KCl, 1·9; and NH_4Cl, 1·6.

It will be recalled that these values represent an allowance for the total ion-solvent interaction; we are claiming that the thermodynamic properties of the solution are the same as those which would be expected if the 'molecule' of solute consisted of two ions, solvated with a total of h molecules of water, rather than asserting that in fact the kinetic entities are (taking lithium chloride as an example) an unhydrated chloride ion and a lithium ion solvated with 7·1 molecules of water. On this basis the high value for the hydration number of the hydrogen ion in hydrochloric acid is not unreasonable. The familiar formula H_3O^+ is no more than a statement that at any given moment the proton must be on one water molecule or another; it is quite likely that its presence would lead to an intensification of the temporary bonds of that molecule to its neighbours, so giving the hydrogen ion a large thermodynamic hydration number.

BASCOMBE and BELL[1a], and WYATT[1b], have found that the variation of the Hammett acidity function with concentration in strong acid solutions (up to 8 M) is consistent with hydration of the proton to $H^+(H_2O)_4$. A similar conclusion is reached by VAN ECK, MENDEL and BOOG[1c] from X-ray diffraction studies of concentrated hydrochloric acid. Such a clustering would not seriously limit the mobility of the ion in conductance or diffusion, since most of the transport of hydrogen ion occurs by a 'jumping' of the proton from one water molecule to another rather than by the bodily motion of the whole cluster (see p. 121). The introduction of some such abnormal transport mechanism for the hydrogen ion is unavoidable if the extremely high mobility of this ion is to be explained. However, it presents an interesting problem: why is the concentration-dependence of both the conductivity and the transport number of hydrochloric acid so successfully accounted for by the theory developed in Chapter 7 for normal electrolytes, in which the transport is by ordinary motion of the ions through the solvent? On examination of the form of the theoretical expression, an answer suggests itself.

The equivalent conductivity of an ion of an electrolyte is given by equation (7.25); $\frac{\Delta X}{X}$ is the relaxation effect, ΔX being the extra field acting on the ion due to the field of the surrounding ions. This is a purely electrostatic effect, and will be just as effective in stimulating proton-jumps as it will in causing normal ionic motion; consequently the factor $\left(1 + \frac{\Delta X}{X}\right)$ will be applicable to hydrochloric acid. The term $\left(\frac{F^2}{6\pi\eta N}\frac{\kappa}{1 + \kappa a}\right)$ gives the electrophoretic effect. This is a hydrodynamic effect, and as such will affect the chloride ions but cannot be expected to apply to the proton-jump part of the motion of hydrogen ions. However, it has never been claimed that the proton-jump mechanism is the only cause contributing to the conductance of the hydrogen ion: if a cluster of water molecules is associated with a proton, that cluster will move in an electrical field even if the proton does not jump. In fact, only relatively few protons need jump to produce the observed conductivity: the rest of the hydrogen ions will be moving in the normal manner, probably with a mobility comparable to that of lithium ions. Now the electrophoretic term in equation (7.25) does not involve the mobility of the ions directly: we might write the first factor as:

$$\left(\lambda_a^0 + \lambda_n^0 - \frac{F^2}{6\pi\eta N}\cdot\frac{\kappa}{1 + \kappa a}\right)$$

where λ_a^0 is the abnormal or proton-jump part of the limiting conductivity, and λ_n^0 is the part contributed by the normal motion; the electrophoretic correction is the same however the total value of λ_{H^+} is distributed between the two processes. This will mean that equation (7.25) is applicable to the hydrogen ion in hydrochloric acid. In fact, the equation, even in the simplified form (7.36) gives an excellent account of the change in equivalent conductivity with concentration up to several tenths molar. The same argument will of course explain the success of equation (7.40) in reproducing the observed transport numbers. The value of the ion-size parameter required in the transport number equation is 4·4 Å and in the conductivity equation it is 4·3 Å; this is very nearly the value (4·47 Å) demanded by the Debye-Hückel equation for the activity coefficient. It is only at fairly high concentrations that the conductivity begins to drop more rapidly than equation (7.36) would predict.

The conductivity of hydrochloric acid solutions has been thoroughly studied by OWEN and SWEETON[1d] over a wide range of concentrations and temperatures (see *Table 13.1*). Below about

Table 13.1

Equivalent Conductivity (Λ) *of Concentrated Aqueous Hydrochloric Acid Solutions*

c	5°	15°	25°	35°	45°	55°	65°
0	297·6	361·9	426·0	489·0	550·2	609·3	666·6
0·25	266·2	322·1	377·4	431·1	482·8	531·9	578·2
1·00	235·2	284·0	332·3	379·4	424·9	468·2	509·2
2·25	192·0	230·9	270·0	308·6	346·1	382·1	416·3
4·00	143·5	171·6	200·1	228·6	256·9	284·2	310·1
6·25	97·9	116·0	134·7	153·6	172·5	191·2	209·5
9·00	61·3	72·2	83·5	94·9	106·6	118·2	130·0

From OWEN, B. B. and SWEETON, F. H., *J. Amer. chem. Soc.*, 63 (1941) 2811.
The Λ^0 values given in the original paper differ by up to 0·2 unit from those in the table above, which are obtained by applying equation (7.37) to measurements on solutions below 0·1 N.

0·1 N their results can be accurately represented by equation (7.36), the parameter *a* taking the value of 4·3 Å at all temperatures from 5° to 65°. In these relatively dilute solutions, therefore, the acid behaves as a normal non-associated electrolyte. At higher concentrations, however, the conductivity falls more rapidly than equation (7.36) predicts. Thus in 4 N solution at 25° the observed equivalent conductivity is $\Lambda = 200{\cdot}1$; even the introduction of the bulk viscosity ($\eta/\eta^0 = 1{\cdot}255$ at 4 N and 25°C) as in equation (11.50), brings the calculated value down only as far as $\Lambda = 258$, still some 25 per cent high. In view of the comparatively good success of equation (11.50) with other concentrated non-associated electrolytes (see *Figure 11.5*), it seems that some special explanation of its failure for hydrochloric acid must be invoked. Association into hydrogen chloride molecules cannot be the explanation, for the vapour pressure of hydrogen chloride over the 4 N solution is far too small to admit of any significant concentration of such molecules in the liquid. However, the special proton-jump mechanism by which the hydrogen ion is mainly transported (see Chapter 6) provides a reasonable explanation: at the high electrolyte concentrations in question, a substantial proportion of the water molecules must be oriented round ions in positions which leave them unable to participate in the normal coordinated or 'hydrogen-bonded' water structure: such molecules would presumably not be available as arrival points for the 'jumping' protons, the mobility of which would therefore be considerably reduced. This suggestion is due to ONSAGER[2], who further points out that the specific resistance of hydrochloric, sulphuric and nitric acids reaches a maximum of about 1·3 Ω-cm at high concentrations: this leads him to estimate the dielectric relaxation time of water as $1{\cdot}45 \times 10^{-12}$ seconds. The value obtained from high radio frequency measurements is of the order of 10^{-11} sec at room temperatures, so that Onsager's estimate is too low. It was, however, obtained by ignoring the contribution of the anion to the conductivity; the effect of this approximation would at least be in the observed direction, although it would be difficult to estimate its magnitude.

SULPHURIC ACID AS AN IONIZING SOLVENT

Sulphuric acid is of exceptional interest in the study of electrolytes. Its behaviour in aqueous solutions is naturally of great practical importance in view of its widespread use in chemical industry; while from a theoretical view-point perhaps even more valuable information has been gained from the study of sulphuric acid as a solvent for electrolytes.

Most of the present extensive knowledge of the properties of solutions in sulphuric acid is due to some recent comprehensive studies by GILLESPIE and his collaborators[3]. They found a freezing-point of 10·36°C for pure sulphuric acid. (KUNZLER and GIAUQUE[4] found 10·35°.) The freezing-point is depressed by both water and sulphur trioxide in excess of the exact stoichiometric composition

H_2SO_4. The pure liquid has a remarkably high electrical conductivity,

$$K_{sp}^{25^\circ} = 0{\cdot}01033\ \Omega^{-1}\ \text{cm}^{-1}$$

$$K_{sp}^{10{\cdot}4^\circ} = 0{\cdot}00580\ \Omega^{-1}\ \text{cm}^{-1}$$

This conductivity is raised by both excess water and excess sulphur trioxide, though according to Kunzler and Giauque the minimum electrical conductivity occurs not quite at the composition of pure sulphuric acid but at 99·996 ± 0·001 per cent H_2SO_4. The dielectric constant has recently been determined[5, 6] as ε_s (25°C) = 101, so that it is one of the few solvents with a dielectric constant higher than that of water. Its viscosity is also unusually high,

$$\eta_{(25^\circ)} = 0{\cdot}2454 \text{ poise,}$$

some twenty-seven times that of water at 25°.

Thus the properties of most direct relevance to the behaviour of dissolved ions, *viz.*, the self-dissociation, the dielectric constant, and the viscosity, are all substantially greater than the corresponding properties for water, and this fact is reflected in a number of interesting ways.

The conductivity of pure sulphuric acid is attributed to the ionization:

$$2H_2SO_4 \rightleftharpoons H_3SO_4^+ + HSO_4^-$$

for which an apparent molal scale ionization constant

$$K = [H_3SO_4^+][HSO_4^-] = 1{\cdot}7 \times 10^{-4}$$

has been estimated. Another reaction,

$$2H_2SO_4 \rightleftharpoons H_3O^+ + HS_2O_7^-$$

is believed to occur simultaneously, with an ionization constant of 8×10^{-5}. This extensive self-dissociation considerably complicates the interpretation of both cryoscopic and conductivity results for solutions in this solvent; the total concentration of self-dissociation products is estimated as 0·043 molal, in striking contrast with the value of only 2×10^{-7} for the sum of the hydrogen and hydroxyl ion concentrations in water.

Sulphuric acid has good solvent powers for both organic and inorganic compounds: sulphuryl chloride and trichloracetic acid, for example, dissolve as non-electrolytes, while alkali and alkaline-earth metal bisulphates and perchlorates, nitric acid, water, sulphur trioxide, *n*-propylamine, benzoic acid, acetone, and alcohols dissolve as electrolytes. An interesting and unusual feature of the electrochemistry of solutions in sulphuric acid is that, because of the strong proton donating character of the solvent, the anion formed in the electrolyte solutions is almost invariably the bisulphate ion, examples of some ionization-reactions being:

$$KHSO_4 \rightarrow K^+ + HSO_4^-$$

$$H_2O + H_2SO_4 \rightarrow H_3O^+ + HSO_4^-$$

$$C_2H_5OH + 2H_2SO_4 \rightarrow C_2H_5HSO_4 + H_3O^+ + HSO_4^-$$

$$HNO_3 + 2H_2SO_4 \rightarrow NO_2^+ + H_3O^+ + 2HSO_4^-$$

$$NH_4ClO_4 + H_2SO_4 \rightarrow NH_4^+ + HClO_4 + HSO_4^-$$

At present the only feasible way of studying the thermodynamics of such solutions is by freezing-point depression measurements, an extensive study of which, with modern experimental techniques, has recently been made by Gillespie and his co-workers. The ionization equations quoted above are derived from their studies. Gillespie's school reached the conclusion that interionic effects were negligible within experimental error, as had been previously suggested by HAMMETT and DEYRUP[7], and postulated an extremely high ('ferroelectric') dielectric constant for sulphuric acid to account for this. No reliable measurements of the dielectric constant were at that time available, but the recent measurements by BRAND, JAMES, and RUTHERFORD[5] by radio frequency methods at wavelengths as low as 10 cm have overcome the experimental difficulties of measuring the dielectric constant of this highly conducting liquid, and show that the value is approximately $\varepsilon_s = 110$ at 20°. Whilst higher than that of water, this value is certainly not of the 'ferroelectric' order of magnitude, being comparable with that of liquid hydrogen cyanide: in the latter solvent (see Chapter 7) interionic attraction effects are by no means negligible. Brand, James and Rutherford point out that a probable explanation of the 'pseudo-ideal' behaviour of electrolytes in sulphuric acid is that the ionic strengths of the solutions used in the cryoscopic studies are necessarily high (greater than 0·05) because of the strong self-dissociation of the solvent; in this region of ionic strength the activity coefficient and osmotic coefficient would be expected to vary only slowly with concentration, as is the case in water. They have, in fact, shown that the osmotic coefficients of a number of electrolytes in sulphuric acid solutions are in very fair accord with Guggenheim's modification of the Debye-Hückel equation (equation 9.13).

CONDUCTIVITIES OF SOLUTIONS IN SULPHURIC ACID

Gillespie and his co-workers have also made important studies of the conductivity of electrolytes in sulphuric acid, supplemented by measurements of transport numbers, viscosities, and densities. Once again the strong self-dissociation precludes the measurements at low ionic strengths which have proved so valuable with aqueous and other solutions; nevertheless some important conclusions have emerged. In spite of the high viscosity of sulphuric acid, equivalent conductivities are of the same order of magnitude as those found in water. This result is explained when the transport numbers are considered: in Hittorf measurements on the alkali and alkaline-earth metal bisulphates, the highest cation transport number found was 0·030 for the potassium ion in 0·6 molar potassium hydrogen sulphate. The equivalent conductivity of this solution was found to be $\Lambda = 78$ (at 25°), so that the cation contribution to the conductivity is only 2·3 units. In aqueous solutions of this concentration the potassium ion contributes about 50 units to the conductivity. The ratio of the mobilities of the potassium ion in water and sulphuric acid is therefore comparable to the inverse ratio of the viscosities of these solvents, and can be regarded as normal. No marked variation of the transport number with temperature in the range 25°–60° was observed, although a slight increase may occur.

The observed high conductivities must therefore be attributed mainly to the abnormal transport mechanism for the anion: as remarked above, the anion in electrolyte solutions in sulphuric acid is nearly always the bisulphate ion, HSO_4^-. A 'proton-jump' mechanism[8] such as almost certainly exists for the hydrogen and hydroxyl ions in water, is the natural assumption, and is consistent with the known association of sulphuric acid molecules through 'hydrogen bonds'. The $H_3SO_4^+$ ion, *i.e.*, the proton solvated with one sulphuric acid molecule, shows a similar high mobility attributable to the same type of mechanism.

Equivalent conductivities in sulphuric acid are strongly concentration-dependent, that of potassium hydrogen sulphate, for example, dropping from $\Lambda = 158$ at 0·1 to 63 at 1 molar. This drop is considerably more rapid than could be accounted for by interionic effects, and occurs even in the case of ammonium hydrogen sulphate where the viscosity of the solution scarcely changes with concentration, so that it cannot be attributed to increasing viscosity. It therefore seems to arise from some effect of the ions on the proton-jump process responsible for the anion mobility. A somewhat similar effect exists in aqueous solutions of hydrochloric acid, but does not become serious until substantially higher concentrations are reached.

The viscosities of solutions of metal bisulphates in sulphuric acid are highly specific properties of the cation: the ammonium ion scarcely alters the viscosity, while the alkali-metal and alkaline-earth metal cations produce an approximately linear increase in viscosity with concentration. The slope of the viscosity-concentration curves increases in the order

$$NH_4^+ < K^+ < Na^+ < Li^+ < Ba^{++} < Sr^{++}$$

being especially great for the last two members, one molal solutions of which have at least seven times the viscosity of the solvent. This suggests strong ion-solvent interactions, further evidence for which is found in the apparent molal volumes of the cations, which are in all cases lower than the volumes estimated from crystallographic radii, and are, indeed, in most cases negative. These apparent volumes are consistent with an increasing amount of solvation, with resulting electrostriction of the sulphuric acid molecules near the ion, in the same order as is suggested by the viscosities. If the ammonium ion is assumed to solvate with one molecule of H_2SO_4, the results lead to solvation numbers of 2, 3, 3, 8 and 8 for K^+, Na^+, Li^+, Ba^{++} and Sr^{++} respectively. The transport numbers are also consistent with this order for the solvation, and it is the same order as is found for the solvation of metal ions in water, thus indicating that the cation solvation is electrostatic in nature.

NITRIC ACID AS A SOLVENT

The depression of the freezing-point of nitric acid on the addition of either water or dinitrogen pentoxide has been studied by GILLESPIE, HUGHES and INGOLD[9]. Their nitric acid had a freezing-point between $-41{\cdot}71°$ and $-41{\cdot}81°$ (FORSYTH and GIAUQUE[10] record $-41{\cdot}65°$). Dinitrogen pentoxide causes about twice the depression due to an equimolar amount of water; the effect of the former is consistent with dissociation into two ions, but water seems to be dissolved in the molecular form. A more accurate representation of the data is obtained by assuming that the ions of dinitrogen pentoxide are solvated with four molecules of nitric acid whilst the water molecule seems to take up only two molecules of nitric acid. It is believed that ionization occurs according to the equation:

$$N_2O_5 \rightarrow NO_2^+ + NO_3^-$$

and these workers suggest that, if the nitrate ion is solvated with two molecules of nitric acid as seems likely from the observations of CHÉDIN and VANDONI[11] on the vapour-pressure lowering of nitric acid solutions of potassium nitrate, then the nitronium ion must also take up two molecules of nitric acid. Further evidence comes from the electrical conductance of nitric acid which is increased very much on the addition of dinitrogen pentoxide[12], whilst the addition of water up to 10 per cent by weight causes very little change in conductance.

Another interesting feature is the marked rounding-off of the freezing-point curve in the vicinity of 100 per cent nitric acid, indicating considerable self-dissociation:

$$2HNO_3 \leftrightharpoons NO_2^+ + NO_3^- + H_2O$$

for which an equilibrium constant,

$$K = m_{NO_2^+} \cdot m_{NO_3^-} \cdot m_{H_2O} = 0{\cdot}020$$

(in mole kg^{-1} concentration units) has been estimated.

RAMAN SPECTRUM OF NITRIC ACID AND ITS AQUEOUS SOLUTIONS

Like the extinction coefficient of a solution for the absorption of light, the intensity of a Raman line should be proportional to the concentration and not to the activity of the molecule or the ion in which the line originates[13]. There is a strong line at 1050 cm^{-1} in the Raman spectrum of aqueous nitric acid which is also found in the spectrum of the alkali nitrates in aqueous solution: the intensity in concentrated acid solutions is, however, less than in a solution of alkali nitrate of the same concentration. It is likely that the line is characteristic of the nitrate ion and the diminished intensity in concentrated nitric acid solution is taken as evidence of the formation of undissociated molecules. In this way an ionization constant of $K = 23{\cdot}5$ has been calculated[14] in good agreement with $K = 22$ from nuclear magnetic resonance measurements[14a]; the acid is about 50 per cent ionized at 11 N. In the same way perchloric acid has been found to be incompletely dissociated[14a, 15] with $K = 38$; it is therefore considerably stronger than nitric acid and its dissociation falls to 50 per cent only at 15 N. The incomplete dissociation of nitric acid is reflected in its activity coefficient; a plot of the stoichiometric activity coefficient against concentration does not fit into the family of curves formed by other 1 : 1 electrolytes, but MCKAY[15a] has shown that the fit can be realized if the proper ionic activity coefficient, with allowance for incomplete dissociation, is used.

The Raman spectrum of pure nitric acid consists of eight more or less sharp lines and a diffuse band. Six of the lines and the band are attributed to the nitric acid molecule and there is general agreement about the assignment of most of them to various vibrational modes. Valuable work has been done by INGOLD and his school[16] and the spectrum can be summarized as follows:

610 cm^{-1}:	bending of the O—N—OH angle
680	bending of the O—N—O angle
925	stretching of the N—OH bond
1300	symmetric stretching of the NO_2 group
1675	anti-symmetric stretching of this group
3400 (band)	OH stretching, the band being diffuse because of intermolecular hydrogen bonding
1535	first overtone of the out-of-plane vibrations of the NO_3 group.

The remaining two lines are not due to the nitric acid molecule: that at 1050 cm^{-1} is assigned to the nitrate ion and that at 1400 cm^{-1} to the nitronium ion, NO_2^+. Both lines are weak and are caused by some self-dissociation of the molecule:

$$2HNO_3 \rightarrow H_2O + NO_2^+ + NO_3^-$$

The assignment of this sharp, highly polarized line to the nitronium ion is supported in several ways. A number of solid nitronium salts have been isolated: ($NO_2^+ ClO_4^-$), ($NO_2^+ HS_2O_7^-$), ($(NO_2^+)_2 S_2O_7^{--}$) and ($NO_2^+ SO_3F^-$), and in each case the Raman spectrum shows a line at 1400 cm^{-1} together with, of course, lines characteristic of the anion. Moreover, the salt-like character of ($NO_2^+ ClO_4^-$) has been confirmed[17] by an x-ray crystallographic study. The Raman spectrum of solid dinitrogen pentoxide gives both lines suggesting that in the solid state this substance has the very interesting salt-like structure, ($NO_2^+ NO_3^-$) analogous to phosphorus pentachloride, ($PCl_4^+ PCl_6^-$) and again x-ray crystallography supports this structure[18]. Ingold has also studied this problem in another way: in pure nitric acid both the lines at 1050 and 1400 cm^{-1} are weak, but on the addition of approximately 10 mole per cent of either perchloric acid or selenic acid, the line at

1400 cm^{-1} was enhanced and that at 1050 cm^{-1} was suppressed. This is exactly what we would expect from the reactions:

$$HNO_3 + 2HClO_4 \rightarrow H_3O^+ + NO_2^+ + 2ClO_4^-$$

$$HNO_3 + 2H_2SeO_4 \rightarrow H_3O^+ + NO_2^+ + 2HSeO_4^-$$

If sulphuric acid were used instead of perchloric or selenic acid, a similar reaction would be expected:

$$HNO_3 + 2H_2SO_4 \rightarrow H_3O^+ + NO_2^+ + 2HSO_4^-$$

but it was observed that both lines were enhanced. At first sight this might seem to be anomalous but it is readily explained when it is realized that the bisulphate ion, HSO_4^-, itself has a Raman line at 1050 cm^{-1}, a fact which has created some confusion in that experiments on HNO_3—H_2SO_4 mixtures have suggested that the two lines are in some way coupled together. It required these experiments in which anions were produced with no Raman line in the region of 1050 cm^{-1} to demonstrate that the two lines had separate origins; indeed, if the nitronium ion is a product of this reaction, it can, because of its centro-symmetric nature, give only one Raman line. It is also significant that if dinitrogen pentoxide is added to nitric acid both lines are enhanced because of the dissociation:

$$N_2O_5 \rightarrow NO_2^+ + NO_3^-$$

The attribution of the 1050 cm^{-1} line to the nitrate ion is amply justified by its occurrence in the spectra of non-associated nitrates in aqueous solution[13], and it was by this line that Redlich arrived at a value for the dissociation constant of nitric acid in aqueous solution.

THE RAMAN SPECTRUM OF SULPHURIC ACID

Ingold *et al.* list seven lines in the Raman spectrum of sulphuric acid at 391, 416, 562, 910, 976 and 1376 cm^{-1} with a broad band at 1125—95 cm^{-1}. The bisulphate ion, HSO_4^-, has lines at 590, 895 and 1050 cm^{-1}, the last being the only one not close to molecular sulphuric acid lines and therefore the most useful one for detecting the bisulphate ion; it is supposed to be due to a stretching of the S—OH bond. The line at 590 cm^{-1} lies close to the 562 cm^{-1} line of sulphuric acid and that at 895 cm^{-1} is close to the 910 cm^{-1} line. It therefore requires very careful examination of the microphotometer records to find evidence for these bisulphate ion lines.

The addition of sulphur trioxide to sulphuric acid leads to a weakening of the molecular acid lines and when the solutions have the composition of disulphuric acid, $H_2S_2O_7$, the molecular sulphuric acid lines are absent and have been replaced by a new set with a strong line at 735 cm^{-1}, useful for characterizing disulphuric acid.

Addition of further sulphur trioxide leads to trisulphuric acid, $H_2S_3O_{10}$ with a strong, characteristic line at 480 cm^{-1} and another at 530 cm^{-1} which is also useful for identification although sulphur trioxide itself has a line close to this. There is some evidence of the existence of tetrasulphuric acid, $H_2S_4O_{13}$ and even higher polymeric forms before sulphur trioxide appears in monomeric and polymeric forms. It is possible to make an assignment of the different Raman lines to various molecular and ionic forms of these acids for details of which the original paper should be consulted.

AQUEOUS SULPHURIC ACID

YOUNG[19] has described the construction of a Raman spectrograph which gives results of high quantitative accuracy. Considerable work has already been done on aqueous sulphuric acid solutions, using the 910 cm^{-1} line to identify the undissociated sulphuric acid molecule, the 1040 cm^{-1} line for the bisulphate ion, and a line at 980 cm^{-1} for the sulphate ion, SO_4^{--}. Thus, by comparing the intensity of the 980 cm^{-1} line in a solution of ammonium sulphate and in a sulphuric acid solution, the SO_4^{--} ion concentration can be calculated, assuming that the ratio of the intensities of the lines is the ratio of the ion concentrations. The HSO_4^- ion concentration is obtained from the 1040 cm^{-1} line and the concentration of H_2SO_4 molecules by difference. This latter concentration should be proportional to the intensity of the 910 cm^{-1} line, and the data did satisfy this severe test. Perhaps the most concise way of representing the results of this work, which is fully supported by recent nuclear magnetic resonance measurements[19a], is in the form of a graph (*Figure 13.1*) from the Record of Chemical Progress[19] which shows that, except in extremely dilute solution, the SO_4^{--} ion is not a major constituent of these solutions; at moderate concentrations it is the HSO_4^- ion which predominates and only above $c = 14$ does the undissociated molecule contribute significantly. The dotted line in the figure is calculated on the assumption that each molecule of water added to pure sulphuric acid reacts according to the equation:

$$H_2SO_4 + H_2O \rightarrow H_3O^+ + HSO_4^-$$

Since the second dissociation of sulphuric acid is relatively weak, it is suppressed by the hydrogen ion resulting from the strong first dissociation. With a degree of ionization α for the bisulphate ion, we have:

$$K = \frac{\gamma_{H^+}\gamma_{SO_4^{--}}}{\gamma_{HSO_4^-}} \frac{\alpha(1+\alpha)m}{1-\alpha} \approx 0{\cdot}01$$

Since *Figure 13.1* indicates that $\alpha \approx 0{\cdot}3$ at 2 M, the activity coefficient part of this expression must have the low value of $\sim 0{\cdot}01$, and

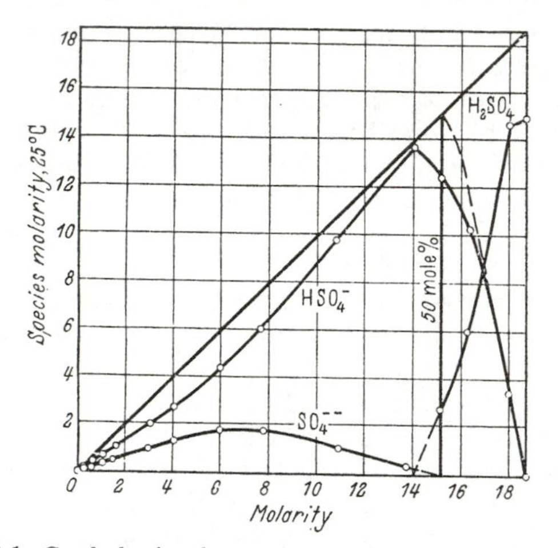

Figure 13.1. Graph showing the proportion of H_2SO_4 *molecules,* HSO_4^- *ions and* SO_4^{--} *ions in aqueous sulphuric acid solution. (From* YOUNG, T. F., *Rec. chem. Progr.*, 12 (1951) 81)

this must be mainly due to the low value of $\gamma_{SO_4^{--}}$ since the ratio $\gamma_{H^+}/\gamma_{HSO_4^-}$ must be near unity.

It is interesting that the thermodynamic behaviour of aqueous sulphuric acid approximates to that of the 1 : 1 electrolyte, hydrochloric acid; a similar effect occurs with ammonium sulphate and ammonium chloride. If we treat a 1 : 2 electrolyte formally as a 1 : 1 electrolyte, its osmotic coefficient becomes $\phi' = 3\phi/2$, ϕ being its value on the basis of a 2 : 1 electrolyte ($\nu = 3$). In *Figure 13.2* this modified osmotic coefficient ϕ' for sulphuric acid is compared with the osmotic coefficient ϕ of the genuine 1 : 1 electrolytes, ammonium chloride and hydrochloric acid, at concentrations up to 6 M. The curve is somewhat higher than that of hydrochloric acid but is clearly of the same type. At its lower extremity it begins to show a rise which is of course due to the increase of the second dissociation with dilution. Above 0·5 M, there is much more similarity between the sulphuric acid and hydrochloric acid curves than there is between hydrochloric acid and ammonium chloride. The difference between the latter is attributable to the great difference in the extent of 'thermodynamic' hydration of the proton and the ammonium ion, while the difference between the two acids is probably mainly due to the fact that the bisulphate ion is larger than the

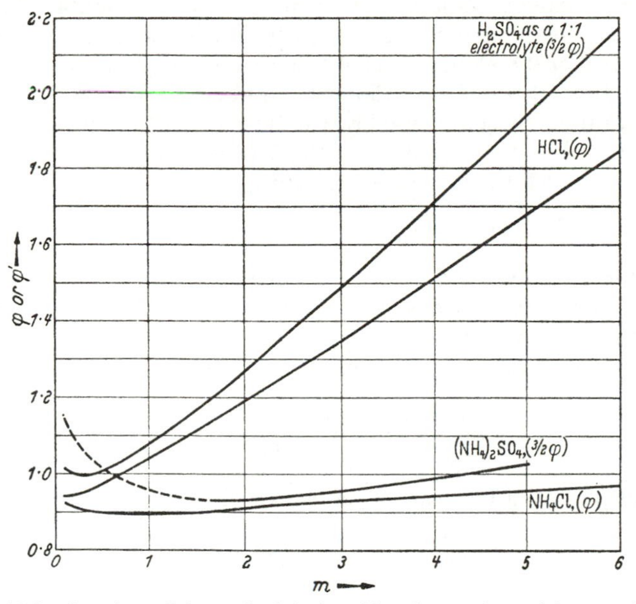

Figure 13.2. Osmotic coefficients of sulphuric acid and ammonium sulphate considered as effectively 1 : 1 *electrolytes. From* WISHAW, B. F. *and* STOKES, R. H., *Trans. Faraday Soc.*, 50 (1954) 954

chloride ion. In passing, it is worth noting that at these high concentrations ammonium sulphate behaves much more like the 1 : 1 electrolyte ammonium chloride than like a fully dissociated 1 : 2 salt: indeed, it is doubtful whether there is such a substance except at great dilution. The ion-pair $NH_4SO_4^-$ is considerably less stable than the covalently-bound particle HSO_4^-, as is evidenced by the dashed part of the ammonium sulphate curve in the figure which shows that dissociation into a 1 : 2 electrolyte is becoming significant below 2 M.

IONIZATION CONSTANT OF THE SECOND STAGE OF DISSOCIATION OF SULPHURIC ACID

Except at very high concentrations sulphuric acid is a non-associated electrolyte in its first stage of dissociation, to which YOUNG and

BLATZ[19] assigned an ionization constant of the order of 10^3. In its second stage of dissociation it is a moderately weak electrolyte with an ionization constant about 0·01. An acid of this strength produces sufficient ions to make the computation much more difficult than that for a much weaker acid like acetic acid. SHERRILL and NOYES[20] made a computation from conductance data which is interesting in being one of the early calculations in which the importance of the Debye-Hückel theory was realized. The equivalent conductivity Λ of a solution of sulphuric acid of molality m, in which all the molecules have lost the first hydrogen ion by dissociation and a fraction, α, have lost the second hydrogen ion to form the SO_4^{--} and a fraction $(1 - \alpha)$ remain in the HSO_4^- state, is:

$$2\Lambda = (1 + \alpha)\lambda_{H^+} + (1 - \alpha)\lambda_{HSO_4^-} + 2\alpha\lambda_{SO_4^{--}}$$

The observed transport number is obtained, at least in principle, by measuring the net transfer of hydrogen ion to the region around an electrode when current is passed. Part of this is due to transport of HSO_4^- ions in the opposite direction so that:

$$t_{H^+} = \frac{(1 + \alpha)\lambda_{H^+} - (1 - \alpha)\lambda_{HSO_4^-}}{2\Lambda}$$

and the two equations can be solved to give

$$\alpha = \frac{(1 + t_{H^+})\Lambda - \lambda_{H^+}}{\lambda_{H^+} + \lambda_{SO_4^{--}}}$$

The transport number is known, λ_{H^+} is found from the conductivity and transport number of hydrochloric acid at a comparable ionic concentration and λ_{SO} - from data for potassium sulphate. A certain amount of successive approximation is necessary because λ_{H^+} and $\lambda_{SO_4^{--}}$ have to be interpolated at ionic strengths not known at the commencement of the calculation.

The two equations of this method can also be solved to give:

$$\lambda_{HSO_4^-} = \frac{(1 - t_{H^+})\Lambda - \alpha\lambda_{SO_4^{--}}}{(1 - \alpha)}$$

Since the conductivity of a solution of sodium hydrogen sulphate is:

$$\Lambda = \lambda_{Na^+} + (1 - \alpha)\lambda_{HSO_4^-} + \alpha\lambda_{H^+} + 2\alpha\lambda_{SO^{--}}$$

and $\lambda_{HSO_4^-}$ is known from the sulphuric acid measurements, a second value of the ionization constant of sulphuric acid can be got from the conductivity of the sodium salt. Sherrill and Noyes arrived at

$K_2 = 0{\cdot}0115$ by both methods, but a recent recalculation[21] has led to the opinion that 0·0102 would be a better value.

A second attack[22] on this problem follows the work of Harned and Ehlers on acetic acid by using the cell:

$$H_2|NaHSO_4(m), Na_2SO_4(m'), NaCl(m'')|AgCl, Ag$$

Whilst this cell gives very reproducible potentials, there are difficulties in the calculation that are met with even in the case of an acid like formic acid, but are enhanced when the acid is polybasic and one of the ionization constants is of the order of 0·01. However, by a tedious set of approximations, Hamer arrived at values of K between 0° and 60°, that at 25° being 0·0120. Hamer's data have been recalculated[21] making allowance for the formation of some $NaSO_4^-$ ions, to give $K = 0{\cdot}0102$ at 25°.

A third method uses the cell:

$$H_2|HCl(m), H_2SO_4(m')|AgCl, Ag,$$

an interesting variant of the Harned-Ehlers cell which avoids the correction for $NaSO_4^-$ ion formation, apart from which the calculation is similar to that used with Hamer's cell. DAVIES, JONES and MONK[21] arrived at $K_2 = 0{\cdot}0103$ at 25°.

Perhaps the most reliable value comes from the spectrophotometric work of YOUNG, KLOTZ and SINGLETERRY[23] using a method which is not unlike that of VON HALBAN[24] *et al.* for picric acid and α-dinitrophenol, but which is not limited to weak acids giving coloured solutions.

Two absorption cells are used, one filled with a 'reference' indicator solution (4×10^{-6} N methyl orange) and between 3×10^{-4} and 6×10^{-4} N hydrochloric acid, so that the pH is about 3·4, and substantial proportions of each of the coloured forms of methyl orange are present. The other cell contains a similar indicator solution to which is added sodium sulphate. The intensity of the light transmitted by this solution for light of wavelength 5200 Å is determined by a photoelectrically registering spectrophotometer. Since both the red and the yellow forms of the indicator are present, by Beer's Law:

$$\log \frac{I_0}{I} = \alpha c l \varepsilon_a + (1 - \alpha) c l \varepsilon_b$$

where I_0 and I are the intensities of the incident and transmitted light, l is the cell length, α is the fraction of the total indicator concentration, c, which is in the yellow form, In^-, this form having an extinction coefficient ε_a, whilst $(1 - \alpha)$ is in the 'red' form, HIn, which has an extinction coefficient ε_b. $l\varepsilon_a$ and $l\varepsilon_b$ are determinable by adding an excess of acid or alkali to the solution so that the measurement of the transmitted light through the stock indicator solution with about 5×10^{-4} N hydrochloric acid and sodium sulphate in amounts up to about 0·04 N is essentially a determination of the indicator ratio, $c_{In^-}/c_{HIn} = \alpha/(1 - \alpha)$. But this ratio occurs in the equilibrium equation:

$$K_{In} = \frac{y_{H^+}y_{In^-}}{y_{HIn}} \frac{c_{H^+}c_{In^-}}{c_{HIn}}$$

or

$$\log c_{H^+} = \log K_{In} - \log R - 2 \log y$$

where R is the ratio c_{In^-}/c_{HIn} and y^2 is an abbreviation for $y_{H^+}y_{In^-}/y_{HIn}$.

Using the same indicator solution but with no added sodium sulphate, let the results of intensity measurements be represented by:

$$\log c^0_{H^+} = \log K_{In} - \log R^0 - 2 \log y^0$$

The addition of sodium sulphate has therefore altered R in two ways, a neutral salt effect on y resulting from change in the total ionic strength, and a change in c_{H^+} resulting from combination of hydrogen and sulphate ions. It is now assumed that the addition of a salt such as sodium chloride changes y but not c_{H^+}; let it be added in such amount that the total ionic strength is increased as it was on the addition of sodium sulphate. Then the new R value is given by:

$$\log c^0_{H^+} = \log K_{In} - \log R' - 2 \log y$$

Hence:

$$\log \frac{c_{H^+}}{c^0_{H^+}} = \log \frac{R^0}{R} + 2 \log \frac{y^0}{y}$$

and

$$\log \frac{R^0}{R'} = -2 \log \frac{y^0}{y}$$

The method therefore gives c_{H^+}, the hydrogen ion concentration of the sodium sulphate solution relative to that of the stock solution, say $c_{H^+} = rc^0_{H^+}$. But the bisulphate ion is subject to the equilibrium equation:

$$K_2 = \frac{y_{H^+}y_{SO_4^{--}}c_{H^+}c_{SO_4^{--}}}{y_{HSO_4^-}c_{HSO_4^-}}$$

or

$$K_2 = \frac{y_{H^+}y_{SO_4^{--}}}{y_{HSO_4^-}} \cdot \frac{r[c - (1 - r)c^0_{H^+}]}{(1 - r)}$$

where c is the stoichiometric sodium sulphate concentration. Fortunately $c^0_{H^+}$ is small and can be determined to a sufficient approximation with a glass electrode. The activity coefficient term is estimated by a Debye-Hückel approximation (equation 9.7) and by an extrapolation to zero concentration, the true value of K_2 is determined. *Table 13.2* gives the mean values quoted by Singleterry.

Table 13.2

Second Ionization Constant of Sulphuric Acid

Temp.	K_2
5°	0·0185 ± 0·0005
15°	0·0139 ± 0·0004
25°	0·0104 ± 0·0003
35°	0·0077 ± 0·0002
45°	0·00565 ± 0·00007
55°	0·00413 ± 0·00001

The limits shown in this table correspond to the agreement between two sets of measurements made by Singleterry, in one of which sodium chloride was used as 'neutral' salt and in the other barium chloride. The ionization constants can be represented by:

$$\log K_2 = -\frac{475{\cdot}14}{T} + 5{\cdot}0435 - 0{\cdot}018222T$$

with the following thermodynamic properties for the dissociation process at 25°.

$$\Delta\bar{H}^0 = -5237 \text{ cal mole}^{-1}$$

$$\Delta\bar{C}^0_P = -49{\cdot}7 \text{ cal deg}^{-1} \text{ mole}^{-1}$$

$$\Delta\bar{S}^0 = -26{\cdot}6 \text{ cal deg}^{-1} \text{ mole}^{-1}$$

Singleterry estimated $\Delta\bar{H}^0 = -5188$ and -5319 cal mole^{-1} from the two sets of measurements; the entropy change is almost the same in each of his calculations, but he derived $-45{\cdot}9$ and -57 cal deg^{-1} mole^{-1} for the partial molal heat capacity so it is evident that a mean value should be used with caution. This equation predicts that K_2 should have a maximum value of 0·14 at $-112°$. It is of course dangerous to extrapolate so far from the range of temperatures in which this equation is valid; nevertheless it is evident from a plot of $\log K_2$ against the temperature that the maximum cannot be attained without a considerable reduction of temperature below 5°.

The high value of this ionization constant results in some anomalous properties of sulphuric acid in comparison with non-associated electrolytes. For example, the apparent molal volume of a simple electrolyte in aqueous solution is usually a linear function of the square root of the volume concentration; this statement, sometimes known as Masson's rule[25], often holds up to surprisingly high concentrations. The behaviour of sulphuric acid is very different, as was shown by KLOTZ and ECKERT[26]. The circles in *Figure 13.3* represent their experimental measurements and the lower straight

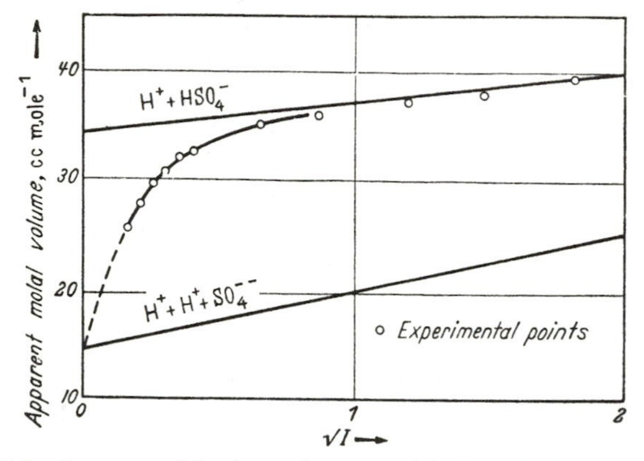

Figure 13.3. Apparent molal volume of aqueous sulphuric acid. I in molarity units

line is the calculated apparent molal volume of the hypothetical fully dissociated ($2H^+ + SO_4^{--}$) electrolyte, obtained by applying the additivity rule to the apparent molal volumes of potassium sulphate, hydrochloric acid and potassium chloride. It is evident that only at the most extreme dilutions will the apparent molal volume of sulphuric acid be at all close to that expected of the completely dissociated acid. At concentrations experimentally accessible the volume is considerably higher and becomes linear in the square root of the concentration at high concentrations when the solution contains effectively only H^+ and HSO_4^- ions. Klotz and Eckert were able, from the known degrees of dissociation, to calculate the apparent molal volume of the hypothetical fully dissociated ($H^+ + HSO_4^-$) electrolyte, shown by the upper straight line of *Figure 13.3*. Thus they have demonstrated that the anomalous position of the experimental points can be resolved by assuming two straight lines to represent the variation of the volume function with $\sqrt{I}$ and apportioning the contribution of each species according to the known fraction present.

The surface tension of aqueous sulphuric acid exhibits greater complexity. The surface tension of a salt solution usually increases linearly with the molality, the slope being characteristic of the salt, but for hydrochloric and nitric acids the surface tension decreases with increasing concentration and the slopes are not quite straight lines.

The curve of the surface tension of sulphuric acid solutions against the concentration is markedly temperature dependent; at 0° and at low concentrations the curve has a negative slope leading to a minimum surface tension at about 0·6 M after which the surface tension increases again to a flat maximum at about 7 M. The minimum is not found at higher temperatures, although the initial branch of the curve at 18° is almost sigmoid in shape, and the maximum occurs at higher concentrations as the temperature is increased. By employing an additivity principle similar to that used by Klotz and Eckert, YOUNG and GRINSTEAD[27] were able to calculate the surface tension of solutions of the hypothetical, fully dissociated acid ($2H^+ + SO_4^{--}$) from data for hydrochloric acid, sodium sulphate and sodium chloride and to show that the surface tension should decrease with increasing concentration. That of the fully dissociated acid ($H^+ + HSO_4^-$), however, should increase with concentration. Qualitatively we can see that the observed minimum may well result from the balance set up between the positive slope of the ($H^+ + HSO_4^-$) curve and the negative slope of the ($2H^+ + SO_4^{--}$) curve. Young and Grinstead were able to go further than this and to show that, from the known degrees of dissociation at various concentrations, it could be predicted that at 0° the minimum should be at 0·65 M (observed 0·5–0·7 M) and the lowering of the surface tension, relative to pure water, at the minimum should be 0·15 (observed 0·21 dyn cm^{-1}). The maximum is more difficult to account for quantitatively; in these solutions the SO_4^{--} ion is negligible in amount, the HSO_4^- is present in considerable quantity but is diminishing in extent relative to the undissociated sulphuric acid molecule. Pure sulphuric acid has a considerably lower surface tension than water and the formation of the undissociated sulphuric acid molecule should lower the surface tension of the solution, *i.e.*, it should act contrary to the elevating effect of the ($H^+ + HSO_4^-$) acid and hence there should be a maximum surface tension. The quantitative calculation is made difficult because the behaviour of two-component liquid mixtures is not yet thoroughly understood, but Young and Grinstead were able to show that the value of the maximum surface tension and the concentration at which it is found are in accord with the idea that it could be compounded of values due to the two solute species, bisulphate ions and undissociated molecules.

It should also be mentioned that the heat of dilution of sulphuric acid to infinite dilution is very large. This is mainly due to the heat liberated on the ionization of the bisulphate ion which is present in considerable amount at ordinary concentrations, but, of course, dissociates completely on sufficient dilution. By a similar process of compounding the contribution of the ($H^+ + H^+ + SO_4^=$) and ($H^+ + HSO_4^-$) species, Young and Blatz were able to give a remarkably good account of the observed heat of dilution of sulphuric acid solution up to about 0·05 M.

Selenic acid seems to be an acid comparable in strength to sulphuric acid, the second ionization constant being 0·0120 at 25° according to PAMFILOV and AGAFONOVA[(28)]. Their measurements extended over the range 0° to 30° and calculations based on their results suggest $\Delta \bar{H}^0 = -2080$ cal mole^{-1}, which is considerably less than Young *et al.* found for sulphuric acid. Telluric acid has very different properties: salts such as Ag_6TeO_6 can be prepared, the first ionization constant is $2{\cdot}31 \times 10^{-8}$ and the second is about 10^{-12}, so that telluric acid is very weak even in its first dissociation[(29)]. Sulphurous acid[(30)] has ionization constants $K_1 = 1{\cdot}72 \times 10^{-2}$ and $K_2 = 6{\cdot}24 \times 10^{-8}$ whilst iodic acid[(31, 32)] ($K = 0{\cdot}168$) and trichloracetic acid[(32)] ($K = 0{\cdot}232$) are two more examples of acids intermediate between non-associated electrolytes and the majority of the weak acids. By contrast, the ionization constants of periodic acid[(33)] are $K_1 = 0{\cdot}028$ and $K_2 = 5{\cdot}38 \times 10^{-9}$.

Finally it may be mentioned that hydrofluoric acid is unlike the other halide acids in being a weak acid with an ionization constant[(34)] of $6{\cdot}7 \times 10^{-4}$ at 25° and with a strong tendency to associate:

$$HF + F^- \rightleftarrows HF_2^-$$

the 'association constant' being 3·9 at 25°. This leads to low values of the stoichiometric activity coefficient as follows:

m	0·001	0·003	0·005	0·01	0·03	0·05	0·1	0·3	0·5	1·0
γ	0·544	0·371	0·300	0·224	0·136	0·106	0·077	0·044	0·031	0·024

REFERENCES

1a BASCOMBE, K. N. and BELL, R. P., *Disc. Faraday Soc.*, 24 (1957) 158

1b WYATT, P. A. H., *ibid.*, 24 (1957) 162

1c VAN ECK, C. L. P. *v.* P., MENDEL, H. and BOOG, W., *ibid.*, 24 (1957) 200

1d OWEN, B. B. and SWEETON, F. H., *J. Amer. chem. Soc.*, 63 (1941) 2811

2 ONSAGER, L., *Ann. N.Y. Acad. Sci.*, 46 (1945) 265

3 GILLESPIE, R. J. with HUGHES, E. D., INGOLD, C. K., GRAHAM, J., PEELING, E. R. A. and WASIF, S., *J. chem. Soc.*, (1950) 2473–2551, 2997; (1953) 204, 964

These papers contain a comprehensive bibliography of earlier work, in particular that of HANTZSCH, A. and of HAMMETT, L. P.

REFERENCES

[4] KUNZLER, J. E. and GIAUQUE, W. F., *J. Amer. chem. Soc.*, 74 (1952) 804

[5] BRAND, J. C. D., JAMES, J. C. and RUTHERFORD, A., *J. chem. Soc.*, (1953) 2447

[6] GILLESPIE, R. J. and COLE, R. H., *Trans. Faraday Soc.*, 52 (1956) 1325

[7] HAMMETT, L. P. and DEYRUP, A. J., *J. Amer. chem. Soc.*, 55 (1933) 1900

[8] HAMMETT, L. P. and LOWENHEIM, F. A., *ibid.*, 56 (1934) 2620

[9] GILLESPIE, R. J., HUGHES, E. D. and INGOLD, C. K., *J. chem. Soc.* (1950) 2552

[10] FORSYTHE, W. R. and GIAUQUE, W. F., *J. Amer. chem. Soc.*, 64 (1942) 48

[11] CHÉDIN, J. and VANDONI, R., *C.R. Acad. Sci., Paris*, 227 (1948) 1232

[12] BERL, E. and SAENGER, H. H., *Monatsh.*, 54 (1929) 1036

[13] REDLICH, O., *Chem. Rev.*, 39 (1946) 333

[14] YOUNG, T. F. and KRAWETZ, A. A., quoted by REDLICH, O. and HOOD, G. C., *Disc. Faraday Soc.*, 24 (1957) 87

[14a] HOOD, G. C., REDLICH, O. and REILLY, C. A., *J. chem. Phys.*, 22 (1954) 2067

[15] REDLICH, O., HOLT, E. K. and BIGELEISEN, J., *J. Amer. chem. Soc.*, 66 (1944) 13

[15a] MCKAY, H. A. C., *Trans. Faraday Soc.*, 52 (1956) 1568

[16] INGOLD, C. K., MILLEN, D. J. and POOLE, H. G., *J. chem. Soc.* (1950) 2576; MILLEN, D. J., *ibid.* (1950) 2589, 2600, 2606; INGOLD, C. K. and MILLEN, D. J., *ibid.* (1950) 2612; GOULDEN, J. D. S. and MILLEN, D. J., *ibid.* (1950) 2620

[17] COX, E. G., JEFFERY, G. A. and TRUTER, M. R., *Nature, Lond.*, 162 (1948) 259

[18] GRISON, E., ERIKS, K. and DE VRIES, J. L., *Acta cryst., Camb.*, 3 (1950) 290

[19] YOUNG, T. F. and BLATZ, L. A., *Chem. Rev.*, 44 (1949) 93; YOUNG, T. F., *Rec. chem. Prog.*, 12 (1951) 81

[19a] REDLICH, O. and HOOD, G. C., *Disc. Faraday Soc.*, 24 (1957) 87

[20] SHERRILL, M. S. and NOYES, A. A., *J. Amer. chem. Soc.*, 48 (1926) 1861

[21] DAVIES, C. W., JONES, H. W. and MONK, C. B., *Trans. Faraday Soc.*, 48 (1952) 921; see also KERKER, M., *J. Amer. chem. Soc.*, 79 (1957) 3664

[22] HAMER, W. J., *J. Amer. chem. Soc.*, 56 (1934) 860

[23] KLOTZ, I. M., SINGLETERRY, C. R., *Theses*, University of Chicago (1940)

[24] HALBAN, H. VON and SIEDENTOPF, K., *Z. phys. Chem.*, 100 (1922) 208; HALBAN, H. VON and EBERT, L., *ibid.*, 112 (1924) 359; HALBAN, H. VON and KORTÜM, G., *ibid.*, 170 A (1934) 351

[25] MASSON, D. O., *Phil. Mag.*, 8 (1929) 218

[26] KLOTZ, I. M. and ECKERT, C. F., *J. Amer. chem. Soc.*, 64 (1942) 1878

[27] YOUNG, T. F. and GRINSTEAD, S. R., *Ann. N.Y. Acad. Sci.*, 51 (1949) 765

[28] PAMFILOV, A. V. and AGAFONOVA, A. L., *Zhur. Fiz. Khim.*, 24 (1950) 1147; *Chem. Abstr.*, 45 (1951) 2293

[29] BLANC, E., *J. Chim. phys.*, 18 (1920) 28; BRITTON, H. T. S. and ROBINSON, R. A., *Trans. Faraday Soc.*, 28 (1932) 531; FOUASSON, F., *Ann. Chim.*, 3 (1948) 594; ANTIKAINEN, P. J., *Suomen Kem.*, 28B (1955) 135; 30B (1957) 201

[30] TARTAR, H. V. and GARRETSON, H. H., *J. Amer. chem. Soc.*, 63 (1941) 808

[31] FUOSS, R. M. and KRAUS, C. A., *ibid.*, 55 (1933) 476

[32] HALBAN, H. VON and BRÜLL, J., *Helv. chim. Acta*, 27 (1944) 1719

[33] NÄSÄNEN, R., *Acta chem. scand.*, 8 (1954) 1587

[34] BROENE, H. H. and DE VRIES, T., *J. Amer. chem. Soc.*, 69 (1947) 1644

14

ION ASSOCIATION

THE concept of ionic association provides a relatively simple and self-consistent method of dealing with the situation which arises when ions of opposite sign are close together. In these circumstances the energy of their mutual electrical attraction may be considerably greater than their thermal energy, so that they form a virtually new entity in the solution, of sufficient stability to persist through a number of collisions with solvent molecules. In the case of a symmetrical electrolyte, such ion-pairs will have no net charge, though they should have a dipole moment. They will therefore make no contribution to the electrical conductivity, while their thermodynamic effects will be those of removing a certain number of ions from the solution and replacing them by half the number of dipolar 'molecules'. With unsymmetrical electrolytes the position will be more complicated, since the simplest and most probable type of ion association, that involving only two particles, will result in the appearance of a new ionic species of a charge type not previously present; this will contribute to the conductivity, though less than would its constituent ions in a free state. In such cases further association to form neutral particles may also be reasonably expected.

The question which immediately presents itself is: when can two adjacent ions be called an ion-pair? This is rather like the other question we have had to consider: when is a water molecule to be regarded as part of the hydration shell of an ion? and we shall give a rather similar answer, *viz.*, that an ion-pair must be long-lived enough to be a recognizable kinetic entity in the solution. We have treated the hydration question by a simplified picture in which different degrees of hydration are smoothed out to an average number of molecules of water of hydration. Similarly we use the idea, due to BJERRUM[1], that the average effects of ion-pair formation may be calculated on the basis that all oppositely charged ions within a certain distance of one another are 'associated' into ion-pairs, though in reality a momentarily fast-moving ion might come within this distance of another and pass by without forming a pair.

Bjerrum proposed that this critical distance, which we shall denote by q, should be chosen as:

$$q = \frac{|z_1 z_2| \boldsymbol{e}^2}{2\varepsilon \boldsymbol{k} T} \qquad \ldots.(14.1)$$

This is seen to be the distance at which the mutual electrical potential energy of the two ions:

$$|z_1 z_2| \boldsymbol{e}^2/(\varepsilon q)$$

is equal to $2\boldsymbol{k}T$. The reason for this particular choice appears from the following argument:

In discussing the Poisson-Boltzmann equation:

$$\nabla^2 \psi_j = -\frac{4\pi}{\varepsilon} \sum_i n_i z_i \boldsymbol{e} \exp\left(-\frac{z_i \boldsymbol{e} \psi_j}{\boldsymbol{k}T}\right)$$

we have said that no self-consistent solution is possible unless the series expansion of the exponential is stopped at the first power of ψ, or at the second for the special case of symmetrical electrolytes, and that pursuing the expansion further, apart from the mathematical complexity, leads to difficulty with the principle of linear superposition. The Bjerrum treatment avoids these difficulties. The density of i-ions around a selected j-ion is given as before by the Boltzmann expression (4.5) and the number in a shell of thickness dr at a distance r is:

$$n_i \exp\left(-\frac{z_i \boldsymbol{e} \psi_j}{\boldsymbol{k}T}\right) 4\pi r^2 \mathrm{d}r$$

When r is small, Bjerrum neglects the effect of interionic forces on the reasonable ground that the potential of the central ion will be dominant and writes:

$$\psi_j = \frac{z_j \boldsymbol{e}}{\varepsilon r}$$

so that the number of i-ions in the shell is:

$$4\pi n_i \exp\left(-\frac{z_i z_j \boldsymbol{e}^2}{\varepsilon \boldsymbol{k} T r}\right) r^2 \mathrm{d}r$$

Considering a series of shells each of equal thickness, dr, the number of ions which on a time average find themselves in each succeeding ring, can be calculated. In *Table 14.1*, we give the results for an aqueous solution at 25° containing a 1 : 1 electrolyte, for the cases where z_i and z_j are of opposite sign and of the same sign. The second column contains the value of the probability factor, the next column the volume of a shell 0·1 Å thick, and the last column the number of ions to be found in each shell. (In making the calculation it has been assumed that the probability factor in the second column is constant in any one shell of thickness 0·1 Å; this, of

Table 14.1

r (Å)	$\exp[\mathbf{e}^2/(\varepsilon \mathbf{k} Tr)]$	$4\pi r^2\,dr \times 10^{23}$ ($dr = 0{\cdot}1 \times 10^{-8}$ cm)	*Number of ions in shell* $\times 10^{22}$	
			Ions of opposite charge	*Ions of like charge*
2	35·57	0·50	1·77 n_i	0·001 n_j
2·5	17·36	0·79	1·37 n_i	0·005 n_j
3	10·78	1·13	1·22 n_i	0·01 n_j
3·57	7·39	1·60	1·18 n_i	0·02 n_j
4	5·95	2·01	1·20 n_i	0·03 n_j
5	4·17	3·14	1·31 n_i	0·08 n_j
6	3·28	4·52	1·48 n_i	0·14 n_j
7	2·77	6·14	1·70 n_i	0·22 n_j
8	2·44	8·04	1·96 n_i	0·33 n_j

course, is not so, but this crude method of calculation suffices for purpose of illustration.) It will be seen that when i and j are ions of opposite sign, then with increasing r there is a decreasing probability of finding an i-ion in any unit of volume, but the volume of the shell increases and the two opposing effects combine to give a distance at which there is minimum probability of finding an i-ion anywhere on a sphere surrounding the central j-ion at this critical distance. The position of minimum probability is:

$$q = \frac{|z_i z_j| \mathbf{e}^2}{2\varepsilon \mathbf{k} T}$$

as can readily be shown by differentiating the function

$$r^2 \exp\left(-\frac{z_i z_j \mathbf{e}^2}{\varepsilon \mathbf{k} Tr}\right)$$

For a 1 : 1 electrolyte in water at 25° $q = 3{\cdot}57$ Å; at distances closer to the central ion the population of oppositely charged ions increases rapidly (see *Figure 14.1*): the population also increases at greater distances but the rate of increase is less. There is no such effect with ions of like charge: there is small probability of finding them close to the central ion and the population shows no minimum. As regards the ions of opposite charge, if the distance of closest approach is $3{\cdot}57|z_1z_2|$ Å or more, it is assumed that there will be no ion-pairs. If the ions can approach closer than this, Bjerrum would regard those within the sphere of radius $3{\cdot}57|z_1z_2|$ Å as 'undissociated' ion-pairs. It is to the ions outside that the Debye-Hückel theory is to be applied. (We use the expression 'number of ions' although it would be closer to physical reality to say, at greater length, the time average probability of finding an oppositely charged ion within this critical distance.) Before we apply these

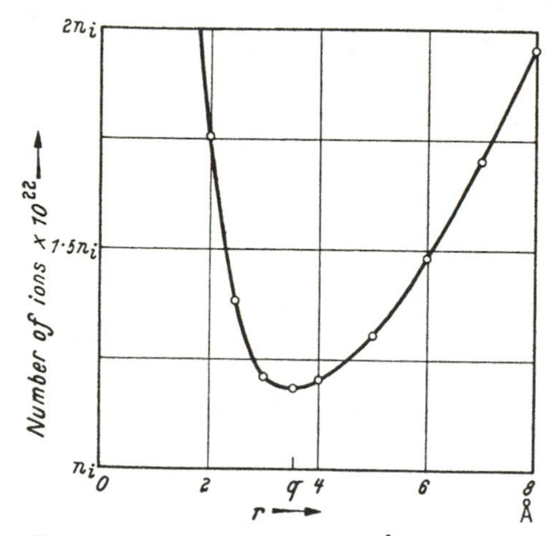

Figure 14.1. Number of ions in a shell 0·1 Å *thick at a distance from a central ion*

considerations to ion-pair formation, let us consider the magnitude of the effect we are discussing. Consider a solution of a 1 : 1 electrolyte, 0·01 N in concentration corresponding to $n_1 = 6 \times 10^{18}$ ions/c.c. Even in the absence of any electrical force exercised by the central ion, the 'normal' distribution would lead to the presence of 0·0127 ions in the shell between 2 and 8 Å, or, putting it more realistically, a single ion has a volume of $1{\cdot}7 \times 10^5$ cu. Å at its disposal. The attractive force of the central ion increases the concentration to an extent given approximately by averaging over the figures in the penultimate column of *Table 14.1*, *viz.*, 0·050 ions in the shell. This figure is probably too large because Bjerrum has simplified the treatment of the problem by subjecting the ions surrounding the central ion to the potential of this ion alone, whereas allowance for the interionic forces would act in the opposite direction.

The degree of association $(1 - \alpha)$ is obtained by integrating the number of ions in all the shells from the distance of closest approach up to the critical Bjerrum distance:

$$(1 - \alpha) = 4\pi n_1 \int_a^q \exp\left(-\frac{z_1 z_2 e^2}{\varepsilon k T r}\right) r^2 \, dr$$

Put

$$x = -\frac{z_1 z_2 e^2}{\varepsilon k T r}$$

so that the integral becomes:

$$-\left(\frac{|z_1 z_2| e^2}{\varepsilon k T}\right)^3 \int_b^2 \frac{e^x}{x^4} \, dx$$

where

$$\frac{|z_1 z_2| e^2}{\varepsilon k T a} = b$$

and

$$\frac{|z_1 z_2| e^2}{\varepsilon k T q} = 2$$

Thus

$$(1 - \alpha) = \frac{4\pi N c}{1000} \left(\frac{|z_1 z_2| e^2}{\varepsilon k T}\right)^3 Q(b)$$

where

$$Q(b) = \int_2^b x^{-4} e^x \, dx$$

Values of the integral, $Q(b)$, have been tabulated[1, 2] (see Appendix 14.1). The law of mass action gives:

$$\frac{\alpha^2 y^2 c}{(1 - \alpha)} = K$$

assuming that the activity coefficient of the ion-pair is unity. The calculation now proceeds in three stages:

1. In very dilute solutions, $\alpha \approx 1, y \approx 1$ and

$$\frac{1}{K} \approx \frac{1 - \alpha}{c} \approx \frac{4\pi N}{1000} \times \left(\frac{|z_1 z_2| e^2}{\varepsilon k T}\right)^3 Q(b) \qquad \ldots . (14.2)$$

For any value of a ($< q$) there are corresponding values of b and $Q(b)$ and hence of $\frac{1}{K}$ so that K is a function of the closest distance of approach of the ions.

2. From the two equations:

$$\frac{\alpha^2 y^2 c}{(1 - \alpha)} = K$$

and

$$-\log f = \frac{A\sqrt{(\alpha c)}}{1 + Bq\sqrt{(\alpha c)}}$$

the degree of association $(1 - \alpha)$ at any value of c can be calculated by successive approximations. As the theory is not expected to apply to solutions that are not dilute the distinction between the activity coefficients f and y can be ignored.

3. From the equation (cf. eq. 2.40):

$$f\alpha c = f_{obs.} c$$

knowing α and f from the second computation, we can calculate $f_{obs.}$, the activity coefficient which should result from experimental measurements, assuming complete dissociation, on an electrolyte

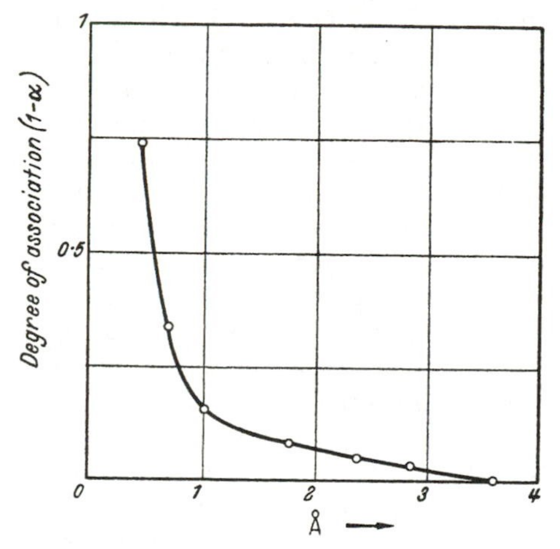

Figure 14.2. Effect of distance of closest approach on the degree of association at 0·1 M *for a* 1 : 1 *electrolyte in water*

possessing the value of a adopted at the commencement of the calculations. Bjerrum has given extensive tables of the degree of association of 1 : 1 electrolytes in water at 18° and the activity coefficients which should be observed on the assumption of complete dissociation. The tables cover the range 0·0001 to 2 N for values of a between 0·47 and 2·82 Å. *Figure 14.2* shows how the degree of association varies with a at $m = 0·1$. At $a = 2$ Å, only about 2·5 per cent of the ions are associated; a has to be reduced to 1·4 Å to increase this to 10 per cent and only at about 0·6 Å do the ion-pairs preponderate over the free ions. Such small ionic radii are unusual and therefore we should not look to aqueous solutions expecting to find outstanding examples of ion-pair formation in 1 : 1 electrolytes.

FUOSS[(2a)] has recently pointed out that a continuous distribution such as that shown in *Figure 14.1* ignores the discrete molecular nature of the solvent: he suggests that two ions should be counted as a pair only if they are in contact, with no solvent molecule intervening. Configurations in which the ions are separated by only a fraction of the diameter of a solvent molecule are highly improbable. On this basis he finds for the dissociation constant K of a 1 : 1 electrolyte the simpler result:

$$\frac{1}{K} = \frac{4\pi N a^3 e^b}{3000} \qquad \ldots.(14.2a)$$

For large values of b—*i.e.*, in solvents of low dielectric constant, this result differs from (14.2) by approximately a factor b, which is of minor importance compared to the large value e^b. Some further discussion of this new theory is given in Appendix 14.3.

One example can be quoted which illustrates Bjerrum's theory in solvents of dielectric constant not less than 57. The dissociation constant of lanthanum ferricyanide, $LaFe(CN)_6$, has been determined recently[3, 4], not only in water as solvent, but also in aqueous mixtures of ethanol, glycol, acetone, dioxan and glycine, the last being used to provide solvents of dielectric constant greater than that of water. The dissociation constants were derived from conductivity measurements in very dilute solution and it was found that $K = 1{\cdot}82 \times 10^{-4}$ in water as solvent, a value comparable with that of formic acid. The critical distance for a 3 : 3 electrolyte is 32·1 Å: calculation shows that a closest distance of approach of 7·2 Å corresponds, on Bjerrum's theory, to a dissociation constant of $1{\cdot}82 \times 10^{-4}$ if we regard any ion distant between 7·2 and 32·1 Å from an oppositely charged ion as forming, temporarily at least, an ion-pair with its neighbour. It was also found that Walden's rule held for these solutions, $\Lambda^0\eta^0$ changing very little from one solvent to another, and it was therefore assumed that this distance of 7·2 Å would not vary with the nature of the solvent. K is then a function of the dielectric constant which appears twice in equation (14.2), in the $\dfrac{|z_1 z_2| e^2}{\varepsilon kT}$ factor and in the $Q(b)$ factor. The continuous line in *Figure 14.3* shows how K should vary with the dielectric constant on Bjerrum's theory, the points being the observed dissociation constants. Considering how difficult to determine are these dissociation constants, requiring accurate measurements at very low concentrations, it is not surprising that there is some scatter of the points, but the dissociation constant decreases with decreasing dielectric constant in a way very close, indeed, to that predicted by Bjerrum's theory.

Solvents of lower dielectric constant should favour ion-pair formation to an even more marked degree. At the critical distance defined by $q = \frac{|z_1 z_2| e^2}{2\varepsilon \boldsymbol{k} T}$, the potential energy of the ion-pair is $2\,\boldsymbol{k}T$, that is to say, the energy necessary to separate the pair is comparable with their energy of thermal motion. Whilst in water as solvent the

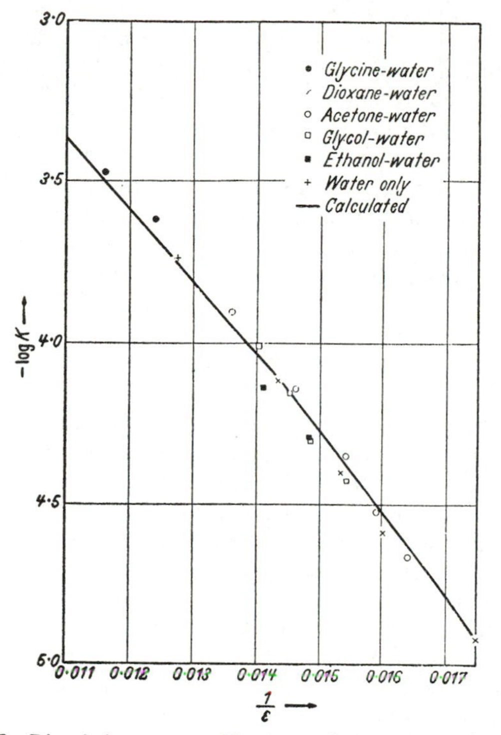

Figure 14.3. Dissociation constant of lanthanum ferricyanide as a function of the dielectric constant of the solvent, compared with the prediction of Bjerrum's equation

majority of ions, especially the solvated ones, cannot approach within their critical distances, q can exceed the ordinary ionic diameter if the dielectric constant is lowered. A convincing proof of this has been advanced by KRAUS and FUOSS[2, 5] using conductivity measurements on tetra*iso*amylammonium nitrate in a series of water–dioxan mixtures covering a wide range of dielectric constant from 2·2 to 79. Solutions of concentration as low as $c = 10^{-5}$ were examined and the spread of the dielectric constant led to a tremendous variation of the equivalent conductivity; for example, at

$c = 0{\cdot}0005$, Λ was 85·1 in water but only 0·000129 in dioxan. With changing electrolyte concentration in any one solvent the conductivity exhibits curious changes. With pure dioxan as solvent there is a minimum in very dilute solution (at $c = 2 \times 10^{-5}$): the curve of conductivity against concentration (best plotted as $\log \Lambda$ versus $\log c$) then shows three points of inflection at higher concentrations. On the addition of water to the solvent, that is, on increasing the dielectric constant, the minimum is found at higher concentrations and becomes less pronounced. Thus, with 4 per cent of water ($\varepsilon = 3{\cdot}5$) the minimum is at $c = 3 \times 10^{-3}$ and disappears if there is 20 per cent of water in the solvent ($\varepsilon = 12$).

It is with the conductivities in very dilute solution that we are now concerned, *i.e.*, the conductivities at concentrations lower than that at which we find the minimum. Very marked departures from the limiting Onsager equation are found in solutions of low dielectric constant; for example, the Onsager equation for a 9·5 per cent water solution ($\varepsilon = 5{\cdot}84$) is:

$$\Lambda = 30 - 473\sqrt{c}$$

predicting $\Lambda = 20{\cdot}5$ at $c = 4 \times 10^{-4}$ whilst the observed figure was only $\Lambda = 2{\cdot}48$. It is now assumed that this indicates ion-pair formation and a series of approximations gives a dissociation constant of the order of 10^{-6}. Fuoss and Kraus had at their disposal dissociation constants of tetra*iso*amylammonium nitrate in nine solutions. From each dissociation constant they were able, using Bjerrum's equation, to calculate that the distance of closest approach was 6·4 Å (the values ranged from 6·01 to 6·70 Å). They plotted a graph of $\log K$ as a function of $\log \varepsilon$, $\log K$ being calculated by Bjerrum's equation for $a = 6{\cdot}4$ Å and on this graph the experimental values of K agreed remarkably well with the predicted curve. Another method of showing this agreement is to calculate (see *Table 14.2*) K for each solvent assuming a constant value of $a = 6{\cdot}4$ Å and compare K with the experimental values. This is a severe test of the theory, because the dielectric constant varies by sixteenfold and the dissociation constant varies over a range of 10^{-15}. Only at the lowest water content is there a difference which could be called significant, and in this solution the minimum in the conductivity curve is found at $c = 0{\cdot}0007$, so that the disturbing factors to which the minimum is due may well have affected the measurements at lower concentrations. This experiment of Kraus and Fuoss must be regarded as establishing the essential soundness of Bjerrum's concept of electrostatic ion-pairs, though Fuoss[2a] now considers that the data are perhaps better represented by equation (14.2a) than by Bjerrum's result (14.2).

Ion-pair formation does indeed occur when most electrolytes are dissolved in any solvent other than one of the few which have high dielectric constants. Water is one of these; as it is also the cheapest

Table 14.2

Dissociation Constant of Tetraisoamylammonium Nitrate in Dioxane–Water Mixtures

$a = 6{\cdot}4$ Å			
% H_2O	ε	K (obs.)	K (calc.)
0·60	2·38	2×10^{-16}	2×10^{-15}
1·24	2·56	1×10^{-14}	2×10^{-14}
2·35	2·90	1×10^{-12}	1×10^{-12}
4·01	3·48	$2{\cdot}5 \times 10^{-10}$	$1{\cdot}4 \times 10^{-10}$
6·37	4·42	3×10^{-8}	$1{\cdot}7 \times 10^{-8}$
9·50	5·84	$1{\cdot}65 \times 10^{-6}$	$1{\cdot}6 \times 10^{-6}$
14·95	8·5	1×10^{-4}	$0{\cdot}9 \times 10^{-4}$
20·2	11·9	9×10^{-4}	7×10^{-4}
53·0	38·0	0·25	0·28

and most accessible of solvents, it is not surprising that much of our information about electrolytic conductivity concerns aqueous solutions. This is fortunate in one way because electrolytes obey comparatively simple laws in solvents of such high dielectric constants, but it should not be allowed to obscure the fact that electrolytes are incompletely dissociated in the majority of solvents. This is illustrated by Appendix 14.2 which lists the limiting equivalent conductivities and dissociation constants of a number of salts in seven solvent media. Even simple salts are weak electrolytes in solvents of low dielectric constant: to emphasize this we quote a few examples from recent work[5a]:

Salt	Solvent	ε	Temp. °C	K
KBr	Acetic acid	6·20	30	$1{\cdot}1 \times 10^{-7}$
KBr	Ammonia	22	−34	$18{\cdot}9 \times 10^{-4}$
CsCl	Ethanol	24·30	25	$6{\cdot}6 \times 10^{-3}$
KI	Acetone	20·70	25	$8{\cdot}02 \times 10^{-3}$
KI	*n*-Propanol	20·1	25	$3{\cdot}0 \times 10^{-3}$
KI	Pyridine	12·0	25	$2{\cdot}1 \times 10^{-4}$
NaI	Ethylenediamine	12·9	25	$6{\cdot}86 \times 10^{-4}$

TRIPLE ION FORMATION

Simple electrostatic theory shows that a system of two charged spheres placed symmetrically on each side of an oppositely charged sphere, all three being of the same size, has an energy 50 per cent greater than that of two oppositely charged spheres. Thus there is reason to believe that triple ions $(+ - +)$ or $(- + -)$, might be formed in solvents of low dielectric constant. The following treatment is taken from a paper by FUOSS and KRAUS[6]. Consider the simplified case of an extremely dilute solution where the activity coefficients can be equated to unity and the limiting conductivity at infinite dilution is a sufficient approximation to the conductivity of a fully dissociated salt solution at this low concentration; let the solvent be one of low dielectric constant so that the degree of dissociation of ion-pairs is very small and $(1 - \alpha) \approx 1$.

Then for the reaction:

$$MX \rightleftharpoons M^+ + X^-$$

$$K \approx \alpha^2 c$$

If there is a possibility of the further equilibria:

$$(MXM)^+ \rightleftharpoons MX + M^+$$

and

$$(XMX)^- \rightleftharpoons MX + X^-$$

let

$$k = \frac{[M^+]\,[MX]}{[MXM^+]} = \frac{[X^-]\,[MX]}{[XMX^-]}$$

an equality which implies that ions M^+ and X^- are equal in size and that there is equal probability of forming $(MXM)^+$ or $(XMX)^-$ triple ions.

The total concentration is:

$$c = [MX] + \tfrac{1}{2}[M^+] + \tfrac{1}{2}[X^-] + \tfrac{3}{2}[MXM^+] + \tfrac{3}{2}[XMX^-]$$

Put

$$\alpha_T = [MXM^+]/c = [XMX^-]/c$$

so that, if α and α_T are small,

$$k \approx \frac{\alpha}{\alpha_T} c \quad \text{and} \quad \alpha_T \approx \frac{\sqrt{(Kc)}}{k}$$

Let Λ^0 be the limiting conductivity at infinite dilution of the simple ions, *i.e.*,

$$\Lambda^0 = \lambda^0_{M^+} + \lambda^0_{X^-}$$

and Λ^0_T that of the triple ions, *i.e.*,

$$\Lambda^0_T = \lambda^0_{MXM^+} + \lambda^0_{XMX^-}$$

Then the observed conductivity will be:

$$\Lambda = \alpha\Lambda^0 + \alpha_T\Lambda_T^0$$
$$= \sqrt{\left(\frac{K}{c}\right)}\,\Lambda^0 + \frac{\sqrt{(Kc)}}{k}\,\Lambda_T^0$$

which is of the form

$$\Lambda = Ac^{-1/2} + Bc^{1/2}$$

This is the equation of a curve with a minimum and, by differentiation, it can be shown that the concentration corresponding to the minimum conductivity is:

$$c_{\text{min}} = \frac{A}{B} = \frac{k\Lambda^0}{\Lambda_T^0} \quad \text{and} \quad \Lambda_{\text{min}} = 2\sqrt{(AB)}$$

a condition which gives three more important relations:

$$K = c_{\text{min}}\left(\frac{\Lambda_{\text{min}}}{2\Lambda^0}\right)^2, \; k = c_{\text{min}}\frac{\Lambda_T^0}{\Lambda^0}$$
$$\Lambda_{\text{min}} = 2\alpha_{\text{min}}\Lambda^0 = 2\alpha_{T(\text{min})}\Lambda_T^0$$

showing that at the minimum, the conductivity is due in equal parts to single and to triple ions. *Figure 14.4* shows a plot of $\Lambda\sqrt{c}$ against

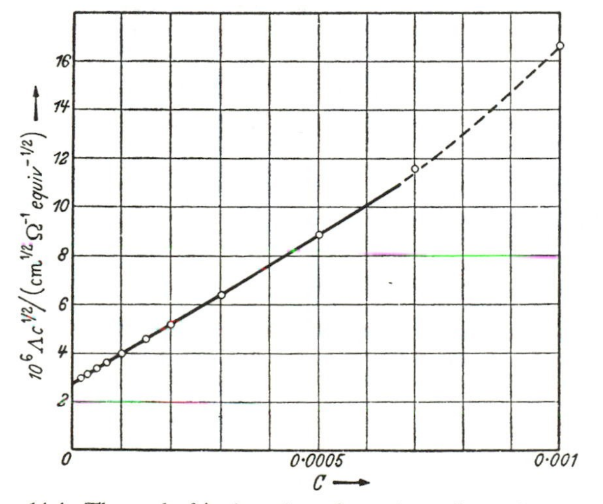

Figure 14.4. The graph of $\Lambda\sqrt{c}$ *against* c *for tetraisoamylammonium nitrate in water–dioxan of dielectric constant* 2·56

c for tetra*iso*amylammonium nitrate in a water–dioxan solvent of dielectric constant 2·56. Up to $c = 0{\cdot}0007$, the points lie on a straight line whose slope is 0·0119 and intercept $2{\cdot}85 \times 10^{-6}$. If

Λ^0 is equated to 30 (by comparison with Λ^0 for this salt in solvents of similar viscosity) and Λ_T^0 is put equal to 10 on the ground that the triple ions will move about three times as slowly, then

$$c_{\min} = A/B = 2{\cdot}85 \times 10^{-6}/0{\cdot}0119 = 2{\cdot}4 \times 10^{-4}$$

$$\Lambda_{\min} = 2\sqrt{(AB)} = 3{\cdot}68 \times 10^{-4}$$

$$K = 9 \times 10^{-15}$$

$$k = 8 \times 10^{-5}$$

To show the magnitude of the two dissociations the values in *Table 14.3* have been calculated.

Table 14.3

$c \times 10^5$	$\alpha \times 10^5$	$\alpha_T \times 10^5$	$\Lambda_{\text{calc.}} \times 10^4$	$\Lambda_{\text{obs.}} \times 10^4$
1·5	2·4	0·5	7·7	7·5
3·0	1·7	0·7	5·8	5·8
8·0	1·1	1·1	4·4	—
10	0·95	1·2	4·05	4·03
24	0·61	1·9	3·68	—
30	0·55	2·1	3·75	3·68
100	0·30	3·8	4·70	5·25

As the concentration increases from very small values, α decreases more rapidly than α_T increases and the conductivity decreases; at $c = 8 \times 10^{-5}$, $\alpha = \alpha_T$, but the conductivity is still decreasing. It is only when $c = 24 \times 10^{-5}$ that the conductivity contributions of the two types of ions are equal and the conductivity has a minimum value, after which the formation of triple ions is dominant and the conductivity increases again.

It must be noted now that, by selecting a solvent of such low dielectric constant (and therefore low values of α and α_T) the calculation has been capable of simplification by neglecting the interionic effects. For a solvent of higher dielectric constant, interionic forces are no longer negligible and the computation is not so straightforward.

Fuoss and Kraus were able to carry the argument one stage further: by treating the approach of a negative ion towards the positive ion of an ion-pair, subject to coulomb forces only, they were able to show that there is a certain value of the distance which is critical: once the approaching ion is within this critical distance it is to be regarded as forming a triple ion. The dissociation constant can be derived in the form of a complicated integral for details of which the original paper should be consulted. However, assuming a critical distance of 9 Å, good agreement with the experimental values of k was found: thus for the water–dioxan mixture of dielectric constant 2·56 for which k was found to be 8×10^{-5}, theory yielded $9 \cdot 3 \times 10^{-5}$. The critical distance of 9 Å may seem very different from 6·4 Å which had to be used in dealing with ion-pair formation. This, however, was a simple case of $(+\ -)$ union, whereas the triple ion formation is an average process involving $(+\ -\ +)$ and $(-\ +\ -)$ in one of which two very large ions participate.

An interesting question arises when two ions are competing to form a triple ion. On a random distribution XMX^- and YMY^- should be present in equal amount and XMY^- at twice this concentration. This has been found[6a] to be true for tetra-*n*-butylammonium chloride and azide in benzene but for the chloride-nitrate, chloride-perchlorate and nitrate-perchlorate mixtures the XMY^- triple ion is favoured.

QUADRUPOLE FORMATION

The existence of a minimum in the conductivity concentration curve of an electrolyte has been explained by the formation of triple ions. At higher concentrations the conductivity changes in a complicated way and it is probable that higher aggregates are formed, for example, quadrupoles $(+\ -\ +\ -)$. Definite evidence for this has been found from the freezing-point measurements of solutions of tri*iso*amylammonium picrate in benzene[7]. At extremely low concentrations the freezing-points can be explained on the basis of an equation[8] for the j function of the freezing-point depression if a reasonable model is assumed for the ion-pair—an ellipsoid with axes in the ratio 2 : 1 containing a point dipole of moment 12·9 Debye units[9]. But at higher concentrations the apparent molecular weight increases. It is assumed that this reaction occurs:

$$2M^+X^- \rightarrow M^+X^-M^+X^-$$

Let a fraction, α, of the M^+X^- ion-pairs associate in this way: then we can write:

$$k_4 = \frac{2(1-\alpha)^2 c}{\alpha}$$

As each ion-pair is replaced by $\frac{\alpha}{2}$ quadrupoles leaving $(1-\alpha)$ ion-pairs, the total number of particles becomes $\left(1 - \frac{\alpha}{2}\right)$. This we equate to the osmotic coefficient and hence to $(1 - j)$ or $j = \frac{\alpha}{2}$ and $(1 - 2j) = (1 - \alpha)$ so that:

$$k_4 = \frac{(1 - 2j)^2 c}{j}$$

or, rearranging:

$$\frac{j}{(1 - 2j)^2} = c/k_4$$

Therefore the function $j/(1 - 2j)^2$, obtained from the experimental data, should be a linear function of the concentration. This is exactly what Fuoss and Kraus[9] found—plotted in this way a straight line was obtained up to a concentration of about 0·03 N, the slope giving a dissociation constant of 0·105 for this particular electrolyte in benzene.

ION-PAIR FORMATION IN WATER AS SOLVENT

The electrolytes so far considered as showing evidence of ion-pair formation do so to a marked extent. The phenomenon therefore occurs to a measurable degree even at low concentrations where the conductivity and the activity coefficient of the few dissociated ions can justifiably be described by equations known to be very good approximations at such high dilutions. It is, however, suspected that ion-pair formation does occur in some electrolytes, for example, with potassium nitrate in aqueous solution, but to an extent much less than in the examples we have already considered. The conductivity of potassium nitrate follows the predictions of the Onsager limiting equation much more closely and to much higher concentrations than we have any right to expect; this can be expressed differently by saying that if the $(1 + \kappa a)$ factor is introduced into the Onsager limiting conductivity equation, the a value required (about 1·9 Å) is just possible if the planar nature of the nitrate ion permits a number of close encounters. The activity coefficient of potassium nitrate is also much lower than we would expect. It is believed by many that the behaviour of potassium nitrate can be explained by postulating a small amount of ion-pair formation; about 3 per cent at 0·1 N would suffice. Unfortunately, whilst it would be comparatively easy to detect the ions if only 3 per cent of the potassium nitrate were dissociated, it is not easy to detect the ion-pairs if only 3 per cent are present in this form, except perhaps in unusual examples where the ion-pairs have characteristic Raman or ultra-violet absorption spectra. We have to measure the diminution from 100 per cent to something of the order of 97 per cent for the proportion of ions present and, to add to the difficulties, this has to be done in a range of comparatively high concentration for the ion-pair effect to be noticeable, assuming that it does exist. We must therefore try to estimate ionic concentrations of such magnitude that our theories are inapplicable with any great assurance; for example, if the ionic concentration is estimated from conductance, it is difficult to prove that the 3 per cent deviation from theoretical prediction is due to ion-pair formation and not to some defect in the theory. The subject has indeed suffered from the difficulty of knowing how to describe the behaviour of the ionized part of the molecule at comparatively high concentrations. DAVIES[10] has used an empirical equation to describe the conductivity of a solution up to about 0·5 N and by comparison with observed conductivities, he has come to the conclusion that many salts, including sodium and potassium nitrate, sodium and potassium iodate, silver nitrate and potassium bromate are only about 97 per cent dissociated at 0·1 N.

We have already seen that the theoretical equation (7.36) works very well for aqueous 1 : 1 electrolytes of the non-associated type. It would therefore seem reasonable to expect it to represent the conductivity of an associated electrolyte if we put a equal to the critical Bjerrum distance: for a temperature of 25°, we write:

$$\Lambda_{\text{calc.}} = \Lambda^0 - (0{\cdot}2300\Lambda^0 + 60{\cdot}65)\frac{\sqrt{(\alpha c)}}{1 + Ba\sqrt{(\alpha c)}}$$

α being the degree of dissociation of the ion-pairs. But if we put $a = 3{\cdot}57$ Å in this equation, we should also use the same value of a in the equation for the activity coefficient:

$$-\log f = \frac{A\sqrt{(\alpha c)}}{1 + Ba\sqrt{(\alpha c)}} \qquad \ldots.(14.3)$$

which is required to calculate the dissociation constant:

$$K = \frac{\alpha^2 y^2 c}{(1 - \alpha)}$$

ignoring the small difference between f and y. The results of calculations along these lines are given in *Table 14.4* for potassium and silver nitrate and for thallous chloride, using the conductivity data of SHEDLOVSKY[11] for the first two, and of GARRETT and VELLENGA[12] and of BRAY and WINNINGHOF[13] for the last salt. The dissociation 'constant' hardly lives up to its name for the two nitrates, since with increasing concentration it increases for potassium nitrate and decreases for silver nitrate, but a reasonably good 'constant' results for thallous chloride. At low concentrations the 'constant' is very sensitive to small changes in either Λ or Λ^0—a change of 0·01 in either at 0·005 N corresponds to only 0·01 per cent in α but to 5 per cent in $(1 - \alpha)$! On Bjerrum's theory dissociation constants of the order of these we have obtained for

Table 14.4

Dissociation Constants of Potassium Nitrate, Silver Nitrate and Thallous Chloride at 25°

c	$\Lambda_{obs.}$	$\Lambda_{calc.}$	$\alpha = \Lambda_{obs.}/\Lambda_{calc.}$	$-2 \log f$	K
		Potassium Nitrate			
0·005	138·48	138·86	0·9973	0·0664	1·42
0·01	135·82	136·61	0·9942	0·0909	1·38
0·02	132·41	133·67	0·9906	0·1230	1·57
0·05	126·31	128·51	0·9829	0·1792	1·87
0·07	123·56	126·23	0·9788	0·2040	1·98
0·1	120·40	123·60	0·9741	0·2327	2·14
					Average 1·73
		Silver Nitrate			
0·005	127·20	127·43	0·9982	0·0664	2·38
0·01	124·76	125·24	0·9962	0·0910	2·12
0·02	121·41	122·38	0·9921	0·1231	1·88
0·05	115·23	117·39	0·9816	0·1791	1·73
0·1	109·13	112·64	0·9688	0·2322	1·76
					Average 1·97
		Thallous Chloride			
0·00507	143·10	144·85	0·9879	0·0665	0·351
0·00604	142·25	144·34	0·9855	0·0721	0·343
0·00750	141·13	143·65	0·9825	0·0794	0·345
0·01	139·00	142·65	0·9744	0·0901	0·302
0·01108	138·35	142·27	0·9724	0·0942	0·306
0·01501	136·03	141·05	0·9644	0·1074	0·306
0·01607	135·40	140·75	0·9620	0·1105	0·303
					Average 0·322

potassium and silver nitrate would be given by electrolytes whose ions could approach within 2 Å, and this may be a reasonable value if it is remembered that the nitrate ion is planar and some of the encounters can be comparatively close ones. For thallous chloride, however, the distance would be only about 1 Å and this is not consistent with the ionic dimensions. The hypothesis of ion-pair formation can be checked in another way, since the product of α and γ, as calculated above, should equal the activity coefficient measured experimentally and computed assuming complete dissociation. For both potassium and sodium nitrate at 0·1 N, γ

calculated by equation (14.3) is 0·765, the $\alpha\gamma$ product is 0·745 and 0·741 respectively compared with the observed activity coefficients, 0·739 and 0·734 (Appendix 8.10). Again the agreement is not good, but for thallous chloride at 0·01 N, $\alpha\gamma$ is 0·878 compared with the observed value of 0·876.

The problem of incomplete dissociation has been approached from another angle by studying reaction rates[14]. In a reaction between a neutral molecule, S, and the ion X^- of an electrolyte MX the rate depends, according to the transition state theory, on the concentrations of S and X^- and the activity coefficients $f_S f_{X^-}/f_{SX^-}$, SX^- being the transition-complex. In dilute solution f_S should be close to unity and X^- and SX^- having similar charges, their activity coefficients should be almost equal. Consequently the rate of reaction should depend on the concentration rather than the activity of X^- and any ion-pair formation between M^+ and X^- should reduce the reaction rate in amount proportional to the number of ion-pairs formed, unless the transition complex, SX^-, can also form an ion-pair with M^+. Whilst this possibility cannot be excluded and, indeed, seems to be realized in the saponification of ethyl acetate[15], it should be negligible if the complex, SX^-, is large as in the catalytic decomposition of diacetone alcohol by hydroxyl ions. Support for this belief comes from further experiments on the hydrolysis of carbethoxymethyltriethylammonium iodide, $EtCO_2 \cdot CH_2N(Et)_3I$, whose transition complex has zero net charge; conclusions drawn from these experiments agreed with those in which diacetone alcohol was used.

In solutions of potassium or rubidium hydroxide this alcohol is decomposed at a rate directly proportional to the stoichiometric alkali concentration, the reaction constant per mole of hydroxide varying only between 0·2165 and 0·2193 up to 0·4 N. Some curious results were obtained with sodium hydroxide; concentrations increasing up to 0·4 N leading to a decrease of the reaction constant from 0·2182 to 0·2051; this may mean that sodium hydroxide is only 94 per cent dissociated at 0·4 N and 98 per cent at 0·1 N, corresponding to a mean ionic diameter of about 3·1 Å if the mechanism of association were Bjerrum ion-pair formation. The 'effective' radius of the hydrated sodium ion of sodium chloride, allowing for a penetration of 0·7 Å (see Chapter 9) is 2·2 Å, so that the hydroxyl ion could approach within 3·1 Å if it had a radius of 0·9 Å, a not impossible value if the radius of the water molecule is 1·4 Å. But there seems to be an objection to this idea. Whatever our doubts about the accuracy of some of these ionic dimensions, the rubidium and potassium ions must be considerably smaller than the hydrated sodium ion and, therefore, rubidium and cesium hydroxides should be even weaker electrolytes. This is contrary to the findings of Bell and Prue and, indeed, contrary to all our ideas about the alkali hydroxides. We will return to this later. (See p. 423.) At present our interest is more with calcium, barium and thallous hydroxide. With each of these the rate-constant falls with increasing hydroxide concentration, showing large departures from the value of 0·218 for the fully dissociated hydroxides and suggesting a considerable degree of association. Making reasonable assumptions about the activity coefficients of the various species, Bell and Prue decided that the dissociation constants of calcium, barium and thallous hydroxide were 0·051, 0·23 and 0·38 respectively, the last corresponding to about 87 per cent dissociation of thallous hydroxide at 0·1 N. Working from these dissociation constants and the Bjerrum equation (14.2) the distances of closest approach can be calculated and compared with the crystallographic radii as follows:

	Bjerrum distance of closest approach	*Sum of crystallographic radii*
$CaOH^+$	2·55 Å	2·52 Å
$BaOH^+$	5·55	2·88
TlOH	1·23	2·97

The *a* values needed for calcium and barium hydroxides are reasonable, although it is curious that barium and hydroxide ions do not approach nearer than 5·55 Å. But the dimensions of the ions of thallous hydroxide are such that they cannot approach closer than 2·97 Å without an interaction more profound than that induced by coulomb forces. Bell and Prue concluded, therefore, that a covalent link must be formed.

Another method of studying the incomplete dissociation of electrolytes depends on measuring the solubility of a sparingly soluble electrolyte in the presence of another electrolyte[16]. Calcium and thallous iodate are examples of salts of conveniently low solubility. From such measurements we derive the activity coefficient in the presence of the added electrolyte because, if s_0 and s are the solubilities in pure water and in the presence of the other electrolyte and f_0 and f are the corresponding activity coefficients, the condition for saturation is $f_0 s_0 = fs$. These activity coefficients are expected to conform to a selected equation considered to be valid for all salts at low concentrations and a lack of such agreement is taken to mean that an incompletely dissociated 'intermediate ion', ion-pair or molecule has been formed. The method can be illustrated by reference to some measurements of the solubility of thallous iodate in potassium chloride solution[17].

In pure water at 25° thallous iodate is soluble to the extent of $1{\cdot}838 \times 10^{-3}$ mole/l rising to $c = 2{\cdot}359 \times 10^{-3}$ in the presence of potassium chloride at a concentration of 0·05422 N. The activity coefficient, f_0, is now calculated by the equation

$$-\log f = \frac{0{\cdot}5\sqrt{I}}{1+\sqrt{I}} - 0{\cdot}1I \qquad \ldots .(14.4)$$

giving $f_0 = 0{\cdot}954$ at a concentration corresponding to the solubility of thallous iodate in water. From the solubility in 0·05422 N potassium chloride, the activity coefficient of thallous iodate can now be calculated as 0·743. This is the stoichiometric activity coefficient assuming that the solubility measurements give ionic concentrations. But using equation (14.4) for this total ionic strength, an activity coefficient of 0·812 is calculated. The ratio of these two activity coefficients is a measure of the amount of thallous ions which have gone to form thallous chloride molecules. Hence the dissociation constant of thallous chloride can be calculated. In practice it is not quite as simple as this because corrections have to be made by successive approximations to get the total ionic strength; thallous iodate and potassium iodate give small amounts of undissociated molecules and, in some experiments, *e.g.*, when thallous iodate is dissolved in potassium sulphate solution, allowance must be made for KSO_4^- ions. Bell and George give the following dissociation constants:

	0°	25°	40°
$TlSO_4^-$	0·042	0·043	0·044
TlCl	0·165	0·210	0·230
TlOH	0·155	0·150	0·142
TlCNS	0·115	0·160	0·230
TlF	—	0·8	—
$TlFe(CN)_6^{---}$	0·00065	0·00060	0·00054
$CaOH^+$	0·043	0·040	0·033
$CaSO_4$	0·0060	0·0049	0·0041

From the solubility of calcium iodate in calcium hydroxide solution, DAVIES and HOYLE[18] obtained 0·050 for the dissociation constant of $CaOH^+$, in good agreement with the value Bell and Prue found from reaction rate experiments, whilst measurements have been made on magnesium and strontium hydroxide which fit in the series: $MgOH^+$ (0·0026), $CaOH^+$ (0·05), $SrOH^+$ (0·11), $BaOH^+$ (0·23).

ION-PAIR FORMATION WITH 2 : 2 ELECTROLYTES

Reference has already been made to the difficulty of finding equations which will describe the behaviour of the ionized portion of a partly dissociated 1 : 1 electrolyte in any but dilute solutions. Two additional difficulties are met with 2 : 2 electrolytes, such as zinc sulphate. First, it is very doubtful if we have an equation which will describe the conductivity of even a hypothetical non-associated 2 : 2 electrolyte because, as we have seen in Chapter 7, it is doubtful if we are justified in taking only the first two electrophoretic terms in equation (7.24), whilst the introduction of higher terms cannot be justified as long as we use the modified Boltzmann distribution given by equation (4.9). In other words we are attempting the problem of a partially dissociated 2 : 2 electrolyte without adequate solutions of the problem of a non-associated 2 : 2 electrolyte. Secondly, we are in considerable difficulty when we try to find Λ^0 values for such electrolytes. This is not a problem of theoretical significance but it does add to the complexities of the task. For two electrolytes, cadmium sulphate and magnesium sulphate, we can circumvent the latter difficulty because for the first salt we can extrapolate the conductivity data[19] at very low concentrations, and for the second salt we know the limiting mobility of the magnesium ion from magnesium chloride data and that of the sulphate ion from sodium sulphate data. Having obtained $\Lambda^0 = 133{\cdot}07$ indirectly, we can use the measurements of DUNSMORE and JAMES[4] at concentrations below 0·001 molar, applying the method already described for potassium and silver nitrate. We write the conductivity equation (7.36) as:

$$\Lambda = 133{\cdot}07 - 484{\cdot}8\frac{\sqrt{(\alpha c)}}{1 + 9{\cdot}378\sqrt{(\alpha c)}} \qquad \ldots.(14.5)$$

and the activity coefficient equation:

$$-\log f = \frac{4{\cdot}074\sqrt{(\alpha c)}}{1 + 9{\cdot}378\sqrt{(\alpha c)}} \qquad \ldots.(14.6)$$

where the figure 9·378 corresponds to an a value of 14·28 Å—the Bjerrum critical distance for a 2 : 2 electrolyte in water at 25°. The equations can be solved for α by successive approximations to give the results in *Table 14.5.* The values of K show a reasonable degree of constancy with an average of $4{\cdot}96 \times 10^{-3}$.

Of the seven measurements made by Deubner and Heise on cadmium sulphate at 18°, those at the four lowest concentrations agree with the predictions of the Onsager limiting law:

$$\Lambda = 113{\cdot}15 - 408{\cdot}1\sqrt{c}$$

The other three points were at very low concentrations so that we can use equation (14.5) without the $(1 + \kappa a)$ term to get 0·0066, 0·0051 and 0·0043 for the dissociation constant.

Table 14.5

Dissociation Constant of Magnesium Sulphate at 25°

$c \times 10^4$	$\Lambda_{obs.}$	$\Lambda_{calc.}$	α	$-2 \log f$	$K \times 10^3$
0·8098	127·31	129·07	0·9864	0·0672	4·96
1·6336	124·27	127·60	0·9739	0·0919	4·81
2·6924	121·34	126·30	0·9607	0·1138	4·87
4·297	117·85	124·86	0·9439	0·1380	4·97
6·006	114·92	123·70	0·9290	0·1575	5·08
8·380	111·61	122·43	0·9116	0·1791	5·21
0·8511	127·11	128·98	0·9855	0·0687	4·87
1·994	123·13	127·11	0·9687	0·1002	4·75
3·090	120·33	125·90	0·9558	0·1205	4·84
4·270	117·80	124·88	0·9433	0·1376	4·88
5·597	115·50	123·95	0·9318	0·1533	5·01
7·197	113·14	123·02	0·9197	0·1689	5·02
8·846	111·02	122·21	0·9084	0·1825	5·23

$K = 4{\cdot}96 \times 10^{-3}$

A new approach has been devised recently by Jones and Monk[20] with the cell:

$$H_2|HCl, MgSO_4|AgCl, Ag$$

which measures the quantity $\gamma_{H^+}\gamma_{Cl^-}m_{H^+}$ for a solution containing H^+, Cl^-, Mg^{++}, HSO_4^- and SO_4^{--} ions as well as undissociated $MgSO_4$. A series of successive approximations, along with a knowledge of the ionization constant of the HSO_4^- ion is sufficient to give the dissociation constant of magnesium sulphate. Jones and Monk determined this over the temperature range 20° to 35°, finding $K = 0{\cdot}0044$ at 25°. A similar method has been used[21] to study the equilibrium between magnesium ions and phosphate, glucose-1-phosphate or glycerol-1-phosphate ions, the dissociation constants at 25° being $1{\cdot}95 \times 10^{-3}$, $3{\cdot}31 \times 10^{-3}$ and $3{\cdot}25 \times 10^{-3}$ respectively; for calcium glucose-1-phosphate[21a] $K = 3{\cdot}20 \times 10^{-3}$ at 25°. These studies cover a wide temperature range and have much biological interest.

Mention should also be made of a spectrophotometric method[22] which takes advantage of the non-associated nature of bivalent metal perchlorates. Copper perchlorate has a characteristic absorption band in the ultra-violet, presumably due to the copper ion. If lithium sulphate (also, probably, a non-associated electrolyte) is added to solutions of copper perchlorate it is found that the extinction coefficient increases as the amount of lithium sulphate is increased. This is ascribed to formation of $CuSO_4$ molecules or ion-pairs. Measurements have also been made[23] in solutions of copper sulphate alone which lead to a dissociation constant of 0·0035 at 25° in agreement with 0·0039 calculated from conductivity data[24] at the same temperature and 0·0033 from cryoscopic experiments[25]: it is shown that much higher values of the dissociation constant result if a smaller distance of closest approach of the ions is assumed, the above values being calculated on the assumption that the correct a value to use in calculating osmotic and activity coefficients of the free ions is the Bjerrum critical distance of 14 Å.

Supporting evidence has been found recently in two different ways. The first is that, as we saw in Chapter 11, the diffusion coefficients of magnesium and zinc sulphate are explicable if it can be assumed that they form ion-pairs with dissociation constants of the order of 0·005. Another piece of evidence comes from measurements of the Wien effect. We omitted all mention of this when discussing conductance: the ONSAGER-WILSON theory[26] of this effect is complicated but, very briefly, the effect is concerned with the motion of ions under very high potentials such that the ions move so quickly that the 'ionic atmosphere' does not have time to build up completely or, at sufficiently high field strengths, the atmosphere does not build up at all. This leads to an increase of the ionic mobility. If the electrolyte is weak, another effect is superimposed: ONSAGER[27] has shown that at high field strength the ionization constant will be increased and he has obtained an equation relating this to the field strength. It is not easy to see why there should be an increase in the ionization constant but, putting it rather crudely, the absence of the 'ionic atmosphere' round the ion reduces the concentration of ions and, by a mass action effect, favours further ionization of molecules. Patterson *et al.* have improved the experimental methods of determining the Wien effect and they find that, for magnesium, zinc and copper sulphate and for lanthanum ferricyanide[28], the observed Wien effect is much larger than that predicted by the Onsager-Wilson theory. Taking reasonable values for the dissociation constants of these electrolytes at very low field strengths, determined by the methods already described, they use Onsager's equation to calculate the increased dissociation constants at high field strengths and hence the increased ionic concentrations which give the high conductances. With this correction good agreement with the Onsager-Wilson theory is found, thereby indirectly supporting the ion-pair theory.

Another method[29] which promises well is that of sound absorption; 2 : 2 electrolytes, in contrast to other valency types, have two maxima in their sound absorption spectra which can be definitely assigned to interaction between the cation and the anion.

ION-PAIR FORMATION WITH UNSYMMETRICAL ELECTROLYTES

The conductivity method should be applied to ion-pair formation with unsymmetrical electrolytes with considerable caution: we have seen in Chapter 7 that in deriving equation (7.36) we have dropped higher-order electrophoretic terms on grounds of self-consistency, although they are of appreciable magnitude. We are therefore less successful in predicting the conductivities of unsymmetrical electrolytes; *Table 7.6* shows that up to $c = 0{\cdot}005$ we can represent conductivities with an average deviation of 0·3 units for calcium chloride and 0·4 units for lanthanum chloride. Indeed, as the calculated values are higher than the observed, it might well be argued that even these salts are incompletely dissociated, provided we could satisfy ourselves that the difference did not arise from the unsatisfactory nature of the theory when applied to unsymmetrical electrolytes. JENKINS and MONK[30] made measurements on sodium sulphate at concentrations as low as $c = 6 \times 10^{-5}$; at the highest concentration, $c = 6 \times 10^{-4}$, the conductivity was 123·57 whilst the limiting law (equation 7.29) gives 123·85. The introduction of a $(1 + \kappa a)$ factor with $a = 4$ Å (equation 7.36) would raise the calculated conductivity to 124·14, 0·57 units higher than the observed—a very narrow margin when we are in doubt about the higher electrophoretic terms in the theoretical equation! These authors also made measurements on lanthanum sulphate. At their highest concentration, $c = 3 \times 10^{-4}$, the observed conductivity was 72·81, the limiting law gives 126·31 and a $(1 + \kappa a)$ factor with $a = 6$ Å would raise this to 129·07. In this case the difference between the observed conductivity and that calculated by the limiting law is substantial and one can accept ion-pair formation for this electrolyte with much more confidence. Jenkins and Monk calculated $K = 2{\cdot}4 \times 10^{-4}$ in good agreement with $2{\cdot}2 \times 10^{-4}$ found by DAVIES[31] from measurements of the solubility of lanthanum iodate in potassium sulphate solution[32]. To take one set of observations from this work, the solubility is $12{\cdot}153 \times 10^{-4}$ mole/l in 20×10^{-4} molar potassium sulphate, whereas the solubility in water is $8{\cdot}9006 \times 10^{-4}$ mole/l. The solubility product is then

$$3^3 f_0^4 (8{\cdot}9006 \times 10^{-4})^4 = 6{\cdot}06 \times 10^{-12}$$

if we calculate the activity coefficient by the Debye-Hückel limiting law (equation 9.10). The solubility product in the potassium sulphate solution is $11{\cdot}64 \times 10^{-12}$ so that it seems that too much lanthanum iodate has dissolved. It is now assumed that the true concentration of lanthanum ions is reduced by formation of $LaSO_4^+$ ions and, by successive approximations to get the total ionic strength at which f is to be calculated, it is found that the correct solubility product is obtained if the concentration of $LaSO_4^+$ ions is $7{\cdot}499 \times 10^{-4}$ mole/l. The law of mass action then gives $K = 2{\cdot}12 \times 10^{-4}$. It is worth while examining this calculation to see what is the nature of the approximation made by using equation (9.10). If we repeat the calculation using equation (14.4), modified for the 3 : 1 electrolyte, lanthanum iodate, we find the solubility product is $6{\cdot}60 \times 10^{-12}$, the concentration of $LaSO_4^+$ ions in the potassium sulphate solution is $7{\cdot}753 \times 10^{-4}$ mole/l and $K = 2{\cdot}15 \times 10^{-4}$. Thus we can derive a dissociation constant which is almost independent of the assumption we make about the equation for the activity coefficient and in good agreement with that deduced by an entirely different argument from conductivity data.

Table 14.6

Dissociation Constants of Electrolytes in Water at 25°

Cation	*Thiosulphate*	*Sulphate*	*Malonate* $\times 10^4$	*Oxalate* $\times 10^4$
H^+	0·035	0·012	0·02	0·52
Na^+	0·21	0·19	—	—
K^+	0·12	0·11	—	—
Mg^{++}	0·0145	0·0070	14·0	3·7
Ca^{++}	0·0104	0·0053	32·0	10·0
Sr^{++}	0·0092	—	—	29·0
Ba^{++}	0·0047	—	196·0	47·0
Mn^{++}	0·0112	0·0052	5·1	1·3
Co^{++}	0·0090	0·0034	1·9	0·20
Ni^{++}	0·0087	0·0040	0·99	0·05
Zn^{++}	0·0040	0·0049	2·1	0·13

(From Denney, T. O. and Monk, C. B., *Trans. Faraday Soc.*, 47 (1951) 992)

It is the derivation of dissociation constants of large magnitudes, *i.e.*, for largely dissociated electrolytes, from conductivity values which are not far from those predicted for a completely dissociated electrolyte, about which we have our doubts, especially if they are unsupported by data based on solubility measurements. Because of the unsatisfactory nature of the present theory of conductivity for unsymmetrical electrolytes, the solubility method seems more soundly based. It has been used extensively and in *Table 14.6* we give some dissociation constants derived mainly by this method, taken from a compilation by DENNEY and MONK[33].

SPECTROPHOTOMETRIC EVIDENCE FOR ION ASSOCIATION

Just as the ionization constants of some weak acids can be derived from measurements of the absorption spectra in the ultraviolet, so can the dissociation constant of an incompletely dissociated salt provided the two species involved absorb at different wavelengths. Mention has already been made of the use of the absorption spectra of copper sulphate solutions and a further example is that of the $PbCl^+$ ion which has maximum absorption at 2380 Å compared with 2080 Å for the Pb^{++} ion.

An example of one method of using such absorption data (the method of 'continuous variations') is as follows[34]: mixtures of lead perchlorate and potassium chloride each 0·0005 molar are made in various proportions with the total molarity constant, *i.e.*, the solutions are xc with respect to lead perchlorate and $(1 - x)c$ to potassium chloride. The optical density, D, is measured at a wavelength near to 2380 Å. Let α be the fraction of the lead which forms a complex, $PbCl_n$ with a charge $(2 - n)$. Then the concentrations of the various species are:

$$c_{Pb^{++}} = (1 - \alpha)xc$$

$$c_{PbCl_n} = \alpha xc$$

$$c_{Cl^-} = (1 - x)c - n\alpha xc$$

and the optical density is:

$$D = \varepsilon_{Pb^{++}}(1 - \alpha)xc + \varepsilon_{PbCl_n}\alpha xc$$

if we omit the length of the cell from this equation, *i.e.*, we assume a cell of unit length. Then

$$D - xc\varepsilon_{Pb^{++}} = \alpha xc(\varepsilon_{PbCl_n} - \varepsilon_{Pb^{++}})$$

The quantity on the left is the 'excess' optical density, the excess of the observed density over that calculated on the assumption that the density is due entirely to lead ions and that there is no interaction with chloride ions. $\varepsilon_{Pb^{++}}$ is obtained from a solution containing no chloride ions ($x = 1$). Usually, if the experiment is made at a wavelength characteristic of the complex, $xc\varepsilon_{Pb^{++}}$ will be much smaller than D and $(D - xc\varepsilon_{Pb^{++}})$ will have a maximum value (or

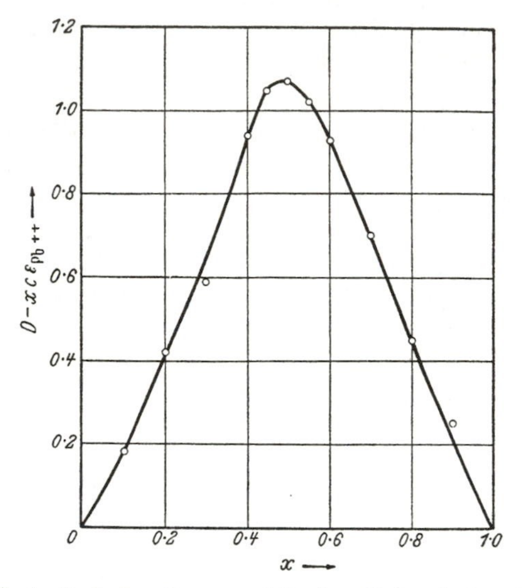

Figure 14.5. The 'method of continuous variations' applied to lead perchlorate–potassium chloride mixtures in 90 *per cent ethanol of total molarity* 0·0005

a minimum value if $\varepsilon_{Pb^{++}} > \varepsilon_{PbCl_n}$) when αx is a maximum. From the law of mass action:

$$K\alpha x = (1 - \alpha)x[(1 - x) - n\alpha x]^n c^n$$

neglecting the activity coefficients, whose variation with x should be small in dilute solution, and αx has a maximum value when

$$x = \frac{1}{n + 1}$$

Thus if the excess optical density $(D - xc\varepsilon_{Pb^{++}})$ is plotted against x, there should be a maximum in the curve at a value of x from which n can be calculated. *Figure 14.5* shows such a graph for lead perchlorate–potassiumchloride mixtures in 90 per cent ethanol which shows clearly that the maximum is at $x = 0{\cdot}5$ and therefore the complex has the formula $PbCl^+$.

This method gives us the composition of the complex but it does not tell us how stable it is. To show how the dissociation constant of a complex may be measured, we refer now to a paper by HERSHENSON, SMITH and HUME[35] on the $PbNO_3^+$ ion. The nitrate ion itself shows maximum absorption about 3000 Å and, whilst the wavelength of maximum absorption for the $PbNO_3^+$ ion has not yet been fixed exactly, it is known that this ion absorbs at 3000 Å. The composition of the complex was determined by the method of continuous variations. Additional measurements were then made on solutions, each of which contained 0·05 N sodium nitrate and lead perchlorate in amounts from 0·1 and 0·6 molar, each solution being brought to a total ionic strength of two by addition of sodium perchlorate. It was hoped that by keeping the total ionic strength constant, changes in the activity coefficients of the various species in these mixtures would be negligible. These solutions contain lead ions, $PbNO_3^+$ ions and nitrate ions each of which absorbs at 3000 Å although the contribution of the lead ions is only that of the foot of the peak at 2080 Å and enters as a minor correction. The optical density, using a cell of unit length, is therefore:

$$D = \varepsilon_{NO_3^-} c_{NO_3^-} + \varepsilon_{PbNO_3^+} c_{PbNO_3^+}$$

If c is the stoichiometric concentration of lead perchlorate, then the concentrations of the various species are:

$$c_{NO_3^-} + c_{PbNO_3^+} = 0{\cdot}05$$

$$c_{Pb^{++}} + c_{PbNO_3^+} = c \approx c_{Pb^{++}}$$

so that:

$$D - 0{\cdot}05\,\varepsilon_{NO_3^-} = (\varepsilon_{PbNO_3^+} - \varepsilon_{NO_3^-}) c_{PbNO}$$

By the law of mass action:

$$c_{PbNO_3^+} = \frac{0{\cdot}05c}{(c + K)}$$

so that:

$$\frac{0{\cdot}05c}{D - 0{\cdot}05\,\varepsilon_{NO_3^-}} = \frac{K}{\varepsilon_{PbNO_3^+} - \varepsilon_{NO_3^-}} + \frac{c}{\varepsilon_{PbNO_3^+} - \varepsilon_{NO_3^-}}$$

The quantity on the left-hand side plotted against the stoichiometric lead perchlorate concentration should give a straight line of slope $(\varepsilon_{PbNO_3^+} - \varepsilon_{NO_3^-})^{-1}$ and intercept $K(\varepsilon_{PbNO_3} - \varepsilon_{NO_3^-})^{-1}$ from which K and $\varepsilon_{PbNO_3^+}$ can be calculated, $\varepsilon_{NO_3^-}$ being found by measuring the solution which contains no lead perchlorate. In *Figure 14.6* a line is drawn through the points calculated from the results of Hershenson, Smith and Hume, using the absorption at 3000 Å. The slope and intercept of this line give $K = 0{\cdot}62$.

ION ASSOCIATION FROM PARTITION EXPERIMENTS

The distribution, or partition, of an electrolyte between two partially miscible liquids is an experiment which can give much information about the state of the electrolyte; it is, however, an experiment which has seldom been performed with the accuracy of the other techniques we have been considering. Of the few accurate measurements which have been made, those on the distribution of the sodium and potassium salts of guaiacol (*o*-methoxyphenol) between water and guaiacol deserve consideration[36]. At first sight, it may seem curious that such a system should be selected, but interest in it arose from the work of Osterhout on the transport of

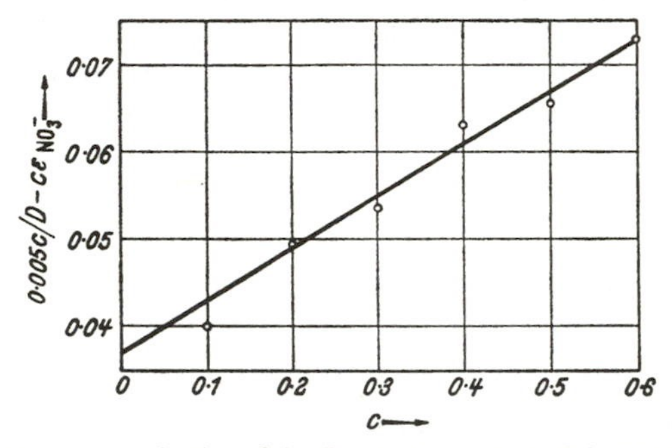

Figure 14.6. Determination of the dissociation constant of the $PbNO_3^+$ *ion*

electrolytes in living cells where a guaiacol–water system was used as a model of the protoplasm-cell sap equilibrium. These salts seem to be fully dissociated electrolytes when dissolved in water (at least up to the highest concentration employed, 0·14 N) whilst they are weak electrolytes, perhaps with ion-pair formation, in the guaiacol solvent, a conclusion which was reached from conductance measurements. The partition experiment itself is simple; the sodium or potassium salt is distributed between the two phases by rotating a 50 ml. glass tube containing the solutions for 15 h. in a thermostat at 25° and the concentrations are determined by titration with hydrochloric acid using a glass electrode differential titration apparatus. In speaking of the water phase or the guaiacol phase it should be understood that we mean the water-saturated-with-guaiacol phase or the guaiacol-saturated-with-water phase respectively. Three moles of guaiacol can dissolve one mole of water and, whilst no direct measurements seem to have been made of the solubility in the other phase, the dielectric constant of the saturated aqueous solution suggests that a mole of guaiacol can be taken up by about two hundred moles of water. From these experiments two quantities can now be derived—the dissociation constant of the salt in the guaiacol phase and the partition coefficient. In the guaiacol phase we have

$$K = \frac{\alpha^2 y^2 c}{1 - \alpha}$$

and for the distribution between the phases:

$$S_0 = \frac{\alpha yc}{y'c'}$$

where c and c' are the concentrations in the guaiacol and the water phases and y and y' are activity coefficients, which can be calculated from the Debye-Hückel equation assuming, on crystallographic evidence, that $a = 7$ Å.

The quantity measured is $S = \frac{c}{c'}$. It can easily be shown that:

$$\alpha = \frac{S_0}{S}\frac{y'}{y} \quad \text{and} \quad S(1 - \alpha) = \frac{S_0^2}{K}c'y'^2$$

A rough value of S_0 is got by plotting S against $c'y'^2$ and extrapolating to $c'y'^2 = 0$; then, by a series of approximations a value of S_0 is found which makes a plot of $S(1 - \alpha)$ against $c'y'^2$ linear and passing through the origin. The slope of this line is S_0^2/K. In this way Shedlovsky and Uhlig found $K = 5{\cdot}5 \times 10^{-5}$ for the potassium salt and $3{\cdot}5 \times 10^{-5}$ for the sodium salt.

SOME GENERAL REMARKS ON ION-PAIR FORMATION IN AQUEOUS SOLUTIONS

We have mentioned two proofs of Bjerrum's hypothesis, one depending on measurements of a high valency type salt in solvents of high dielectric constant and the other relying on experiments in extremely dilute solutions of low dielectric constant. These examples are very convincing, but it by no means follows that such demonstrations validate the case for ion-pair formation with salts of a simpler nature in solvents of high dielectric constant. There are two reasons why we think it would be well to be cautious. When lanthanum ferricyanide is dissolved in water it is believed that the ions cannot approach one another closer than 7·2 Å, but that those separated by distances between 7·2 and 32·1 Å are to be regarded as forming ion-pairs. It should not be forgotten that in this region there are a large number of water molecules. Assuming that the solvent molecules occupy the volume they do in the pure solvent

(30 cu. Å/molecule) there are nearly five thousand water molecules in the shell. There are also large numbers of solvent molecules round an ion in the solutions with which Kraus and Fuoss worked. For example, in a solvent of 4·01 per cent water and 95·99 per cent dioxan, the Bjerrum critical distance is much larger than it is in water. It is about 80 Å for a 1 : 1 electrolyte. We have seen that for one electrolyte 6·4 Å is a reasonable value to assume as the 'exclusion' distance within which other ions cannot penetrate: it is the shell between 6·4 and 80 Å which is concerned with ion-pair formation. In this shell there are about 17,000 solvent molecules. Of all the ions which penetrate into this shell and form, temporarily at least, ion-pairs, a few will get very close to the central ion. But in general the partners in the ion-pair will be held together by electrostatic forces operating through a large number of solvent molecules—large enough to justify us in considering that the solvent in this shell will have the properties, and particularly the dielectric constant, of the solvent in bulk. The critical distance for a 2 : 2 electrolyte in water is 14·28 Å and a sphere of this radius can hold about 400 water molecules of which only a few will be firmly attached to the cation. It is very different for a 1 : 1 electrolyte in water where the Bjerrum critical distance is only 3·57 Å and the *total* volume of the shell in which ion-pair formation is expected is only 190 cu. Å. This volume has to contain the two ions forming the ion-pair so that there is no room left for more than about four water molecules. Are we justified in using the bulk dielectric constant in a region where there are so few solvent molecules and even these must be subject to dielectric saturation?

The second consideration to be advanced concerns the nature of the anions of electrolytes where ion-pair formation is suspected. In very few examples is the anion a simple one. Thallous chloride could be quoted. A dissociation constant of about 0·3 has been calculated from conductivity measurements and about 0·2 from solubility measurements, and on Bjerrum's theory it would be necessary for the ions to approach within about 1 Å, whereas the sum of the crystallographic radii is 3·26 Å. There does not seem to be conclusive proof from Raman or absorption spectra that thallous chloride forms covalently bound molecules, although this does seem to be the explanation of this anomaly of the very close approach of the ions. In the case of lead chloride very convincing evidence for the covalent nature of the bonding of the $PbCl^+$ intermediate ion is provided by ultra-violet absorption spectra[37], which give a dissociation constant of the order of 0·03. Lead nitrate[35] also gives evidence of $PbNO_3^+$ formation, and this is one of the few examples known where covalent bond formation is suspected with the nitrate ion. To return to thallous chloride, the point we want to emphasize is that this 'simple' electrolyte is a dangerous example to quote in favour of Bjerrum's theory; as Bell and George have pointed out, it is difficult to reconcile the behaviour of this salt, if it is to be explained by ion-pair formation, with the crystallographic radii: covalent bond formation is an attractive alternative but not yet a proved fact.

All other cases where ion-pair formation is suspected with 1 : 1 electrolytes in water concern polyatomic anions–nitrates, chlorates, perchlorates, bromates, $H_2PO_4^{--}$ anions. Whilst it may be possible that the planar configuration of the nitrate ion allows a cation to approach in one direction within distances less than 3·57 Å, it is difficult to see how this can happen with the bulky tetrahedral perchlorate ion. Moreover, these suspected examples of ion-pair formation are all found with cations which are either unsolvated or, at the most, contain only a few solvated molecules. The heavily hydrated lithium ion does not associate with anions, whilst the perchlorates of the hydrated bivalent metal cations seem to be non-associated electrolytes (the heavily hydrated calcium ion behaves differently in $CaOH^+$ formation where the hydroxyl radical seems to replace a water molecule). In general, the cations which enter into ion-pair formation are those whose electrical forces are not satisfied by hydration and are therefore free to produce polarization in oxyacid anions. Some aqueous 1 : 1 electrolytes are peculiarly liable to this polarization effect. The conditions for it are a cation with little or no hydration, an anion with an inherently polarizable structure and the possibility of approach within distances much less than is found with hydrated cations. So many examples of so-called ion-pair formation satisfy these conditions that one must ask if the picture of ion-pairs is not too simple? To treat them as ion-pairs may be a first approximation, but we suggest that it would be closer to the truth to say they are examples of interaction between the cation and an induced dipole in the anion.

THE HYPOTHESIS OF 'LOCALIZED' HYDROLYSIS

An inspection of the activity coefficient data in Appendix 8.10 shows that at a given concentration the values for most electrolytes of the alkali metal family are in the order:

$$Li > Na > K > Rb > Cs$$

This is true for the chlorides, bromides, iodides, nitrates, chlorates, perchlorates, *etc.*, and is consistent with the increasing hydration of the cation from caesium to lithium. But the reverse is true for the *hydroxides* of the alkali metals, where we find the order:

$$\mathrm{Cs} > \mathrm{K} > \mathrm{Na} > \mathrm{Li}$$

and, as we have said earlier in this chapter, this is not explicable by ion-pair formation: the low activity coefficients found when ion-pairs are formed, are now found with lithium hydroxide and its conductivity is so low that a dissociation constant of 1·2 has been ascribed to it[37a] although the cation is hydrated and too large to permit ion association. To account for this ROBINSON and HARNED[38] introduced the idea of 'localized hydrolysis'. In the hydration shell around a cation the water molecules must be highly polarized with the positive charges directed away from the cation:

$$\overset{+}{\mathrm{Na}} \text{----} \overset{-}{\mathrm{OH}} \text{----} \overset{+}{\mathrm{H}}$$

It is possible that the bound 'hydrogen ion' can exert sufficient force on a comparatively small ion like the hydroxyl to lead to a short-life binding:

$$\overset{+}{\mathrm{Na}} \text{----} \overset{-}{\mathrm{OH}} \text{----} \overset{+}{\mathrm{H}} \text{----} \overset{-}{\mathrm{OH}}$$

with the formation of a kind of ion-pair but differing in that the water molecule acts as intermediary. The smaller the cation the more polarized will be the solvent molecules so that the effect would decrease from lithium to caesium. As such interaction would lead to reduced activity coefficients, it would explain why lithium hydroxide has a low and caesium hydroxide a high activity coefficient. It would also explain the observation of BELL and PRUE[14] that the catalytic effect of sodium hydroxide on the decomposition of diacetone alcohol is low compared with that of potassium or rubidium hydroxide. The catalytic effect is even less with lithium hydroxide[39].

This effect should operate not only for hydroxyl ions but for any anion which is a proton acceptor. We do indeed find that this reversal of order with $\mathrm{Li} < \mathrm{Na} < \mathrm{K} < \mathrm{Rb} < \mathrm{Cs}$ holds for the formates and acetates and perhaps for the fluorides. It holds also[40] for the osmotic coefficients of magnesium and barium acetates up to 1 M, the magnesium ion being more heavily hydrated and therefore more disposed to this localized hydrolysis. Above 1 M the osmotic coefficient of the barium salt is less than that of the magnesium salt; this is due perhaps to the greater tendency of barium ions to enter into Bjerrum's type of ion-pair formation. This is a good example of the complex behaviour which is met in solutions when we study any concentration region except the dilute. The reversal of order is also found with mixtures; the activity coefficient of the hydrogen and acetate ions of acetic acid is less in a solution of sodium chloride than it is in one of potassium chloride although no such reversal of order is found for the activity coefficient of hydrochloric acid in these salt solutions. The reversal is also found for the ionic activity coefficient of water itself, $\gamma_{H^+}\gamma_{OH^-}$, in these salt solutions. In brief, it seems to happen whenever the cation is small and hydrated and there is an anion which is a proton acceptor.

It has been suggested recently[49] that the hypothesis can be extended to the alkali halides, the effect being comparatively large with the lithium halides and negligible with the caesium halides; thus the order of activity coefficients, $Cl^- > Br^- > I^-$, observed with the rubidium and caesium halides is taken as the norm in the absence of 'localized hydrolysis' and the reversal of the order in the lithium, sodium and potassium halides is ascribed to such hydrolysis; the hydrated cations should be more effective in this respect but for a given cation the degree of hydrolysis should be greater with the smaller anions.

COMPLEX IONS

It is not our intention to discuss in any detail the subject of complex ions[41] but mention should be made of some of the complications introduced by the less stable complex ions. These are found among the halides of the transition metals, particularly the halides of cadmium and zinc, where ion-pair formation is followed by further association into neutral molecules and negatively charged anions until the coordination shell is completed. Bates[42] has discussed the interpretation of electromotive force measurements on solutions of cadmium iodide to which either cadmium sulphate or potassium iodide is added and he has obtained values for the stability constants of the CdI^+, CdI_2 and CdI_3^- species. *Figure 14.7* illustrates his conclusions. Up to about 0·005 M, most of the cadmium is present as Cd^{++} ions although substantial proportions of the CdI^+ ion are found even at 0·001 M. With increasing concentration the proportion of CdI^+ ions increases to a maximum of 45 per cent at 0·01 M and then decreases, the cadmium iodide molecule now becoming important and accounting for 46 per cent of the cadmium at 0·5 M. The complex ion, CdI_3^-, is formed to a lesser extent although at 0·5 M it is present to the extent of 24 per cent. It is possible that further association occurs to CdI_4^{--} ions in more concentrated solutions. This behaviour is typical of the halides of zinc and cadmium although the ease of formation of complex ions is in the order:

$$ZnCl_2 > ZnBr_2 > ZnI_2$$

for the zinc salts and in the reverse order for the cadmium salts:

$$CdI_2 > CdBr_2 > CdCl_2$$

This is also the order of the activity coefficient curves. Indeed, the tendency to complex ion formation with zinc iodide, in contrast to cadmium iodide, is so small that up to about 0·3 M it behaves as a typical non-associated electrolyte, but complex ions do occur at high concentrations as shown by its negative transport number above 3·5 M (see *Table 7.9*). That zinc bromide forms complex ions more readily and that this is even easier with zinc chloride is

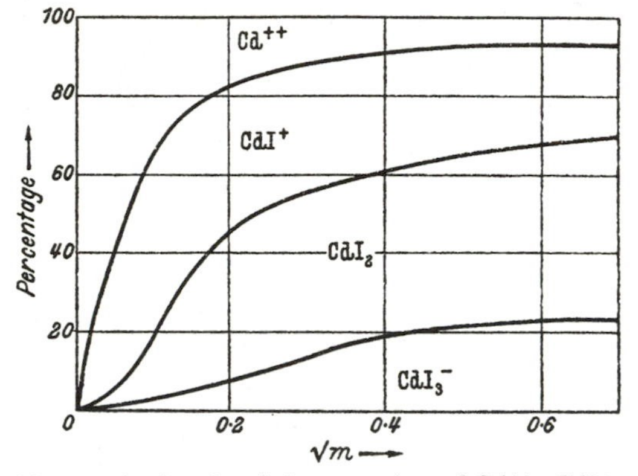

Figure 14.7. Diagram showing the relative proportions of Cd^{++}, CdI^+ *and* CdI_3^- *ions and* CdI_2 *molecules in cadmium iodide solutions up to* 0·5 M

shown not only by the order of the activity coefficient curves, but also by the occurrence of negative transport numbers at about 2·8 M with zinc bromide and at 2 M with zinc chloride.

Evidence for the formula of the complex ions of zinc halides in concentrated solution has come from vapour pressure measurements[43]. Mixtures of a magnesium halide and the corresponding zinc halide of constant total molality were made with different Mg : Zn ratios and the vapour pressures measured with the results shown in *Figure 14.8*. The magnesium halides are non-associated electrolytes and give large vapour pressure lowerings. Moreover, zinc perchlorate has activity coefficients close to those of magnesium perchlorate, so that the hydrated zinc and magnesium ions must be about the same size. It would be expected, therefore, that if no complex ions were formed with a zinc halide, its vapour pressure lowering would be about the same as that of the magnesium halide and the graphs in *Figure 14.8* would be almost horizontal straight lines. Instead, a sharp decrease of the vapour pressure lowering occurs until the zinc and magnesium ions are present in equal amounts and further replacement of magnesium by zinc causes very little change. These results are consistent with ZnX_4^{--} ion formation. For example, at a total molality of 4, the vapour pressure

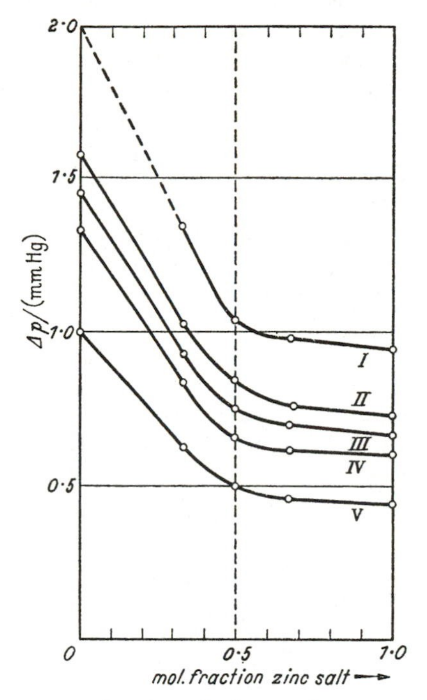

Figure 14.8. Vapour pressure lowerings of mixtures of zinc and magnesium halides at constant total molality

I = $ZnCl_2$—$MgCl_2$ at M = 7.
II = ZnI_2—MgI_2 at M = 5.
III = $ZnBr_2$—$MgBr_2$ at M = 5.
IV = $ZnCl_2$—$MgCl_2$ at M = 5.
V = $ZnCl_2$—$MgCl_2$ at M = 4.

(*From* STOKES, R. H., *Trans. Faraday Soc.*, 44 (1948) 137)

lowering due to 4 M $MgCl_2$ is that due to 4 M—Mg^{++} and 8 M—Cl^-. With magnesium and zinc present in equal amounts, the effect is that due to 2 M—Mg^{++} and 2 M—$ZnCl_4^{--}$, an effect which should be much smaller; with zinc chloride alone the vapour pressure lowering is ascribed to 2 M—Zn^{++} and 2 M—$ZnCl_4^{--}$, so that there should be little difference after the Zn : Mg ratio exceeds

1·0. As *Figure 14.8* shows, this is exactly what is observed with all three zinc halides.

Further evidence for the $ZnCl_4^{--}$ ion has been found from an x-ray study[44] of the complex salt, $(NH_4)_3ZnCl_5$, in the solid state where the $ZnCl_4^{--}$ ions form one unit of the lattice.

Cupric chloride readily forms the $CuCl^+$ ion as spectrophotometric evidence shows[45] with additional evidence that $CuCl_2$ molecules and $CuCl_3^-$ and even $CuCl_4^{--}$ ions also exist[46].

Scandinavian workers have been particularly active in studies aimed at the elucidation of the compositions and stabilities of complex ions in solution. The work of Bjerrum[47] on the ammonia and ethylenediamine complexes of a large number of metal ions and the studies by Sillén[48] and his associates of the complexes of zinc and mercury ions with anions are of great value. Sillén's technique is essentially one of potentiometric titration using a cell with liquid junction. The solution being titrated is kept at a relatively high and constant total ionic strength by the presence of large amounts of electrolytes such as perchloric acid or sodium perchlorate; the ions whose interactions are being studied, *e.g.*, the Zn^{++} and Cl^- ions, are at relatively much lower concentrations. In this way the uncertain effects of the variation of ionic activity coefficients with concentration are eliminated, and the changes in the *concentration* of the Zn^{++} ion as the chloride content of the solution is varied can be followed by measuring the potential of a zinc amalgam electrode relative to a reference calomel electrode. Sillén and Liljeqvist thus arrived at the following estimates of the molar scale constants of various stages of the complex formation between zinc and halide ions in 3 N sodium perchlorate solution at 25°:

$$Zn^{++} + Cl^- \rightleftharpoons ZnCl^+, \qquad K_1 = 0{\cdot}65$$

$$Zn^{++} + 2Cl^- \rightleftharpoons ZnCl_2, \qquad K_2 = 0{\cdot}25$$

$$Zn^{++} + 3Cl^- \rightleftharpoons ZnCl_3^-, \qquad K_3 = 1{\cdot}4$$

$$Zn^{++} + Br^- \rightleftharpoons ZnBr^+, \qquad K_1 = 0{\cdot}25$$

(further stages also occur)

$$Zn^{++} + I^- : \text{for all stages, } K < 0{\cdot}05.$$

This work reinforces the conclusion that the complex formation between zinc and halide ions is least in the iodide and greatest in the chloride, as indicated by the other considerations discussed above. The experiments were not, however, made at high enough concentrations of the zinc and halide ions to reveal the stage

$Zn^{++} + 4X^- \rightleftharpoons ZnX_4^{--}$, for which there is a good deal of evidence from the vapour pressure data for mixed solutions.

Similar work on the mercuric complexes has led to an evaluation of the much larger equilibrium constants for mercuric–halide ion interactions, as follows:

	X =	Cl	Br	I
$Hg^{++} + X^- \rightleftharpoons HgX^+$,	$\log K_1 =$	6·74	9·05	12·87
$Hg^{++} + 2X^- \rightleftharpoons HgX_2$,	$\log K_2 =$	13·22	17·33	23·82
$Hg^{++} + 3X^- \rightleftharpoons Hg\,X_3^-$,	$\log K_3 =$	14·07	19·74	27·60
$Hg^{++} + 4X^- \rightleftharpoons HgX_4^{--}$,	$\log K_4 =$	15·07	21·10	29·83

With mercury, as with cadmium, the iodide complexes are the most stable and the chlorides the least stable, though all are of very high stability compared with those of zinc and cadmium.

REFERENCES

[1] BJERRUM, N., *K. danske vidensk. Selsk.*, 7 (1926) No. 9; 'Selected Papers, p. 108, Einar Munksgaard, Copenhagen (1949)

[2] FUOSS, R. M. and KRAUS, C. A., *J. Amer. chem. Soc.*, 55 (1933) 1019

[2a] FUOSS, R. M., *ibid.*, 80 (1958) 5059

[3] DAVIES, C. W. and JAMES, J. C., *Proc. roy. Soc.*, 195 A (1948) 116; JAMES, J. C., *J. chem. Soc.*, (1950) 1094

[4] DUNSMORE, H. S. and JAMES, J. C., *ibid.*, (1951) 2925

[5] KRAUS, C. A. and FUOSS, R. M., *J. Amer. chem. Soc.*, 55 (1933) 21

[5a] JONES, M. M. and GRISWOLD, E., *ibid.*, 76 (1954) 3247: acetic acid; HNIZDA, V. F. and KRAUS, C. A., *ibid.*, 71 (1949) 1565: liquid ammonia; EL-AGGAN, A. M., BRADLEY, D. C. and WARDLAW, W., *J. chem. Soc.*, (1958) 2092: ethanol; GOVER, T. A. and SEARS, P. G., *J. phys. Chem.*, 60 (1956) 330: *n*-propanol; HIBBARD, B. B. and SCHMIDT, F. C., *J. Amer. chem. Soc.*, 77 (1955) 225: ethylenediamine; REYNOLDS, M. B. and KRAUS, C. A., *ibid.*, 70 (1948) 1709: acetone; BURGESS, D. S. and KRAUS, C. A., *ibid.*, 70 (1948) 706: pyridine

[6] FUOSS, R. M. and KRAUS, C. A., *ibid.*, 55 (1933) 2387

[6a] HUGHES, E. D., INGOLD, C. K., PATAI, S. and POCKER, Y., *J. chem. Soc.*, (1957) 1206

[7] BATSON, F. M. and KRAUS, C. A., *J. Amer. chem. Soc.*, 56 (1934) 2017

[8] FUOSS, R. M., *ibid.*, 56 (1934) 1027

[9] — and KRAUS, C. A., *ibid.*, 57 (1935) 1

[10] DAVIES, C. W., *Trans. Faraday Soc.*, 23 (1927) 351; ROBINSON, R. A. and DAVIES, C. W., *J. chem. Soc.* (1937) 574

[11] SHEDLOVSKY, T., *J. Amer. chem. Soc.*, 54 (1932) 1411

[12] GARRETT, A. B. and VELLENGA, S. J., *ibid.*, 67 (1945) 225

[13] BRAY, W. C. and WINNINGHOF, W. J., *ibid.*, 33 (1911) 1663

[14] BELL, R. P. and PRUE, J. E., *J. chem. Soc.* (1949) 362

[15] — and WAIND, G. M., *ibid.* (1950) 1979

[16] DAVIES, C. W., *ibid.* (1930) 2410

[17] BELL, R. P. and GEORGE, J. H. B., *Trans. Faraday Soc.*, 49 (1953) 619

[18] DAVIES, C. W. and HOYLE, B. E., *J. chem. Soc.*, (1951) 233; STOCK, D. I. and DAVIES, C. W., *Trans. Faraday Soc.*, 44 (1948) 856; COLMAN-PORTER, C. A. and MONK, C. B., *J. chem. Soc.* (1952) 1312

[19] DEUBNER, A. and HEISE, E., *Ann. Phys., Lpz.*, 9 (1951) 213

[20] JONES, H. W. and MONK, C. B., *Trans. Faraday Soc.*, 48 (1952) 929

[21] CLARKE, H. B., CUSWORTH, D. C. and DATTA, S. P., *Biochem. J.*, 58 (1954) 146

[21a] — and DATTA, S. P., *Biochem. J.*, 64 (1956) 604

[22] NÄSÄNEN, R., *Acta chem. scand.*, 3 (1949) 179

[23] DAVIES, W. G., OTTER, R. J. and PRUE, J. E., *Disc. Faraday Soc.*, 24 (1957) 103

[24] OWEN, B. B. and GURRY, R. W., *J. Amer. chem. Soc.*, 60 (1938) 3074

[25] BROWN, P. G. M. and PRUE, J. E., *Proc. Roy. Soc.*, 232A (1955) 320

[26] WILSON, W. S., *Thesis*, Yale University (1936): this thesis does not appear to have been published in a journal but a full account of this and of the paper referred to in (27) is given by HARNED, H. S. and OWEN, B. B., 'Physical Chemistry of Electrolytic Solutions,' p. 128 *et seq.*, Reinhold Publishing Corp. (1958)

[27] ONSAGER, L., *J. chem. Phys.*, 2 (1934) 599

[28] GLEDHILL, J. A. and PATTERSON, A., *J. phys. Chem.*, 56 (1952) 999; BAILEY, F. E. and PATTERSON, A., *J. Amer. chem. Soc.*, 74 (1952) 4428; BERG, D. and PATTERSON, A., *ibid.*, 74 (1952) 4704; *ibid.*, 75 (1953) 1484

[29] EIGEN, M., *Disc. Faraday Soc.*, 24 (1957) 25

[30] JENKINS, I. L. and MONK, C. B., *J. Amer. chem. Soc.*, 72 (1950) 2695

[31] DAVIES, C. W., *J. chem. Soc.*, (1930) 2421

[32] LAMER, V. K. and GOLDMAN, F. H., *J. Amer. chem. Soc.*, 51 (1929) 2632

[33] DENNEY, T. O. and MONK, C. B., *Trans. Faraday Soc.*, 47 (1951) 992

[34] JOB, P., *Ann. chim.*, 9 (1928) 113; VOSBURGH, W. C. and COOPER, G. R., *J. Amer. chem. Soc.*, 63 (1941) 437; FOLEY, R. T. and ANDERSON, R. C., *ibid.*, 71 (1949) 909

[35] HERSHENSON, H. M., SMITH, M. E. and HUME, D. N., *ibid.*, 75 (1953) 507

[36] SHEDLOVSKY, T. and UHLIG, H. H., *J. gen. Physiol.*, 17 (1934) 549, 563

[37] FROMHERZ, H. and LIH, K. H., *Z. phys. Chem.*, 153 A (1931) 321; BIGGS, A. I., PANCKHURST, M. H. and PARTON, H. N., *Trans. Faraday Soc.*, 51 (1955) 802; PANCKHURST, M. H. and PARTON, H. N., *ibid.*, 51 (1955) 806; BIGGS, A. I., PARTON, H. N. and ROBINSON, R. A., *J. Amer. chem. Soc.*, 77 (1955) 5844

[37a] DARKEN, L. S. and MEIER, H. F., *ibid.*, 64 (1942) 621

[38] ROBINSON, R. A. and HARNED, H. S., *Chem. Rev.*, 28 (1941) 419

[39] ÅKERLÖF, G., *J. Amer. chem. Soc.*, 49 (1927) 2955

[40] STOKES, R. H., *ibid.*, 75 (1953) 3856

[41] See, for example, MARTELL, A. E. and CALVIN, M., 'Chemistry of the metal chelate compounds,' Prentice-Hall, Inc., New York (1952)

[42] BATES, R. G. and VOSBURGH, W. C., *J. Amer. chem. Soc.*, 60 (1938) 137

[43] STOKES, R. H., *Trans. Faraday Soc.*, 44 (1948) 137

[44] KLUG, H. P. and ALEXANDER, L., *J. Amer. chem. Soc.*, 66 (1944) 1056

[45] BROWN, J. B., *Proc. roy. Soc.*, New Zealand, 77 (1948) 19; NÄSÄNEN, R., *Acta chem. scand.*, 4 (1950) 140; MCCONNELL, H. and DAVIDSON, N.. *J. Amer. chem. Soc.*, 72 (1950) 3164

REFERENCES

BJERRUM, J., *K. danske vidensk. Selsk.*, 22 (1946) No. 18

— 'Metal Ammine Formation in Aqueous Solutions,' P. Haase and Son, Copenhagen (1941); BJERRUM, J. and NIELSEN, E. J., *Acta chem. scand.*, 2 (1948) 297 and earlier references there cited

SILLÉN, L. G. and LILJEQVIST, B., *Svensk kem. Tidskr.*, 56 (1944) 85; SILLÉN, L. G., *Acta chem. scand.*, 3 (1949) 539 and earlier references there cited

DIAMOND, R. M., *J. Amer. chem. Soc.*, 80 (1958) 4808

15

THE THERMODYNAMICS OF MIXED ELECTROLYTES

A STUDY of the chemical potentials and transport properties o single electrolyte solutions is very important in providing a criterion for judging theoretical predictions. The interactions prevailing in an electrolyte solution are, as we have seen in earlier chapters, of several kinds, and theory provides us with only a partial explanation of them. It is not surprising therefore that mixtures of electrolytes present even more difficult problems to solve. Mixed electrolytes are, however, very important; they are found in numerous processes in chemical industry, they occur in enormous quantities in the water of the oceans and have an important role in the physiological processes of body fluids and cell equilibria. It is also likely that ion exchange resins can be treated as mixed electrolytes. A start has been made with the study of conductance and diffusion processes in mixtures of electrolytes, but it is the thermodynamics of these mixtures which has been studied in the greater detail.

We commence with the system hydrochloric acid–sodium chloride: the top curve of *Figure 15.1* represents the activity coefficient of hydrochloric acid in aqueous solution at 25° in the absence of any other solute. We have already seen, in Chapter 9, that the shape of the curve, the minimum at $\gamma = 0{\cdot}755$ when $m = 0{\cdot}4$ and the rapid increase of the activity coefficient at high concentrations, can be accounted for by postulating that the ions have a mean diameter of 4·47 Å and that the cation is associated with an average of 8 water molecules. The lowest curve of *Figure 15.1* represents the activity coefficient of sodium chloride as a single electrolyte in aqueous solution at 25°; there is a minimum at $\gamma = 0{\cdot}654$, $m = 1{\cdot}2$; a mean ionic diameter of 3·97 Å and a hydration number of 3·5 suffice to represent the activity coefficient up to high concentrations. We could, however, measure the activity coefficient of hydrochloric acid in the presence of sodium chloride: we could make a solution containing two parts of the acid to one part of salt and study the variation of the activity coefficient of the acid as the total concentration is changed: if we were to make such measurements, we would get a curve like the one in *Figure 15.1* which is marked $x_{HCl} = 0{\cdot}667$. In general the shape of the curve is similar to that for hydrochloric acid by itself, but the curve is somewhat lower: thus 3 M—HCl has an activity coefficient of 1·316 which is reduced to 1·225 for a solution of 2 M—HCl + 1 M—NaCl. It is more difficult to determine the activity coefficient of sodium chloride in such mixtures, but it has been done, and the

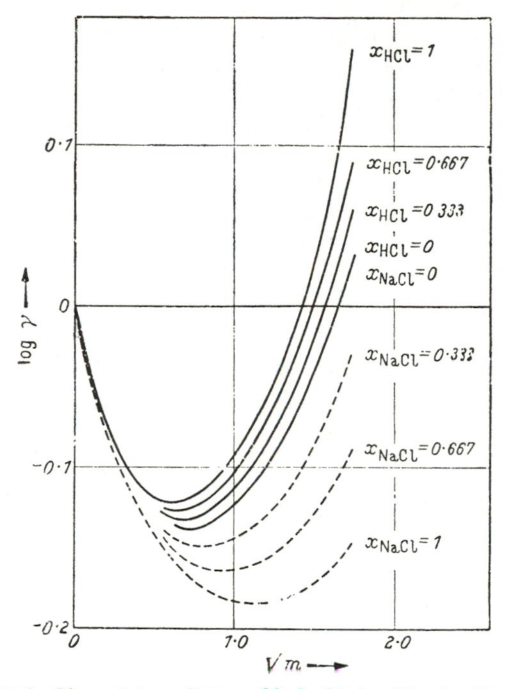

Figure 15.1. The activity coefficients of hydrochloric acid and sodium chloride in mixed electrolyte solution

curve marked $x_{NaCl} = 0{\cdot}667$ shows that the activity coefficient of the salt is raised by the substitution of hydrochloric acid for some of the sodium chloride in the solution. Thus the activity coefficient of sodium chloride as a single electrolyte at 3 M is 0·714 but this becomes 0·816 in a solution of 2 M—NaCl + 1 M—HCl. This behaviour is typical of these mixtures: as the proportion of sodium chloride is increased the activity coefficient of hydrochloric acid decreases so that the curve for $x_{HCl} = 0{\cdot}333$ is even lower than that for $x_{HCl} = 0{\cdot}667$. But starting with solutions containing only sodium chloride and then increasing the proportion of hydrochloric acid in the mixture, the activity coefficient of sodium chloride increases so that the curve for $x_{NaCl} = 0{\cdot}333$ lies above that for $x_{NaCl} = 0{\cdot}667$. It is possible by a method of extrapolation to determine the activity coefficient of hydrochloric acid present in vanishingly small amounts in a solution in which the electrolyte is virtually all sodium chloride. Likewise the activity coefficient of sodium chloride can be determined for the limiting case of zero salt concentration in a solution of hydrochloric acid. If this is done, the curious result emerges that the activity coefficient of hydrochloric acid in a solution containing virtually nothing but sodium chloride is almost identical with that of sodium chloride in a solution containing nothing but hydrochloric acid. These limiting cases, of great theoretical interest, have to be represented by a single curve in *Figure 15.1*, marked $x_{HCl} = 0$, $x_{NaCl} = 0$, because it would require a large scale graph to show the difference between them.

We will represent the activity coefficient of an electrolyte B in a solution containing only this electrolyte by $\gamma_{B(0)}$ and the activity coefficient of B in the limiting case where the electrolyte has been replaced entirely by a second electrolyte C, by $\gamma_{(0)B}$. (These correspond to $\gamma_{1(0)}$ and $\gamma_{(0)1}$ in the literature on the subject; we have made the change because we have used γ_1 to denote the activity coefficient of the ion 1.) Thus the central curve of *Figure 15.1* represents both $\gamma_{(0)HCl}$ and $\gamma_{(0)NaCl}$ whilst the top and bottom curves represent $\gamma_{HCl(0)}$ and $\gamma_{NaCl(0)}$ respectively. The following table shows how close $\gamma_{(0)HCl}$ and $\gamma_{(0)NaCl}$ are.

m	$\gamma_{HCl(0)}$	$\gamma_{(0)HCl}$	$\gamma_{(0)NaCl}$	$\gamma_{NaCl(0)}$	$\gamma_{mean} = \surd[\gamma_{HCl(0)}\,\gamma_{NaCl(0)}]$
0·5	0·757	0·726	0·727	0·681	0·718
1·0	0·809	0·752	0·751	0·657	0·729
2·0	1·009	0·875	0·873	0·668	0·821
3·0	1·316	1·063	1·066	0·714	0·969

It also shows that they are very different from the activity coefficient of either the acid or the salt as a single electrolyte. Moreover the last column, which represents a mean of $\gamma_{HCl(0)}$ and $\gamma_{NaCl(0)}$, $\log \gamma_{mean} = \frac{1}{2}[\log \gamma_{HCl(0)} + \log \gamma_{NaCl(0)}]$, shows that $\gamma_{0(HCl)}$ and $\gamma_{0(NaCl)}$ are closer to $\gamma_{HCl(0)}$ than to $\gamma_{NaCl(0)}$.

This description of the hydrochloric acid–sodium chloride system is typical of mixed electrolyte solutions except that the near identity of $\gamma_{(0)HCl}$ and $\gamma_{(0)NaCl}$ is not found to be quite as close in other cases. Thus in the system hydrochloric acid–potassium chloride at a total concentration of 3 M, $\gamma_{(0)HCl} = 0{\cdot}858$, $\gamma_{(0)KCl} = 0{\cdot}845$, and in the hydrochloric acid–caesium chloride system at the same concentration $\gamma_{(0)HCl} = 0{\cdot}669$ and $\gamma_{(0)CsCl} = 0{\cdot}634$.

Before we commence a study of the methods available for

measuring these activity coefficients, we may consider some theoretical implications of the Debye-Hückel theory. In extremely dilute aqueous solution where the effect of the ionic diameter is negligible, the limiting law (9.10) is applicable and consequently for a mixture of electrolytes of the same valency type a difference between $\gamma_{(0)B}$ and $\gamma_{(0)C}$ at the same total molality can originate only in a difference between the ionic strengths as measured on the molarity and the molality scales, and this difference must be negligible in such dilute solutions. In solutions not so dilute, equation (9.7) should be applicable, but we must be careful what meaning we give to a. Taking as an example a mixture of hydrochloric acid and sodium chloride in the molal ratio $x : (1 - x)$, it is tempting to speculate that the activity coefficient γ_{HCl} can be compounded of two terms, one for the hydrogen ions with an a_{H-Cl} dependent mainly on the sizes of the hydrogen and chloride ions, because encounters between oppositely charged ions are more frequent. The other parameter, for the chloride ion, would be more complicated because we would need something like $[xa_{H-Cl} + (1 - x)a_{Na-Cl}]$ to account for the interactions of the chloride ion with each of the cations. The latter quantity might also be used for the contribution of the chloride ion to γ_{NaCl} but in addition we would need the a_{Na-Cl} term to account for the sodium ion contribution. In any such speculation it should be remembered that, as a simple consequence of the chemical potential being a partial differential coefficient of the total free energy with respect to a concentration, it is necessary that, for a mixture of two 1 : 1 electrolytes[1]: (see p. 441)

$$\left(\frac{\partial \ln \gamma_B}{\partial m_C}\right)_{m_B} = \left(\frac{\partial \ln \gamma_C}{\partial m_B}\right)_{m_C} \qquad \ldots.(15.1)$$

It is therefore difficult to satisfy both this equation and equation (9.7) unless a is almost the same for each electrolyte in the mixture; the only permissible difference in the a values for each of the components would enter because the concentrations in (9.7) are in volume units and those in (15.1) are in molalities, but this would not allow for any great difference in a. An extended equation of the form of (9.11) gives us greater freedom to vary a, but there will still be some restrictions on the a and b values.

GUGGENHEIM'S TREATMENT OF MIXED ELECTROLYTE SOLUTIONS

Starting with equation (9.13), the a parameter being the same for all electrolytes and the term linear in the concentration accounting for the specific interionic effects, GUGGENHEIM[2] has built up a theory of mixed electrolyte solutions consistent with equation (15.1). Following the BRÖNSTED[3] Principle of the Specific Interaction of Ions: 'in a dilute salt solution of constant total concentration, ions will be uniformly influenced by ions of their own sign and specific effects are to be sought in interactions between oppositely charged ions,' Guggenheim introduces specific interaction coefficients for the activity coefficient of an electrolyte, B, in the presence of another electrolyte, C:

$$\ln \gamma_B = -\frac{\alpha\sqrt{I}}{1+\sqrt{I}} + [2xb_{\mathrm{M^+X^-}} + (b_{\mathrm{N^+X^-}} + b_{\mathrm{M^+Y^-}})(1-x)]m$$

Here $\mathrm{M^+}$, $\mathrm{X^-}$ are the ions of B, and $\mathrm{N^+}$, $\mathrm{Y^-}$ the ions of C; for simplicity we consider 1 : 1 electrolytes in aqueous solution at 25°. m is the total molality and xm and $(1-x)m$ are the molalities of B and C respectively. We deviate slightly from Guggenheim's treatment in using activity coefficients and concentrations on the molality scale. It will be noted that in this equation there is no b term for the interaction of an ion with another of its own sign; this accords with Brönsted's Principle. For the other electrolyte we have:

$$\ln \gamma_C = -\frac{\alpha\sqrt{I}}{1+\sqrt{I}} + [2(1-x)b_{\mathrm{N^+Y^-}} + x(b_{\mathrm{N^+X^-}} + b_{\mathrm{M^+Y^-}})]m$$

and it is easy to show that these two equations for the activity coefficients γ_B and γ_C are consistent with equation (15.1). Moreover, the first term on the right-hand side of these equations is not subject to variation with a change in x so we let γ' and ϕ' refer to contributions resulting from the second term.

When $x = 0$ we get:

$$\ln \gamma'_{(0)B} = (b_{\mathrm{M^+Y^-}} + b_{\mathrm{N^+X^-}})m, \quad \ln \gamma'_{C(0)} = 2b_{\mathrm{N^+Y^-}}m$$

and when $x = 1$:

$$\ln \gamma'_{B(0)} = 2b_{\mathrm{M^+X^-}}m, \quad \ln \gamma'_{(0)C} = (b_{\mathrm{N^+X^-}} + b_{\mathrm{M^+Y^-}})m$$

so that:

$$\ln \gamma_B = \ln \gamma_{(0)B} + (\ln \gamma_{B(0)} - \ln \gamma_{(0)B})x$$

with a similar equation for C:

$$\ln \gamma_C = \ln \gamma_{(0)C} + (\ln \gamma_{C(0)} - \ln \gamma_{(0)C})(1-x)$$

The logarithm of the activity coefficient of either component in a mixture maintained at constant total molality is therefore a linear function of the composition. Furthermore,

$$\ln \gamma_{(0)B} = \ln \gamma_{(0)C}$$

These equations can be recast in the form:

$$\ln \gamma_B = \ln \gamma_{(0)B} + (2b_{M^+X^-} - b_{N^+X^-} - b_{M^+Y^-})xm$$

$$\ln \gamma_C = \ln \gamma_{(0)C} + (2b_{N^+Y^-} - b_{N^+X^-} - b_{M^+Y^-})(1 - x)m$$

showing that whilst a plot of $\log \gamma_B$ or $\log \gamma_C$ against x should give straight lines, their slopes will in general be different. Only if the two electrolytes have a cation or an anion in common, $M^+ = N^+$ or $X^- = Y^-$, are the slopes equal in magnitude and opposite in sign. It can also be shown, by the Gibbs-Duhem equation, that:

$$\phi' = m\{b_{M^+X^-}x^2 + (b_{N^+X^-} + b_{M^+Y^-})x(1 - x) + b_{N^+Y^-}(1 - x)^2\}$$

so that, in general, ϕ is not a linear function of the composition: only if the two electrolytes have a common ion, *e.g.*, $X^- = Y^-$, does this reduce to:

$$\phi' = m\{b_{M^+X^-}x + b_{N^+X^-}(1 - x)\}$$

or

$$\phi = \phi_C + (\phi_B - \phi_C)x$$

and the osmotic coefficient is now a linear function of the composition.

EXPERIMENTAL METHODS FOR THE MEASUREMENT OF THE ACTIVITY COEFFICIENTS OF ELECTROLYTES IN MIXED SOLUTIONS

One very accurate method uses cells of the type:

$$H_2|HCl(m_B), NaCl(m_C)|AgCl, Ag$$

whose potential is:

$$E = E^0 - k \log \gamma^2_{HCl} m_B(m_B + m_C)$$

Since E^0 is known from measurements on cells containing hydrochloric acid only, the potential of this cell gives the activity coefficient of hydrochloric acid in the presence of sodium chloride. The method can be used for any electrolyte provided that electrodes are available reversible to each of the ions of the electrolyte. One of the earliest studies of this nature was made by GÜNTELBERG[4] for solutions of hydrochloric acid and lithium, sodium, potassium or caesium chloride, the concentrations being varied in such a way that the total molality remained constant at 0·1 M. This research is a model of experimental skill and accuracy.

Most of the work on these cells has been concentrated on either (1) keeping m_B constant and varying m_C or (2) allowing both m_B and m_C to vary subject to the condition that $(m_B + m_C)$ = constant. Very extensive measurements have been made by the Yale school on both these types. Reference has already been made (Chapter 12) to some of these when we were describing the determination of the ionization constant of water. Solutions of hydrochloric acid + alkali metal chloride and of alkali metal chloride + alkali metal hydroxide (and the corresponding bromide cells) have been studied in this way. Measurements have also been made for sulphuric acid in lithium, sodium and potassium sulphate solution[5].

SYSTEMS AT CONSTANT TOTAL MOLALITY

If we were to draw a vertical line for any one total molality across the four upper curves of *Figure 15.1*, we would get four values for the activity coefficient of hydrochloric acid in the presence of sodium chloride subject to the condition that the molalities of both solutes could vary, but the total molality remained constant. Measurements at constant total molality give more detailed information about γ_{HCl} as it changes from its value from $\gamma_{HCl(0)}$, in a solution containing acid only, to its limiting value $\gamma_{(0)HCl}$, when the solution contains only salt. The work of Güntelberg has been referred to earlier: Harned and his co-workers have made numerous measurements under this condition of constant total molality. From this work has emerged what has been called Harned's rule: the logarithm of the activity coefficient of one electrolyte in a mixture of constant total molality is directly proportional to the molality of the other component. Or:

$$\log \gamma_B = \log \gamma_{B(0)} - \alpha_B m_C \qquad \ldots.(15.2)$$

and when $m_C = m$ = total molality,

$$\log \gamma_{(0)B} = \log \gamma_{B(0)} - \alpha_B m \qquad \ldots.(15.3)$$

so that:

$$\log \gamma_B = \log \gamma_{(0)B} + \alpha_B m_B \qquad \ldots.(15.4)$$

and for the other component:

$$\log \gamma_C = \log \gamma_{C(0)} - \alpha_C m_B = \log \gamma_{(0)C} + \alpha_C m_C \qquad \ldots.(15.5)$$

α_B and α_C being functions of the total molality, m, but not of the individual molalities, m_B, m_C. These equations are contained in the theoretical deductions of Guggenheim, but they are found to be valid over a wider concentration range than one would expect and, as we shall see, at these higher concentrations the osmotic coefficient does not behave as simply as Guggenheim's equations for more dilute solutions predict.

To illustrate these equations we refer to some recent work[6] on the subject in which the activity coefficient of hydrochloric acid in the presence of potassium chloride at a total molality of $m = 2$ was measured by the cell:

$$H_2|HCl(m_B), KCl(m_C)|AgCl, Ag,$$

In the following table we compare the observed values of γ_{HCl} with those calculated by the equation:

$$\log \gamma_{HCl} = 0{\cdot}00358 - 0{\cdot}0580\, m_{KCl}$$

m_{HCl}	0·1	0·5	1·0	2·0
$\gamma_{obs.}$	0·7838	0·8243	0·8822	1·008
$\gamma_{calc.}$	0·7823	0·8252	0·8822	1·008

The agreement is within the experimental error of the measurements. The rule has also been confirmed for potassium chloride as added salt at total molalities of $m = 0{\cdot}1$ to 3, for sodium chloride at $m = 0{\cdot}1$, 1 and 3 and for lithium chloride at these total molalities. Hawkins[7] has shown that the rule is valid for the hydrocholric acid–potassium chloride system at $m = 4$ and 5 and for the hydrochloric acid–lithium chloride and hydrochloric acid–sodium chloride systems up to 6 M (although the acid–lithium chloride system exhibits the puzzling behaviour that the α coefficient for hydrochloric acid is negative, which means that the activity coefficient of the acid is raised by lithium chloride). The rule holds for the HCl—$NaClO_4$[8], HCl—$HClO_4$[9] systems and for the HCl—$Na_2S_2O_6$[9], HCl—$BaCl_2$[10], HCl—$AlCl_3$[11], and HCl—$CeCl_3$[12] systems provided that it is the total ionic strength which is kept constant.

Among the few known exceptions[13] to this rule are the electrolyte pairs $NaOH$—$NaCl$ and KOH—KCl at high concentrations, although the greatest error introduced by using equation (15.2) is only 3·9 per cent in the activity coefficient. Harned and Cook[14] were able to measure the activity coefficients of both the hydroxide and the chloride independently by means of the cells:

$$H_2|MOH(m_B), MCl(m_C)|AgCl, Ag$$

$$M_xHg|MOH(m_B), MCl(m_C)|AgCl, Ag$$

with $\qquad M = K, (m_B + m_C) = 1{\cdot}0$

and $\qquad M = Na, (m_B + m_C) = 0{\cdot}5$ and $1{\cdot}0$

In these cases it was found necessary to use the equations:

$$\log \gamma_B = \log \gamma_{B(0)} - \alpha_B m_C - \beta_B m_C^2 \qquad \ldots.(15.6)$$

$$\log \gamma_C = \log \gamma_{C(0)} - \alpha_C m_B - \beta_C m_B^2 \qquad \ldots.(15.7)$$

Another example of the failure of equation (15.2) is the salt pair $CaCl_2$—$ZnCl_2$, where extensive complex ion formation occurs[15]. Except for the hydroxide-chloride mixtures, the activity of only one of the electrolytes has been measured; as will be shown later, even if Harned's rule holds for one electrolyte it does not necessarily hold for the other electrolyte in the mixture.

VAPOUR PRESSURE MEASUREMENTS ON ELECTROLYTE MIXTURES

If B and C are non-volatile solutes, the vapour pressure of a solution of a particular total molality and particular values of m_B and m_C can be measured by the isopiestic vapour pressure method. OWEN and COOKE[16] were the first to make such measurements, on the potassium chloride–lithium chloride salt pair.

Equations (15.2) and (15.6) are special cases of the general expansion of $\log \gamma_B$ in a series in m_C. However, no case has yet been observed where more than a term in the square of m_C is necessary to express the variation of $\log \gamma_B$ and therefore we can start with equation (15.6) as giving sufficient generality for practical purposes. Putting $m_B = xm$ and $m_C = (1 - x)m$, where m is the total molality, then if m remains constant, the Gibbs-Duhem equation gives for an aqueous solution of two 1 : 1 electrolytes:

$$\begin{aligned} -\ & 55{\cdot}51 \,\mathrm{d} \log a_w \\ &= 2m_B \,\mathrm{d} \log m_B\gamma_B + 2m_C \,\mathrm{d} \log m_C\gamma_C \\ &= 2xm \,\mathrm{d} \log \gamma_B + 2(1 - x)m \,\mathrm{d} \log \gamma_C \\ &= 2xm[\alpha_B m + 2\beta_B(1 - x)m^2]\mathrm{d}x \\ &\qquad\qquad - 2(1 - x)m[\alpha_C m + 2\beta_C x m^2]\mathrm{d}x \\ &= m^2\{4x^2m(\beta_C - \beta_B) + 2[(\alpha_B + \alpha_C) - 2m(\beta_C - \beta_B)]x - 2\alpha_C\}\mathrm{d}x \end{aligned}$$

Integrating from $x = 0$ when $\log a_w = \log a_{w(C)}$, the value in a solution containing only electrolyte C at a molality m:

$$-\frac{55{\cdot}51}{xm^2} \log \frac{a_{w(x)}}{a_{w(C)}} = \tfrac{4}{3}x^2m(\beta_C - \beta_B) + x[(\alpha_B + \alpha_C) - 2m(\beta_C - \beta_B)] - 2\alpha_C \qquad \ldots(15.8)$$

where $a_{w(x)}$ is the water activily of a solution of molality xm of B and $(1 - x)m$ of C. The left-hand side of this equation contains only experimentally ascertainable quantities; it should be a quadratic function of the composition x. If $\beta_B = \beta_C$, then:

$$-\frac{55{\cdot}51}{xm^2} \log \frac{a_{w(x)}}{a_{w(C)}} = x(\alpha_B + \alpha_C) - 2\alpha_C = 0{\cdot}8686 \frac{\phi - \phi_C}{xm} \quad \ldots .(15.9)$$

ϕ being the osmotic coefficient of the mixed solution and ϕ_C that of a solution containing only electrolyte C at a concentration m.

A plot of $-\left(\frac{55{\cdot}51}{xm^2} \log \frac{a_{w(x)}}{a_{w(C)}}\right)$ against x should be a straight line of slope $(\alpha_B + \alpha_C)$ and intercept $-2\alpha_C$.

If $x = 1$, $\log a_{w(x)} = \log a_{w(B)}$, $\phi = \phi_B$ and:

$$-\frac{55{\cdot}51}{m^2} \log \frac{a_{w(B)}}{a_{w(C)}} = \alpha_B - \alpha_C = 0{\cdot}8686 \frac{\phi_B - \phi_C}{m} \quad \ldots .(15.10)$$

a relation between α_B and α_C which is true if Harned's rule holds for both electrolytes.

Before we consider the application of equations (15.8) and (15.9), it would be well to consider some restrictions on the properties of these α and β coefficients.

RELATIONS BETWEEN THE α AND β COEFFICIENTS

Equation (15.1) is a general result of the property of a chemical potential of being a partial differential coefficient of the total free energy with respect to concentration and imposes certain restrictions on the coefficients of equations (15.6) and (15.7). For clearly, in the case of 1 : 1 electrolytes:

$$\left(\frac{\partial \log \gamma_B}{\partial m_C}\right)_{m_B} = \left(\frac{\partial \log \gamma_{B(0)}}{\partial m_C}\right)_{m_B} - \alpha_B - m_C\left(\frac{\partial \alpha_B}{\partial m_C}\right)_{m_B} - 2m_C\beta_B - m_C^2\left(\frac{\partial \beta_B}{\partial m_C}\right)_{m_B}$$

$$\left(\frac{\partial \log \gamma_C}{\partial m_B}\right)_{m_C} = \left(\frac{\partial \log \gamma_{C(0)}}{\partial m_B}\right)_{m_C} - \alpha_C - m_B\left(\frac{\partial \alpha_C}{\partial m_B}\right)_{m_C} - 2m_B\beta_C - m_B^2\left(\frac{\partial \beta_C}{\partial m_B}\right)_{m_C}$$

None of the quantities $\gamma_{B(0)}$, $\gamma_{C(0)}$, α_B, α_C, β_B, β_C being functions of x:

$$\frac{\mathrm{d}\log\gamma_{B(0)}}{\mathrm{d}m} - \alpha_B - (1-x)m\frac{\mathrm{d}\alpha_B}{\mathrm{d}m} - 2(1-x)m\beta_B - (1-x)^2m^2\frac{\mathrm{d}\beta_B}{\mathrm{d}m}$$
$$= \frac{\mathrm{d}\log\gamma_{C(0)}}{\mathrm{d}m} - \alpha_C - xm\frac{\mathrm{d}\alpha_C}{\mathrm{d}m} - 2xm\beta_C - x^2m^2\frac{\mathrm{d}\beta_C}{\mathrm{d}m}$$

Since this must be true for all values of x:

$$xm\frac{\mathrm{d}\alpha_B}{\mathrm{d}m} + 2xm\beta_B + x(2-x)m^2\frac{\mathrm{d}\beta_B}{\mathrm{d}m}$$
$$= -xm\frac{\mathrm{d}\alpha_C}{\mathrm{d}m} - 2xm\beta_C - x^2m^2\frac{\mathrm{d}\beta_C}{\mathrm{d}m}$$

In all the cases so far examined, even if β_B and β_C terms are required to represent the experimental results, it is found that they are extremely small and, moreover, any variation of β_B and β_C with m is beyond experimental detection. For the purpose of dealing with experimental data of the accuracy now available, it is therefore justifiable to write:

$$\frac{\mathrm{d}\alpha_B}{\mathrm{d}m} + 2\beta_B = -\frac{\mathrm{d}\alpha_C}{\mathrm{d}m} - 2\beta_C$$

or $$(\alpha_B + \alpha_C) = \text{constant} - 2m(\beta_B + \beta_C) \quad \ldots .(15.11)$$

and in the even simpler case where $\beta_B = \beta_C = 0$:

$$(\alpha_B + \alpha_C) = \text{constant independent of } m$$

a result deduced by GLUECKAUF, MCKAY and MATHIESON(17).

If Harned's rule applies to electrolyte B, $\beta_B = 0$, then:

$$\left[\frac{\partial}{\partial m}(\alpha_C m_B + \beta_C m_B^2)\right]_{m_C} = \frac{\mathrm{d}}{\mathrm{d}m}\log\frac{\gamma_{C(0)}}{\gamma_{B(0)}} + \alpha_B + m_C\frac{\mathrm{d}\alpha_B}{\mathrm{d}m} \quad \ldots .(15.12)$$

This can be integrated between the limits $m = m_C$ and $m = m$ with respect to m at constant m_C to give:

$$(\alpha_C + \beta_C m_B)m_B = \left[\log\frac{\gamma_{C(0)}}{\gamma_{B(0)}}\right]_{m_C}^{m} + m_C\left[\alpha_B\right]_{m_C}^{m} + \int_{m_C}^{m}\alpha_B\mathrm{d}m \quad \ldots .(15.13)$$

When $m_B \rightarrow m$, $m_C \rightarrow 0$, then from (15.13):

$$(\alpha_C + \beta_C m)m = \log \frac{\gamma_{C(0)}}{\gamma_{B(0)}} + \int_0^m \alpha_B \mathrm{d}m$$

where $\gamma_{C(0)}$ and $\gamma_{B(0)}$ are values at the concentration m. When $m_C \rightarrow m$, $m_B \rightarrow 0$, then from (15.12):

$$\alpha_C = \frac{\mathrm{d}}{\mathrm{d}m} \log \frac{\gamma_{C(0)}}{\gamma_{B(0)}} + \alpha_B + m \frac{\mathrm{d}\alpha_B}{\mathrm{d}m}$$

These last three results are useful in calculating α_C and β_C for an electrolyte C when it is known that Harned's rule applies to the other electrolyte. Similar equations have been deduced by McKay[18]. As an illustration we use the data of Harned and Gancy[6] for mixtures of $B = \mathrm{HCl}$ and $C = \mathrm{KCl}$ at $m = 2$. At $m_C = 0{\cdot}5$, $1{\cdot}0$ and $1{\cdot}5$, the three terms on the right of equation (15.13) have the following values:

m_C	*1st term*	*2nd term*	*3rd term*	$\alpha_C + \beta_C m_B$	$\gamma_C(1)$	$\gamma_B(2)$
0·5	0·1788	0·0020	0·0842	−0·0617	0·7098	0·7092
1·0	0·1187	0·0027	0·0568	−0·0592	0·6608	0·6569
1·5	0·0592	0·0020	0·0288	−0·0568	0·6154	0·6132

from which we conclude that $\alpha_C = -0{\cdot}0543$ and $\beta_C = -0{\cdot}0050$. The last two columns give γ_C (1) calculated from these α_C and β_C values and γ_C (2) calculated on the assumption that Harned's rule holds for C with $\alpha_C = -0{\cdot}0619$ from equation (15.10). The difference is small but sufficient to show that a β_C term is necessary.

ANOTHER METHOD OF USING VAPOUR PRESSURE MEASUREMENTS

Starting with equation (15.1) McKay and Perring[19] have obtained a number of useful transforms, one of which, useful for isopiestic results, is:

$$0{\cdot}002\, W_A \left(\frac{\partial \ln \gamma_C m}{\partial \ln a_A}\right)_{m_B/m_C} = -\frac{1}{m^2}\left(\frac{\partial m}{\partial \ln x}\right)_{aA} - \frac{1}{m}$$

In this equation, applicable to 1 : 1 electrolytes, the left-hand side gives the change in the activity of electrolyte C as the solvent activity increases, *i.e.*, as the total molality changes subject to constancy of x. The right-hand side contains a term for the change in the total molality which it is necessary to make if the solvent activity is to remain constant while the ratio of the molalities of the two electrolytes is changed; it is related therefore to the conditions which make a series of solutions isopiestic with one another. This equation can be integrated:

$$0{\cdot}002\,W_A \ln \gamma_C m = -\int \left[\frac{1}{m^2}\left(\frac{\partial m}{\partial \ln x}\right)_{a_A} + \frac{1}{m}\right] \mathrm{d} \ln a_A \quad \ldots.(15.14)$$

the integration to be performed at constant x. There will, of course, be an integration constant which is eliminated by noting the limiting condition when $x = 0$, *i.e.*, when the solution contains only one electrolyte, C. Now let M and Γ_C be the molality and activity coefficient of electrolyte C in a solution from which electrolyte B is absent but at the same a_A as the mixed solution of total molality m. Then by equation (15.14):

$$0{\cdot}002\,W_A \ln \Gamma_C M = -\int \left[\frac{1}{M^2} \operatorname*{Lt}_{x\to 0} \left(\frac{\partial m}{\partial \ln x}\right)_{a_A} + \frac{1}{M}\right] \mathrm{d} \ln a_A$$

By the Gibbs-Duhem equation:

$$0{\cdot}002\,W_A \ln \Gamma_C M = -\int \frac{1}{M} \mathrm{d} \ln a_A$$

Therefore:

$$0{\cdot}002\,W_A \ln \frac{\gamma_C m}{\Gamma_C M} = -\int_{a_A=1}^{a_A} \left[\frac{1}{m^2}\left(\frac{\partial m}{\partial \ln x}\right)_{a_A} + \frac{1}{m} - \frac{1}{M}\right] \mathrm{d} \ln a_A \quad \ldots.(15.15)$$

It would be well to reiterate the meaning of the symbols in this equation. γ_C is the activity coefficient of electrolyte C in a solution containing both electrolytes at total molality m and having a particular solvent activity a_A; in the absence of electrolyte B, this particular solvent activity is associated with a solution of electrolyte C of molality M and activity coefficient Γ_C; the latter is not the same as $\gamma_{C(0)}$ because $\gamma_{C(0)}$ is the activity coefficient of electrolyte C at a concentration m, electrolyte B being absent. The integration is to be performed at a constant value of x, that associated with the γ_C and m on the left-hand side which it is desired to evaluate; thus m, $\left(\frac{\partial m}{\partial \ln x}\right)_{a_A}$ and M are functions of x and a_A but they are to be assigned their values for a particular x during the integration from $a_A = 1$ up to the value of a_A corresponding to the solution in question. This equation is cast in a form particularly suitable for isopiestic vapour pressure measurements, because the quantity $\left(\frac{\partial m}{\partial \ln x}\right)_{aA}$ can be evaluated as a function of x and a_A by isopiestic measurements using a desiccator with a large number of dishes containing solutions of electrolytes B and C.

An alternative form of equation (15.15):

$$\ln \gamma_C = \ln \Gamma_C + \ln R + \int_0^{M\phi} \left[\frac{1}{m^2} \left(\frac{\partial m}{\partial \ln x} \right)_{M\phi} + \frac{R-1}{M} \right] \mathrm{d}\,(M\phi)$$

where $R = M/m$, will be found useful.

Some isopiestic measurements[20] are available on the system $B = \mathrm{NaCl}$, $C = \mathrm{KCl}$ from which we quote the following results for $m = 4$, calculated by the McKay-Perring method:

x	0·0	0·1	0·3	0·5	0·7	0·9	1·0
$-\log \gamma_B$	0·1965	0·1882	0·1711	0·1533	0·1349	0·1158	0·1061
$-\log \gamma_C$	0·2390	0·2354	0·2282	0·2210	0·2138	0·2066	0·2030

Equation (15.5) with $\alpha_C = -0{\cdot}0090$ represents the results within the limits of experimental error, but the data for sodium chloride need a small β term in equation (15.6); $\alpha_B = 0{\cdot}0246$, $\beta_B = -0{\cdot}0005$. The importance of the method derived by McKay and Perring lies in the evaluation of α_C at a series of x values for a particular value of m without at any time assuming that equation (15.5) is true. The results of the calculation made by the McKay-Perring method can therefore be used as a direct test of the validity of equation (15.5) for any particular salt pair without any prior assumption of its validity.

DISCUSSION OF THE ACTIVITY COEFFICIENTS OF MIXED ELECTROLYTES

We have seen that there is reason to believe that equations like (15.6) and (15.7) are necessary to represent the activity coefficients of a number of electrolytes in mixed solutions. The simpler equations (15.2) and (15.5), however, are valid in some cases and are a fair approximation in others. HARNED[21] has developed the consequences of these equations in considerable detail and his discussion is of such importance as to warrant some recapitulation.

1. The most general case, subject to the limitation that $\beta_B = \beta_C = 0$, occurs when $\alpha_B \neq -\alpha_C$ and $\gamma_{(0)B} \neq \gamma_{(0)C}$. The first inequality necessitates, by equation (15.9), that the osmotic coefficient is a quadratic function of x. The system $B = \mathrm{HCl}$, $C = \mathrm{CsCl}$, approximates to this although it is probable that a small β_C term is needed. E.m.f. measurements of cells without transport made by HARNED and SCHUPP[22] give $\alpha_B = 0{\cdot}098$ and $\alpha_C = -0{\cdot}041$ at 3 M. The activity coefficient of hydrochloric acid by itself at 3 M is $\gamma_{B(0)} = 1{\cdot}316$ that of caesium chloride is $\gamma_{C(0)} = 0{\cdot}478$. Clearly

they are very different. The α coefficients now lead to $\gamma_{(0)B} = 0{\cdot}669$ for the limiting case of hydrochloric acid in a solution containing only 3 M caesium chloride and $\gamma_{(0)C} = 0{\cdot}634$ for the corresponding caesium chloride activity coefficient in 3 M hydrochloric acid. $\gamma_{(0)B}$ and $\gamma_{(0)C}$ therefore do differ considerably although not as much as do $\gamma_{B(0)}$ and $\gamma_{C(0)}$. The inequality of α_B and α_C leads by equation (15.9) to $a_{w(x)} = 0{\cdot}8908$ at $x = 0{\cdot}5$ compared with $a_{w(x)} = 0{\cdot}8868$ if $\log a_w$ were linear in x. This difference may seem small, but the osmotic coefficient of 3 M hydrochloric acid is 1·348, that of 3 M caesium chloride is 0·879 and if the osmotic coefficient were linear in x, it would be 1·114 at $x = 0{\cdot}5$, whereas the observed value is 1·070.

To illustrate further that the osmotic coefficient is far from being a linear function of the composition, a comparison can be made between the observed osmotic coefficients of the NaCl—CsCl system[23] at 3 M and those calculated if the variation were proportional to the composition.

Fraction of CsCl in mixture	0	0·1335	0·2698	0·3689	0·4989	0·6354	0·7978	1·0
ϕ (obs.)	1·045	1·008	0·976	0·953	0·929	0·910	0·895	0·87
ϕ (calc.)	—	1·023	1·000	0·984	0·963	0·940	0·913	—

$$\alpha_B = \alpha_{\text{NaCl}} = 0{\cdot}0429, \ \alpha_C = \alpha_{\text{CsCl}} = -0{\cdot}0048$$

2. $\alpha_B \neq -\alpha_C$ but $\gamma_{(0)B} = \gamma_{(0)C}$. Again the osmotic coefficient is not a linear function of x but:

$$\log \frac{\gamma_{B(0)}}{\gamma_{C(0)}} = (\alpha_B - \alpha_C)m$$

From (2.28) and (15.10) it follows that:

$$\log \frac{\gamma_{B(0)}}{\gamma_{C(0)}} = 0{\cdot}8686(\phi_B - \phi_C) = \frac{2}{m}\int_0^m m \, \mathrm{d} \log \frac{\gamma_{B(0)}}{\gamma_{C(0)}}$$

and this can hold only if:

$$\log \frac{\gamma_{B(0)}}{\gamma_{C(0)}} = Km$$

K being a constant independent of m, an equation proposed by Åkerlöf and Thomas[24]. These conditions are almost true for the system: B = HCl, C = NaCl.[25] We have seen at the beginning of this chapter that $\gamma_{(0)B} = 1{\cdot}063$ and $\gamma_{(0)C} = 1{\cdot}066$ at 3 M but $\alpha_B = 0{\cdot}031$, $\alpha_C = -0{\cdot}058$; to show that the Åkerlöf-Thomas rule is very nearly valid we quote the following figures for the ratio of the two activity coefficients over a range of concentration:

m	1	2	3	4
$\gamma_{HCl(0)}$	0·809	1·009	1·316	1·762
$\gamma_{NaCl(0)}$	0·657	0·668	0·714	0·783
$\frac{1}{m}\log\frac{\gamma_{HCl(0)}}{\gamma_{NaCl(0)}}$	0·0903	0·0896	0·0885	0·0881

3. $\alpha_B = -\alpha_C$ but $\gamma_{(0)B} \neq \gamma_{(0)C}$. The osmotic coefficient is now linear in x and from equation (15.9):

$$\phi_B - \phi_C = -2{\cdot}303\, m\alpha_C = 2{\cdot}303\, m\alpha_B$$

so that for any pair of electrolytes for which $\alpha_B = -\alpha_C$, the values of α_B and α_C are very simply related to the osmotic coefficients of the electrolytes. This condition rarely holds: the system B = KCl, C = CsCl[26] approximates to this behaviour with $\alpha_B = 0{\cdot}011$ and $\alpha_C = -0{\cdot}005$ at 3 M. That the osmotic coefficient is very nearly a linear function of the composition is shown by the following comparison of the observed osmotic coefficients with those calculated on the assumption of a linear variation:

Fraction of CsCl *in mixture*	0	0·1411	0·3025	0·4007	0·6443	0·7726	1·0
ϕ (obs.)	0·937	0·927	0·916	0·910	0·896	0·890	0·879
ϕ (calc.)	—	0·929	0·919	0·914	0·900	0·892	—

The activity coefficient of 3 M KCl is 0·569 and that of 3 M CsCl is 0·478; with $\alpha_B = 0{\cdot}011$ and $\alpha_C = -0{\cdot}005$, we calculate $\gamma_{(0)B} = 0{\cdot}527$ and $\gamma_{(0)C} = 0{\cdot}494$ so that $\gamma_{0(B)}$ and $\gamma_{0(C)}$ are by no means the same.

4. $\alpha_B = -\alpha_C$ and $\gamma_{0(B)} = \gamma_{0(C)}$. These conditions are approximated by the salt pair B = LiCl and C = NaCl.[27] At $m = 3$, $\phi_B = 1{\cdot}286$ and $\phi_C = 1{\cdot}045$ so that if $\alpha_B = -\alpha_C$, it is necessary that $\alpha_B = 0{\cdot}035$ and $\alpha_C = -0{\cdot}035$. These values have been found experimentally, although there are small β terms in equations (15.7) and (15.8) as a result of which this example can be quoted only as an approximation to the case where $\alpha_B = -\alpha_C$. That the osmotic coefficient is a linear function of the molality can be seen from the following comparison of the observed osmotic coefficients with those calculated on the assumption of a linear variation:

Fraction of LiCl *in mixture*	0	0·3392	0·5167	0·6699	1·0
ϕ (obs.)	1·045	1·125	1·170	1·207	1·286
ϕ (calc.)	—	1·127	1·170	1·206	—

Moreover, the activity coefficients of lithium chloride and sodium chloride, each in the absence of the other electrolyte, are 1·156 and 0·714 respectively and if $\alpha_{LiCl} = -\alpha_{NaCl} = 0{\cdot}035$, it follows that $\gamma_{(0)LiCl} = 0{\cdot}908$ and $\gamma_{(0)NaCl} = 0{\cdot}909$. If these conditions hold over a range of values of m, it follows that the Åkerlöf-Thomas rule is applicable and indeed it is found by experiment to be a good approximation over a considerable concentration range as the following figures show:

m	1	2	3	4	5	6
γ_{LiCl}	0·774	0·921	1·156	1·510	2·02	2·72
γ_{NaCl}	0·657	0·668	0·714	0·783	0·874	0·986
$\frac{1}{m}\log\frac{\gamma_{LiCl}}{\gamma_{NaCl}}$	0·0711	0·0698	0·0698	0·0713	0·0728	0·0734

We can summarize this as follows: $\beta_B = \beta_C = 0$ but:

1. $\alpha_B \neq -\alpha_C$, $\gamma_{(0)B} \neq \gamma_{(0)C}$, the osmotic coefficient is a quadratic in x. Example: the HCl—CsCl system.

2. $\alpha_B \neq -\alpha_C$, $\gamma_{(0)B} = \gamma_{(0)C}$, the osmotic coefficient is a quadratic in x and the Åkerlöf-Thomas rule applies. Example: the HCl—NaCl system.

3. $\alpha_B = -\alpha_C$, $\gamma_{(0)B} \neq \gamma_{(0)C}$, the osmotic coefficient is linear in x and the α_B and α_C coefficients are determined in terms of $(\phi_B - \phi_C)$. Example: the system KCl—CsCl.

4. $\alpha_B = -\alpha_C$, $\gamma_{(0)B} = \gamma_{(0)C}$, the osmotic coefficient is linear in x, the Åkerlöf-Thomas rule holds and α_B and α_C are determined in terms of $(\phi_B - \phi_C)$. Example: the system LiCl—NaCl.

There are few electrolyte pairs, however, to which Harned's rule is strictly applicable; these include the systems: B = HCl, C = LiCl (up to 3 M[28]); B = HCl, C = NaCl[25]; B = NaCl, C = KCl[20]; B = NaCl, C = CsCl[23]; B = KCl, C = KBr[29]; B = KCl, C = CsCl[26].

Several types of system are possible if Harned's rule does not hold:

1. $\beta_B = 0$, $\beta_C \neq 0$. The system B = HCl, C = KCl has already been quoted as an example and, from the calculations of ARGERSINGER and MOHILNER[30], this seems to be true of mixtures of hydrochloric acid with barium chloride, strontium chloride, aluminium chloride and cerium chloride.

2. $\beta_B \approx \beta_C \neq 0$. A system of this type, $B = \text{LiCl}$, $C = \text{NaCl}$[27], has been investigated by vapour pressure measurements. For any one total molality the function on the left of equation (15.8) gave a good straight line graph when plotted against x, indicating that $\beta_B \approx \beta_C$, but the slopes of this graph were not the same at all values of the total molality; indeed, to a first approximation the slopes were proportional to m:

$$(\alpha_{\text{NaCl}} + \alpha_{\text{LiCl}}) = -0{\cdot}013 + 0{\cdot}004\, m$$

It follows from equation (15.11) that $(\beta_B + \beta_C)$ cannot be neglected but must be of the order of $-0{\cdot}002$ and since $\beta_B \approx \beta_C$, each must be about $-0{\cdot}001$. Small β_B and β_C values are also found for the KCl—LiCl and LiCl—LiNO_3 systems.

3. $\beta_B \neq \beta_C \neq 0$. An example[26] is the complicated CsCl—LiCl system, also studied by vapour pressure measurements. At no value of the total molality between $m = 0{\cdot}5$ and $m = 6$ did the plot of the left-hand side of equation (15.8) against x give a straight line, so that $\beta_B \neq \beta_C$, and at least one of them has a non-zero value. To explore this system thoroughly would need much tedious and very precise work; a preliminary survey has been made as follows. At one particular total molality, $m = 5$, vapour pressure measurements were made at many values of x so that the curvature of the graph corresponding to equation (15.8) could be ascertained with some accuracy and the values of α_{CsCl}, α_{LiCl} and $(\beta_{\text{CsCl}} - \beta_{\text{LiCl}})$ necessary to represent this graph were evaluated for $m = 5$. It was then assumed that β_{CsCl} and β_{LiCl} were independent of m and equation (15.8) was used with these β quantities together with the less extensive experimental data at other total molalities, to evaluate the two α coefficients. The sum of these coefficients $(\alpha_{\text{CsCl}} + \alpha_{\text{LiCl}})$, was found to be a linear function of the total molality. Thus there were obtained the two equations:

$$\beta_{\text{LiCl}} - \beta_{\text{CsCl}} = -0{\cdot}0058$$

valid at $m = 5$ and assumed valid at other values of m, and:

$$\begin{aligned}(\alpha_{\text{CsCl}} + \alpha_{\text{LiCl}}) &= \text{constant} - 2\,m\,(\beta_{\text{LiCl}} + \beta_{\text{CsCl}}) \\ &= 0{\cdot}082 + 0{\cdot}009\, m,\end{aligned}$$

whence $\beta_{\text{CsCl}} = 0{\cdot}001$ and $\beta_{\text{LiCl}} = -0{\cdot}005$

A reinvestigation of these systems by the McKay-Perring method is needed. For the system, p-toluenesulphonic acid and its sodium salt[31], even equations (15.6) and (15.7) give only an approximation to the observed behaviour.

CALCULATION OF α COEFFICIENTS FROM OTHER DATA

The α coefficients of equations (15.2) to (15.5) determine a number of important properties of mixed electrolyte solutions and it is clear that much experimental work could be saved if we had some sound method of calculating these coefficients from the properties of single electrolyte solutions. This is possible for dilute solutions if equations of the type of (9.13) are valid. For example, if the activity coefficient of one electrolyte can be written:

$$\ln \gamma_{B(0)} = -\frac{\alpha\sqrt{I}}{1+\sqrt{I}} + 2b_{M^+X^-}m$$

and a similar equation with $b_{N^+X^-}$ is valid for an electrolyte C with the same anion X^- then, from Guggenheim's equation for a mixed solution:

$$\ln \gamma_B = \ln \gamma_{(0)B} + (b_{M^+X^-} - b_{N^+X^-})xm$$

and therefore the α_B of equation (15.2) is:

$$\alpha_B = 0{\cdot}4343\ (b_{M^+X^-} - b_{N^+X^-})$$

that is to say, the α_B coefficient is predictable from the properties of single electrolyte solutions. Using the activity coefficients of hydrochloric acid, sodium, potassium and caesium chloride at 0·1 M to calculate b and putting $B = MX = HCl$, $C = NX = LiCl$, NaCl, KCl or CsCl, we can calculate the following α_B coefficients for hydrochloric acid in alkali halide solution and compare them with the values deduced from Güntelberg's work:

Electrolyte	$-\log \gamma$	$0{\cdot}4343\ b_{N^+X^-}$	α_B (*calc.*)	α_B (*obs.*)
HCl	0·0991	(0·116)	—	—
LiCl	0·1024	0·100	0·016	0·009
NaCl	0·1090	0·067	0·049	0·043
KCl	0·1135	0·044	0·072	0·077
CsCl	0·1215	0·004	0·112	0·143

For this calculation to be exact it is of course necessary that $\gamma_{(0)B} = \gamma_{(0)C}$, so that there are very few systems at higher concentrations to which this method of prediction can be applied. Again, if $\alpha_B = -\alpha_C$ we can use the relation $(\phi_B - \phi_C) = 2{\cdot}303m\alpha_B$; this would lead for the HCl—CsCl system to $\alpha_B = -\alpha_C = 0{\cdot}068$ at 3 M whereas the experimental value is $\alpha_B = 0{\cdot}098$. Clearly, such calculations are not likely to give more than an order of magnitude. To emphasize this, we give a few comparisons at 1 M:

Electrolyte		$\alpha_B = -\alpha_C = \dfrac{\phi_B - \phi_C}{2{\cdot}303}$	*Observed*
B	*C*		α_B
HCl	NaCl	0·045	0·032
HBr	NaBr	0·050	0·038
HCl	KCl	0·061	0·056
HBr	KBr	0·072	0·080

This important problem of calculating properties of mixtures from those of the components is still far from solution.

A SIMPLE ADDITIVITY RULE FOR THE VAPOUR PRESSURE LOWERING OF A MIXED ELECTROLYTE SOLUTION

For some purposes, when the highest accuracy is not required, these somewhat complicated considerations of mixtures can be overlooked and a simple additivity rule used. The vapour pressures of solutions containing electrolytes such as ($2KCl + MgCl_2$) have been measured[32] over a wide range of concentration at 25°. A solution 0·5 M with respect to the double salt, K_2MgCl_4, has a vapour pressure lowering of $\Delta p/p^0 = 0{\cdot}06040$: the potassium chloride concentration is 1 M at which concentration and in the absence of other solutes the vapour pressure lowering, $\Delta p/p^0$, is 0·03182. Similarly for 0·5 M magnesium chloride, $\Delta p/p^0 = 0{\cdot}02525$. If we add these two contributions to get a calculated vapour pressure lowering, we find $\Delta p/p^0 = 0{\cdot}05707$, a value which differs by 5·5 per cent from the observed. Better agreement can be obtained by a slight elaboration which is illustrated as follows: this solution of ($2KCl + MgCl_2$) has a total ionic strength of 2·5 and we use the molal vapour pressure lowerings of the components at this total ionic strength, $\Delta p/(mp^0) = 0{\cdot}03195$ for potassium chloride and 0·05530 for magnesium chloride; the contribution to $\Delta p/p^0$ of the mixture is 0·03195 for potassium chloride and 0·02765 for magnesium chloride with a total of 0·05960 which differs by only 1·3 per cent from the observed value. Agreement of this order is found with a number of these mixtures up to a total molality of unity and even in the more searching example of a solution of lithium chloride and calcium chloride, agreement within 5 per cent can be obtained even with solutions of 4 M—$CaCl_2$ + 8 M—LiCl. Thus at a concentration of 3·833 M—Li_2CaCl_4 the observed relative vapour pressure lowering is 0·7698 and the calculated 0·7379, a difference of only 4·2 per cent.

To take a third example, good agreement can be obtained by applying this empirical rule to mixtures of lithium chloride and lithium nitrate[19]. A mixture of 4·662 M—$LiNO_3$ and 5·338 M—LiCl is known to have $\Delta p/p^0 = 0{\cdot}5141$; calculating from the known data for the component salts we find $\Delta p/p^0 = 0{\cdot}5215$, a difference of only 1·4 per cent. This empirical rule is almost equivalent to assuming that the osmotic coefficient is a linear function of the fraction of lithium nitrate in the mixture, and hence to assuming that:

$$\alpha_{LiCl} = -\alpha_{LiNO_3} = \frac{1}{2{\cdot}303m}(\phi_{LiCl} - \phi_{LiNO_3}) = 0{\cdot}036$$

which is very different to the values found by more detailed study of the system, $\alpha_{LiCl} = 0{\cdot}050$ and $\alpha_{LiNO_3} = -0{\cdot}023$. The former value of α_{LiCl} gives $\gamma_{LiCl} = 6{\cdot}39$ and the latter gives $\gamma_{LiCl} = 5{\cdot}50$ for the activity coefficient of lithium chloride in the mixture. We emphasize this point because, whilst the empirical rule is a very useful one for calculating properties of the solvent, it can be a dangerous rule if applied to properties of the component solutes.

THE SOLVATION OF MIXED ELECTROLYTES

We next inquire whether the 'hydration' equation developed in Chapter 9 assists us in explaining some of the peculiarities of mixed electrolyte solutions. Suppose we had S moles of water containing one mole of electrolyte B and ζ moles of electrolyte C. For simplicity we consider only 1 : 1 electrolytes. Let h_B and h_C be the hydration numbers of the electrolytes. It can be shown that equation (9.16) becomes:

$$\ln\frac{f'_B}{f_B} + \zeta\ln\frac{f'_C}{f_C} = \frac{h_B + \zeta h_C}{2}\ln a_w + (1+\zeta)\ln\frac{S + 2(1+\zeta) - h_B - \zeta h_C}{S + 2(1+\zeta)}$$

or, in terms of the molalities, m_B and m_C:

$$\ln\frac{f'_B}{f_B} + \zeta\ln\frac{f'_C}{f_C} = \frac{h_B + \zeta h_C}{2}\ln a_w + (1+\zeta)\ln\frac{1 + 0{\cdot}018(2m - h_B m_B - h_C m_C)}{1 + 0{\cdot}036m}$$

where $m = m_B + m_C$. Converting now into molal activity coefficients:

$$\ln\frac{f'_B}{\gamma_B} + \zeta\ln\frac{f'_C}{\gamma_C} = \frac{h_B + \zeta h_C}{2}\ln a_w + (1+\zeta)\ln\left[1 + 0{\cdot}018(2m - h_B m_B - h_C m_C)\right]$$

If we can decompose this into two equations:

$$\ln \gamma_B = \ln f'_B - \frac{h_B}{2} \ln a_w - \ln [1 + 0{\cdot}018(2m - h_B m_B - h_C m_C)]$$

$$\ln \gamma_C = \ln f'_C - \frac{h_C}{2} \ln a_w - \ln [1 + 0{\cdot}018(2m - h_B m_B - h_C m_C)]$$

and assume that f'_B is independent of the composition of the solution at constant total molality, then:

$$\log \gamma_{B(0)} = \log f'_B - \frac{h_B}{2} \log a_{w(B)} - \log [1 + 0{\cdot}018(2m - h_B m)]$$

and:

$$\log \gamma_{(0)B} = \log f'_B - \frac{h_B}{2} \log a_{w(C)} - \log [1 + 0{\cdot}018(2m - h_C m)]$$

Therefore, if equation (15.2) is to apply, we have:

$$\alpha_B m = 0{\cdot}0078 h_B m(\phi_B - \phi_C) + \log \frac{1 + 0{\cdot}018(2 - h_C)m}{1 + 0{\cdot}018(2 - h_B)m}$$

or, to a close approximation:

$$\alpha_B = 0{\cdot}0078 h_B(\phi_B - \phi_C) + 0{\cdot}0078(h_B - h_C)$$

and $$\alpha_C = 0{\cdot}0078 h_C(\phi_C - \phi_B) + 0{\cdot}0078(h_C - h_B)$$

Calculations of some α values have been made using these equations and *Table 15.1* gives the results for a total molality of unity.

Table 15.1

Comparison of Observed and Calculated α Coefficients at a Total Molality of Unity

Electrolyte		α_B		$-\alpha_C$	
B	*C*	*obs.*	*calc.*	*obs.*	*calc.*
HCl	LiCl	0·005	0·008	0·012	0·008
KCl	CsCl	0·016	0·015	0·019	0·015
NaCl	CsCl	0·021	0·029	0·047	0·027
HCl	NaCl	0·032	0·041	0·058	0·038
HCl	KCl	0·056	0·056	0·072	0·050

Although the agreement with the experimental values is not good, the crude theory we have developed does at least predict the sign and the magnitude of the effect and puts the coefficients in the right order. The average deviation is only 0·01 in α_B or α_C. More cannot be expected until we have a much clearer picture of the multiform complications of these solutions.

We have not considered the possible variation of f'_B or f'_C with composition at constant total molality. This alone is a complicated problem, requiring a survey of the modifications of the Debye-Hückel theory for the interaction of ions of different sizes. We know very nearly nothing at all about the effect of a size-difference between the ions. Again, as was pointed out in Chapter 9, the 'hydration' equation ignores 'non-electrolyte' effects so that the hydration number, h, is made to account not only for the hydration effect, but also for the effects of the free volume ratio and of the heat of mixing of the hydrated ions with the solvent. Our picture of even a single electrolyte in solution is far from complete and we need not be surprised to find that the delicate interactions between the electrolytes in a mixture are even less well understood. It is, indeed, encouraging to find that we can ignore these finer details of the picture and get even such qualitative agreement as that shown in *Table 15.1*. In the above argument the consequences of the 'hydration' equation have been carried to an extreme with the hope that this crude picture may form at least a basis for improvement and that α_B and α_C will become calculable. The consequences of a successful theory would be important. At present the thermodynamic properties of a comparatively simple system like sea-water are known only as a result of tedious experiments; simple as this system is, it has many degrees of freedom and questions such as, for example, the effect of a change in the sodium chloride–magnesium chloride ratio on the water activity cannot be answered today and would necessitate considerable experimental work. The properties of sea-water should be calculable from the properties of a few solutions each containing a single salt but, with our present theory, we can make only the most approximate estimate of the interactions of these salts[33]. The various physiological fluids can be quoted as another example where a theory of mixed electrolyte solutions would lead to progress whilst the problem of the activity coefficient of a weak acid in the presence of one of its salts, *i.e.*, in a buffer solution, does not seem to be completely soluble until we know much more about the interactions of two electrolytes in a solution.

REFERENCES

1 GUGGENHEIM, E. A., 'Thermodynamics, an advanced treatment for chemists and physicists,' p. 307, North-Holland Publishing Co., Amsterdam (1949)
2 — *Phil. Mag.*, 19 (1935) 588
3 BRÖNSTED, J. N., *J. Amer. chem. Soc.*, 44 (1922) 877
4 GÜNTELBERG, E., *Z. phys. Chem.*, 123 (1926) 199; 'Studier over Elektrolyt-Activiteter,' G.E.C. Gads Forlag, Copenhagen (1938)
5 ÅKERLÖF, G., *J. Amer. chem. Soc.*, 48 (1926) 1160
6 HARNED, H. S. and GANCY, A. B., *J. phys. Chem.*, 62 (1958) 627; see also HARNED, H. S., *J. Amer. chem. Soc.*, 48 (1926) 326; HARNED, H. S. and ÅKERLÖF, G., *Phys. Z.*, 27 (1926) 411
7 HAWKINS, J. E., *J. Amer. chem. Soc.*, 54 (1932) 4480
8 BATES, S. J. and URMSTON, J. W., *ibid.*, 55 (1933) 4068
9 MURDOCK, P. G. and BARTON, R. C., *ibid.*, 55 (1933) 4074
10 HARNED, H. S. and GARY, R., *ibid.*, 76 (1954) 5924
11 HARNED, H. S. and MASON, C. M., *ibid.*, 53 (1931) 3377
12 MASON, C. M. and KELLAM, D. B., *J. phys. Chem.*, 38 (1934) 689
13 HARNED, H. S. and HARRIS, J. M., *J. Amer. chem. Soc.*, 50 (1928) 2633
14 — and COOK, M. A., *ibid.*, 59 (1937) 1890
15 ROBINSON, R. A. and FARRELLY, R. O., *J. phys. Chem.*, 51 (1947) 704
16 OWEN, B. B. and COOKE, T. F., *J. Amer. chem. Soc.*, 59 (1937) 2273
17 GLUECKAUF, E., MCKAY, H. A. C. and MATHIESON, A. R., *J. chem. Soc.* (1949) S 299
18 MCKAY, H. A. C., *Trans. Faraday Soc.*, 51 (1955) 903
19 MCKAY, H. A. C. and PERRING, J. K., *ibid.*, 49 (1953) 163
20 ROBINSON, R. A., *J. phys. Chem.*, 65 (1961) 662; the system sodium chloride–barium chloride has also been studied by ROBINSON, R. A. and BOWER, V. E., *J. Res. nat. Bur. Stand.*, 69A (1965) 19
21 HARNED, H. S., *J. Amer. chem. Soc.*, 57 (1935) 1865; HARNED, H. S. and OWEN, B. B., 'The Physical Chemistry of Electrolytic Solutions,' chap. 14, pp. 602 *et seq.*, Reinhold Publishing Corporation, New York (1958)
22 HARNED, H. S. and SCHUPP, O. E., *J. Amer. chem. Soc.*, 52 (1930) 3892
23 ROBINSON, R. A., *ibid.*, 74 (1952) 6035
24 ÅKERLÖF, G. and THOMAS, H. C., *ibid.*, 56 (1934) 593
25 HARNED, H. S., *ibid.*, 57 (1935) 1865
26 ROBINSON, R. A., *Trans. Faraday Soc.*, 49 (1953) 1147
27 ROBINSON, R. A. and LIM, C. K., *ibid.*, 49 (1953) 1144
28 HARNED, H. S. and COPSON, H. R., *J. Amer. chem. Soc.*, 55 (1933) 2206
29 MCCOY, W. H. and WALLACE, W. E., *ibid.*, 78 (1956) 1830
30 ARGERSINGER, W. J. and MOHILNER, D. M., *J. phys. Chem.*, 61 (1957) 99
31 BONNER, O. D. and HOLLAND, V. F., *J. Amer. chem. Soc.*, 77 (1955) 5828
32 ROBINSON, R. A. and STOKES, R. H., *Trans. Faraday Soc.*, 41 (1945) 752
33 — *J. Mar. biol. Ass.* U.K., 33 (1954) 449

APPENDIX 1.1

Physical Properties of Water

Temp. °C	*Density* g/ml	*Specific volume* ml/g	*Vapour Pressure* mm Hg	*Dielectric constant*	*Viscosity centipoise*
0	0·99987	1·00013	4·580	87·740	1·787
5	0·99999	1·00001	6·538	85·763	1·516
10	0·99973	1·00027	9·203	83·832	1·306
15	0·99913	1·00087	12·782	81·945	1·138
18	0·99862	1·00138	15·471	80·835	1·053
20	0·99823	1·00177	17·529	80·103	1·002
25	0·99707	1·00293	23·753	78·303	0·8903
30	0·99568	1·00434	31·824	76·546	0·7975
35	0·99406	1·00598	42·180	74·823	0·7194
38	0·99299	1·00706	49·702	73·817	0·6783
40	0·99224	1·00782	55·338	73·151	0·6531
45	0·99024	1·00985	71·90	71·511	0·5963
50	0·98807	1·01207	92·56	69·910	0·5467
55	0·98573	1·01448	118·11	68·344	0·5044
60	0·98324	1·01705	149·47	66·813	0·4666
65	0·98059	1·01979	187·65	65·319	0·4342
70	0·97781	1·02270	233·81	63·855	0·4049
75	0·97489	1·02576	289·22	62·425	0·3788
80	0·97183	1·02899	355·31	61·027	0·3554
85	0·96865	1·03237	433·64	59·657	0·3345
90	0·96534	1·03590	525·92	58·317	0·3156
95	0·96192	1·03959	634·04	57·005	0·2985
100	0·95838	1·04343	760·00	55·720	0·2829

REFERENCES

[1] Density and Specific Volume: *Int. crit. Tab.*, Vol. III, pp. 25–26; see also OWEN, B. B., WHITE, J. R. and SMITH, J. S., *J. Amer. chem. Soc.*, 78 (1956) 3561

[2] Vapour Pressure: KEYES, F. G., *J. chem. Phys.*, 15 (1947) 602

[3] Dielectric constant: MALMBERG, C. G. and MARYOTT, A. A., *J. Res. nat. Bur. Stand.*, 56 (1956) 1.

[4] Viscosity: SWINDELLS, J. F., COE, J. R. and GODFREY, T. B., *ibid.*, 48 (1952) 1; COE, J. R. and GODFREY, T. B., *J. appl. Phys.*, 15 (1944) 625; WEBER, W., *Z. angew. Phys.*, 7 (1955) 96.

APPENDIX 1.2

DENSITIES, dielectric constant and viscosities of some electrolytic solvents. (Temperature 25°C unless otherwise noted.)

Solvent	*Density* (g/ml)	*Dielectric Constant*	*Viscosity* (*centipoise*)
Water	0·99707	78·30	0·8903
Acetone	0·7850	20·70	0·3040
Acetonitrile	0·7768	36·7	0·344
Ammonia (−34°)	0·6826	22	0·2558
Benzene	0·87368	2·273	0·6028
o-Dichlorobenzene	1·3003	9·93	1·96
1 : 1 Dichloroethane	1·1667	10·00	0·466
1 : 2 Dichloroethane	1·2453	10·36	0·787
Dimethylacetamide	0·9366	37·78	0·919
Dimethylformamide	0·9443	36·71	0·796
Dimethylpropionamide	0·9205	32·9	0·935
Dimethylsulphoxide	1·0958	46·7	1·96
Dioxan	1·0269	2·209	1·196
Ethanol	0·7851	24·30	1·078
Ethylenediamine	0·8922	12·9	1·54
Formamide	1·1292	109·5	3·302
Glycerol	1·2583	42·5	945
Hydrogen cyanide (18°)	0·6900	118·3	0·206
Hydrogen peroxide (20°)	1·4489	74	1·24
Methanol	0·7868	32·63	0·5445
N-Methylacetamide (40°)	0·9420	165·5	3·020
N-Methylbutyramide (30°)	0·9068	124·7	7·472
N-Methylformamide	0·9976	182·4	1·65
N-Methylpropionamide (30°)	0·9269	164·3	4·568
Nitrobenzene	1·1986	34·82	1·811
n-Propanol	0·7995	20·1	2·004
Pyridine	0·9779	12·0	0·8824
Sulphuric acid	1·8255	101	24·54
Sulpholane (30°)	1·2623	$43{\cdot}3_3$	10·29

The above values are selected from a wide variety of sources. Extensive references may be found in TIMMERMANS, J., 'Physiochemical Constants of Pure Organic Compounds,' Elsevier (1950), in WEISSBERGER, A., and PROSKAUER, E., 'Organic Solvents,' Interscience Publishers Inc., New York (1955) and in numerous papers by KRAUS, C. A., and collaborators (see refs. to Appendix 14.2) and WALDEN, P., and collaborators.

In most cases the viscosities quoted have been obtained by viscometers calibrated using the older value of the viscosity of water; on the new basis, the values would be 0.3% lower.

The density of ethylenediamine was determined for us by Dr. P. W. Brewster in the laboratories of Prof. F. C. Schmidt, University of Indiana.

APPENDIX 2.1

RELATION between molality, mean molality, activity and mean activity coefficient for various valency types; m = molality of solute; a_B = activity of solute.

Subscripts 1, 2, refer to cation and anion respectively.

Type	*Example*	$\gamma_\pm$	$m_\pm = Qm$	$a_B = (m_\pm\gamma_\pm)^\nu$
Non-electrolyte	Sucrose	—	—	$m\gamma$
1 : 1, 2 : 2, 3 : 3	KCl, $ZnSO_4$, $LaFe(CN)_6$	$(\gamma_1\gamma_2)^{1/2}$	m	$m^2\gamma_\pm^2$
2 : 1	$CaCl_2$	$(\gamma_1\gamma_2^2)^{1/3}$	$4^{1/3}m$	$4m^3\gamma_\pm^3$
1 : 2	Na_2SO_4	$(\gamma_1^2\gamma_2)^{1/3}$	$4^{1/3}m$	$4m^3\gamma_\pm^3$
3 : 1	$LaCl_3$	$(\gamma_1\gamma_2^3)^{1/4}$	$27^{1/4}m$	$27m^4\gamma_\pm^4$
1 : 3	$K_3Fe(CN)_6$	$(\gamma_1^3\gamma_2)^{1/4}$	$27^{1/4}m$	$27m^4\gamma_\pm^4$
4 : 1	$Th(NO_3)_4$	$(\gamma_1\gamma_2^4)^{1/5}$	$256^{1/5}m$	$256m^5\gamma_\pm^5$
1 : 4	$K_4Fe(CN)_6$	$(\gamma_1^4\gamma_2)^{1/5}$	$256^{1/5}m$	$256m^5\gamma_\pm^5$
3 : 2	$Al_2(SO_4)_3$	$(\gamma_1^2\gamma_2^3)^{1/5}$	$108^{1/5}m$	$108m^5\gamma_\pm^5$

$$Q = (\nu_1^{\nu_1}\,\nu_2^{\nu_2})^{1/\nu}$$

APPENDIX 2.2

The function $\sigma(x) = \frac{3}{x^3}\left[1 + x - \frac{1}{1+x} - 2\ln(1+x)\right]$

x	$\sigma(x)$	x	$\sigma(x)$	x	$\sigma(x)$	x	$\sigma(x)$
0	1·0000	0·50	0·5377	1·00	0·3411	2·0	0·17604
0·05	0·9293	0·55	·5108	1·10	0·3154	2·25	0·15407
0·10	0·8662	0·60	·4860	1·20	0·2926	2·50	0·13608
0·15	0·8097	0·65	·4631	1·30	0·2723	2·75	0·12115
0·20	0·7588	0·70	·4418	1·40	0·2541	3·00	0·10860
0·25	0·7129	0·75	·4220	1·50	0·2377	3·25	0·09796
0·30	0·6712	0·80	·4035	1·60	0·2229	3·50	0·08884
0·35	0·6332	0·85	·3863	1·70	0·2095	3·75	0·08096
0·40	0·5986	0·90	·3703	1·80	0·1973	4·00	0·07412
0·45	0·5668	0·95	·3553	1·90	0·1862	4·25	0·06812

APPENDIX 2.3

Values of the function $\phi^0 = 1 - 1{\cdot}352\sqrt{m}\,\sigma(\beta\sqrt{m})$ applicable to 2:1 and 1:2 electrolytes at 25° in water. See Equations 8.4a and 8.4b and Appendix 2.2.

β \ m =	0·1	0·2	0·3	0·4
1·8	0·7858	0·7571	0·7445	0·7381
2·0	0·7986	0·7751	0·7656	0·7613
2·2	0·8104	0·7911	0·7841	0·7814
2·4	0·8210	0·8053	0·8004	0·7991
2·6	0·8308	0·8181	0·8149	0·8146
2·8	0·8398	0·8296	0·8278	0·8283
3·0	0·8479	0·8401	0·8393	0·8405
3·2	0·8554	0·8496	0·8497	0·8515
3·4	0·8626	0·8582	0·8592	0·8612

[1] GUGGENHEIM, E. A. and STOKES, R. H., *Trans. Faraday Soc.*, 54 (1958) 1646

APPENDIX 3.1

Table of Ionic Radii (in Ångström units)

Li^{+}	0·60	Be^{++}	0·31					F^{-}	1·36
Na^{+}	0·95	Mg^{++}	0·65	Al^{+++}	0·50			Cl^{-}	1·81
K^{+}	1·33	Ca^{++}	0·99					Br^{-}	1·95
NH_4^{+}	(1·48)								
Rb^{+}	1·48	Sr^{++}	1·13					I^{-}	2·16
Cs^{+}	1·69	Ba^{++}	1·35	La^{+++}	1·15				
						Mn^{++}	(0·80)		
						Fe^{++}	(0·75)		
						Co^{++}	(0·72)		
						Ni^{++}	(0·70)		
		Zn^{++}	0·74						
Ag^{+}	1·26	Cd^{++}	0·97						
Tl^{+}	(1·44)	Hg^{++}	1·10	Tl^{+++}	0·95				

Data from PAULING, L., 'The Nature of the Chemical Bond,' Chap. X, Cornell University Press (1940). The figures in brackets were obtained by Pauling from a set of data given by GOLDSCHMIDT, V. M., 'Geochemische Verteilungsgesetze der Elemente,' Skrifter det Norske Videnskaps. Akad. Oslo I. Matem-Naturvid Klasse (1926); *idem, Trans. Faraday Soc.*, 25 (1929) 253.

APPENDIX 5.1

Specific Conductivities of Potassium Chloride Solutions (ref. 5.6)

Solution (demal)	g KCl/1000 g *solution* (*in vacuo*)	K_{sp} (int. ohm^{-1} cm^{-1})		
		0°	18°	25°
1 D	71·1352	$0{\cdot}06517_6$	$0{\cdot}09783_8$	$0{\cdot}11134_2$
0·1 D	7·41913	$0{\cdot}007137_9$	$0{\cdot}011166_7$	$0{\cdot}012856_0$
0·01 D	0·745263	$0{\cdot}0007736_4$	$0{\cdot}0012205_2$	0·0014087

APPENDIX 6.1

Limiting Equivalent Conductivities of Ions at 25° in water

λ^0 in cm^2 Int. Ω^{-1} equiv.$^{-1}$

Ion	$\lambda°$*	*Reference*	*Ion*	$\lambda°$	*Reference*
H^+	$349{\cdot}8_1$	1, 2	OH^-	$199{\cdot}1_8$	3
Li^+	$38{\cdot}6_8$	2	F^-	$55{\cdot}4$	4
Na^+	$50{\cdot}10$	2, 5	Cl^-	$76{\cdot}35$	5, 6
K^+	$73{\cdot}50$	5, 6	Br^-	$78{\cdot}14$	5, 6
Rb^+	$77{\cdot}8_1$	7	I^-	$76{\cdot}8_4$	6
Cs^+	$77{\cdot}2_6$	7	N_3^-	69	8
Ag^+	$61{\cdot}9_0$	2	NO_3^-	$71{\cdot}46$	2
Tl^+	$74{\cdot}7$	9	ClO_3^-	$64{\cdot}6$	10
NH_4^+	$73{\cdot}5_5$	11	BrO_3^-	$55{\cdot}7_4$	12
$CH_3NH_3^+$	$58{\cdot}7_2$	11a	IO_3^-	$40{\cdot}5_4$	12c
$(CH_3)_2NH_2^+$	$51{\cdot}8_7$	11a	ClO_4	$67{\cdot}3_6$	13
$(CH_3)_3NH^+$	$47{\cdot}2_5$	11a	IO_4^-	$54{\cdot}5_5$	10
NMe_4^+	$44{\cdot}9_2$	12	ReO_4^-	$54{\cdot}9_7$	10
NEt_4^+	$32{\cdot}6_6$	12	HCO_3^-	$44{\cdot}5_0$	14
NPr_4^+	$23{\cdot}4_2$	12	Formate	$54{\cdot}5_9$	15
NBu_4^+	$19{\cdot}4_7$	12	Acetate	$40{\cdot}9_0$	16
NAm_4^+	$17{\cdot}4_7$	12	Bromoacetate	$39{\cdot}2_2$	16a
$(CH_3)_3(C_6H_5)N^+$	$34{\cdot}6_5$	12a	Chloroacetate	$42{\cdot}2_0$	16a
$CH_2OH{\cdot}CH_2{\cdot}NH_3^+$	$42{\cdot}2_3$	12b	Cyanoacetate	$43{\cdot}4_3$	16b
Be^{++}	45	4	Fluoroacetate	$44{\cdot}3_9$	16a
Mg^{++}	$53{\cdot}0_5$	17	Iodoacetate	$40{\cdot}6_0$	16b
Ca^{++}	$59{\cdot}50$	17, 19	Propionate	$35{\cdot}8$	18
Sr^{++}	$59{\cdot}4_5$	17	Butyrate	$32{\cdot}6$	18
Ba^{++}	$63{\cdot}6_3$	17	Benzoate	$32{\cdot}3_8$	20
Cu^{++}	$53{\cdot}6$	21	Picrate	$30{\cdot}39$	12
Zn^{++}	$52{\cdot}8$	21	SO_4^{--}	$80{\cdot}0_2$	22
Co^{++}	55	24	$C_2O_4^{--}$	$74{\cdot}1_5$	23
Pb^{++}	$69{\cdot}5$	24a	CO_3^{--}	$69{\cdot}3$	25
La^{+++}	$69{\cdot}7$	26	$Fe(CN)_6^{---}$	$100{\cdot}9$	27
Ce^{+++}	$69{\cdot}8$	26	$P_3O_9^{---}$	$83{\cdot}6$	28
Pr^{+++}	$69{\cdot}6$	26	$Fe(CN)_6^{----}$	$110{\cdot}_5$	29
Nd^{+++}	$69{\cdot}4$	26	$P_4O_{12}^{----}$	$93{\cdot}_7$	28
Sm^{+++}	$68{\cdot}5$	26	$P_2O_7^{----}$	$95{\cdot}_9$	30
Eu^{+++}	$67{\cdot}8$	26	$P_3O_{10}^{5-}$	109	31
Gd^{+++}	$67{\cdot}3$	26	$[Co(NH_3)_6]^{+++}$	$101{\cdot}9$	27
Dy^{+++}	$65{\cdot}6$	26	$[Co_2\,tri\text{-}en_3]^{6+}$	$68{\cdot}_7$	32
Ho^{+++}	$66{\cdot}3$	26	$[Ni_2\,tri\text{-}en_3]^{4+}$	$52{\cdot}_5$	33
Er^{+++}	$65{\cdot}9$	26			
Tm^{+++}	$65{\cdot}4$	26			
Yb^{+++}	$65{\cdot}6$	26			

* The number of significant figures given is such that the last figure, if printed normally, is considered reliable within 1 or 2; and if printed below the line of the other figures, within about 5.

APPENDIX 6.1

REFERENCES TO APPENDIX 6.1

[1] OWEN, B. B. and SWEETON, F. H., *J. Amer. chem. Soc.*, 63 (1941) 2811
[2] SHEDLOVSKY, T., *ibid.*, 54 (1932) 1411
[3] MARSH, K. N. and STOKES, R. H., *Aust. J. Chem.* 17 (1964) 740
[4] WALDEN, P., LANDOLT-BÖRNSTEIN, 'Tabellen', Eg. IIIc, p. 2059, Julius Springer, Berlin (1936)
[5] BENSON, G. C. and GORDON, A. R., *J. chem. Phys.*, 13 (1945) 473
[6] OWEN, B. B. and ZELDES, H., *ibid.*, 18 (1950) 1083
[7] VOISENET, W. E., Thesis, Yale (1951); OWEN, B. B., *J. chim. phys.*, 49 (1952) C-72
[8] SEMENCHENKO, V. and SERPINSKII, V. V., *Z. phys. Chem.*, 167 A (1933) 197
[9] ROBINSON, R. A. and DAVIES, C. W., *J. chem. Soc.*, (1937) 574
[10] MONK, C. B., *J. Amer. chem. Soc.*, 70 (1948) 3281
[11] LONGSWORTH, L. G., *ibid.*, 57 (1935) 1185
[11a] JONES, J. H., SPUHLER, F. J. and FELSING, W. A., *ibid.*, 64 (1942) 965
[12] DAGGETT, H. M., BAIR, E. J. and KRAUS, C. A., *ibid.*, 73 (1951) 799
[12a] MCDOWELL, M. J. and KRAUS, C. A., *ibid.*, 73 (1951) 2170; SEARS, P. G., WILHOIT, E. D. and DAWSON, L. R., *J. chem. Phys,*. 23 (1955) 1274
[12b] SIVERTZ, V., REITMEIER, R. E. and TARTAR, H. V., *J. Amer. chem. Soc.*, 62 (1940) 1379
[12c] SPIRO, M., *J. phys. Chem.*, 60 (1956) 976; KRIEGER, K. A. and KILPATRICK, M., *J. Amer. chem. Soc.*, 64 (1942) 7
[13] JONES, J. H., *ibid.*, 67 (1945) 855
[14] SHEDLOVSKY, T. and MACINNES, D. A., *ibid.*, 57 (1935) 1705
[15] SAXTON, B. and DARKEN, L. S., *ibid.*, 62 (1940) 846
[16] MACINNES, D. A. and SHEDLOVSKY, T., *ibid.*, 54 (1932) 1429
[16a] IVES, D. J. G. and PRYOR, J. H., *J. chem. Soc.*, (1955) 2104; values of Λ^0 are given from 15° to 35°
[16b] FEATES, F. S. and IVES, D. J. G., *ibid.*, (1956) 2798; values of Λ^0 are given from 5° to 45°
[17] SHEDLOVSKY, T. and BROWN, A. S., *J. Amer. chem. Soc.*, 56 (1934) 1066
[18] BELCHER, D., *ibid.*, 60 (1938) 2744
[19] BENSON, G. C. and GORDON, A. R., *J. chem. Phys.*, 13 (1945) 470
[20] BROCKMAN, F. G. and KILPATRICK, M., *J. Amer. chem. Soc.*, 56 (1934) 1483
[21] OWEN, B. B. and GURRY, R. W., *ibid.*, 60 (1938) 3074
[22] JENKINS, I. L. and MONK, C. B., *ibid.*, 72 (1950) 2695
[23] DARKEN, L. S., *ibid.*, 63 (1941) 1007
[24] CANTELLO, R. C. and BERGER, A. J., *ibid.*, 52 (1930) 2648
[24a] NANCOLLAS, G. H., *J. chem. Soc.*, (1955) 1458
[25] MONK, C. B., *ibid.*, (1949) 429
[26] SPEDDING, F. H., PORTER, P. E. and WRIGHT, J. M., *J. Amer. chem. Soc.*, 74 (1952) 2055; SPEDDING, F. H. and YAFFE, I. S., *ibid.*, 74 (1952) 4751; SPEDDING, F. H. and DYE, J. L., *ibid.*, 76 (1954) 879
[27] HARTLEY, G. S. and DONALDSON, G. W., *Trans. Faraday Soc.*, 33 (1937) 457
[28] DAVIES, C. W. and MONK, C. B., *J. chem. Soc.* (1949) 413
[29] OWEN, B. B., *J. Amer. chem. Soc.*, 61 (1939) 1393
[30] MONK, C. B., *J. chem. Soc.* (1949) 423
[31] — *ibid.*, (1949) 427
[32] JAMES, J. C., *Trans. Faraday Soc.*, 47 (1951) 392
[33] DAVIES, C. W. and OWEN, B. D. R., *ibid.*, 52 (1956) 998

APPENDIX 6.2

Limiting Equivalent Conductivities, λ^0, of Ions in Water at Various Temperatures

Ion	0°	5°	15°	18°	25°	35°	45°	55°	100°
H^+	225	250·1	300·6	315	349·8$_1$	397·0	441·4	483·1	630
OH^-	105	—	165·9$_3$	175·8$_2$	199·1$_8$	233·0$_4$	267·2$_3$	301·4$_4$	450
Li^+	19·4	22·7$_6$	30·2$_0$	32·8	38·6$_8$	48·0$_0$	58·0$_4$	68·7$_4$	115
Na^+	26·5	30·3$_0$	39·7$_7$	42·8	50·10	61·5$_4$	73·7$_3$	86·8$_8$	145
K^+	40·7	46·7$_5$	59·6$_6$	63·9	73·50	88·2$_1$	103·4$_9$	119·2$_9$	195
Rb^+	43·9	50·1$_3$	63·4$_4$	66·5	77·8$_1$	92·9$_1$	108·5$_5$	124·2$_5$	—
Cs^+	44	50·0$_3$	63·1$_6$	67	77·2$_6$	92·1$_0$	107·5$_3$	123·6$_6$	—
Ag^+	33·1	—	—	53·5	61·9$_0$	—	—	—	175
NH_4^+	40·2	—	—	63·9	73·5$_5$	88·7$_3$	—	—	180
$N(CH_3)_4^+$	24·1	—	—	40·0	44·9$_2$	—	—	—	—
$N(C_2H_5)_4^+$	16·4	—	—	28·2	32·6$_6$	—	—	—	—
$N(C_3H_7)_4^+$	11·5	—	—	20·9	23·4$_2$	—	—	—	—
$N(C_4H_9)_4^+$	9·6	—	—	—	19·4$_7$	—	—	—	—
$N(C_5H_{11})_4^+$	8·8	—	—	—	17·4$_7$	—	—	—	—
F^-	—	—	—	47·3	55·4	—	—	—	—
Cl^-	41·0	47·5$_1$	61·4$_1$	66·0	76·35	92·2$_1$	108·9$_2$	126·4$_0$	212
Br^-	42·6	49·2$_5$	63·1$_5$	68·0	78·1$_4$	94·0$_3$	110·6$_8$	127·8$_6$	—
I^-	41·4	48·5$_7$	62·1$_7$	66·5	76·8$_4$	92·3$_9$	108·6$_4$	125·4$_4$	—
NO_3^-	40·0	—	—	62·3	71·46	85·4$_8$	—	—	195
ClO_4^-	36·9	—	—	58·8	67·3$_6$	—	—	—	185
Acetate	20·1	—	—	35	40·9$_0$	—	—	—	—
Mg^{++}	28·9	—	—	44·9	53·0$_5$	—	—	—	165
Ca^{++}	31·2	—	46·9$_8$	50·7	59·50	73·2$_6$	88·2$_1$	—	180
Sr^{++}	31	—	—	50·9	59·4$_5$	—	—	—	—
Ba^{++}	34·0	—	—	54·6	63·6$_3$	—	—	—	195
Cd^{++}	—	—	—	44·8	—	—	—	—	—
La^{+++}	34·$_4$	—	—	59·5	69·7$_5$	—	—	—	215
SO_4^{--}	41	—	—	68·4	80·0$_2$	—	—	—	260
Viscosity of water (centipoises)	1·787	1·516	1·138	1·053	0·8903	0·7194	0·5963	0·5044	0·2829

Note: The data at 25° are from references in Appendix 6.1. Those for OH^- from 15 to 55° are from the paper by Marsh, K. N. and Stokes, R. H., *Aust. J. Chem.*, 17 (1964) 740; λ^0 (OH^-)=368·8$_2$ at 75°; otherwise the data at 0° and 18° are from the limiting conductivities given in Landolt-Börnstein, based on $\lambda^0_{Cl^-}$(0°) = 41·0, $\lambda^0_{Cl^-}$(18°) = 66·0.

Data at 5°, 15°, 35°, 45° and 55° from papers of Gordon and his co-workers, and Owen *et al.* Those for the NH_4^+ and NO_3^- ions at 35° are due to Campbell, A. N. and Bock, E., *Canad. J. Chem.* 36 (1958) 330.

Data at 100° from Landolt-Börnstein, readjusted to base $\lambda^0_{Cl^-}$(100°) = 212.

The values at 100° are reliable only within several units; those at 5°, 15°, 25°, 35°, 45° to within the last figure given; those at 0° and 18° to within two or three units in the last figure given.

APPENDIX 6.3

Table 1

Equivalent Conductivity of Typical Electrolytes up to High Concentrations, in Aqueous Solution at 25°; c *in* mole/l; Λ *in* cm² int Ω^{-1} equiv^{-1}

c	NaCl	KCl	$BaCl_2$	$LaCl_3$
0·0	126·45	149·85	139·98	146·0
0·0005	124·51	147·81	134·34	135·21
0·001	123·74	146·95	132·27	131·16
0·005	120·64	143·55	123·94	118·11
0·01	118·53	141·27	119·09	111·25
0·02	115·76	138·34	—	—
0·05	111·06	133·37	105·19	94·95
0·1	106·74	128·96	98·68	87·89
0·2	101·71	124·08	—	—
0·5	93·62	117·27	80·60	66·68
1	85·76	111·87	68·98	51·15
1·5	79·86	108·27	—	—
2	74·71	105·23	—	—
3	65·57	99·46	—	—
4	57·23	93·46	—	—
5	49·46	—	—	—
Ref.	1, 2	1, 2	3, 4, 7	5, 6

[1] SHEDLOVSKY, T., *J. Amer. chem. Soc.*, 54 (1932) 1411
[2] CHAMBERS, J. F., STOKES, J. M. and STOKES, R. H., *J. phys. Chem.*, 60 (1956) 985
[3] JONES, G. and DOLE, M., *J. Amer. chem. Soc.*, 52 (1930) 2245
[4] SHEDLOVSKY, T. and BROWN, A. S., *ibid.*, 56 (1934) 1066
[5] JONES, G. and BICKFORD, C. F., *ibid.*, 56 (1934) 602
[6] LONGSWORTH, L. G. and MACINNES, D. A., *ibid.*, 60 (1938) 3070
[7] CALVERT, R., CORNELIUS, J. A., GRIFFITHS, V. S. and STOCK, D. I., *J. phys. Chem.*, 62 (1958) 47

APPENDIX 6.3

Table 2

Bibliography Recent Conductance Measurements in Concentrated Aqueous Solutions

Solute	*Max. conc. mole/l.*	*Temp.* °C	*Ref.*	*Solute*	*Max. conc. mole/l.*	*Temp.* °C	*Ref.*
HCl	9–12	5–65	1	KBr	3·75	0, 25	9*
$LiClO_3$	19	25	2*	KI	6	25, 50	6
$LiClO_3$	23·11†	131·8	2*	KH_2PO_4	1·9	25	8
$LiNO_3$	13·6	25	3*	NH_4Cl	5	25	10
$LiNO_3$	14·4	110	3*	NH_4NO_3	8	25	10
NaCl	5	25	5	NH_4NO_3	11	25, 35	11*
NaCl	5	50	6	NH_4NO_3	15	95	16*
NaI	10	0, 30, 50	7*	NH_4NO_3	18·0†	180	12*
$NaClO_3$	10	0, 30, 50	7*	$AgNO_3$	8	25, 35	11*
NaCNS	10	0, 30, 50	7*	$AgNO_3$	14	95	16*
Na_2HPO_4	3·9	25	8	$AgNO_3$	23·19†	221·7	12*
KCl	4	25	5	H_2SO_4	18	50, 75	13*
$LiNO_3$ in EtOH and EtOH—H_2O	3–11	25	4*	H_2SO_4	18	25–155	14
				H_3PO_4	18	25	8*
				$K_3Fe(CN)_6$	1	25	15
				$K_4Fe(CN)_6$	0·7	25	15
				$MgSO_4$	2·9	25	15
				HCOOK	6·5	50·5	17*
				HCOOCs	10	50·5	17*

* These papers include viscosity data. † Fused salt.

For work at lower concentrations, see references to *Table 1;* for extensive earlier work, usually of lower accuracy, see *Int. crit. Tab.*, Vol. VI, pp. 230–256.

[1] Owen, B. B. and Sweeton, F. H., *J. Amer. chem. Soc.*, 63 (1941) 2811

[2] Campbell, A. N. and Patterson, W. G., *Canad. J. Chem.*, 36 (1958) 1004

[3] Campbell, A. N., Debus, G. H. and Kartzmark, E. M., *ibid.*, 33 (1955) 1508

[4] Campbell, A. N., Debus, G. H. and Kartzmark, E. M., *ibid.*, 34 (1956) 1232

[5] Chambers, J. F., Stokes, J. M. and Stokes, R. H., *J. phys. Chem.*, 60 (1956) 985

[6] Chambers, J. F., *ibid.*, 62 (1958) 1136

[7] Miller, M. L., *ibid.*, 60 (1956) 189

[8] Mason, C. M. and Culvern, J. B., *J. Amer. chem. Soc.*, 71 (1949) 2387

[9] Jones, G. and Bickford, C. E., *ibid.*, 56 (1934) 602

[10] Wishaw, B. F. and Stokes, R. H., *ibid.*, 76 (1954) 2065

[11] Campbell, A. N. and Kartzmark, E. M., *Canad. J. Res.*, 28B (1950) 43; Campbell, A. N., Gray, A. P. and Kartzmark, E. M., *Canad. J. Chem.*, 31 (1953) 617

[12] Campbell, A. N., Kartzmark, E. M., Bednas, M. E. and Herron, J. T., *ibid.*, 32 (1954) 1051

[13] Campbell, A. N., Kartzmark, E. M., Bisset, D. and Bednas, M. E., *ibid.*, 31 (1953) 303

[14] Roughton, J. E., *J. appl. Chem.*, 1 (1951) S141

[15] Calvert, R., Cornelius, J. A., Griffiths, V. S. and Stock, D. I., *J. phys. Chem.*, 62 (1958) 47

[16] Campbell, A. N., Kartzmark, E. M., *Can. J. Chem.*, 30 (1952) 128

[17] Rice, M. J. and Kraus, C. A., *Proc. Nat. Acad. Sci. (U.S.A.)*, 39 (1953) 802

APPENDIX 7.1

Values of Parameters in the Debye-Hückel-Onsager equations for Aqueous 1 : 1 Electrolytes

°C	$10^8 \frac{e^2}{\varepsilon kT}$ (cm)	A	10^{-8} B	B_1	B_2
0	6·971	0·4918	0·3248	0·2211	29·82
5	7·004	·4952	·3256	·2227	35·23
10	7·039	·4989	·3264	·2243	41·00
15	7·076	·5028	·3273	·2261	47·18
18	7·099	·5053	·3278	·2271	51·07
20	7·115	·5070	·3282	·2280	53·73
25	7·156	·5115	·3291	·2300	60·65
30	7·200	·5161	·3301	·2321	67·91
35	7·246	·5211	·3312	·2343	75·52
38	7·274	·5242	·3318	·2357	80·25
40	7·294	·5262	·3323	·2366	83·46
45	7·344	·5317	·3334	·2391	91·72
50	7·396	·5373	·3346	·2416	100·4
55	7·450	·5432	·3358	·2443	109·2
60	7·506	·5494	·3371	·2470	118·5
65	7·564	·5558	·3384	·2499	127·8
70	7·625	·5625	·3397	·2529	137·6
75	7·688	·5695	·3411	·2560	147·7
80	7·752	·5767	·3426	·2593	158·1
85	7·820	·5842	·3440	·2627	168·7
90	7·889	·5920	·3456	·2662	179·6
95	7·961	·6001	·3471	·2698	190·8
100	8·036	·6086	·3488	·2736	202·2

For polyvalent electrolytes the following factors are useful:

Type	*Example*	$\frac{\lvert z_1 z_2 \rvert \sqrt{I}}{\sqrt{c}}$	$\frac{\sqrt{I}}{\sqrt{c}}$
1 : 1	KCl	1	1
2 : 1	$CaCl_2$	$2\sqrt{3} =$ 3·464	$\sqrt{3} =$ 1·732
2 : 2	$ZnSO_4$	8·000	2·000
3 : 1	$LaCl_3$	$3\sqrt{6} =$ 7·350	$\sqrt{6} =$ 2·450
4 : 1	$Th(NO_3)_4$	$4\sqrt{10} =$ 12·65	$\sqrt{10} =$ 3·162
3 : 2	$Al_2(SO_4)_3$	$6\sqrt{15} =$ 23·24	$\sqrt{15} =$ 3·873
3 : 3	$LaFe(CN)_6$	27·00	3·000

The units of A of equation (9.7) are mole$^{-\frac{1}{2}}$ $l^{\frac{1}{2}}$, those of B in equation (9.7) are cm^{-1} mole$^{-\frac{1}{2}}$ $l^{\frac{1}{2}}$, those of B_1 of equation (7.35) are mole$^{-\frac{1}{2}}$ $l^{\frac{1}{2}}$ and those of B_2 of equation (7.32) are cm^2 Ω^{-1} equiv.$^{-1}$ $(l/\text{mole})^{\frac{1}{2}}$.

APPENDIX 8.1

Table 1

Values of $k = 2{\cdot}3026\ \boldsymbol{RT/F}$ *and of* $\boldsymbol{RT/F^2}$

°C	k (abs. v)	$10^7\ \boldsymbol{RT/F^2}$
0	0·054197	2·4381
5	0·055189	2·4827
10	0·056182	2·5273
15	0·057173	2·5719
18	0·057768	2·5987
20	0·058165	2·6166
25	0·059158	2·6612
30	0·060149	2·7058
35	0·061141	2·7505
38	0·061736	2·7772
40	0·062133	2·7951
45	0·063126	2·8397
50	0·064117	2·8843
55	0·065109	2·9290
60	0·066102	2·9736
65	0·067093	3·0182
70	0·068085	3·0629
75	0·069078	3·1075
80	0·070069	3·1521
85	0·071061	3·1967
90	0·072054	3·2414
95	0·073046	3·2860
100	0·074037	3·3306

Values of k in international volts were given by MANOV, G. G., BATES, R. G., HAMER, W. J. and ACREE, S. F., *J. Amer. chem. Soc.*, 65(1943) 1765.

The values of $\boldsymbol{RT/F^2}$ in the third column are given in int. Ω equiv. sec^{-1} in order that they may be used with the usual cm^2 int. Ω^{-1} equiv^{-1} units of λ to give diffusion coefficients in cm^2 sec^{-1}, e.g., in equations (11.3) and (11.49).

APPENDIX 8.2

Standard cell potentials, E^0. The sign convention used is that the potential is recorded as positive if the right-hand electrode of the cell, as written, is positive with respect to the left-hand electrode. (Molal concentration scale.)

1. The cell: $H_2|HCl|AgCl, Ag$ in various solvents.

Table 1

Water		*Water (high temp.)*		°C	*Dioxan*				*Methanol*						
°C		°C			20%	45%	70%	82%	10%	20%	43·3%	64%	84·2%	94·2%	100%
0	0·23655	55	0·20056	0	0·21975	0·18938	0·10584	—	0·22762	0·22022	—	—	—	—	—
5	0·23413	60	0·19649	5	0·21677	0·18468	0·09784	− 0·0130	0·22547	0·21837	—	—	—	—	—
10	0·23142	70	0·18782	10	0·21362	0·17972	0·08970	− 0·0246	0·22328	0·21631	—	—	—	—	—
15	0·22857	80	0·1787	15	0·21025	0·17454	0·08123	− 0·0370	0·22085	0·21405	0·2010	0·1864	0·1452	0·0979	0·0014
20	0·22557	90	0·1695	20	0·20674	0·16916	0·07267	− 0·0487	0·21821	0·21155	0·1975	0·1813	0·1384	0·0908	− 0·0044
25	0·22234	95	0·1651	25	0·20303	0·16358	0·06395	− 0·0614	0·21535	0·20881	0·1939	0·1765	0·1319	0·0838	− 0·0103
30	0·21904			30	0·19914	0·15778	0·05500	− 0·0738	0·21220	0·20567	0·1901	0·1717	0·1252	0·0768	− 0·0164
35	0·21565			35	0·19505	0·15182	0·04587	− 0·0871	0·20892	0·20246	0·1860	0·1668	0·1184	0·0693	− 0·0228
40	0·21208			40	0·19080	0·14560	0·03661	− 0·1012	0·20550	0·19910	0·1818	0·1620	0·1114	0·0618	(− 0·0293)
45	0·20835			45	0·18634	0·13925	0·02705	− 0·1172	—	—	0·1771	0·1563	0·1039	0·0539	(− 0·0361)
50	0·20449			50	0·18171	0·13282	0·01746	—	—	—	—	—	—	—	—
Ref.	1				2	2	3	4	5	5	6	6	6	6	6

Table 1—(Contd.)

°C	*Ethanol*		Iso*propanol*			*Glycerol*			50% *Glycerol* (*high temp.*)	
	10%	20%	5%	10%	20%	10%	30%	50%	°C	
0	0·22726	0·21606	0·23106	0·22543	0·21612	0·23075	0·21684	0·20065	55	0·15890
5	—	—	0·22892	0·22365	0·21492	0·22824	0·21421	0·19760	60	0·15420
10	0·22328	0·21367	0·22654	0·22158	0·21336	0·22557	0·21141	0·19441	65	0·14936
15	—	—	0·22390	0·21922	0·21138	0·22274	0·20851	0·19103	70	0·14437
20	0.21901	0·21013	0·22107	0·21667	0·20906	0·21970	0·20545	0·18760	75	0·13912
25	0·21467	0·20757	0·21807	0·21383	0·20637	0·21650	0·20221	0·18398	80	0·13394
30	0·21383	0·20587	0·21494	0·21081	0·20341	0·21315	0·19882	0·18015	85	0·12838
35	—	—	0·21164	0·20754	0·20009	0·20965	0·19521	0·17618	90	0·12280
40	0·20783	0·19962	0·20809	0·20410	0·19652	0·20600	0·19140	0·17202		
45	—	—	—	—	—	—	—	0·16780		
50	—	—	—	—	—	—	—	0·16341		
	7	7	8	8	8	9	9	10		

Table 1—(Contd.)

The cell: $H_2|HCl|AgCl, Ag$ *at* 25°C

% Solvent	Acetone	2:3 Butylene glycol	Ethanol	Ethylene glycol	Fructose	Glucose	Glycerol	Methanol	n-Propanol	Propylene glycol
4·92	—	—	—	—	—	—	0·21960	—	—	—
5	0·2190	—	—	$0{\cdot}2190_5$	$0{\cdot}2190_0$	$0{\cdot}2186_3$	—	—	—	—
10	$0{\cdot}2156_5$	0·2144	—	$0{\cdot}2163_5$	$0{\cdot}2150_2$	$0{\cdot}2141_9$	—	—	0·2141	$0{\cdot}2150_5$
15	—	—	—	$0{\cdot}2133_0$	—	—	—	—	—	—
20	$0{\cdot}2079_5$	0·2063	—	$0{\cdot}2102_0$	—	$0{\cdot}2045_1$	—	0·2094	0·2066	$0{\cdot}2077_5$
21·2	—	—	—	—	—	—	$0{\cdot}2082_5$	—	—	—
30	—	—	$0{\cdot}2003_3$	$0{\cdot}2036_0$	—	$0{\cdot}1935_5$	—	—	—	—
40	$0{\cdot}1859_5$	—	$0{\cdot}1945_4$	$0{\cdot}1972_0$	—	—	—	0·1968	—	—
50	—	—	$0{\cdot}1858_8$	—	—	—	—	—	—	—
60	—	—	—	$0{\cdot}1807_0$	—	—	—	0·1818	—	—
80	—	—	—	—	—	—	—	0·1492	—	—
90	—	—	—	—	—	—	—	0·1135	—	—
100	—	—	$-0{\cdot}0813_8$	—	—	—	—	−0·0099	—	—
Ref.	11	12	13, 14	12, 15	16	17	18	19, 20	21	12

Methylethyl ketone (10%) $0{\cdot}2153_5$ V, (20%) 0·2078 V. Ref. 11a
Triethylene glycol (10%) $0{\cdot}2161_5$ V, (20%) $0{\cdot}209_4$ V. Ref. 11a
Fructose (17%) $0{\cdot}2088_6$ V, (25%) $0{\cdot}2020_0$ V. Ref. 16a
*Iso*propanol (100%) −0·109 V, *n*-butanol (100%) −0·132 V, *iso*butanol (100%) −0·134 V, *iso*amyl alcohol (100%) −0·134 V, Benzyl alcohol (100%) −0·163 V. Ref. 40
Formamide (100%) 0·204 V. Ref. 41
Acetic acid (100%) at 35°, $-0{\cdot}6208_4$ V. Formic acid (100%) at 35°, $-0{\cdot}1302_2$ V. Ref. 22

2. The cell: $H_2|HBr|AgBr, Ag, (E_1^0)$; or $H_2|HI|AgI, Ag, (E_2^0)$.

°C	0	5	10	15	20	25
E_1^0	0·08163	0·07991	0·07802	0·07595	0·07372	0·07131
E_2^0	—	− 0·14712	− 0·14805	− 0·14920	− 0·15062	− 0·15225
°C	30	35	40	45	50	
E_1	0·06872	0·06597	0·06304	0·05995	0·05667	—
E_2^0	− 0·15396	− 0·15586	− 0·15787	—	—	—

Ref. (23)

For the first cell in methanol at 25°, $E^0 = -0{\cdot}1328$ ref. (24) and in ethanol at 35°,

$$E_1^0 = -0{\cdot}06895 \text{ and } E_2^0 = -0{\cdot}2404_7 \text{ ref. (25).}$$

3. The cell: $H_2|HX|HgX, Hg$ where X = Cl, Br or I.

$$X = Cl, \quad E^0 = 0{\cdot}26796 \text{ at } 25° \text{ ref. (26).}$$

Over the temperature range 0° to 60°:

$$E^0 = 0{\cdot}26647 - 3{\cdot}465 \times 10^{-4} (t - 30) - 2{\cdot}87 \times 10^{-6} (t - 30)^2$$
$$8{\cdot}5 \times 10^{-9} (t - 30)^3 \quad \text{ref. (27).}$$

In water-methanol mixtures (x per cent methanol by weight):

x	20·22	43·12	68·33	97·29
E^0	0·2545	0·2415	0·2173	0·1027

Ref. (28)

$$X = Br, \quad E^0 = 0{\cdot}13956 \text{ at } 25° \text{ ref. (29).}$$

$$X = I, \quad E^0 = -0{\cdot}0405 \text{ at } 25° \text{ ref. (30).}$$

4. The cell: $H_2|H_2SO_4|Hg_2SO_4, Hg$.

°C	0	5	10	15	20	25	30
E^0	0·63495	0·63097	0·62704	0·62307	0·61930	0·61515	0·61107
°C	35	40	45	50	55	60	—
E^0	0·60701	0·60305	0·59900	0·59487	0·59051	0·58659	—

Ref. (31)

For this cell in methanol: ref. (32)

°C	20	25	30	35	—	—	—
E^0	0·5443	0·5392	0·5351	0·5318	—	—	—

in water-ethylene glycol mixtures (x per cent glycol by weight):

Ref. (33)	x	5	10	20	30
	E^0	0·6095	0·6077	0·6026	0·5982

5. The cell: $H_2|H_2SO_4|PbSO_4, PbO_2$ (Pt).*

°C	0	5	10	15	20	25	30
E^0	1·67694	1·67846	1·67998	1·68159	1·68322	1·68488	1·68671
°C	35	40	45	50	55	60	—
E^0	1·68847	1·69036	1·69231	1·69436	1·69649	1·69861	—

Ref. (34)

6. Cells: $M_xHg|MX_2|AgX$, Ag, M = Zn or Cd; X = Cl, Br or I.

°C	$ZnCl_2$	$ZnBr_2$	ZnI_2	$CdCl_2$	$CdBr_2$
0	—	—	—	0·58151	—
5	—	—	0·6176	0·58039	0·4250
10	0·99617	—	0·6161	0·57900	0·4248
15	0·99192	—	0·6145	0·57755	0·4243
20	0·98849	0·83684	0·6126	0·57581	0·4236
25	0·98485	0·83388	0·6105	0·57390	0·4227
30	0·98103	0·83084	0·6083	0·57175	0·4215
35	0·97702	0·82766	0·6059	0·56955	0·4201
40	0·97281	0·82430	0·6033	0·56730	0·4185
Ref.	35	36	37	38	39

* See also BECK, W. H., SINGH, K. P. and WYNNE-JONES, W. F. K., *Trans. Faraday Soc.* 55 (1959) 331.

APPENDIX 8·2

Data from:

[1] BATES, R. G. and BOWER, V. E., *J. Res. nat. Bur. Stand.*, 53 (1954) 283; see also HARNED, H. S. and EHLERS, R. W., *J. Amer. chem. Soc.*, 54 (1932) 1350. Their results, recalculated with Birge's values of the constants in k, differ on the average by less than 0·1 mV from those of BATES and BOWER; those of HARNED, H. S. and PAXTON, T. R., *J. phys. Chem.*, 57 (1953) 531 by less than 0·05 mV.

[2] HARNED, H. S., *J. Amer. chem. Soc.*, 60 (1938) 336

[3] HARNED, H. S. and CALMON, C., *ibid.*, 60 (1938) 2130

[4] HARNED, H. S., WALKER, F. and CALMON, C., *ibid.*, 61 (1939) 44; DANYLUK, S. S., TANIGUCHI, H. and JANZ, G. J., *J. phys. Chem.* 61 (1957) 1679

[5] HARNED, H. S. and THOMAS, H. C., *J. Amer. chem. Soc.*, 57 (1935) 1666

[6] AUSTIN, J. M., HUNT, A. H., JOHNSON, F. A. and PARTON, H. N., private communication

[7] PATTERSON, A. and FELSING, W. A., *J. Amer. chem. Soc.*, 64 (1942) 1478

[8] MOORE, R. L. and FELSING, W. A., *ibid.*, 69 (1947) 1076; HARNED, H. S. and ALLEN, D. S., *J. phys. Chem.*, 58 (1954) 191, have found $E^0 = 0{\cdot}2060$V for 20% *iso*propanol at 25°

[9] KNIGHT, S. B., CROCKFORD, H. D. and JAMES, F. W., *ibid.*, 57 (1953) 463

[10] HARNED, H. S. and NESTLER, F. H. M., *J. Amer. chem. Soc.*, 68 (1946) 665

[11] FEAKINS, D. and FRENCH, C. M., *J. chem. Soc.*, (1956) 3168

[11a] FEAKINS, D. and FRENCH, C. M., *J. chem. Soc.*, (1957) 2284

[12] CLAUSSEN, B. H. and FRENCH, C. M., *Trans. Faraday Soc.*, 51 (1955) 1124

[13] HARNED, H. S. and ALLEN, D. S., *J. phys. Chem.*, 58 (1954) 191

[14] TANIGUCHI, H. and JANZ, G. J., *ibid.*, 61 (1957) 688; MUKHERJEE, L. M., *J. phys. Chem.*, 58 (1954) 1042 found $E^0 = +\,0{\cdot}00977$ v. in 100% ethanol at 35°

[15] KNIGHT, S. B., MASI, J. F. and ROESEL, D., *J. Amer. chem. Soc.*, 68 (1946) 661; CROCKFORD, H. D., KNIGHT, S. B. and STATON, H. A., *ibid.*, 72 (1950) 2164

[16] CROCKFORD, H. D. and SAKHNOVSKY, A. A., *ibid.*, 73 (1951) 4177

[16a] CROCKFORD, H. D., LITTLE, W. F. and WOOD, W. A., *J. phys. Chem.*, 61 (1957) 1674

[17] WILLIAMS, J. P., KNIGHT, S. B. and CROCKFORD, H. D., *J. Amer. chem. Soc.*, 72 (1950) 1277

[18] LUCASSE, W. W., *Z. phys. Chem.*, 121 (1926) 254

[19] OJWA, I. T., *J. phys. Chem.*, 60 (1956) 754

[20] KOSKIKALLIO, J., *Suomen Kem.*, 30B (1957) 38, 43, 111

[21] CLAUSSEN, B. H. and FRENCH, C. M., *Trans. Faraday Soc.*, 51 (1955) 708

[22] MUKHERJEE, L. M., *J. Amer. chem. Soc.*, 79 (1957) 4040

[23] HARNED, H. S. and DONELSON, J. G., *ibid.*, 59 (1937) 1280; OWEN, B. B., *ibid.*, 57 (1935) 1526

[24] KANNING, E. W. and CAMPBELL, A. W., *ibid.*, 64 (1942) 517

[25] MUKHERJEE, L. M., *J. phys. Chem.*, 60 (1956) 974

[26] HILLS, G. J. and IVES, D. J. G., *J. chem. Soc.*, (1951) 318

[27] GRZYBOWSKI, A. K., *J. phys. Chem.*, 62 (1958) 550

[28] SCHWABE, K. and ZIEGENBALG, S., *Z. Elektrochem.*, 62 (1958) 172

[29] LARSON, W. D., *J. Amer. chem. Soc.*, 62 (1940) 765; DAKIN, T. W. and EWING, D. T., *ibid.*, 62 (1940) 2280

[30] BATES, R. G. and VOSBURGH, W. C., *ibid.*, 59 (1937) 1188

[31] HARNED, H. S. and HAMER, W. J., *ibid.*, 57 (1935) 27

[32] KANNING, E. W. and BOWMAN, M. G., *ibid.*, 68 (1946) 2042

[33] FRENCH, C. M. and HUSSAIN, Ch. F., *J. chem. Soc.*, (1955) 2211

[34] HAMER, W. J., *J. Amer. chem. Soc.*, 57 (1935) 9

[35] ROBINSON, R. A. and STOKES, R. H., *Trans. Faraday Soc.*, 36 (1940) 740

[36] STOKES, R. H. and STOKES, J. M., *ibid.*, 41 (1945) 688

[37] BATES, R. G., *J. Amer. chem. Soc.*, 60 (1938) 2983

[38] HARNED, H. S. and FITZGERALD, M. E., *ibid.*, 58 (1936) 2624; TREUMANN, W. B. and FERRIS, L. M., *ibid.*, 80 (1958) 5048

[39] BATES, R. G., *ibid.*, 61 (1939) 308

[40] IZMAILOV, N. A. and ALEXANDROV, V. V., *Zhur. fiz. khim.* 31 (1957) 2619

[41] MANDEL, M. and DECROLY, P., *Trans. Faraday Soc.*, 56 (1960) 29

APPENDIX 8.3

Table 1

Water Activities, Osmotic Coefficients, Activity Coefficients and Relative Molal Vapour Pressure Lowerings of Sodium and Potassium Chloride Solutions at 25°

m	Sodium chloride				Potassium chloride			
	a_w[a]	ϕ	$1 + \log \gamma$	$\frac{p^0 - p}{Mp^0}$	a_w	ϕ	$1 + \log \gamma$	$\frac{p^0 - p}{Mp^0}$
0·1	0·996646	0·9324	0·8912	0·03354	0·996668	0·9266	0·8864	0·03332
0·2	0·993360	0·9245	0·8661	0·03320	0·993443	0·9130	0·8562	0·03279
0·3	0·99009	0·9215	0·8511	0·03303	0·99025	0·9063	0·8373	0·03250
0·4	0·98682	0·9203	0·8406	0·03295	0·98709	0·9017	0·8233	0·03228
0·5	0·98355	0·9209	0·8332	0·03290	0·98394	0·8989	0·8124	0·03212
0·6	0·98025	0·9230	0·8278	0·03292	0·98078	0·8976	0·8038	0·03203
0·7	0·97692	0·9257	0·8240	0·03296	0·97763	0·8970	0·7967	0·03196
0·8	0·97359	0·9288	0·8211	0·03301	0·97448	0·8970	0·7907	0·03190
0·9	0·97023	0·9320	0·8190	0·03308	0·97133	0·8971	0·7854	0·03186
1·0	0·96686	0·9355	0·8175	0·03314	0·96818	0·8974	0·7809	0·03182
1·2	0·9601	0·9428	0·8158	0·03325	0·9619	0·8986	0·7733	0·03175
1·4	0·9532	0·9513	0·8159	0·03343	0·9556	0·9010	0·7676	0·03171
1·6	0·9461	0·9616	0·8178	0·03369	0·9492	0·9042	0·7634	0·03175
1·8	0·9389	0·9723	0·8208	0·03394	0·9428	0·9081	0·7603	0·03178
2·0	0·9316	0·9833	0·8245	0·03420	0·9364	0·9124	0·7580	0·03180
2·2	0·9242	0·9948	0·8291	0·03445	0·9299	0·9168	0·7564	0·03186
2·4	0·9166	1·0068	0·8344	0·03475	0·9234	0·9214	0·7554	0·03192
2·6	0·9089	1·0192	0·8402	0·03504	0·9169	0·9264	0·7549	0·03198
2·8	0·9011	1·0321	0·8466	0·03532	0·9103	0·9315	0·7548	0·03204
3·0	0·8932	1·0453	0·8535	0·03560	0·9037	0·9367	0·7550	0·03210
3·2	0·8851	1·0587	0·8608	0·03591	0·8971	0·9421	0·7557	0·03216
3·4	0·8769	1·0725	0·8684	0·03621	0·8904	0·9477	0·7567	0·03223
3·6	0·8686	1·0867	0·8766	0·03650	0·8837	0·9531	0·7578	0·03230
3·8	0·8600	1·1013	0·8852	0·03684	0·8770	0·9588	0·7593	0·03237
4·0	0·8515	1·1158	0·8939	0·03713	0·8702	0·9647	0·7610	0·03245
4·2	0·8428	1·1306	0·9029	0·03743	0·8634	0·9707	0·7629	0·03252
4·4	0·8339	1·1456	0·9122	0·03775	0·8566	0·9766	0·7649	0·03259
4·6	0·8250	1·1608	0·9218	0·03804	0·8498	0·9824	0·7670	0·03266
4·8	0·8160	1·1761	0·9315	0·03833	0·8429	0·9883	0·7693	0·03273
5·0	0·8068	1·1916	0·9415	0·03864	—	—	—	—
5·2	0·7976	1·2072	0·9517	0·03892	—	—	—	—
5·4	0·7883	1·2229	0·9620	0·03920	—	—	—	—
5·6	0·7788	1·2389	0·9726	0·03950	—	—	—	—
5·8	0·7693	1·2548	0·9833	0·03977	—	—	—	—
6·0	0·7598	1·2706	0·9940	0·04003	—	—	—	—

[a] Vapour pressures in columns 2, 5, 6 and 9 are tabulated relative to $p^0 = 23{\cdot}753$ mm for pure water at 25°.

APPENDIX 8.4

Table 1

Water Activities, Osmotic Coefficients and Activity Coefficients of Sulphuric Acid Solutions at 25°

m	a_w	ϕ	γ	m	a_w	ϕ	γ
0·1	0·99633	0·680	0 2655	19·0	0·0925	2·318	1·771
0·2	0·99281	0·668	0·2090	20·0	0·0796	2·341	1·940
0·3	0·98923	0·668	0·1826	21·0	0·0686	2·361	2·114
0·5	0·98190	0·676	0·1557	22·0	0·0589	2·381	2·300
0·7	0·97427	0·689	0·1417	23·0	0·0506	2·401	2·495
1·0	0·96176	0·721	0·1316	24·0	0·0441	2·407	2·666
1·5	0·93872	0·780	0·1263	26·0	0·0331	2·426	3·040
2·0	0·91261	0·846	0·1276	28·0	0·0250	2·438	3·423
2·5	0·8836	0·916	0·1331	30·0	0·0191	2·441	3·792
3·0	0·8516	0·991	0·1422	32·0	0·01472	2·439	4·152
3·5	0·8166	1·071	0·1547	34·0	0·01148	2·431	4·493
4·0	0·7799	1·150	0·1700	36·0	0·00900	2·421	4·828
4·5	0·7422	1·226	0·1875	38·0	0·00711	2·408	5·145
5·0	0·7032	1·303	0·2081	40·0	0·00575	2·386	5·406
5·5	0·6643	1·376	0·2312	42·0	0·00467	2·364	5·656
6·0	0·6259	1·445	0·2567	44·0	0·00381	2·342	5·891
6·5	0·5879	1·512	0·2852	46·0	0·00315	2·317	6·097
7·0	0·5509	1·576	0·3166	48·0	0·00262	2·291	6·278
7·5	0·5152	1·636	0·350	50·0	0·00220	2·265	6·443
8·0	0·4814	1·691	0·386	52·0	0·001855	2·238	6·586
8·5	0·4488	1·744	0·426	54·0	0·001585	2·209	6·700
9·0	0·4180	1·793	0·467	56·0	0·001355	2·182	6·817
9·5	0·3886	1·841	0·512	58·0	0·001168	2·154	6·906
10·0	0·3612	1·884	0·559	60·0	0·001010	2·127	6·982
11·0	0·3111	1·964	0·661	62·0	0·000882	2·099	7·045
12·0	0·2681	2·030	0·770	64·0	0·000774	2·071	7·091
13·0	0·2306	2·088	0·888	66·0	0·000684	2·043	7·125
14·0	0·1980	2·140	1·017	68·0	0·000606	2·016	7·153
15·0	0·1698	2·187	1·154	70·0	0·000537	1·990	7·171
16·0	0·1456	2·228	1·300	72·0	0·000480	1·964	7·181
17·0	0·1252	2·262	1·450	74·0	0·000430	1·938	7·184
18·0	0·1076	2·292	1·608	76·0	0·000387	1·913	7·182

APPENDIX 8.5

Table 1

Water Activities, Osmotic Coefficients and Activity Coefficients of Calcium Chloride Solutions at 25°

m	a_w	ϕ	γ	m	a_w	ϕ	γ
0·1	0·99540	0·854	0·518	3·0	0·7494	1·779	1·483
0·2	0·99073	0·862	0·472	3·5	0·6875	1·981	2·078
0·3	0·98590	0·876	0·455	4·0	0·6239	2·182	2·934
0·4	0·98086	0·894	0·448	4·5	0·5602	2·383	4·17
0·5	0·97552	0·917	0·448	5·0	0·4988	2·574	5·89
0·6	0·96998	0·940	0·453	5·5	0·4425	2·743	8·18
0·7	0·96423	0·963	0·460	6·0	0·3916	2·891	11·11
0·8	0·95818	0·988	0·470	6·5	0·3482	3·003	14·53
0·9	0·95174	1·017	0·484	7·0	0·3117	3·081	18·28
1·0	0·94504	1·046	0·500	7·5	0·2815	3·127	22·13
1·2	0·93072	1·107	0·539	8·0	0·2561	3·151	26·02
1·4	0·91521	1·171	0·587	8·5	0·2337	3·165	30·1
1·6	0·8986	1·237	0·644	9·0	0·2139	3·171	34·2
1·8	0·8808	1·305	0·712	9·5	0·1963	3·171	38·5
2·0	0·8618	1·376	0·792	10·0	0·1804	3·169	43·0
2·5	0·8091	1·568	1·063				

APPENDIX 8.6

Table 1

Water Activities, Osmotic Coefficients and Activity Coefficients of Sucrose Solutions at 25°

m	a_w	ϕ	$\log \gamma$	m	a_w	ϕ	$\log \gamma$
0·1	0·99819	1·0072	0·0062	1·6	0·96755	1·1447	0·1197
0·2	0·99635	1·0148	0·0127	1·8	0·96292	1·1652	0·1365
0·3	0·99449	1·0226	0·0193	2·0	0·95818	1·1857	0·1535
0·4	0·99260	1·0308	0·0262	2·5	0·94582	1·2369	0·1961
0·5	0·99068	1·0393	0·0333	3·0	0·93284	1·2863	0·2382
0·6	0·98874	1·0480	0·0405	3·5	0·91938	1·3331	0·2792
0·7	0·98676	1·0569	0·0479	4·0	0·90560	1·3761	0·3185
0·8	0·98475	1·0661	0·0554	4·5	0·8916	1·4152	0·3557
0·9	0·98272	1·0754	0·0631	5·0	0·8776	1·4500	0·3906
1·0	0·98065	1·0849	0·0709	5·5	0·8635	1·4809	0·4233
1·2	0·97641	1·1044	0·0868	6·0	0·8496	1·5084	0·4541
1·4	0·97204	1·1244	0·1031				

APPENDIX 8.7

Table 1

Molal Activity Coefficients of Salts at the Temperature of the Freezing Point of the Solution

m	LiCl	LiBr	$LiNO_3$	$LiClO_3$	$LiClO_4$	LiFor	LiAc	NaCl	NaBr	$NaNO_3$	$NaClO_3$	$NaClO_4$	NaFor	NaAc	KCl	KBr	KNO_3	$KClO_3$
0·001	0·964	0·967	0·967	0·967	0·968	0·967	0·967	0·967	0·968	0·967	0·967	0·967	0·967	0·967	0·967	0·967	0·966	0·968
0·002	0·950	0·955	0·955	0·955	0·956	0·954	0·954	0·955	0·957	0·954	0·954	0·953	0·954	0·954	0·954	0·954	0·953	0·956
0·005	0·924	0·934	0·932	0·933	0·935	0·931	0·931	0·932	0·937	0·931	0·930	0·929	0·932	0·932	0·930	0·931	0·927	0·933
0·01	0·899	0·891	0·910	0·911	0·915	0·907	0·908	0·909	0·917	0·906	0·905	0·904	0·909	0·910	0·905	0·906	0·899	0·908
0·02	0·870	0·886	0·882	0·884	0·890	0·876	0·878	0·880	0·891	0·874	0·873	0·873	0·880	0·882	0·874	0·876	0·862	0·874
0·05	0·826	0·847	0·839	0·843	0·853	0·827	0·832	0·831	0·847	0·818	0·819	0·821	0·833	0·841	0·821	0·824	0·794	0·810
0·1	0·791	0·817	0·804	0·810	0·825	0·784	0·794	0·787	0·808	0·765	0·769	0·773	0·792	0·808	0·773	0·778	0·726	0·745
0·2	0·760	0·793	0·772	0·782	0·805	0·739	0·758	0·740	0·765	0·701	0·711	0·720	0·748	0·778	0·720	0·726	0·642	0·665
0·3	0·748	0·787	0·758	0·771	0·804	0·714	0·742	0·712	0·741	0·659	0·673	0·686	0·722	0·766	0·687	0·694	0·584	—
0·4	0·744	0·788	0·752	0·768	0·810	0·697	0·733	0·693	0·726	0·627	0·645	0·662	0·704	0·760	0·663	0·670	0·540	—
0·5	0·745	0·795	0·751	0·769	0·821	0·686	0·729	0·678	0·714	0·601	0·621	0·640	0·691	0·759	0·645	0·652	0·504	—
0·6	0·748	0·805	0·752	0·773	0·835	0·677	0·728	0·666	0·705	0·578	0·602	0·623	0·681	0·760	0·630	0·638	0·473	—
0·7	0·755	0·818	0·755	0·780	0·852	0·672	0·730	0·657	0·699	0·559	0·585	0·609	0·673	0·762	0·618	0·625	0·446	—
0·8	0·763	0·832	0·760	0·788	0·871	0·667	0·732	0·650	0·695	0·542	0·569	0·597	0·666	0·766	0·607	0·615	0·423	—
0·9	0·773	0·849	0·766	0·798	0·891	0·663	0·736	0·643	0·691	0·526	0·556	0·586	0·661	0·774	0·597	0·606	0·403	—
1·0	0·784	0·867	0·772	0·809	0·913	0·660	0·742	0·639	0·688	0·512	0·544	0·576	0·656	0·781	0·589	0·598	0·393	—
1·1	0·797	0·886	0·780	0·820	0·936	0·658	0·749	0·634	0·686	0·499	0·534	0·567	0·653	0·789	0·594	0·590	—	—
Ref.	1	1	2	3	3	4	4	1	1	2	3	3	4	4	1	1	2	3

m	$KClO_4$	KFor	KAc	NH_4Cl	NH_4Br	NH_4I	NH_4NO_3	$CH_3 \cdot NH_3Cl$	$(CH_3)_2NH_2Cl$	$(CH_3)_3NHCl$	$Mg(ClO_4)_2$	$Ca(ClO_4)_2$	$Sr(ClO_4)_2$	$Ba(ClO_4)$
0·001	0·965	0·967	0·967	0·961	0·964	0·962	0·959	0·961	0·959	0·955	0·901	0·900	0·900	0·897
0·002	0·951	0·955	0·955	0·944	0·936	0·946	0·942	0·943	0·941	0·932	0·869	0·869	0·871	0·864
0·005	0·923	0·933	0·932	0·911	0·901	0·917	0·912	0·909	0·904	0·888	0·814	0·814	0·815	0·805
0·01	0·893	0·911	0·910	0·880	0·870	0·889	0·882	0·877	0·869	0·842	0·764	0·763	0·763	0·751
0·02	0·852	0·882	0·882	0·845	0·834	0·856	0·844	0·838	0·827	0·789	0·707	0·706	0·706	0·689
0·05	—	0·837	0·841	0·790	0·780	0·804	0·783	0·778	0·762	0·715	0·633	0·633	0·630	0·602
0·1	—	0·796	0·810	0·742	0·733	0·760	0·726	0·729	0·708	0·661	0·587	0·587	0·580	0·541
0·2	—	0·755	0·783	0·689	0·683	0·711	0·660	0·677	0·654	0·606	0·567	0·554	0·543	0·489
0·3	—	0·732	0·774	0·658	0·653	0·683	0·616	0·649	0·622	0·572	0·575	0·551	0·526	0·463
0·4	—	0·717	0·772	0·636	0·633	0·662	0·584	0·627	0·601	0·550	0·598	0·561	0·534	0·449
0·5	—	0·707	0·775	0·619	0·617	0·646	0·557	0·611	0·584	0·533	0·630	0·579	0·542	0·441
0·6	—	0·699	0·781	0·606	0·605	0·634	0·534	0·599	0·571	0·520	0·668	0·603	0·555	0·438
0·7	—	0·693	0·788	0·596	0·593	0·623	0·515	0·590	0·563	0·509	0·713	0·633	0·572	0·437
0·8	—	0·689	0·797	0·587	0·586	0·615	0·497	0·581	0·553	0·499	0·767	0·668	0·594	0·438
0·9	—	0·686	0·808	0·579	0·578	0·607	0·482	0·575	0·546	0·492	0·828	0·710	0·619	0·442
1·0	—	0·684	0·819	0·572	0·572	0·600	0·467	0·569	0·542	0·486	0·898	0·763	0·648	0·449
1·1	—	0·680	0·831	0·567	0·566	0·594	0·454	—	—	—	—	—	—	—
Ref.	3	4	4	5	5	5	5	6	6	6	7	7	7	7

For = formate Ac = acetate

Data from: [1] SCATCHARD, G. and PRENTISS, S. S., *J. Amer. chem. Soc.*, 55 (1933) 4355; [2] SCATCHARD, G., PRENTISS, S. S. and JONES, P. T., *ibid.*, 54 (1932) 2690; [3] SCATCHARD, G., PRENTISS S. S. and JONES, P. T., *ibid.*, 56 (1934) 805; [4] SCATCHARD, G., and PRENTISS, S. S., *ibid.*, 56 (1934) 807; [5] SCATCHARD, G., and PRENTISS, S. S., *ibid.*, 54 (1932) 2696; [6] JONES, J. H., SPUHLER, F. J. and FELSING, W. A., *ibid.*, 64 (1942) 965; [7] NICHOLSON, D. E. and FELSING, W. A., *ibid.*, 72 (1950) 4469; 73 (1951) 3520

APPENDIX 8.8

Osmotic and Activity Coefficients of Sodium Chloride and Potassium Bromide between 60° and 100° from Measurement of the Boiling Point Elevation

Table 1

Sodium Chloride

m	*Molal Osmotic Coefficient*					*Molal Activity Coefficient*				
	60°	70°	80°	90°	100°	60°	70°	80°	90°	100°
0·05	0·940	0·939	0·938	0·936	0·935	0·811	0·807	0·803	0·799	0·794
0·1	0·929	0·927	0·926	0·924	0·923	0·766	0·762	0·757	0·752	0·746
0·2	0·921	0·919	0·918	0 916	0·914	0·721	0·717	0·711	0·705	0·698
0·3	0·919	0·917	0·916	0·914	0·911	0·697	0·691	0·686	0·679	0·672
0·4	0·921	0·919	0·917	0·915	0·912	0·682	0·676	0·671	0·663	0·655
0·5	0·923	0·921	0·920	0·917	0·915	0·672	0·667	0·660	0·653	0·644
0·6	0·927	0·925	0·923	0·921	0·918	0·665	0·659	0·653	0·645	0·636
0·7	0·931	0·928	0·927	0·924	0·921	0·661	0·654	0·648	0·640	0·631
0·8	0·935	0·934	0·931	0·929	0·926	0·657	0·651	0·644	0·636	0·627
1·0	0·944	0·942	0·940	0·937	0·935	0·655	0·648	0·641	0·632	0·622
1·5	0·968	0·968	0·966	0·963	0·960	0·662	0·656	0·646	0·638	0·629
2·0	0·999	0·998	0·995	0·991	0·986	0·683	0·672	0·663	0·651	0·641
2·5	1·031	1·029	1·026	1·022	1·016	0·707	0·697	0·685	0·674	0·659
3·0	1·061	1·059	1·057	1·053	1·048	0·736	0·724	0·709	0·700	0·687
3·5	1·092	1·090	1·086	1·082	1·077	0·771	0·758	0·742	0·730	0·716
4·0	1·130	1·127	1·120	1·113	1·105	0·811	0·794	0·777	0·763	0·746

Data from SMITH, R. P., *J. Amer. chem. Soc.*, 61 (1939) 500; SMITH, R. P. and HIRTLE, D. S., *ibid.*, 61 (1939) 1123.

Table 2

Potassium Bromide

m	*Molal Osmotic Coefficient*					*Molal Activity Coefficient*				
	60°	70°	80°	90°	100°	60°	70°	80°	90°	100°
0·1	0·924	0·922	0·922	0·920	0·920	0·759	0·756	0·752	0·748	0·744
0·2	0·913	0·912	0·912	0·911	0·910	0·711	0·708	0·704	0·699	0·695
0·3	0·909	0·909	0·909	0·908	0·907	0·684	0·681	0·677	0·673	0·668
0·4	0·908	0·908	0·908	0·907	0·906	0·666	0·663	0·660	0·655	0·650
0·5	0·909	0·909	0·909	0·908	0·906	0·653	0·649	0·647	0·643	0·637
0·6	0·911	0·911	0·910	0·909	0·908	0·644	0·641	0·638	0·633	0·628
0·8	0·915	0·915	0·914	0·914	0·912	0·632	0·628	0·625	0·620	0·615
1·0	0·920	0·920	0·920	0·920	0·919	0·623	0·621	0·618	0·613	0·607
1·5	0·936	0·937	0·937	0·937	0·936	0·615	0·613	0·611	0·606	0·600
2·0	0·953	0·955	0·956	0·955	0·955	0·616	0·615	0·613	0·608	0·602
2·5	0·971	0·973	0·974	0·974	0·973	0·622	0·622	0·620	0·615	0·609
3·0	0·988	0·991	0·992	0·992	0·992	0·631	0·631	0·629	0·625	0·619
3·5	1·005	1·008	1·010	1·010	1·010	0·642	0·642	0·641	0·637	0·630
4·0	1·022	1·026	1·028	1·028	1·028	0·654	0·655	0·654	0·650	0·644

Data from JOHNSON, G. C. and SMITH, R. P., *J. Amer. chem. Soc.*, 63 (1941) 1351.

APPENDIX 8.9

Osmotic and Activity Coefficients of Electrolytes at Concentrations up to 0·1 M and at temperatures other than the freezing point

Table 1

Molal Activity Coefficients at 25°

m	HCl	HBr	HI	TlCl	H_2SO_4	$LaCl_3$
0·001	0·966	0·966	0·966	0·962	0·830	0·790
0·002	0·952	0·952	0·953	0·946	0·757	0·729
0·005	0·929	0·930	0·931	0·912	0·639	0·636
0·01	0·905	0·906	0·908	0·876	0·544	0·560
0·02	0·876	0·879	0·882	—	0·453	0·483
0·05	0·830	0·838	0·845	—	0·340	0·388

$\sqrt{m}$	ϕ_{NaCl}				γ_{NaCl}			
	15°	25°	35°	45°	15°	25°	35°	45°
0·04	—	—	—	—	0·9576	0·9570	0·9563	0·9554
0·08	0·9740	0·9737	0·9732	0·9727	0·9211	0·9198	0·9185	0·9171
0·12	—	—	—	—	0·8892	0·8878	0·8859	0·8839
0·16	0·9552	0·9548	0·9541	0·9532	0·8614	0·8598	0·8576	0·8551
0·20	—	—	—	—	0·8368	0·8352	0·8327	0·8297
0·24	0·9416	0·9414	0·9406	0·9395	0·8147	0·8134	0·8108	0·8074
0·28	—	—	—	—	0·7956	0·7940	0·7914	0·7876
0·32	0·9319	0·9320	0·9312	0·9298	0·7782	0·7768	0·7741	0·7702

$\sqrt{m}$	ϕ_{KCl}				γ_{KCl}			
	15°	25°	35°	45°	15°	25°	35°	45°
0·04	—	—	—	—	0·9572	0·9568	0·9561	0·9552
0·08	0·9736	0·9733	0·9728	0·9723	0·9202	0·9192	0·9177	0·9162
0·12	—	—	—	—	0·8876	0·8861	0·8843	0·8823
0·16	0·9538	0·9533	0·9526	0·9517	0·8588	0·8570	0·8549	0·8523
0·20	—	—	—	—	0·8329	0·8312	0·8285	0·8259
0·24	0·9385	0·9381	0·9372	0·9362	0·8097	0·8078	0·8052	0·8021
0·28	—	—	—	—	0·7887	0·7869	0·7840	0·7807
0·32	0·9266	0·9264	0·9255	0·9243	0·7697	0·7679	0·7649	0·7617

$\sqrt{m}$	ϕ_{CaCl_2}			γ_{CaCl_2}		
	15°	25°	35°	15°	25°	35°
0·04	0·9545	0·9538	0·9530	0·8660	0·8640	0·8618
0·08	0·9215	0·9203	0·9188	0·7700	0·7667	0·7631
0·12	0·8975	0·8960	0·8940	0·6986	0·6947	0·6901
0·16	0·8805	0·8788	0·8763	0·6443	0·6397	0·6346
0·20	0·8687	0·8669	0·8640	0·6021	0·5974	0·5914
0·24	0·8612	0·8594	0·8560	0·5689	0·5642	0·5576
0·28	0·8571	0·8554	0·8514	0·5424	0·5377	0·5305

APPENDIX 8.9

The logarithm of the activity coefficient of a number of rare earth halides can be represented, up to 0·03 M, by: $3{\cdot}745\sqrt{c}/(1 + 0{\cdot}8049 \times 10^8 a\sqrt{c})$ using the following values of a

Salt	$10^8\,a$ (cm)	*Salt*	$10^8\,a$ (cm)
$LaCl_3$	5·75	$LaBr_3$	6·20
$CeCl_3$	5·75	—	—
$PrCl_3$	5·73	$PrBr_3$	6·10
$NdCl_3$	5·92	$NdBr_3$	6·06
$SmCl_3$	5·63	—	—
$EuCl_3$	5·60	—	—
$GdCl_3$	5·63	$GdBr_3$	5·72
—	—	$HoBr_3$	6·42
$ErCl_3$	5·65	$ErBr_3$	5·90
$YbCl_3$	5·65	—	—

REFERENCES

HCl: HARNED, H. S. and EHLERS, R. W., *J. Amer. chem. Soc.*, 55 (1933) 2179; ROBINSON, R. A. and HARNED, H. S., *Chem. Rev.*, 28 (1941) 419

HBr: HARNED, H. S., KESTON, A. S. and DONELSON, J. G., *J. Amer. chem. Soc.*, 58 (1936) 989

HI: Extrapolated from data of HARNED, H. S. and ROBINSON, R. A., *Trans. Faraday Soc.*, 37 (1941) 302

TlCl: COWPERTHWAITE, I. A., LAMER, V. K. and BARKSDALE, J., *J. Amer. chem. Soc.*, 56 (1934) 544

H_2SO_4: HARNED, H. S. and HAMER, W. J., *ibid.*, 57 (1935) 27

$LaCl_3$: SHEDLOVSKY, T., *ibid.*, 72 (1950) 3680
This paper contains results for HCl, NaCl, KCl and $CaCl_2$.

NaCl: JANZ, G. J. and GORDON, A. R., *ibid.*, 65 (1943) 218

KCl: HORNIBROOK, W. J., JANZ, G. J. and GORDON, A. R., *ibid.*, 64 (1942) 513

$CaCl_2$: MCLEOD, H. G. and GORDON, A. R., *ibid.*, 68 (1946) 58

Rare earth chlorides and bromides: SPEDDING, F. H., PORTER, P. E. and WRIGHT, J. M., *ibid.*, 74 (1952) 2781; SPEDDING, F. H. and YAFFE, I. S., *ibid.*, 74 (1952) 4751

APPENDIX 8.10

Osmotic Coefficients of Electrolytes at 25°

Table 1

m	HCl	HBr	HI	$HClO_4$	HNO_3	LiOH	LiCl	LiBr	LiI	$LiClO_4$	$LiNO_3$	LiAc	LiTol	NaOH	NaF	NaCl	NaBr	NaI	$NaClO_3$	$NaClO_4$	$NaBrO_3$
0·1	0·943	0·948	0·953	0·947	0·940	0·894	0·939	0·943	0·952	0·951	0·938	0·935	0·928	0·925	0·924	0·932	0·934	0·938	0·927	0·930	0·918
0·2	0·945	0·954	0·969	0·951	0·935	0·889	0·939	0·944	0·966	0·959	0·935	0·928	0·917	0·925	0·908	0·925	0·928	0·936	0·913	0·920	0·896
0·3	0·952	0·964	0·984	0·958	0·936	0·881	0·945	0·952	0·980	0·971	0·940	0·929	0·912	0·929	0·898	0·922	0·928	0·939	0·904	0·915	0·883
0·4	0·963	0·978	1·001	0·966	0·940	0·874	0·954	0·960	0·995	0·985	0·946	0·931	0·908	0·933	0·891	0·920	0·929	0·945	0·897	0·912	0·873
0·5	0·974	0·993	1·019	0·976	0·944	0·870	0·963	0·970	1·008	0·999	0·954	0·935	0·906	0·937	0·886	0·921	0·933	0·952	0·892	0·910	0·865
0·6	0·986	1·007	1·038	0·988	0·950	0·865	0·973	0·981	1·022	1·013	0·962	0·940	0·906	0·941	0·882	0·923	0·937	0·959	0·888	0·909	0·857
0·7	0·998	1·023	1·057	1·000	0·957	0·862	0·984	0·993	1·034	1·027	0·970	0·945	0·905	0·945	0·879	0·926	0·942	0·967	0·885	0·910	0·851
0·8	1·011	1·038	1·075	1·013	0·964	0·860	0·995	1·007	1·049	1·043	0·978	0·951	0·905	0·949	0·876	0·929	0·947	0·975	0·883	0·911	0·845
0·9	1·025	1·054	1·094	1·026	0·971	0·858	1·006	1·021	1·063	1·058	0·987	0·956	0·905	0·953	0·874	0·932	0·953	0·983	0·882	0·912	0·839
1·0	1·039	1·072	1·113	1·041	0·979	0·857	1·018	1·035	1·080	1·072	0·997	0·962	0·905	0·958	0·872	0·936	0·958	0·991	0·880	0·913	0·833
1·2	1·067	1·111	1·153	1·072	0·994	0·861	1·041	1·067	1·111	1·104	1·015	0·975	0·904	0·969	—	0·943	0·969	1·007	0·878	0·916	0·824
1·4	1·096	1·147	1·193	1·106	1·009	0·864	1·066	1·098	1·143	1·137	1·033	0·988	0·901	0·980	—	0·951	0·983	1·025	0·876	0·920	0·815
1·6	1·126	1·184	1·233	1·141	1·025	0·868	1·091	1·130	1·176	1·170	1·052	1·001	0·899	0·991	—	0·962	0·997	1·043	0·874	0·925	0·808
1·8	1·157	1·222	1·273	1·175	1·042	0·871	1·116	1·163	1·212	1·204	1·070	1·014	0·894	1·002	—	0·972	1·012	1·061	0·875	0·930	0·804
2·0	1·188	1·261	1·315	1·210	1·060	0·874	1·142	1·196	1·250	1·238	1·088	1·027	0·893	1·015	—	0·983	1·028	1·079	0·876	0·934	0·800
2·5	1·266	1·365	1·424	1·305	1·106	0·881	1·212	1·278	1·351	1·328	1·134	1·061	0·899	1·054	—	1·013	1·067	1·129	0·879	0·947	0·792
3·0	1·348	1·475	1·535	1·406	1·154	0·885	1·286	1·364	1·467	1·419	1·181	1·093	0·912	1·094	—	1·045	1·107	1·188	0·881	0·960	—
3·5	1·431	—	—	1·511	—	0·888	1·366	1·467	—	1·512	1·227	1·123	0·930	1·139	—	1·080	1·150	1·243	0·886	0·975	—
4·0	1·517	—	—	1·622	—	0·891	1·449	1·578	—	1·595	1·270	1·153	0·951	1·195	—	1·116	1·199	—	—	0·991	—
4·5	1·598	—	—	1·738	—	—	1·533	1·687	—	—	1·312	—	0·972	1·255	—	1·153	—	—	—	1·008	—
5·0	1·680	—	—	1·860	—	—	1·619	1·793	—	—	1·352	—	—	1·314	—	1·192	—	—	—	1·025	—
5·5	1·763	—	—	1·981	—	—	1·705	1·891	—	—	1·387	—	—	1·374	—	1·231	—	—	—	1·042	—
6·0	1·845	—	—	2·106	—	—	1·791	1·989	—	—	1·420	—	—	1·434	—	1·271	—	—	—	1·060	—

Ac = acetate

Tol = p-toluenesulphonate

APPENDIX 8.10

Osmotic Coefficients of Electrolytes at 25°

Table 2

m	$NaNO_3$	Na formate	Na acetate	Na propionate	Na butyrate	Na valerate	Na caproate	Na heptylate	Na caprylate	Na pelargonate	Na caprate	Na H malonate	Na H succinate	Na H adipate
0·1	0·921	0·931	0·940	0·944	0·944	0·944	0·946	0·946	0·947	—	—	0·923	0·924	0·931
0·2	0·902	0·924	0·939	0·947	0·949	0·951	0·952	0·953	0·955	—	—	0·907	0·910	0·921
0·3	0·890	0·922	0·945	0·955	0·961	0·963	0·964	0·966	0·968	0·812	—	0·896	0·898	0·917
0·4	0·881	0·923	0·951	0·965	0·975	0·983	0·984	0·978	0·960	0·607	0·448	0·888	0·892	0·914
0·5	0·873	0·925	0·959	0·975	0·991	0·998	0·999	0·987	0·882	0·521	0·370	0·880	0·888	0·912
0·6	0·867	0·927	0·967	0·986	1·009	1·013	1·017	0·986	0·802	0·458	0·326	0·876	0·885	0·911
0·7	0·862	0·929	0·977	0·997	1·027	1·028	1·034	0·982	0·722	0·416	0·293	0·872	0·882	0·911
0·8	0·858	0·932	0·986	1·008	1·043	1·042	1·046	0·958	0·644	0·385	0·270	0·869	0·880	—
0·9	0·854	0·936	0·994	1·020	1·060	1·057	1·052	0·892	0·568	0·362	0·251	0·866	0·879	—
1·0	0·851	0·939	1·002	1·032	1·076	1·071	1·054	0·833	0·535	0·343	0·235	0·863	0·878	—
1·2	0·845	0·945	1·018	1·058	1·106	1·101	1·050	0·740	0·486	0·316	0·230	0·860	0·877	—
1·4	0·839	0·951	1·038	1·083	1·132	1·126	1·033	0·683	0·448	0·304	0·231	0·858	0·878	—
1·6	0·835	0·957	1·057	1·107	1·157	1·142	1·000	0·635	0·420	0·300	0·234	0·857	0·880	—
1·8	0·830	0·964	1·074	1·131	1·182	1·152	0·960	0·596	0·394	0·298	0·236	0·856	0·883	—
2·0	0·826	0·970	1·092	1·151	1·203	1·159	0·925	0·562	0·386	0·296	—	0·856	0·887	—
2·5	0·817	0·988	1·137	1·205	1·241	1·124	0·827	0·505	0·387	0·289	—	0·855	0·895	—
3·0	0·810	1·005	1·181	1·252	1·272	1·065	0·770	0·492	0·394	—	—	0·855	0·907	—
3·5	0·804	1·022	1·223	—	1·297	0·980	0·746	0·495	—	—	—	0·855	0·917	—
4·0	0·797	—	—	—	—	—	0·745	0·502	—	—	—	0·856	0·929	—
4·5	0·792	—	—	—	—	—	0·761	0·511	—	—	—	0·857	0·942	—
5·0	0·788	—	—	—	—	—	—	0·523	—	—	—	0·858	0·958	—
5·5	0·787	—	—	—	—	—	—	—	—	—	—	—	—	—
6·0	0·788	—	—	—	—	—	—	—	—	—	—	—	—	—

m	NaTol	NaCNS	NaH_2PO_4	KOH	KF	KCl	KBr	KI	$KClO_3$	$KBrO_3$	KNO_3	KAc
0·1	0·924	0·937	0·911	0·933	0·930	0·927	0·928	0·932	0·913	0·910	0·906	0·943
0·2	0·907	0·934	0·884	0·930	0·919	0·913	0·916	0·922	0·887	0·881	0·873	0·944
0·3	0·897	0·935	0·864	0·934	0·915	0·906	0·910	0·918	0·867	0·858	0·851	0·951
0·4	0·887	0·938	0·847	0·941	0·914	0·902	0·906	0·917	0·849	0·837	0·833	0·958
0·5	0·880	0·943	0·832	0·951	0·915	0·899	0·904	0·917	0·832	0·816	0·817	0·968
0·6	0·874	0·948	0·819	0·960	0·916	0·898	0·904	0·918	0·816	—	0·802	0'977
0·7	0·867	0·953	0·808	0·970	0·919	0·897	0·904	0·919	0·802	—	0·790	0·987
0·8	0·861	0·958	0·798	0·982	0·923	0·897	0·905	0·922	—	—	0·778	0·997
0·9	0·855	0·963	0·789	0·992	0·926	0·897	0·906	0·924	—	—	0·767	1·007
1·0	0·849	0·969	0·780	1·002	0·931	0·897	0·907	0·926	—	—	0·756	1·017
1·2	0·837	0·979	0·765	1·025	0·941	0·899	0·910	0·931	—	—	0·736	1·038
1·4	0·824	0·990	0·751	1·050	0·951	0·901	0·914	0·937	—	—	0·718	1·060
1·6	0·811	1·002	0·739	1·075	0·962	0·904	0·917	0·943	—	—	0·700	1·081
1·8	0·799	1·014	0·729	1·099	0·973	0·908	0·922	0·950	—	—	0·684	1·103
2·0	0·787	1·025	0·721	1·124	0·984	0·912	0·927	0·957	—	—	0·669	1·123
2·5	0·763	1·055	0·705	1·183	1·014	0·924	0·941	0·974	—	—	0·631	1·177
3·0	0·748	1·086	0·696	1·248	1·048	0·937	0·955	0·990	—	—	0·602	1·228
3·5	0·738	1·118	0·691	1·317	1·084	0·950	0·969	1·006	—	—	0·577	1·274
4·0	0·733	1·150	0·691	1·387	1·124	0·965	0·984	1·021	—	—	—	—
4·5	—	—	0·694	1·459	—	0·980	1·000	1·032	—	—	—	—
5·0	—	—	0·699	1·524	—	—	1·015	—	—	—	—	—
5·5	—	—	0·706	1·594	—	—	1·028	—	—	—	—	—
6·0	—	—	0·713	1·661	—	—	—	—	—	—	—	—

APPENDIX 8.10

Osmotic Coefficients of Electrolytes at 25°

Table 3

m	KH malonate	KH succinate	KH adipate	KTol	KCNS	KH_2PO_4	NH_4Cl	NH_4NO_3	RbCl	RbBr	RbI
0·1	0·920	0·922	0·928	0·921	0·926	0·901	0·927	0·911	0·923	0·922	0·921
0·2	0·903	0·904	0·917	0·901	0·911	0·868	0·913	0·890	0·907	0·905	0·904
0·3	0·891	0·892	0·909	0·886	0·904	0·843	0·906	0·876	0·898	0·897	0·896
0·4	0·877	0·882	0·904	0·873	0·900	0·823	0·901	0·864	0·893	0·892	0·890
0·5	0·866	0·875	0·900	0·860	0·897	0·805	0·899	0·855	0·889	0·888	0·886
0·6	0·856	0·870	0·899	0·847	0·896	0·789	0·897	0·847	0·887	0·886	0·884
0·7	0·847	0·867	0·898	0·834	0·895	0·773	0·896	0·840	0·886	0·884	0·881
0·8	0·840	0·862	0·898	0·822	0·895	0·760	0·896	0·834	0·886	0·882	0·880
0·9	0·835	0·859	0·898	0·809	0·894	0·747	0·896	0·829	0·885	0·881	0·879
1·0	0·829	0·856	0·899	0·798	0·894	0·736	0·897	0·823	0·885	0·881	0·878
1·2	0·820	0·851	—	0·775	0·893	0·716	0·898	0·813	0·886	0·880	0·878
1·4	0·811	0·848	—	0·751	0·892	0·698	0·900	0·803	0·888	0·881	0·878
1·6	0·805	0·846	—	0·732	0·892	0·683	0·903	0·793	0·890	0·882	0·880
1·8	0·802	0·845	—	0·715	0·893	0·669	0·906	0·785	0·893	0·884	0·882
2·0	0·799	0·845	—	0·700	0·894	—	0·909	0·776	0·896	0·887	0·886
2·5	0·792	0·848	—	0·664	0·898	—	0·918	0·758	0·905	0·893	0·893
3·0	0·785	0·854	—	0·637	0·903	—	0·926	0·743	0·916	0·899	0·901
3·5	0·778	0·865	—	0·615	0·908	—	0·936	0·728	0·928	0·907	0·911
4·0	0·771	0·870	—	—	0·912	—	0·945	0·715	0·941	0·916	0·921
4·5	0·764	0·876	—	—	0·917	—	0·953	0·702	0·952	0·924	0·931
5·0	0·757	—	—	—	0·921	—	0·958	0·690	0·966	0·934	0·940
5·5	—	—	—	—	—	—	0·963	0·679	—	—	—
6·0	—	—	—	—	—	—	0·969	0·670	—	—	—

m	$RbNO_3$	RbAc	CsOH	CsCl	CsBr	CsI	$CsNO_3$	CsĀc	$AgNO_3$	$TlClO_4$	$TlNO_3$	TlĀc
0·1	0·903	0·943	0·954	0·917	0·917	0·916	0·902	0·945	0·903	0·900	0·881	0·913
0·2	0·871	0·945	0·945	0·897	0·896	0·895	0·869	0·947	0·870	0·867	0·833	0·891
0·3	0·847	0·952	0·945	0·885	0·882	0·880	0·842	0·954	0·847	0·842	0·800	0·876
0·4	0·826	0·961	0·942	0·875	0·873	0·870	0·820	0·964	0·827	0·821	0·775	0·865
0·5	0·809	0·971	0·962	0·869	0·865	0·863	0·802	0·975	0·811	0·804	—	0·855
0·6	0·794	0·981	0·972	0·864	0·861	0·858	0·787	0·986	0·795	—	—	0·849
0·7	0·781	0·992	0·984	0·861	0·857	0·855	0·774	0·996	0·779	—	—	0·843
0·8	0·768	1·002	0·994	0·859	0·854	0·852	0·761	1·006	0·766	—	—	0·838
0·9	0·756	1·013	1·004	0·858	0·852	0·849	0·748	1·016	0·754	—	—	0·833
1·0	0·745	1·023	1·014	0·857	0·850	0·846	0·736	1·026	0·742	—	—	0·829
1·2	0·725	1·046	—	0·856	0·849	0·842	0·715	1·049	0·720	—	—	0·823
1·4	0·706	1·068	—	0·856	0·848	0·839	0·695	1·072	0·699	—	—	0·818
1·6	0·689	1·091	—	0·857	0·848	0·836	—	1·095	0·680	—	—	0·814
1·8	0·673	1·114	—	0·859	0·850	0·834	—	1·119	0·662	—	—	0·810
2·0	0·656	1·137	—	0·864	0·852	0·832	—	1·142	0·646	—	—	0·807
2·5	0·620	1·192	—	0·871	0·859	0·827	—	1·196	0·609	—	—	0·801
3·0	0·588	1·248	—	0·880	0·866	0·822	—	1·251	0·576	—	—	0·796
3·5	0·561	1·302	—	0·891	0·874	—	—	1·306	0·550	—	—	0·789
4·0	0·538	—	—	0·901	0·884	—	—	—	0·523	—	—	0·783
4·5	0·516	—	—	0·913	0·892	—	—	—	0·502	—	—	0·777
5·0	—	—	—	0·923	0·901	—	—	—	0·483	—	—	0·772
5·5	—	—	—	0·934	—	—	—	—	0·467	—	—	0·766
6·0	—	—	—	0·945	—	—	—	—	0·452	—	—	0·760

APPENDIX 8.10

Osmotic Coefficients of Electrolytes at 25°

Table 4

m	$MgCl_2$	$MgBr_2$	MgI_2	$Mg(ClO_4)_2$	$Mg(NO_3)_2$	$Mg\bar{A}c_2$	$CaCl_2$	$CaBr_2$	CaI_2	$Ca(ClO_4)_2$
0·1	0·861	0·874	0·892	0·898	0·857	0·797	0·854	0·863	0·880	0·883
0·2	0·877	0·898	0·921	0·935	0·869	0·793	0·862	0·878	0·906	0·911
0·3	0·895	0·928	0·957	0·974	0·890	0·795	0·876	0·900	0·935	0·942
0·4	0·919	0·963	0·998	1·016	0·914	0·800	0·894	0·927	0·969	0·976
0·5	0·947	1·004	1·044	1·062	0·940	0·807	0·917	0·958	1·008	1·014
0·6	0·976	1·042	1·090	1·108	0·967	0·816	0·940	0·990	1·044	1·051
0·7	1·004	1·082	1·139	1·158	0·991	0·826	0·963	1·022	1·083	1·089
0·8	1·036	1·127	1·192	1·211	1·017	0·838	0·988	1·057	1·128	1·131
0·9	1·071	1·172	1·249	1·267	1·046	0·850	1·017	1·093	1·173	1·175
1·0	1·108	1·218	1·306	1·323	1·074	0·861	1·046	1·131	1·217	1·219
1·2	1·184	1·314	1·421	1·437	1·134	0·886	1·107	1·207	1·310	1·310
1·4	1·264	1·410	1·537	1·558	1·192	0·910	1·171	1·286	1·407	1·405
1·6	1·347	1·510	1·660	1·683	1·251	0·935	1·237	1·370	1·504	1·503
1·8	1·434	1·610	1·784	1·815	1·311	0·961	1·305	1·455	1·605	1·605
2·0	1·523	1·715	1·912	1·945	1·372	0·987	1·376	1·547	1·710	1·710
2·5	1·762	1·999	2·25	2·306	1·535	1·049	1·568	1·790	—	1·992
3·0	2·010	2·29	2·60	2·667	1·710	1·109	1·779	2·048	—	2·261
3·5	2·264	2·59	2·96	3·036	1·878	1·159	1·981	2·297	—	2·521
4·0	2·521	2·89	3·34	3·397	2·043	1·207	2·182	2·584	—	2·769
4·5	2·783	3·19	3·72	—	2·209	—	2·383	2·908	—	3·005
5·0	3·048	3·50	4·11	—	2·376	—	2·574	3·239	—	3·233
5·5	—	—	—	—	—	—	2·743	3·564	—	3·454
6·0	—	—	—	—	—	—	2·891	3·880	—	3·655

m	$Ca(NO_3)_2$	$SrCl_2$	$SrBr_2$	SrI_2	$Sr(ClO_4)_2$	$Sr(NO_3)_2$	$BaCl_2$	$BaBr_2$	BaI_2	$Ba(ClO_4)_2$
0·1	0·827	0·850	0·859	0·876	0·864	0·816	0·843	0·851	0·869	0·857
0·2	0·819	0·854	0·871	0·899	0·886	0·796	0·837	0·857	0·891	0·868
0·3	0·818	0·864	0·888	0·925	0·915	0·785	0·843	0·869	0·918	0·884
0·4	0·821	0·880	0·908	0·955	0·947	0·778	0·853	0·884	0·949	0·905
0·5	0·825	0·899	0·932	0·987	0·982	0·773	0·864	0·906	0·985	0·929
0·6	0·831	0·918	0·957	1·021	1·017	0·769	0·877	0·926	1·017	0·954
0·7	0·837	0·937	0·983	1·056	1·052	0·765	0·890	0·945	1·050	0·977
0·8	0·843	0·959	1·011	1·095	1·090	0·762	0·904	0·965	1·085	1·000
0·9	0·850	0·983	1·042	1·136	1·130	0·760	0·918	0·989	1·122	1·024
1·0	0·859	1·009	1·074	1·177	1·170	0·757	0·934	1·013	1·159	1·046
1·2	0·879	1·061	1·142	1·264	1·249	0·754	0·967	1·063	1·231	1·094
1·4	0·898	1·116	1·210	1·352	1·329	0·754	1·002	1·112	1·308	1·141
1·6	0·917	1·173	1·284	1·443	1·413	0·754	1·034	1·162	1·388	1·188
1·8	0·934	1·232	1·360	1·540	1·492	0·755	1·065	1·212	1·470	1·233
2·0	0·953	1·292	1·440	1·641	1·577	0·758	—	1·263	1·599	1·279
2·5	1·001	1·454	—	—	1·789	0·768	—	—	—	1·394
3·0	1·051	1·631	—	—	2·008	0·783	—	—	—	1·509
3·5	1·103	1·802	—	—	2·196	0·800	—	—	—	1·619
4·0	1·157	1·966	—	—	2·372	0·818	—	—	—	1·713
4·5	1·210	—	—	—	2·538	—	—	—	—	1·791
5·0	1·263	—	—	—	2·693	—	—	—	—	1·862
5·5	1·313	—	—	—	2·834	—	—	—	—	1·945
6·0	1·361	—	—	—	2·962	—	—	—	—	—

APPENDIX 8.10

Osmotic Coefficients of Electrolytes at 25°

Table 5

m	$Ba(NO_3)_2$	$Ba\bar{A}c_2$	$MnCl_2$	$FeCl_2$	$CoCl_2$	$CoBr_2$	CoI_2	$Co(NO_3)_2$	$NiCl_2$	$CuCl_2$	$Cu(NO_3)_2$
0·1	0·771	0·800	0·853	0·854	0·857	0·871	0·89	0·854	0·857	0·845	0·847
0·2	0·724	0·807	0·859	0·863	0·869	0·894	0·92	0·861	0·868	0·843	0·849
0·3	0·687	0·817	0·872	0·877	0·886	0·922	0·96	0·875	0·885	0·848	0·860
0·4	0·659	0·828	0·889	0·896	0·907	0·955	1·00	0·892	0·907	0·860	0·875
0·5	—	0·841	0·908	0·920	0·932	0·992	1·05	0·914	0·934	0·876	0·895
0·6	—	0·849	0·929	0·943	0·959	1·029	1·10	0·936	0·960	0·892	0·914
0·7	—	0·857	0·950	0·968	0·982	1·068	1·15	0·958	0·987	0·908	0·934
0·8	—	0·864	0·971	0·994	1·011	1·109	1·21	0·981	1·016	0·922	0·955
0·9	—	0·869	0·995	1·024	1·043	1·152	1·26	1·007	1·048	0·938	0·978
1·0	—	0·873	1·022	1·055	1·075	1·196	1·32	1·033	1·082	0·952	1·001
1·2	—	0·881	1·072	1·117	1·141	1·286	1·43	1·087	1·150	0·978	1·046
1·4	—	0·884	1·124	1·180	1·208	1·382	1·54	1·143	1·221	1·000	1·087
1·6	—	0·885	1·173	1·244	1·274	1·482	1·65	1·199	1·293	1·022	1·131
1·8	—	0·884	1·221	1·307	1·339	1·580	1·78	1·258	1·366	1·043	1·177
2·0	—	0·878	1·264	1·371	1·404	1·678	1·90	1·317	1·442	1·062	1·224
2·5	—	0·856	1·366	—	1·564	1·921	2·24	1·468	1·633	1·100	1·339
3·0	—	0·832	1·454	—	1·711	2·149	2·56	1·620	1·816	1·131	1·480
3·5	—	0·804	1·528	—	1·821	2·358	2·87	1·769	1·969	1·160	1·610
4·0	—	—	1·584	—	1·896	2·564	3·17	1·913	2·100	1·183	1·732
4·5	—	—	1·634	—	—	2·737	3·41	2·053	2·202	1·201	1·841
5·0	—	—	1·671	—	—	2·880	3·59	2·196	2·292	1·219	1·940
5·5	—	—	1·704	—	—	2·990	3·63	2·323	—	—	2·035
6·0	—	—	1·735	—	—	—	3·61	—	—	—	2·125

m	$ZnCl_2$	$ZnBr_2$	ZnI_2	$Zn(ClO_4)_2$	$Zn(NO_3)_2$	$CdCl_2$	$CdBr_2$	CdI_2	$Cd(NO_3)_2$	$Pb(ClO_4)_2$
0·1	0·847	0·869	0·893	0·893	0·862	0·622	0·592	0·416	0·850	0·858
0·2	0·845	0·886	0·924	0·928	0·873	0·571	0·533	0·390	0·852	0·870
0·3	0·842	0·911	0·957	0·966	0·890	0·542	0·502	0·371	0·861	0·886
0·4	0·838	0·937	0·994	1·010	0·909	0·522	0·480	0·365	0·873	0·907
0·5	0·833	0·962	1·038	1·056	0·934	0·506	0·466	0·366	0·888	0·930
0·6	0·829	0·984	1·083	1·105	0·958	0·492	0·455	0·370	0·903	0·954
0·7	0·824	1·002	1·124	1·157	0·982	0·481	0·449	0·376	0·917	0·976
0·8	0·817	1·018	1·163	1·212	1·009	0·473	0·445	0·383	0·931	1·002
0·9	0·810	1·032	1·194	1·269	1·037	0·465	0·441	0·388	0·946	1·031
1·0	0·805	1·039	1·220	1·328	1·064	0·458	0·439	0·394	0·962	1·060
1·2	0·792	1·047	1·260	1·450	1·120	0·448	0·439	0·405	0·995	1·118
1·4	0·782	1·049	1·283	1·578	1·180	0·440	0·444	0·419	1·025	1·179
1·6	0·781	1·047	1·291	1·708	1·238	0·435	0·449	0·436	1·057	1·240
1·8	0·785	1·043	1·292	1·843	1·296	0·431	0·455	0·451	1·085	1·301
2·0	0·792	1·042	1·282	1·986	1·355	0·428	0·462	0·463	1·114	1·363
2·5	0·820	1·048	1·262	2·358	1·510	0·430	0·483	0·507	1·182	1·521
3·0	0·858	1·066	1·262	2·739	1·664	0·434	0·504	—	—	1·693
3·5	0·903	1·100	1·278	3·117	1·814	0·442	0·527	—	—	1·853
4·0	0·955	1·143	1·297	3·494	1·957	0·454	0·548	—	—	1·999
4·5	1·022	1·195	1·335	—	2·098	0·466	—	—	—	2·137
5·0	1·091	1·253	1·381	—	2·235	0·482	—	—	—	2·271
5·5	1·160	1·314	1·436	—	2·366	0·497	—	—	—	2·399
6·0	1·229	1·379	1·487	—	2·489	0·514	—	—	—	2·516

APPENDIX 8.10

Osmotic Coefficients of Electrolytes at 25°

Table 6

m	$Pb(NO_3)_2$	UO_2Cl_2	$UO_2(ClO_4)_2$	$UO_2(NO_3)_2$	Li_2SO_4	Na_2SO_4	Na_2CrO_4	$Na_2S_2O_3$
0·1	0·746	0·870	0·921	0·875	0·818	0·793	0·814	0·805
0·2	0·692	0·890	0·972	0·899	0·792	0·753	0·788	0·774
0·3	0·656	0·914	1·028	0·928	0·780	0·725	0·769	0·753
0·4	0·628	0·942	1·089	0·960	0·775	0·705	0·759	0·741
0·5	0·606	0·975	1·152	0·996	0·772	0·690	0·751	0·731
0·6	0·588	1·005	1·216	1·031	0·773	0·678	0·747	0·724
0·7	0·572	1·034	1·284	1·067	0·775	0·667	0·743	0·719
0·8	0·558	1·066	1·356	1·105	0·778	0·658	0·740	0·713
0·9	0·545	1·097	1·430	1·143	0·782	0·650	0·738	0·709
1·0	0·533	1·128	1·507	1·182	0·787	0·642	0·737	0·707
1·2	0·511	1·188	1·661	1·258	0·800	0·631	0·737	0·707
1·4	0·494	1·247	1·823	1·330	0·815	0·625	0·743	0·713
1·6	0·481	1·303	1·990	1·398	0·832	0·621	0·751	0·719
1·8	0·472	1·356	2·165	1·466	0·847	0·620	0·763	0·728
2·0	0·465	1·406	2·354	1·532	0·867	0·621	0·780	0·739
2·5	—	1·530	2·818	1·673	0·923	0·635	0·838	0·779
3·0	—	1·655	3·284	1·764	0·984	0·661	0·905	0·829
3·5	—	—	3·721	1·807	—	0·696	0·992	0·890
4·0	—	—	4·152	1·809	—	0·740	1·100	—
4·5	—	—	4·561	1·790	—	—	—	—
5·0	—	—	4·907	1·765	—	—	—	—
5·5	—	—	5·220	1·753	—	—	—	—
6·0	—	—	—	—	—	—	—	—

m	Na_2 fumarate	Na_2 maleate	K_2SO_4	K_2CrO_4	$K_2Cr_2O_7$	$(NH_4)_2SO_4$	Rb_2SO_4	Cs_2SO_4
0·1	0·812	0·770	0·779	0·805	0·868	0·767	0·799	0·804
0·2	0·801	0·744	0·742	0·774	0·813	0·731	0·764	0·772
0·3	0·797	0·731	0·721	0·753	0·779	0·707	0·740	0·751
0·4	0·797	0·726	0·703	0·741	0·753	0·690	0·724	0·739
0·5	0·803	0·726	0·691	0·733	0·735	0·677	0·714	0·731
0·6	0·816	0·729	0·679	0·727	—	0·667	0·705	0·725
0·7	0·827	0·733	0·670	0·722	—	0·658	0·698	0·721
0·8	0·841	0·738	—	0·718	—	0·652	0·691	0·717
0·9	0·855	0·742	—	0·714	—	0·646	0·686	0·714
1·0	0·870	0·747	—	0·711	—	0·640	0·681	0·712
1·2	0·894	0·760	—	0·709	—	0·632	0·677	0·711
1·4	0·920	0·773	—	0·711	—	0·628	0·677	0·713
1·6	0·949	0·788	—	0·716	—	0·624	0·679	0·716
1·8	0·972	0·804	—	0·722	—	0·623	0·684	0·722
2·0	0·993	0·820	—	0·730	—	0·623	—	—
2·5	—	0·863	—	0·757	—	0·626	—	—
3·0	—	0·910	—	0·794	—	0·635	—	—
3·5	—	—	—	0·830	—	0·647	—	—
4·0	—	—	—	—	—	0·660	—	—
4·5	—	—	—	—	—	0·673	—	—
5·0	—	—	—	—	—	0·686	—	—
5·5	—	—	—	—	—	0·699	—	—
6·0	—	—	—	—	—	—	—	—

APPENDIX 8.10

Osmotic Coefficients of Electrolytes at 25°

Table 7

m	$BeSO_4$	$MgSO_4$	$MnSO_4$	$NiSO_4$	$CuSO_4$	$ZnSO_4$	$CdSO_4$	UO_2SO_4	$AlCl_3$	$ScCl_3$
0·1	0·582	0·606	0·587	0·581	0·561	0·590	0·565	0·529	0·819	0·797
0·2	0·560	0·562	0·538	0·533	0·515	0·533	0·513	0·488	0·841	0·827
0·3	0·544	0·540	0·516	0·508	0·494	0·506	0·490	0·463	0·889	0·868
0·4	0·536	0·529	0·501	0·488	0·478	0·492	0·476	0·460	0·947	0·917
0·5	0·535	0·522	0·490	0·475	0·469	0·483	0·466	0·462	1·008	0·969
0·6	0·540	0·518	0·481	0·465	0·462	0·476	0·458	0·465	1·074	1·027
0·7	0·548	0·517	0·475	0·458	0·458	0·473	0·452	0·470	1·145	1·090
0·8	0·555	0·518	0·472	0·456	0·457	0·473	0·450	0·477	1·220	1·156
0·9	0·565	0·520	0·472	0·456	0·458	0·474	0·449	0·485	1·299	1·222
1·0	0·580	0·525	0·475	0·459	0·461	0·478	0·452	0·495	1·382	1·291
1·2	0·608	0·542	0·485	0·472	0·473	0·489	0·461	0·517	1·560	1·430
1·4	0·640	0·567	0·504	0·492	0·491	0·508	0·476	0·542	1·749	1·572
1·6	0·676	0·597	0·527	0·517	—	0·533	0·496	0·571	1·951	1·718
1·8	0·715	0·630	0·556	0·551	—	0·566	0·522	0·600	2·175	1·869
2·0	0·757	0·666	0·588	0·589	—	0·602	0·551	0·628	—	—
2·5	0·889	0·780	0·677	0·708	—	0·717	0·632	0·710	—	—
3·0	1·019	0·922	0·782	—	—	0·861	0·726	0·792	—	—
3·5	1·171	—	0·909	—	—	1·033	0·832	0·873	—	—
4·0	1·327	—	1·048	—	—	—	—	0·951	—	—
4·5	—	—	—	—	—	—	—	1·025	—	—
5·0	—	—	—	—	—	—	—	1·092	—	—
5·5	—	—	—	—	—	—	—	1·150	—	—
6·0	—	—	—	—	—	—	—	1·198	—	—

m	$CrCl_3$	YCl_3	$LaCl_3$	$CeCl_3$	$PrCl_3$	$NdCl_3$	$SmCl_3$	$EuCl_3$	$Cr(NO_3)_3$
0·1	0·811	0·789	0·788	0·782	0·784	0·783	0·789	0·794	0·795
0·2	0·833	0·810	0·800	0·805	0·801	0·801	0·809	0·812	0·818
0·3	0·875	0·847	0·833	0·835	0·830	0·832	0·841	0·842	0·860
0·4	0·926	0·892	0·871	0·872	0·866	0·871	0·879	0·882	0·906
0·5	0·983	0·939	0·912	0·914	0·905	0·913	0·921	0·926	0·953
0·6	1·045	0·989	0·955	0·955	0·945	0·954	0·964	0·971	1·003
0·7	1·111	1·042	0·998	1·007	0·996	1·006	1·019	1·027	1·055
0·8	1·181	1·100	1·052	1·057	1·046	1·056	1·074	1·082	1·111
0·9	1·250	1·161	1·102	1·107	1·100	1·110	1·128	1·137	1·168
1·0	1·319	1·223	1·154	1·158	1·154	1·165	1·186	1·193	1·227
1·2	1·443	1·354	1·266	1·264	1·271	1·283	1·302	1·310	1·343
1·4	—	1·491	1·384	1·387	1·388	1·404	1·427	1·438	1·456
1·6	—	1·631	1·502	1·504	1·507	1·527	1·554	1·570	—
1·8	—	1·780	1·623	1·638	1·631	1·656	1·686	1·707	—
2·0	—	1·940	1·748	1·777	1·759	1·789	1·824	1·853	—
2·5	—	—	—	—	—	—	—	—	—
3·0	—	—	—	—	—	—	—	—	—
3·5	—	—	—	—	—	—	—	—	—
4·0	—	—	—	—	—	—	—	—	—
4·5	—	—	—	—	—	—	—	—	—
5·0	—	—	—	—	—	—	—	—	—
5·5	—	—	—	—	—	—	—	—	—
6·0	—	—	—	—	—	—	—	—	—

APPENDIX 8.10

Osmotic Coefficients of Electrolytes at 25°

Table 8

m	$K_3Fe(CN)_6$	$K_4Fe(CN)_6$	$Al_2(SO_4)_3$	$Cr_2(SO_4)_3$	$Th(NO_3)_4$
0·1	0·727	0·595	0·420	0·414	0·675
0·2	0·695	0·556	0·390	0·401	0·685
0·3	0·682	0·535	0·391	0·412	0·705
0·4	0·678	0·518	0·421	0·437	0·734
0·5	0·676	0·506	0·477	0·473	0·770
0·6	0·676	0·498	0·545	0·524	0·807
0·7	0·679	0·494	0·625	0·585	0·846
0·8	0·685	0·494	0·718	0·657	0·885
0·9	0·694	0·501	0·809	0·740	0·925
1·0	0·705	—	0·922	0·832	0·965
1·2	0·727	—	—	1·031	1·044
1·4	0·750	—	—	—	1·120
1·6	—	—	—	—	1·192
1·8	—	—	—	—	1·259
2·0	—	—	—	—	1·325
2·5	—	—	—	—	1·455
3·0	—	—	—	—	1·546
3·5	—	—	—	—	1·616
4·0	—	—	—	—	1·659
4·5	—	—	—	—	1·688
5·0	—	—	—	—	1·706

APPENDIX 8.10

Activity Coefficients of Electrolytes at 25°

Table 9

m	HCl	HBr	HI	$HClO_4$	HNO_3	LiOH	LiCl	LiBr	LiI	$LiClO_4$	$LiNO_3$
0·1	0·796	0·805	0·818	0·803	0·791	0·718	0·790	0·796	0·815	0·812	0·788
0·2	0·767	0·782	0·807	0·778	0·754	0·663	0·757	0·766	0·802	0·794	0·752
0·3	0·756	0·777	0·811	0·768	0·735	0·628	0·744	0·756	0·804	0·792	0·736
0·4	0·755	0·781	0·823	0·766	0·725	0·603	0·740	0·752	0·813	0·798	0·728
0·5	0·757	0·789	0·839	0·769	0·720	0·583	0·739	0·753	0·824	0·808	0·726
0·6	0·763	0·801	0·860	0·776	0·717	0·566	0·743	0·758	0·838	0·820	0·727
0·7	0·772	0·815	0·883	0·785	0·717	0·553	0·748	0·767	0·852	0·834	0·729
0·8	0·783	0·832	0·908	0·795	0·718	0·541	0·755	0·777	0·870	0·852	0·733
0·9	0·795	0·850	0·935	0·808	0·721	0·532	0·764	0·789	0·888	0·869	0·737
1·0	0·809	0·871	0·963	0·823	0·724	0·523	0·774	0·803	0·910	0·887	0·743
1·2	0·840	0·917	1·027	0·858	0·734	0·512	0·796	0·837	0·955	0·931	0·757
1·4	0·876	0·969	1·098	0·900	0·745	0·503	0·823	0·874	1·007	0·979	0·774
1·6	0·916	1·029	1·175	0·947	0·758	0·496	0·853	0·917	1·063	1·034	0·792
1·8	0·960	1·094	1·260	0·998	0·775	0·489	0·885	0·964	1·127	1·093	0·812
2·0	1·009	1·168	1·356	1·055	0·793	0·485	0·921	1·015	1·198	1·158	0·835
2·5	1·147	1·389	1·641	1·227	0·846	0·475	1·026	1·161	1·418	1·350	0·896
3·0	1·316	1·674	2·015	1·448	0·909	0·467	1·156	1·341	1·715	1·582	0·966
3·5	1·518	—	—	1·726	—	0·460	1·317	1·584	—	1·866	1·044
4·0	1·762	—	—	2·08	—	0·454	1·510	1·897	—	2·18	1.125
4·5	2·04	—	—	2·53	—	—	1·741	2·28	—	—	1·215
5·0	2·38	—	—	3·11	—	—	2·02	2·74	—	—	1·310
5·5	2·77	—	—	3·83	—	—	2·34	3·27	—	—	1·407
6·0	3·22	—	—	4·76	—	—	2·72	3·92	—	—	1·506

APPENDIX 8.10

Activity Coefficients of Electrolytes at 25°

Table 10

m	LiAc	LiTol	NaOH	NaF	NaCl	NaBr	NaI	$NaClO_3$	$NaClO_4$	$NaBrO_3$	$NaNO_3$	Na formate	Na acetate	Na propionate
0·1	0·784	0·772	0·764	0·765	0·778	0·782	0·787	0·772	0·775	0·758	0·762	0·778	0·791	0·800
0·2	0·742	0·723	0·725	0·710	0·735	0·741	0·751	0·720	0·729	0·696	0·703	0·734	0·757	0·772
0·3	0·721	0·695	0·706	0·676	0·710	0·719	0·735	0·688	0·701	0·657	0·666	0·710	0·744	0·763
0·4	0·709	0·674	0·695	0·651	0·693	0·704	0·727	0·664	0·683	0·628	0·638	0·696	0·737	0·762
0·5	0·700	0·659	0·688	0·632	0·681	0·697	0·723	0·645	0·668	0·605	0·617	0·685	0·735	0·764
0·6	0·691	0·647	0·683	0·616	0·673	0·692	0·723	0·630	0·656	0·585	0·599	0·676	0·736	0·769
0·7	0·689	0·638	0·680	0·603	0·667	0·689	0·724	0·617	0·648	0·569	0·583	0·671	0·740	0·777
0·8	0·688	0·630	0·677	0·592	0·662	0·687	0·727	0·606	0·641	0·554	0·570	0·667	0·745	0·787
0·9	0·688	0·623	0·676	0·582	0·659	0·687	0·731	0·597	0·635	0·541	0·558	0·664	0·752	0·797
1·0	0·689	0·617	0·677	0·573	0·657	0·687	0·736	0·589	0·629	0·528	0·548	0·661	0·757	0·808
1·2	0·693	0·605	0·679	—	0·654	0·692	0·747	0·575	0·622	0·507	0·530	0·658	0·769	0·833
1·4	0·700	0·595	0·684	—	0·655	0·699	0·763	0·563	0·616	0·489	0·514	0·657	0·789	0·864
1·6	0·709	0·586	0·690	—	0·657	0·706	0·780	0·553	0·613	0·473	0·501	0·656	0·809	0·897
1·8	0·719	0·575	0·698	—	0·662	0·718	0·799	0·545	0·611	0·461	0·489	0·657	0·829	0·932
2·0	0·729	0·568	0·707	—	0·668	0·731	0·820	0·538	0·609	0·450	0·478	0·658	0·851	0·966
2·5	0·762	0·558	0·741	—	0·688	0·768	0·883	0·525	0·609	0·426	0·455	0·667	0·914	1·061
3·0	0·798	0·556	0·782	—	0·714	0·812	0·963	0·515	0·611	—	0·437	0·678	0·982	1·160
3·5	0·837	0·559	0·833	—	0·746	0·865	1·053	0·508	0·617	—	0·422	0·691	1·057	—
4·0	0·877	0·566	0·901	—	0·783	0·929	—	—	0·626	—	0·408	—	—	—
4·5	—	0·575	0·982	—	0·826	—	—	—	0·637	—	0·396	—	—	—
5·0	—	—	1·074	—	0·874	—	—	—	0·649	—	0·386	—	—	—
5·5	—	—	1·178	—	0·928	—	—	—	0·662	—	0·378	—	—	—
6·0	—	—	1·296	—	0·986	—	—	—	0·677	—	0·371	—	—	—

Table 10—(Contd.)

m	Na butyrate	Na valerate	Na caproate	Na heptylate	Na caprylate	Na pelargonate	Na caprate	Na H malonate	Na H succinate	Na H adipate	NaTol	NaCNS	NaH_2PO_4
0·1	0·800	0·800	0·803	0·803	—	—	—	0·764	0·765	0·776	0·765	0·787	0·744
0·2	0·774	0·776	0·779	0·780	—	—	—	0·709	0·712	0·730	0·709	0·750	0·675
0·3	0·769	0·771	0·775	0·777	—	—	—	0·674	0·677	0·703	0·674	0·731	0·629
0·4	0·774	0·780	0·783	0·780	—	—	—	0·647	0·653	0·683	0·648	0·720	0·593
0·5	0·782	0·790	0·794	0·783	0·693	0·390	0·285	0·626	0·635	0·670	0·627	0·715	0·563
0·6	0·795	0·805	0·810	0·781	0·621	0·335	0·244	0·609	0·618	0·658	0·609	0·712	0·539
0·7	0·812	0·817	0·826	0·775	0·553	0·295	0·212	0·595	0·607	0·650	0·593	0·710	0·517
0·8	0·830	0·835	0·841	0·754	0·491	0·264	0·184	0·582	0·596	—	0·579	0·710	0·499
0·9	0·848	0·852	0·851	0·700	0·434	0·239	0·169	0·572	0·586	—	0·566	0·711	0·483
1·0	0·868	0·868	0·858	0·650	0·401	0·219	0·147	0·563	0·579	—	0·554	0·712	0·468
1·2	0·908	0·907	0·865	0·562	0·349	0·189	0·120	0·546	0·565	—	0·532	0·716	0·442
1·4	0·952	0·945	0·855	0·512	0·309	0·168	0·107	0·533	0·556	—	0·511	0·723	0·420
1·6	0·992	0·984	0·830	0·468	0·279	0·152	0·097	0·523	0·548	—	0·493	0·730	0·401
1·8	1·036	1·012	0·799	0·430	0·253	0·140	0·089	0·514	0·543	—	0·476	0·737	0·385
2·0	1·083	1·030	0·763	0·398	0·236	0·130	—	0·507	0·538	—	0·460	0·744	0·371
2·5	1·182	1·027	0·673	0·340	0·206	0·126	—	0·490	0·529	—	0·427	0·779	0·343
3·0	1·278	0·982	0·612	0·306	0·185	—	—	0·477	0·526	—	0·402	0·814	0·320
3·5	1·368	0·901	0·576	0·284	—	—	—	0·467	0·524	—	0·383	0·854	0·305
4·0	—	—	0·556	0·267	—	—	—	0·458	0·525	—	0·368	0·897	0·293
4·5	—	—	0·542	0·255	—	—	—	0·451	0·528	—	—	—	0·283
5·0	—	—	—	0·245	—	—	—	0·445	0·534	—	—	—	0·276
5·5	—	—	—	—	—	—	—	—	—	—	—	—	0·270
6·0	—	—	—	—	—	—	—	—	—	—	—	—	0·265

APPENDIX 8.10

Activity Coefficients of Electrolytes at 25°

Table 11

m	KOH	KF	KCl	KBr	KI	$KClO_3$	$KBrO_3$	KNO_3	KAc	KH malonate	KH succinate	KH adipate	KTol	KCNS
0·1	0·776	0·775	0·770	0·772	0·778	0·749	0·745	0·739	0·796	0·759	0·762	0·772	0·762	0·769
0·2	0·739	0·727	0·718	0·722	0·733	0·681	0·674	0·663	0·766	0·702	0·705	0·724	0·702	0·716
0·3	0·721	0·700	0·688	0·693	0·707	0·635	0·625	0·614	0·754	0·665	0·668	0·693	0·662	0·685
0·4	0·713	0·682	0·666	0·673	0·689	0·599	0·585	0·576	0·750	0·634	0·640	0·669	0·632	0·663
0·5	0·712	0·670	0·649	0·657	0·676	0·568	0·552	0·545	0·751	0·610	0·619	0·654	0·605	0·646
0·6	0·712	0·661	0·637	0·646	0·667	0·541	—	0·519	0·754	0·588	0·602	0·642	0·582	0·633
0·7	0·715	0·654	0·626	0·636	0·660	0·518	—	0·496	0·759	0·570	0·588	0·631	0·560	0·623
0·8	0·721	0·650	0·618	0·629	0·654	—	—	0·476	0·766	0·554	0·575	0·622	0·541	0·614
0·9	0·728	0·646	0·610	0·622	0·649	—	—	0·459	0·774	0·541	0·564	0·615	0·523	0·606
1·0	0·735	0·645	0·604	0·617	0·645	—	—	0·443	0·783	0·528	0·553	0·609	0·506	0·599
1·2	0·754	0·643	0·593	0·608	0·640	—	—	0·414	0·803	0·507	0·536	—	0·476	0·587
1·4	0·778	0·644	0·586	0·602	0·637	—	—	0·390	0·827	0·488	0·521	—	0·448	0·577
1·6	0·804	0·647	0·580	0·598	0·636	—	—	0·369	0·854	0·472	0·510	—	0·424	0·569
1·8	0·832	0·652	0·576	0·595	0·636	—	—	0·350	0·881	0·460	0·501	—	0·404	0·562
2·0	0·863	0·658	0·573	0·593	0·637	—	—	0·333	0·910	0·450	0·493	—	0·386	0·556
2·5	0·947	0·678	0·569	0·593	0·644	—	—	0·297	0·995	0·427	0·478	—	0·347	0·546
3·0	1·051	0·705	0·569	0·595	0·652	—	—	0·269	1·086	0·408	0·468	—	0·317	0·538
3·5	1·181	0·738	0·572	0·600	0·662	—	—	0·246	1·181	0·392	0·463	—	0·292	0·533
4·0	1·314	0·779	0·577	0·608	0·673	—	—	—	—	0·377	0·457	—	—	0·529
4·5	1·49	—	0·583	0·616	0·683	—	—	—	—	0·365	0·453	—	—	0·526
5·0	1·67	—	—	0·626	—	—	—	—	—	0·353	—	—	—	0·524
5·5	1·90	—	—	0·636	—	—	—	—	—	—	—	—	—	—
6·0	2·14	—	—	—	—	—	—	—	—	—	—	—	—	—

Table 11—(Contd.)

m	KH_2PO_4	NH_4Cl	NH_4NO_3	RbCl	RbBr	RbI	$RbNO_3$	RbAc	CsOH	CsCl	CsBr	CsI	$CsNO_3$	CsAc
0·1	0·731	0·770	0·740	0·764	0·763	0·762	0·734	0·796	0·809	0·756	0·754	0·754	0·733	0·799
0·2	0·653	0·718	0·677	0·709	0·706	0·705	0·658	0·767	0·774	0·694	0·694	0·692	0·655	0·771
0·3	0·602	0·687	0·636	0·675	0·673	0·671	0·606	0·756	0·757	0·656	0·654	0·651	0·602	0·761
0·4	0·561	0·665	0·606	0·652	0·650	0·647	0·565	0·753	0·752	0·628	0·626	0·621	0·561	0·759
0·5	0·529	0·649	0·582	0·634	0·632	0·629	0·534	0·755	0·752	0·606	0·603	0·599	0·528	0·762
0·6	0·501	0·636	0·562	0·620	0·617	0·614	0·508	0·759	0·755	0·589	0·586	0·581	0·501	0·768
0·7	0·477	0·625	0·545	0·608	0·605	0·602	0·485	0·766	0·761	0·575	0·571	0·567	0·478	0·776
0·8	0·456	0·617	0·530	0·599	0·595	0·591	0·465	0·773	0·767	0·563	0·558	0·554	0·458	0·783
0·9	0·438	0·609	0·516	0·590	0·586	0·583	0·446	0·782	0·775	0·553	0·547	0·543	0·439	0·792
1·0	0·421	0·603	0·504	0·583	0·578	0·575	0·430	0·792	0·785	0·544	0·538	0·533	0·422	0·802
1·2	0·393	0·592	0·483	0·572	0·565	0·562	0·402	0·815	—	0·529	0·523	0·516	0·393	0·826
1·4	0·369	0·584	0·464	0·563	0·556	0·551	0·377	0·840	—	0·518	0·510	0·501	0·368	0·853
1·6	0·348	0·578	0·447	0·556	0·547	0·544	0·356	0·869	—	0·509	0·500	0·489	—	0·883
1·8	0·332	0·574	0·433	0·551	0·541	0·537	0·338	0·900	—	0·501	0·493	0·479	—	0·916
2·0	—	0·570	0·419	0·546	0·536	0·533	0·321	0·933	—	0·496	0·486	0·470	—	0·950
2·5	—	0·564	0·391	0·539	0·526	0·524	0·285	1·023	—	0·485	0·474	0·450	—	1·041
3·0	—	0·561	0·368	0·536	0·520	0·518	0·257	1·126	—	0·479	0·465	0·434	—	1·145
3·5	—	0·560	0·348	0·536	0·516	0·516	0·234	1·240	—	0·475	0·460	—	—	1·263
4·0	—	0·560	0·331	0·538	0·514	0·515	0·216	—	—	0·474	0·457	—	—	—
4·5	—	0·561	0·316	0·541	0·514	0·516	0·200	—	—	0·474	0·455	—	—	—
5·0	—	0·562	0·302	0·546	0·515	0·517	—	—	—	0·475	0·453	—	—	—
5·5	—	0·563	0·290	—	—	—	—	—	—	0·477	—	—	—	—
6·0	—	0·564	0·279	—	—	—	—	—	—	0·480	—	—	—	—

APPENDIX 8.10

Activity Coefficients of Electrolytes at 25°

Table 12

m	$AgNO_3$	$TlClO_4$	$TlNO_3$	$Tl\bar{A}c$
0·1	0·734	0·730	0·702	0·750
0·2	0·657	0·652	0·606	0·686
0·3	0·606	0·599	0·545	0·644
0·4	0·567	0·559	0·500	0·614
0·5	0·536	0·527	—	0·589
0·6	0·509	—	—	0·570
0·7	0·485	—	—	0·553
0·8	0·464	—	—	0·539
0·9	0·446	—	—	0·526
1·0	0·429	—	—	0·515
1·2	0·399	—	—	0·496
1·4	0·374	—	—	0·480
1·6	0·352	—	—	0·466
1·8	0·333	—	—	0·454
2·0	0·316	—	—	0·444
2·5	0·280	—	—	0·422
3·0	0·252	—	—	0·405
3·5	0·229	—	—	0·389
4·0	0·210	—	—	0·376
4·5	0·194	—	—	0·364
5·0	0·181	—	—	0·354
5·5	0·169	—	—	0·344
6·0	0·159	—	—	0·335

APPENDIX 8.10

Activity Coefficients of Electrolytes at 25°

Table 13

m	$MgCl_2$	$MgBr_2$	MgI_2	$Mg(ClO_4)_2$	$Mg(NO_3)_2$	$MgAc_2$	$CaCl_2$	$CaBr_2$	CaI_2	$Ca(ClO_4)_2$	$Ca(NO_3)_2$
0·1	0·528	0·542	0·571	0·577	0·522	0·459	0·518	0·532	0·552	0·557	0·488
0·2	0·488	0·512	0·550	0·565	0·480	0·397	0·472	0·491	0·524	0·532	0·429
0·3	0·476	0·511	0·558	0·576	0·467	0·366	0·455	0·481	0·524	0·532	0·397
0·4	0·474	0·520	0·575	0·599	0·465	0·347	0·448	0·482	0·535	0·544	0·378
0·5	0·480	0·538	0·605	0·633	0·469	0·335	0·448	0·490	0·553	0·564	0·365
0·6	0·490	0·564	0·643	0·673	0·478	0·326	0·453	0·504	0·576	0·589	0·356
0·7	0·505	0·591	0·688	0·723	0·488	0·320	0·460	0·521	0·605	0·618	0·349
0·8	0·521	0·627	0·742	0·780	0·501	0·316	0·470	0·542	0·641	0·654	0·344
0·9	0·543	0·668	0·805	0·849	0·518	0·314	0·484	0·567	0·682	0·695	0·340
1·0	0·569	0·714	0·879	0·925	0·536	0·313	0·500	0·596	0·731	0·743	0·338
1·2	0·630	0·826	1·053	1·112	0·580	0·314	0·539	0·664	0·840	0·853	0·337
1·4	0·708	0·962	1·272	1·355	0·631	0·316	0·587	0·746	0·978	0·992	0·337
1·6	0·802	1·128	1·556	1·667	0·691	0·321	0·644	0·846	1·148	1·161	0·339
1·8	0·914	1·333	1·928	2·08	0·758	0·327	0·712	0·968	1·356	1·372	0·342
2·0	1·051	1·593	2·39	2·59	0·835	0·336	0·792	1·119	1·617	1·634	0·347
2·5	1·538	2·56	4·27	4·78	1·088	0·358	1·063	1·654	—	2·62	0·362
3·0	2·32	4·20	7·81	8·99	1·449	0·386	1·483	2·53	—	4·21	0·382
3·5	3·55	7·06	14·8	17·26	1·936	0·414	2·08	3·88	—	6·76	0·407
4·0	5·53	12·0	28·6	33·3	2·59	0·445	2·93	6·27	—	10·77	0·438
4·5	8·72	20·8	56·7	—	3·50	—	4·17	10·64	—	17·02	0·472
5·0	13·92	36·1	113·	—	4·74	—	5·89	18·43	—	26·7	0·510
5·5	—	—	—	—	—	—	8·18	31·7	—	41·7	0·551
6·0	—	—	—	—	—	—	11·11	55·7	—	63·7	0·596

Table 13—(Contd.)

m	$SrCl_2$	$SrBr_2$	SrI_2	$Sr(ClO_4)_2$	$Sr(NO_3)_2$	$BaCl_2$	$BaBr_2$	BaI_2	$Ba(ClO_4)_2$	$Ba(NO_3)_2$	$BaAc_2$
0·1	0·515	0·527	0·549	0·528	0·478	0·508	0·517	0·536	0·524	0·431	0·462
0·2	0·466	0·483	0·516	0·494	0·410	0·450	0·269	0·503	0·481	0·345	0·406
0·3	0·446	0·468	0·513	0·488	0·373	0·425	0·450	0·496	0·464	0·295	0·380
0·4	0·436	0·465	0·520	0·494	0·348	0·411	0·440	0·504	0·459	0·262	0·366
0·5	0·433	0·467	0·532	0·507	0·329	0·403	0·438	0·517	0·462	—	0·356
0·6	0·434	0·473	0·551	0·525	0·314	0·397	0·442	0·534	0·469	—	0·349
0·7	0·437	0·484	0·573	0·546	0·302	0·397	0·446	0·556	0·477	—	0·344
0·8	0·445	0·497	0·603	0·573	0·292	0·397	0·452	0·581	0·487	—	0·340
0·9	0·453	0·515	0·637	0·604	0·283	0·397	0·462	0·610	0·500	—	0·337
1·0	0·465	0·535	0·675	0·638	0·275	0·401	0·473	0·642	0·513	—	0·334
1·2	0·493	0·583	0·767	0·718	0·262	0·411	0·500	0·716	0·545	—	0·329
1·4	0·528	0·643	0·878	0·812	0·253	0·424	0·533	0·805	0·581	—	0·323
1·6	0·570	0·715	1·013	0·928	0·244	0·439	0·571	0·914	0·622	—	0·319
1·8	0·619	0·800	1·181	1·060	0·238	0·455	0·614	1·043	0·674	—	0·314
2·0	0·675	0·906	1·396	1·220	0·232	—	0·661	1·208	0·718	—	0·309
2·5	0·862	—	—	1·755	0·223	—	—	—	0·868	—	0·294
3·0	1·135	—	—	2·57	0·217	—	—	—	1·047	—	0·278
3·5	1·504	—	—	3·68	0·214	—	—	—	1·287	—	0·263
4·0	1·993	—	—	5·20	0·212	—	—	—	1·545	—	—
4·5	—	—	—	7·30	—	—	—	—	1·826	—	—
5·0	—	—	—	10·09	—	—	—	—	2·13	—	—
5·5	—	—	—	13·73	—	—	—	—	—	—	—
6·0	—	—	—	18·43	—	—	—	—	—	—	—

APPENDIX 8.10

Activity Coefficients of Electrolytes at 25°

Table 14

m	$MnCl_2$	$FeCl_2$	$CoCl_2$	$CoBr_2$	CoI_2	$Co(NO_3)_2$	$NiCl_2$	$CuCl_2$	$Cu(NO_3)_2$	$ZnCl_2$	$ZnBr_2$	ZnI_2
0·1	0·518	0·520	0·523	0·540	0·56	0·521	0·523	0·510	0·512	0·518	0·547	0·572
0·2	0·471	0·475	0·479	0·507	0·54	0·474	0·479	0·457	0·461	0·465	0·510	0·550
0·3	0·452	0·456	0·463	0·503	0·55	0·455	0·463	0·431	0·440	0·435	0·502	0·555
0·4	0·444	0·450	0·459	0·511	0·57	0·448	0·460	0·419	0·430	0·413	0·504	0·573
0·5	0·442	0·452	0·462	0·526	0·60	0·448	0·464	0·413	0·427	0·396	0·511	0·605
0·6	0·445	0·456	0·470	0·548	0·64	0·451	0·471	0·411	0·428	0·382	0·519	0·635
0·7	0·450	0·465	0·479	0·574	0·69	0·458	0·482	0·411	0·432	0·371	0·528	0·672
0·8	0·457	0·475	0·492	0·605	0·74	0·468	0·496	0·412	0·438	0·359	0·537	0·713
0·9	0·468	0·490	0·511	0·641	0·81	0·480	0·515	0·415	0·446	0·350	0·547	0·750
1·0	0·481	0·508	0·531	0·682	0·88	0·493	0·536	0·419	0·456	0·341	0·552	0·788
1·2	0·509	0·549	0·578	0·780	1·05	0·526	0·586	0·427	0·479	0·325	0·561	0·857
1·4	0·544	0·598	0·634	0·904	1·27	0·566	0·647	0·436	0·504	0·311	0·567	0·914
1·6	0·583	0·656	0·699	1·057	1·54	0·613	0·720	0·446	0·534	0·302	0·569	0·957
1·8	0·626	0·722	0·773	1·241	1·89	0·668	0·805	0·457	0·570	0·296	0·570	0·991
2·0	0·671	0·797	0·860	1·462	2·3	0·730	0·906	0·468	0·610	0·291	0·572	1·012
2·5	0·796	—	1·120	2·23	4·3	0·926	1·236	0·496	0·728	0·287	0·581	1·053
3·0	0·938	—	1·458	3·38	7·4	1·189	1·692	0·522	0·905	0·289	0·598	1·106
3·5	1·088	—	1·832	5·04	13·2	1·535	2·26	0·549	1·120	0·297	0·626	1·170
4·0	1·240	—	2·22	7·54	23	1·984	2·96	0·575	1·384	0·309	0·664	1·239
4·5	1·401	—	—	10·90	39	2·57	3·76	0·599	1·693	0·330	0·714	1·336
5·0	1·56	—	—	15·19	60	3·33	4·69	0·623	2·05	0·356	0·774	1·453
5·5	1·72	—	—	—	80	—	—	0·650	2·48	0·385	0·845	1·588
6·0	1·89	—	—	—	99	—	—	0·676	2·99	0·420	0·930	1·747

Table 14—(Contd.)

m	$Zn(ClO_4)_2$	$Zn(NO_3)_2$	$CdCl_2$	$CdBr_2$	CdI_2	$Cd(NO_3)_2$	$Pb(ClO_4)_2$	$Pb(NO_3)_2$	UO_2Cl_2	$UO_2(ClO_4)_2$	$UO_2(NO_3)_2$
0·1	0·568	0·530	0·2280	0·1900	0·1060	0·516	0·525	0·405	0·539	0·604	0·543
0·2	0·552	0·487	0·1638	0·132	0·0685	0·467	0·483	0·316	0·505	0·612	0·512
0·3	0·560	0·472	0·1329	0·105	0·0523	0·445	0·467	0·267	0·497	0·646	0·510
0·4	0·583	0·467	0·1139	0·089	0·0433	0·433	0·462	0·234	0·500	0·698	0·518
0·5	0·615	0·471	0·1006	0·0780	0·0376	0·428	0·465	0·210	0·512	0·762	0·534
0·6	0·655	0·478	0·0905	0·0699	0·0337	0·426	0·471	0·192	0·527	0·841	0·555
0·7	0·704	0·487	0·0827	0·0638	0·0307	0·426	0·479	0·176	0·544	0·935	0·578
0·8	0·763	0·499	0·0765	0·0591	0·0285	0·428	0·491	0·164	0·565	1·049	0·608
0·9	0·831	0·516	0·0713	0·0551	0·0267	0·431	0·506	0·154	0·589	1·183	0·641
1·0	0·909	0·533	0·0669	0·0518	0·0251	0·436	0·523	0·145	0·614	1·341	0·679
1·2	1·102	0·572	0·0599	0·0468	0·0228	0·449	0·563	0·130	0·671	1·741	0·761
1·4	1·356	0·623	0·0546	0·0431	0·0214	0·463	0·613	0·118	0·737	2·30	0·855
1·6	1·681	0·677	0·0504	0·0402	0·0199	0·481	0·669	0·109	0·808	3·06	0·943
1·8	2·11	0·741	0·0469	0·0380	0·0189	0·498	0·734	0·102	0·885	4·14	1·083
2·0	2·68	0·814	0·0441	0·0361	0·0180	0·518	0·809	0·095	0·968	5·70	1·218
2·5	5·04	1·045	0·0389	0·0328	0·0168	0·573	1·045	—	1·216	12·90	1·602
3·0	9·77	1·358	0·0352	0·0305	—	—	1·386	—	1·535	29·8	2·00
3·5	19·17	1·766	0·0325	0·0290	—	—	1·831	—	—	67·9	2·37
4·0	37·9	2·30	0·0306	0·0278	—	—	2·39	—	—	154·6	2·64
4·5	—	2·98	0·0291	—	—	—	3·22	—	—	345	2·85
5·0	—	3·86	0·0279	—	—	—	4·05	—	—	724	3·01
5·5	—	5·07	0·0270	—	—	—	5·23	—	—	1457	3·20
6·0	—	6·38	0·0263	—	—	—	6·67	—	—	—	—

APPENDIX 8.10

Activity Coefficients of Electrolytes at 25°

Table 15

m	Li_2SO_4	Na_2SO_4	Na_2CrO_4	$Na_2S_2O_3$	Na_2fumarate	Na_2maleate	K_2SO_4	K_2CrO_4	$(NH_4)_2SO_4$	Rb_2SO_4	Cs_2SO_4
0·1	0·478	0·452	0·479	0·466	0·468	0·427	0·436	0·466	0·423	0·460	0·464
0·2	0·406	0·371	0·407	0·390	0·405	0·352	0·356	0·390	0·343	0·382	0·390
0·3	0·369	0·325	0·364	0·347	0·372	0·312	0·313	0·347	0·300	0·338	0·345
0·4	0·344	0·294	0·337	0·319	0·350	0·287	0·283	0·320	0·270	0·308	0·317
0·5	0·326	0·270	0·317	0·298	0·337	0·270	0·261	0·298	0·248	0·285	0·297
0·6	0·313	0·252	0·301	0·282	0·330	0·260	0·243	0·282	0·231	0·269	0·279
0·7	0·303	0·237	0·289	0·267	0·325	0·248	0·229	0·269	0·218	0·254	0·267
0·8	0·295	0·225	0·278	0·256	0·322	0·241	—	0·259	0·206	0·243	0·256
0·9	0·288	0·213	0·269	0·247	0·321	0·234	—	0·248	0·198	0·233	0·247
1·0	0·283	0·204	0·261	0·239	0·321	0·229	—	0·240	0·189	0·224	0·240
1·2	0·277	0·1890	0·249	0·226	0·322	0·222	—	0·228	0·175	0·211	0·226
1·4	0·273	0·1774	0·240	0·218	0·325	0·217	—	0·219	0·165	0·200	0·218
1·6	0·271	0·1680	0·234	0·211	0·332	0·214	—	0·212	0·156	0·193	0·211
1·8	0·270	0·1605	0·231	0·206	0·338	0·213	—	0·205	0·149	0·186	0·205
2·0	0·269	0·1544	0·229	0·202	0·345	0·212	—	0·200	0·144	—	—
2·5	0·280	0·1441	0·232	0·199	—	0·213	—	0·194	0·132	—	—
3·0	0·294	0·1387	0·244	0·203	—	0·218	—	0·194	0·125	—	—
3·5	—	0·1367	0·263	0·211	—	—	—	0·195	0·119	—	—
4·0	—	0·1376	0·294	—	—	—	—	—	0·116	—	—

APPENDIX 8.10

Activity Coefficients of Electrolytes at 25°

Table 16

m	$BeSO_4$	$MgSO_4$	$MnSO_4$	$NiSO_4$	$CuSO_4$	$ZnSO_4$	$CdSO_4$	UO_2SO_4	$AlCl_3$	$ScCl_3$
0·1	(0·150)	(0·150)	(0·150)	(0·150)	(0·150)	(0·150)	(0·150)	(0·150)	(0·337)	(0·320)
0·2	0·109	0·107	0·105	0·105	0·104	0·104	0·103	0·102	0·305	0·288
0·3	0·0885	0·0874	0·0848	0·0841	0·0829	0·0835	0·0822	0·0807	0·302	0·282
0·4	0·0769	0·0756	0·0725	0·0713	0·0704	0·0714	0·0699	0·0689	0·313	0·287
0·5	0·0692	0·0675	0·0640	0·0627	0·0620	0·0630	0·0615	0·0611	0·331	0·298
0·6	0·0639	0·0616	0·0578	0·0562	0·0559	0·0569	0·0553	0·0566	0·356	0·316
0·7	0·0600	0·0571	0·0530	0·0515	0·0512	0·0523	0·0505	0·0515	0·388	0·339
0·8	0·0570	0·0536	0·0493	0·0478	0·0475	0·0487	0·0468	0·0483	0·429	0·369
0·9	0·0546	0·0508	0·0463	0·0448	0·0446	0·0458	0·0438	0·0458	0·479	0·405
1·0	0·0530	0·0485	0·0439	0·0425	0·0423	0·0435	0·0415	0·0439	0·539	0·443
1·2	0·0506	0·0453	0·0404	0·0390	0·0388	0·0401	0·0379	0·0409	0·701	0·544
1·4	0·0493	0·0434	0·0380	0·0368	0·0365	0·0378	0·0355	0·0391	0·936	0·677
1·6	0·0488	0·0423	0·0365	0·0353	—	0·0363	0·0338	0·0379	1·284	0·853
1·8	0·0490	0·0417	0·0356	0·0345	—	0·0356	0·0327	0·0372	1·819	1·089
2·0	0·0497	0·0417	0·0351	0·0343	—	0·0357	0·0321	0·0367	—	—
2·5	0·0538	0·0439	0·0349	0·0357	—	0·0367	0·0317	0·0370	—	—
3·0	0·0613	0·0492	0·0373	—	—	0·0408	0·0329	0·0383	—	—
3·5	0·0724	—	0·0413	—	—	0·0480	0·0356	0·0401	—	—
4·0	0·0875	—	0·0473	—	—	—	—	0·0433	—	—
4·5	—	—	—	—	—	—	—	0·0465	—	—
5·0	—	—	—	—	—	—	—	0·0500	—	—
5·5	—	—	—	—	—	—	—	0·0536	—	—
6·0	—	—	—	—	—	—	—	0·0571	—	—

Activity Coefficients of Electrolytes at 25°

Table 17

m	$CrCl_3$	YCl_3	$LaCl_3$	$CeCl_3$	$PrCl_3$	$NdCl_3$	$SmCl_3$	$EuCl_3$	$Cr(NO_3)_3$
0·1	(0·331)	(0·314)	(0·314)	(0·309)	(0·311)	(0·310)	(0·314)	(0·318)	(0·319)
0·2	0·298	0·278	0·274	0·273	0·273	0·272	0·278	0·282	0·285
0·3	0·294	0·269	0·263	0·261	0·260	0·261	0·267	0·270	0·279
0·4	0·300	0·271	0·261	0·260	0·258	0·259	0·266	0·270	0·281
0·5	0·314	0·278	0·266	0·264	0·262	0·264	0·271	0·276	0·291
0·6	0·335	0·291	0·274	0·272	0·268	0·272	0·280	0·286	0·304
0·7	0·362	0·307	0·285	0·286	0·281	0·284	0·296	0·303	0·322
0·8	0·397	0·329	0·302	0·302	0·297	0·301	0·314	0·322	0·344
0·9	0·436	0·355	0·321	0·320	0·316	0·321	0·336	0·345	0·371
1·0	0·481	0·385	0·342	0·342	0·338	0·344	0·362	0·371	0·401
1·2	0·584	0·462	0·398	0·395	0·395	0·403	0·424	0·436	0·474
1·4	—	0·566	0·470	0·469	0·467	0·480	0·509	0·525	0·565
1·6	—	0·701	0·561	0·559	0·558	0·577	0·616	0·641	—
1·8	—	0·884	0·677	0·684	0·675	0·704	0·756	0·792	—
2·0	—	1·136	0·825	0·847	0·825	0·867	0·940	0·995	—

APPENDIX 8.10

Activity Coefficients of Electrolytes at 25°

Table 18

m	$K_3Fe(CN)_6$	$K_4Fe(CN)_6$	$Al_2(SO_4)_3$	$Cr_2(SO_4)_3$	$Th(NO_3)_4$
0·1	(0·268)	(0·139)	(0·0350)	(0·0458)	0·279
0·2	0·212	0·0993	0·0225	0·0300	0·225
0·3	0·184	0·0808	0·0176	0·0238	0·203
0·4	0·167	0·0693	0·0153	0·0207	0·192
0·5	0·155	0·0614	0·0143	0·0190	0·189
0·6	0·146	0·0556	0·0140	0·0182	0·188
0·7	0·140	0·0512	0·0142	0·0181	0·191
0·8	0·135	0·0479	0·0149	0·0185	0·195
0·9	0·131	0·0454	0·0159	0·0194	0·201
1·0	0·128	—	0·0175	0·0208	0·207
1·2	0·124	—	—	0·0250	0·224
1·4	0·122	—	—	—	0·246
1·6	—	—	—	—	0·269
1·8	—	—	—	—	0·296
2·0	—	—	—	—	0·326
2·5	—	—	—	—	0·405
3·0	—	—	—	—	0·486
3·5	—	—	—	—	0·568
4·0	—	—	—	—	0·647
4·5	—	—	—	—	0·722
5·0	—	—	—	—	0·791

APPENDIX 8.10

Data at High Concentrations

Table 19

m	HCl		$HClO_4$		LiCl		LiBr		$LiNO_3$		NaOH		KOH		NH_4NO_3		CsCl		$AgNO_3$		$Pb(ClO_4)_2$	
	ϕ	γ	ϕ	γ	ϕ	γ	ϕ	γ	ϕ	γ	ϕ	γ	ϕ	γ	ϕ	γ	ϕ	γ	ϕ	γ	ϕ	γ
7	2·008	4·37	2·365	7·44	1·965	3·71	2·206	5·76	1·485	1·723	1·567	1·599	1·81	2·80	0·653	0·2605	0·966	0·486	0·426	0·142	2·737	10·69
8	2·163	5·90	2·629	11·83	2·143	5·10	2·432	8·61	1·541	1·952	1·707	2·00	1·96	3·66	0·639	0·2451	0·989	0·496	0·408	0·129	2·915	16·31
9	2·315	7·94	2·901	19·11	2·310	6·96	2·656	12·92	1·591	2·19	1·853	2·54	2·09	4·72	0·627	0·2318	1·004	0·503	0·393	0·118	3·057	23·7
10	2·444	10·44	3·167	30·9	2·464	9·40	2·902	19·92	1·633	2·44	1·993	3·22	2·22	6·05	0·616	0·2205	1·013	0·508	0·378	0·109	3·194	34·1
11	2·559	13·51	3·433	50·1	2·607	12·55	3·150	31·0	1·668	2·69	2·131	4·09	2·36	7·87	0·607	0·2104	1·018	0·512	0·371	0·102	3·297	46·8
12	2·663	17·25	3·688	80·8	2·730	16·41	3·356	46·3	1·700	2·95	2·262	5·18	2·50	10·2	0·598	0·2016	—	—	0·363	0·096	3·365	61·4
13	2·760	21·8	3·935	129·5	2·830	20·9	3·581	70·6	1·727	3·20	2·382	6·48	2·60	12·8	0·591	0·1936	—	—	0·356	0·090	—	—
14	2·853	27·3	4·166	205	2·915	26·2	3·776	104·7	—	—	2·488	8·02	2·66	15·4	0·583	0·1864	—	—	—	—	—	—
15	2·944	34·1	4·393	322	2·978	31·9	3·912	146·0	—	—	2·574	9·71	2·76	19·1	0·576	0·1797	—	—	—	—	—	—
16	3·033	42·4	4·608	500	3·023	37·9	4·025	198·0	—	—	2·643	11·55	2·87	23·9	0·569	0·1736	—	—	—	—	—	—
17	—	—	—	—	3·044	43·8	4·110	260·	—	—	2·694	13·43	—	—	0·562	0·1679	—	—	—	—	—	—
18	—	—	—	—	3·057	49·9	4·173	331·	—	—	2·730	15·37	—	—	0·556	0·1628	—	—	—	—	—	—
19	—	—	—	—	3·066	56·3	4·216	411·	—	—	2·756	17·33	—	—	0·550	0·1579	—	—	—	—	—	—
20	—	—	—	—	3·063	62·4	4·217	485·	—	—	2·772	19·28	—	—	0·544	0·1534	—	—	—	—	—	—

These osmotic and activity coefficients are molal coefficients; they are taken mainly from Stokes, R. H., *Trans. Faraday Soc.*, 44 (1948) 295; Robinson, R. A. and Stokes, R. H., *ibid.*, 45 (1949) 612, with the following additions:

Smith, E. R. B. and Robinson, R. A., *Trans. Faraday Soc.*, 38 (1942) 70: Sodium salts of fatty acids.
Stokes, J. M., *J. Amer. chem. Soc.*, 70 (1948) 1944: Primary sodium and potassium salts of malonic, succinic and adipic acid.
Wishaw, B. F. and Stokes, R. H., *Trans. Faraday Soc.*, 49 (1953) 27: Ammonium chloride and nitrate, *ibid.*, 50 (1954) 952: Ammonium sulphate.
Robinson, R. A., McCoach, H. J. and Lim, C. K., *J. Amer. chem. Soc.*, 72 (1950) 5783: Cobalt bromide and iodide.
Stokes, R. H., *ibid.*, 75 (1953) 3856: Magnesium and barium acetate.
Robinson, R. A., Lim, C. K. and Ang, K. P., *ibid.*, 75 (1953) 5130: Calcium, strontium and barium perchlorate.
Biggs, A. I., Parton, H. N. and Robinson, R. A., *ibid.*, 77 (1955) 5844: Lead perchlorate and nitrate.
Robinson, R. A. and Lim, C. K., *J. chem. Soc.*, (1951) 1840: Uranyl chloride, perchlorate and nitrate.
Robinson, R. A., *ibid.* (1952) 4543: Beryllium and uranyl sulphate.

APPENDIX 8.10

Table 20

Some Recent Osmotic and Activity Coefficient Data at 25°

A. *Osmotic Coefficients*

m	*A*	*B*	*C*	*D*	*E*	NaH_2AsO_4	KH_2AsO_4	Na_2HPO_4	Na_2HAsO_4
0·1	0·922	0·914	0·913	0·857	0·891	0·924	0·913	0·802	0·820
0·2	0·899	0·890	0·885	0·813	0·905	0·902	0·883	0·754	0·785
0·3	0·887	0·870	0·864	0·800	0·937	0·887	0·861	0·720	0·761
0·4	0·877	0·851	0·844	0·792	0·968	0·874	0·842	0·693	0·742
0·5	0·869	0·831	0·826	0·786	0·997	0·862	0·827	0·670	0·726
0·6	0·861	0·812	0·811	0·782	1·027	0·852	0·813	0·651	0·712
0·7	0·854	0·793	0·795	0·784	1·057	0·842	0·801	0·634	0·700
0·8	0·849	0·774	0·782	0·796	1·088	0·833	0·790	0·620	0·689
0·9	0·843	0·755	0·771	0·807	1·120	0·825	0·781	0·608	0·679
1·0	0·838	0·738	0·758	0·822	1·151	0·817	0·772	0·596	0·670
1·2	0·830	0·708	0·740	0·857	1·222	0·802	0·757	—	—
1·4	0·824	0·687	0·728	0·899	1·293	—	—	—	—
1·6	0·817	0·673	0·721	0·944	1·362	—	—	—	—
1·8	0·812	0·664	0·718	0·992	1·437	—	—	—	—
2·0	0·809	0·659	0·713	1·042	—	—	—	—	—
2·5	0·806	0·662	0·709	—	—	—	—	—	—
3·0	0·816	0·680	0·711	—	—	—	—	—	—
3·5	0·837	0·705	0·718	—	—	—	—	—	—
4·0	0·867	0·734	0·734	—	—	—	—	—	—
4·5	0·899	0·774	0·752	—	—	—	—	—	—
5·0	0·936	0·809	0·774	—	—	—	—	—	—

A = *p*-Toluenesulphonic acid
F = *p*-Ethylbenzenesulphonic acid
C = 2:5-Dimethylbenzenesulphonic acid
D = 4:4′-Bibenzyldisulphonic acid
E = *m*-Benzenedisulphonic acid

APPENDIX 8.10

Table 20—(Contd.)

A. Osmotic Coefficients

m	K_2HPO_4	K_2HAsO_4	$Ga(ClO_4)_3$	$Co(en)_3Cl_3$	Na_3PO_4	Na_3AsO_4	K_3PO_4	K_3AsO_4	$ThCl_4$	$K_4Mo(CN)_8$	$Pt(en)_3Cl_4$	$Pt(en)_3(ClO_4)_4$
0·1	0·805	0·833	0·867	0·627	0·678	0·689	0·709	0·738	0·736	0·603	0·536	0·777
0·2	0·764	0·811	0·903	0·575	0·618	0·640	0·678	0·724	0·776	0·560	0·491	0·698
0·3	0·739	0·799	0·971	0·541	0·579	0·612	0·665	0·724	0·840	0·529	0·478	—
0·4	0·722	0·790	1·051	0·515	0·550	0·593	0·658	0·726	0·906	0·508	—	—
0·5	0·708	0·784	1·139	0·500	0·527	0·579	0·655	0·730	0·974	0·497	—	—
0·6	0·698	0·779	1·233	0·489	0·508	0·569	0·654	0·734	1·048	0·489	—	—
0·7	0·690	0·775	1·332	0·481	0·492	0·561	0·653	0·738	1·129	0·485	—	—
0·8	0·684	0·771	1·436	0·474	—	—	—	—	1·214	0·483	—	—
0·9	0·679	0·769	1·545	0·471	—	—	—	—	1·302	0·483	—	—
1·0	0·674	0·766	1·661	0·469	—	—	—	—	1·390	0·485	—	—
1·2	—	—	1·912	—	—	—	—	—	1·536	0·494	—	—
1·4	—	—	2·185	—	—	—	—	—	1·665	0·506	—	—
1·6	—	—	2·479	—	—	—	—	—	1·847	—	—	—
1·8	—	—	2·774	—	—	—	—	—	—	—	—	—
2·0	—	—	3·068	—	—	—	—	—	—	—	—	—
2·5	—	—	—	—	—	—	—	—	—	—	—	—
3·0	—	—	—	—	—	—	—	—	—	—	—	—
3·5	—	—	—	—	—	—	—	—	—	—	—	—
4·0	—	—	—	—	—	—	—	—	—	—	—	—
4·5	—	—	—	—	—	—	—	—	—	—	—	—
5·0	—	—	—	—	—	—	—	—	—	—	—	—

B. Activity Coefficients at 25°

m	A	B	C	D	E	NaH_2AsO_4	KH_2AsO_4	Na_2CO_3	Na_2HPO_4	Na_2HAsO_4	K_2HPO_4	K_2HAsO_4
0·1	0·759	0·758	0·749	0·640	0·686	0·767	0·750	0·466	0·467	0·488	0·469	0·501
0·2	0·703	0·685	0·679	0·545	0·646	0·708	0·679	0·394	0·381	0·411	0·387	0·432
0·3	0·660	0·635	0·634	0·500	0·647	0·667	0·630	0·356	0·331	0·366	0·342	0·395
0·4	0·630	0·603	0·596	0·465	0·657	0·637	0·593	0·332	0·297	0·334	0·310	0·369
0·5	0·608	0·570	0·565	0·442	0·674	0·611	0·562	0·313	0·269	0·310	0·288	0·349
0·6	0·589	0·542	0·537	0·426	0·694	0·589	0·537	0·301	0·249	0·290	0·270	0·334
0·7	0·573	0·516	0·513	0·412	0·715	0·569	0·515	0·290	0·232	0·274	0·256	0·322
0·8	0·559	0·492	0·493	0·403	0·738	0·552	0·495	0·281	0·217	0·260	0·243	0·311
0·9	0·546	0·469	0·474	0·398	0·763	0·537	0·479	0·272	0·205	0·249	0·234	0·301
1·0	0·535	0·449	0·456	0·397	0·790	0·522	0·463	0·264	0·195	0·238	0·225	0·294
1·2	0·515	0·415	0·429	0·399	0·856	0·498	0·438	0·250	—	—	—	—
1·4	0·498	0·387	0·406	0·410	0·931	—	—	0·238	—	—	—	—
1·6	0·483	0·366	0·389	0·424	1·015	—	—	0·227	—	—	—	—
1·8	0·469	0·348	0·374	0·443	1·108	—	—	—	—	—	—	—
2·0	0·459	0·335	0·362	0·467	—	—	—	—	—	—	—	—
2·5	0·439	0·311	0·338	—	—	—	—	—	—	—	—	—
3·0	0·427	0·299	0·321	—	—	—	—	—	—	—	—	—
3·5	0·425	0·292	0·310	—	—	—	—	—	—	—	—	—
4·0	0·430	0·289	0·303	—	—	—	—	—	—	—	—	—
4·5	0·437	0·292	0·300	—	—	—	—	—	—	—	—	—
5·0	0·448	0·296	0·299	—	—	—	—	—	—	—	—	—

A = *p*-Toluenesulphonic acid
B = *p*-Ethylbenzenesulphonic acid
C = 2:5-Dimethylbenzenesulphonic acid
D = 4:4-Bibenzyldisulphonic acid
E = *m*-Benzenedisulphonic acid

Table 20—(Contd.)

B. Activity Coefficients at 25°

m	$Ga(ClO_4)_3$	$Co(en)_3Cl_3$	Na_3PO_4	Na_3AsO_4	K_3PO_4	K_3AsO_4	$ThCl_4$	$K_4Mo(CN)_8$	$Pt(en)_3Cl_4$	$Pt(en)_3(ClO_4)_4$
0·1	0·443	0·221	0·293	0·299	0·312	0·331	0·292	0·145	0·117	0·308
0·2	0·422	0·164	0·216	0·225	0·244	0·270	0·257	0·104	0·0806	0·239
0·3	0·439	0·135	0·177	0·188	0·211	0·242	0·253	0·0831	0·0652	—
0·4	0·477	0·115	0·151	0·165	0·190	0·224	0·261	0·0713	—	—
0·5	0·532	0·101	0·134	0·148	0·175	0·212	0·275	0·0632	—	—
0·6	0·604	0·0906	0·120	0·136	0·164	0·202	0·297	0·0567	—	—
0·7	0·697	0·0832	0·109	0·126	0·156	0·195	0·327	0·0521	—	—
0·8	0·814	0·0774	—	—	—	—	0·364	0·0488	—	—
0·9	0·961	0·0728	—	—	—	—	0·409	0·0459	—	—
1·0	1·150	0·0690	—	—	—	—	0·463	0·0436	—	—
1·2	1·704	—	—	—	—	—	0·583	0·0400	—	—
1·4	2·63	—	—	—	—	—	0·729	0·0376	—	—
1·6	4·21	—	—	—	—	—	0·966	—	—	—
1·8	6·85	—	—	—	—	—	—	—	—	—
2·0	11·20	—	—	—	—	—	—	—	—	—
2·5	—	—	—	—	—	—	—	—	—	—
3·0	—	—	—	—	—	—	—	—	—	—
3·5	—	—	—	—	—	—	—	—	—	—
4·0	—	—	—	—	—	—	—	—	—	—
4·5	—	—	—	—	—	—	—	—	—	—
5·0	—	—	—	—	—	—	—	—	—	—

APPENDIX 8.10

REFERENCES TO TABLE 20

BONNER, O. D., EASTERLING, G. D., WEST, D. L. and HOLLAND, V. F., *J. Amer. chem. Soc.*, 77 (1955) 242: p-Toluenesulphonic acid and p-ethylbenzenesulphonic acid

BONNER, O. D., HOLLAND, V. F. and SMITH, L. L., *J. phys. Chem.*, 60 (1956) 1102: 2 : 5-Dimethylbenzenesulphonic acid, 4 : 4′-bibenzylsulphonic acid and m-benzenesulphonic acid

BRUBAKER, C. H., *J. Amer. chem. Soc.*, 78 (1956) 5762; 79 (1957) 4274 Potassium octacyanomolybdate, tris-(ethylenediamine)-cobalt (III) chloride and tris-(ethylenediamine)-platinum (IV) chloride and perchlorate

PATTERSON, C. S., TYREE, S. Y. and KNOX, K., *ibid.*, 77 (1955) 2195: Gallium perchlorate

ROBINSON, R. A., *ibid.*, 77 (1955) 6200: Thorium chloride

SCATCHARD, G. and BRECKENRIDGE, R. C., *J. phys. Chem.*, 58 (1954) 596: Sodium and potassium phosphates and arsenates

TAYLOR, C. E., *ibid.*, 59 (1955) 653: Sodium carbonate—e.m.f. measurements were made between 15° and 65° and v.p. measurements between 65° and 95°

The following may also be noted:

JOHNSON, J. S. and KRAUS, K. A., *J. Amer. chem. Soc.*, 74 (1952) 4436; JOHNSON, J. S., KRAUS, K. A. and YOUNG, T. F., *ibid.*, 76 (1954) 1436: Uranyl fluoride by the freezing point method and also at 30°, the latter being expressed relative to $\gamma = 1$ at m = 0·15

LIETZKE, M. H. and STOUGHTON, R. W., *ibid.*, 78 (1956) 4520: Indium sulphate relative to $\gamma = 1$ at 0·1 M

APPENDIX 8.11

Concentrations of Solutions giving Specified Water Activities at 25°

Table 1

a_w	H_2SO_4		NaOH		$CaCl_2$	
	m	%	*m*	%	*m*	%
0·95	1·263	11·02	1·465	5·54	0·927	9·33
0·90	2·224	17·91	2·726	9·83	1·584	14·95
0·85	3·025	22·88	3·840	13·32	2·118	19·03
0·80	3·730	26·79	4·798	16·10	2·579	22·25
0·75	4·398	30·14	5·710	18·60	2·995	24·95
0·70	5·042	33·09	6·565	20·80	3·400	27·40
0·65	5·686	35·80	7·384	22·80	3·796	29·64
0·60	6·341	38·35	8·183	24·66	4·188	31·73
0·55	7·013	40·75	8·974	26·42	4·581	33·71
0·50	7·722	43·10	9·792	28·15	4·990	35·64
0·45	8·482	45·41	10·64	29·86	5·431	37·61
0·40	9·304	47·71	11·54	31·58	5·912	39·62
0·35	10·21	50·04	12·53	33·38	6·478	41·83
0·30	11·25	52·45	13·63	35·29	7·183	44·36
0·25	12·47	55·01	14·96	37·45	—	—
0·20	13·94	57·76	16·67	40·00	—	—
0·15	15·81	60·80	19·10	43·32	—	—
0·10	18·48	64·45	23·05	47·97	—	—
0·05	23·17	69·44	—	—	—	—

Table 2

Water Activity of Saturated Solutions at 25°

Solid phase	a_w	*Solid phase*	a_w
$K_2Cr_2O_7$	0·9800	$SrCl_2 \cdot 6H_2O$	0·7083
KNO_3	0·9248	NH_4NO_3	0·6183
$BaCl_2 \cdot 2H_2O$	0·9019	$NaBr \cdot 2H_2O$	0·5770
$3CdSO_4 \cdot 8H_2O$	0·8891	$Mg(NO_3)_2 \cdot 6H_2O$	0·5286
$ZnSO_4 \cdot 7H_2O$	0·8710	$Ca(NO_3)_2 \cdot 4H_2O$	0·4997
KCl	0·8426	$LiNO_3 \cdot 3H_2O$	0·4706
KBr	0·8071	$K_2CO_3 \cdot 2H_2O$	0·4276
$(NH_4)_2SO_4$	0·7997	$MgCl_2 \cdot 6H_2O$	0·3300
NH_4Cl	0·7710	$K(C_2H_3O_2)\ 1 \cdot 5H_2O$	0·2245
NaCl	0·7528	$LiCl \cdot H_2O$	0·1105
$NaNO_3$	0·7379	$NaOH \cdot H_2O$	0·0703

From STOKES, R. H. and ROBINSON, R. A., *Ind. Eng. Chem.*, 41 (1949) 2013.

APPENDIX 10.1

Values of e^{z^2} Corresponding to Round Values of the Function $f(z)$ in Equation (10.20)

$f(z)$	e^{z^2}	$f(z)$	e^{z^2}	$f(z)$	e^{z^2}	$f(z)$	e^{z^2}
0	1·0000	0·25	1·8336	0·50	3·2645	0·75	7·7985
0·01	1·0593	0·26	1·8725	0·51	3·3522	0·76	8·1907
0·02	1·0968	0·27	1·9123	0·52	3·4448	0·77	8·6192
0·03	1·1303	0·28	1·9531	0·53	3·5417	0·78	9·0909
0·04	1·1620	0·29	1·9949	0·54	3·6425	0·79	9·6071
0·05	1·1923	0·30	2·0378	0·55	3·7491	0·80	10·1820
0·06	1·2220	0·31	2·0819	0·56	3·8592	0·81	10·819
0·07	1·2517	0·32	2·1274	0·57	3·9760	0·82	11·530
0·08	1·2806	0·33	2·1743	0·58	4·0980	0·83	12·330
0·09	1·3099	0·34	2·2226	0·59	4·2264	0·84	13·240
0·10	1·3394	0·35	2·2723	0·60	4·3622	0·85	14·023
0·11	1·3690	0·36	2·3235	0·61	4·5045	—	—
0·12	1·3987	0·37	2·3764	0·62	4·6561	—	—
0·13	1·4287	0·38	2·4310	0·63	4·8144	—	—
0·14	1·4592	0·39	2·4875	0·64	4·9831	—	—
0·15	1·4902	0·40	2·5459	0·65	5·1624	—	—
0·16	1·5216	0·41	2·6065	0·66	5·3530	—	—
0·17	1·5535	0·42	2·6690	0·67	5·5549	—	—
0·18	1·5860	0·43	2·7338	0·68	5·7707	—	—
0·19	1·6191	0·44	2·8011	0·69	6·0010	—	—
0·20	1·6529	0·45	2·8710	0·70	6·2480	—	—
0·21	1·6874	0·46	2·9437	0·71	6·5121	—	—
0·22	1·7227	0·47	3·0192	0·72	6·7967	—	—
0·23	1·7588	0·48	3·0978	0·73	7·1063	—	—
0·24	1·7957	0·49	3·1795	0·74	7·4360	—	—
0·25	1·8336	0·50	3·2645	0·75	7·7985	—	—

Linear interpolation is accurate to 1 part in 5,000 for $f(z) > 0{\cdot}03$; below this value a graph is preferable.

APPENDIX 10.2

Airy Integral Refinement of the 'Quarter Wave' Approximation for the Interference Conditions

M_j and Z_j correspond to maxima and minima, respectively, in the light-intensity.

j	M_j	Z_j
0	0·21822	0·75867
1	1·24229	1·75395
2	2·24565	2·75254
3	3·24698	3·75187
4	4·24769	4·75148
5	5·24813	5·75122
6	6·24843	6·75104
7	7·24864	7·75091
8	8·24881	8·75080
9	9·24894	9·75072
10	10·24904	10·75065
12	12·24920	12·75055
14	14·24931	14·75048
16	16·24939	16·75042
18	18·24946	18·75038
20	20·24951	20·75034
22	22·24956	22·75031
24	24·24959	24·75028
26	26·24962	26·75026
28	28·24965	28·75024
30	30·24967	30·75023
40	40·24976	40·75017

The table shows that for fringe-numbers $j > 10$, the approximations $M_j = j + \frac{1}{4}$, $Z_j = j + \frac{3}{4}$, as given in equation (10.20), are adequate.

From Gosting, L. J. and Morris, M. S., *J. Amer. chem. Soc.*, 71 (1949) 1998.

APPENDIX 11.1

Diffusion Coefficients of Dilute Aqueous Electrolyte Solutions

Salt	*Temp.*	$c = 0$*	0·001	0·002	0·003	0·005	0·007	0·01	*Ref.*
LiCl	25°	1·366	1·345	1·337	1·331	1·323	1·318	1·312	(1)
NaCl	25°	1·610	1·585	1·576	1·570	1·560	1·555	1·545	(1)
KCl	20°	1·763	1·739	1·729	1·722	1·708	—	1·692	(2)
KCl	25°	1·993	1·964	1·954	1·945	1·934	1·925	1·917	(3)
KCl	30°	2·230	—	—	2·174	2·161	2·152	2·144	(2)
RbCl	25°	2·051	—	2·011	2·007	1·995	1·984	1·973	(4)
CsCl	25°	2·044	2·013	2·000	1·992	1·978	1·969	1·958	(16)
$LiNO_3$	25°	1·336	—	—	1·296	1·289	1·283	1·276	(17)
$NaNO_3$	25°	1·568	—	1·535	—	1·516	1·513	1·503	(17)
$KClO_4$	25°	1·871	1·845	1·841	1·835	1·829	1·821	1·790	(18)
KNO_3	25°	1·928	1·899	1·884	1·879	1·866	1·857	1·846	(5)
$AgNO_3$	25°	1·765	—	—	1·719	1·708	1·698	—	(6)
$MgCl_2$	25°	1·249	1·187	1·169	1·158	—	—	—	(7)
$CaCl_2$	25°	1·335	1·249	1·225	1·201	1·179	—	—	(8)
$CaCl_2$	25°	1·335	1·263	1·243	1·230	1·213	1·201	1·188	(19)
$SrCl_2$	25°	1·334	1·269	1·248	1·236	1·219	1·209	—	(9)
$BaCl_2$	25°	1·385	1·320	1·298	1·283	1·265	—	—	(7)
Li_2SO_4	25°	1·041	0·990	0·974	0·965	0·950	—	—	(10)
Na_2SO_4	25°	1·230	1·175	1·160	1·147	1·123	—	—	(10)
Cs_2SO_4	25°	1·569	1·489	1·454	1·437	1·420	—	—	(11)
$MgSO_4$	25°	0·849	0·768	0·740	0·727	0·710	(0·704 at $c =$ 0·006)	—	(12)
$ZnSO_4$	25°	0·846	0·748	0·733	0·724	0·705	—	—	(13)
$LaCl_3$	25°	1·293	1·175	1·145	1·126	1·105	1·084	(1·021 at $c =$ 0·026)	(14)
$K_4Fe(CN)_6$	25°	1·468	—	—	1·213	1·183	—	—	(15)

All the above results were obtained by Harned's conductimetric method.

* Values in this column are Nernst limiting values derived from the limiting mobilities of the ions.

c in mole/l

D in $cm^2 \, sec^{-1} \times 10^{-5}$

The D values tabulated at round concentrations have been interpolated graphically from the original data in the references.

REFERENCES

1 Harned, H. S. and Hildreth, C. L., *J. Amer. chem. Soc.*, 73 (1951) 650

2 — and Nuttall, R. L., *ibid.*, 71 (1949) 1460

3 — — *ibid.*, 69 (1947) 736

4 — and Blander, M., *ibid.*, 75 (1953) 2853

5 — and Hudson, R. M., *ibid.*, 73 (1951) 652

6 — and Hildreth, C. L., *ibid.*, 73 (1951) 3292

7 — and Polestra, F. M., *ibid.*, 76 (1954) 2064

8 — and Levy, A. L., *ibid.*, 71 (1949) 2781

9 — and Polestra, F. M., *ibid.*, 75 (1953) 4168

10 — and Blake, C. A., *ibid.*, 73 (1951) 2448

[11] — — *ibid.*, 73 (1951) 5882
[12] — and HUDSON, R. M., *ibid.*, 73 (1951) 5880
[13] — — *ibid.*, 73 (1951) 3781
[14] — and BLAKE, C. A., *ibid.*, 73 (1951) 4255
[15] — and HUDSON, R. M., *ibid.*, 73 (1951) 5083
[16] HARNED, H. S., BLANDER, M. and HILDRETH, C. L., *ibid.*, 76 (1954) 4219
[17] HARNED, H. S. and SHROPSHIRE, J. A., *ibid.*, 80 (1958) 2618, 2967
[18] HARNED, H. S., PARKER, H. W. and BLANDER, M., *ibid.*, 77 (1955) 2071
[19] HARNED, H. S. and PARKER, H. W., *ibid.*, 77 (1955) 265

APPENDIX 11.2

Diffusion Coefficients of Concentrated Aqueous Electrolyte Solutions at 25°

c	HCl	HBr	LiCl	LiBr	NaCl	CsCl	NaBr	NaI	KCl	KBr	KI	NH_4Cl	NH_4NO_3	$LiNO_3$	$CaCl_2$	$(NH_4)_2SO_4$	$BaCl_2$
0*	3·336	3·400	1·366	1·377	1·610	2·044	1·625	1·614	1·993	2·016	1·999	1·994	1·929	1·336	1·335	1·530	1·385
0·05	3·07$_3$	3·15$_6$	1·28$_0$	1·30$_0$	1·507	—	1·53$_3$	1·52$_7$	1·864	1·89$_2$	1·89$_1$	—	1·788	—	1·121	0·802	1·179
0·1	3·05$_0$	3·14$_6$	1·26$_9$	1·27$_9$	1·483	1·871	1·51$_7$	1·52$_0$	1·844	1·87$_4$	1·86$_5$	1·838	1·769	1·240	1·110	0·825	1·159
0·2	3·06$_4$	3·19$_0$	1·26$_7$	1·28$_5$	1·475	1·857	1·50$_7$	1·53$_2$	1·838	1·87$_0$	1·85$_9$	1·836	1·749	1·243	1·107	0·867	1·150
0·3	3·09$_3$	3·24$_9$	1·26$_9$	1·29$_6$	1·475	1·855	1·51$_5$	1·54$_7$	1·838	1·87$_2$	1·88$_4$	1·841	1·739	1·248	1·116	0·897	1·151
0·5	3·18$_4$	3·38$_8$	1·27$_8$	1·32$_8$	1·474	1·860	1·54$_2$	1·58$_0$	1·850	1·88$_5$	1·95$_5$	1·861	1·724	1·260	1·140	0·938	1·160
0·7	3·28$_6$	3·55$_2$	1·28$_8$	1·36$_0$	1·475	1·871	1·56$_9$	1·61$_2$	1·866	1·91$_7$	2·00$_1$	1·883	1·709	1·274	1·168	0·972	1·168
1·0	3·43$_6$	3·87	1·30$_2$	1·40$_4$	1·484	1·902	1·59$_6$	1·66$_2$	1·892	1·97$_5$	2·06$_5$	1·921	1·690	1·293	1·203	1·011	1·179
1·5	3·74$_3$	—	1·33$_1$	1·47$_3$	1·495	—	1·62$_9$	1·75$_1$	1·943	2·06$_2$	2·16$_6$	1·986	1·661	1·317	1·263	1·047	1·180
2·0	4·04$_6$	—	1·36$_3$	1·54$_2$	1·516	2·029	1·66$_8$	1·84$_6$	1·999	2·13$_2$	2·25$_4$	2·051	1·633	1·332	1·307	1·069	—
2·5	4·33$_7$	—	1·39$_7$	1·59$_7$	—	—	1·70$_2$	1·92$_5$	2·057	2·19$_9$	2·34$_7$	2·113	1·605	1·336	1·306	1·088	—
3·0	4·65$_8$	—	1·43$_0$	1·65$_0$	1·565	2·175	—	1·99$_2$	2·112	2·28$_0$	2·44$_0$	2·164	1·578	1·332	1·265	1·106	—
3·5	4·92	—	1·46$_4$	1·69$_3$	—	—	—	—	2·160	2·35$_4$	2·53$_3$	2·203	—	—	1·195	1·122	—
4·0	5·17	—	—	—	1·594	2·291	—	—	2·196 (c = 3·9)	2·43	—	2·235	1·524	1·292	—	1·135	—
4·5	—	—	—	—	—	—	—	—	—	—	—	2·257	—	—	—	—	—
5·0	—	—	—	—	1·590	2·364	—	—	—	—	—	2·264	1·472	1·238	—	—	—
6·0	—	—	—	—	—	2·335	—	—	—	—	—	—	1·421	1·157	—	—	—
7·0	—	—	—	—	—	—	—	—	—	—	—	—	1·370	—	—	—	—
8·0	—	—	—	—	—	—	—	—	—	—	—	—	1·320	—	—	—	—
Method	*A*	*A*	*A*	*A*	*A C*	*C*	*A*	*A*	*B, C*	*A*	*A*	*C*	*C*	*C*	*C*	*C*	*C*
Reference	1	1	1	1	1 8	7	1	2	3, 4	1	2	5	6	6	5, 7	6	8

* Nernst limiting values. Methods *A*—magnetically-stirred diaphragm cell; *B*—conductimetric; *C*—Gouy interference.

Units 10^{-5} cm^2 sec^{-1}

REFERENCES

[1] STOKES, R. H., *J. Amer. chem. Soc.*, 72 (1950) 2243
[2] DUNLOP, P. J. and STOKES, R. H., *ibid.*, 73 (1951) 5456
[3] HARNED, H. S. and NUTTALL, R. L., *ibid.*, 71 (1949) 1460
[4] GOSTING, L. J., *ibid.*, 72 (1950) 4418
[5] HALL, J. R., WISHAW, B. F. and STOKES, R. H., *ibid.*, 75 (1953) 1556
[6] WISHAW, B. F. and STOKES, R. H., *ibid.*, 76 (1954) 2065
[7] LYONS, P. A. and RILEY, J. F., *ibid.*, 76 (1954) 5216
[8] VITAGLIANO, V. and LYONS, P. A., *ibid.*, 78 (1956) 1549

APPENDIX 11.3

Individual ionic values of the coefficient A_2 in the viscosity equation:

$$\eta/\eta^\circ = 1 + A_1\sqrt{c} + A_2 c$$

c in mole/l.

°C	Li^+	Na^+	K^+	Rb^+	Cs^+	NH_4^+	Ag^+	H^+		
15	0·1615	0·0860	−0·0200			−0.0137				
25	0·1495	0·0863	−0·0070	(−0·030)	(−0·045)	−0·0074	(0·091)	(0·069)		
35	0·1385	0·0851	+0·0049			−0·0027				
42·5	0·1310	0·0861	+0·0121			+0·0018				
°C	Be^{2+}	Mg^{2+}	Ca^{2+}	Sr^{2+}	Ba^{2+}	Fe^{2+}	Co^{2+}	Ni^{2+}	La^{3+}	Ce^{3+}
15	0·4345	0·4091				[0·4372]$^{15\cdot5^\circ}$				0·5841
25	0·3923	0·3852	(0·285)	(0·265)	0·220	0·4160			(0·588)	0·5765
35	0·3444	0·3625			(0·276)	0·3955	0·360	0·306		0·5573
42·5	0·3105	0·3472				[0·3950]$^{40^\circ}$				0·5427
°C	Cl^-	Br^-	I^-	OH^-	IO_3^-	BrO_3^-	ClO_3^-	NO_3^-	MnO_4^-	SO_4^{2-}
15	−0·0200		[−0·0880]$^{18^\circ}$	(0·109)$^{18^\circ}$	(0·125)$^{18^\circ}$		(−0·041)$^{18^\circ}$	(−0·055)$^{18^\circ}$		0·1889
25	−0·0070	(−0·042)	−0·0685		(0·140)	(0·0062)	(0·0240)	(−0·0460)	(−0·059)	0·2085
35	+0·0049		−0·0536				(−0·0084)			0·2277
42·5	+0·0121		[−0·0490]$^{40^\circ}$							0·2399

KAMINSKY, M., *Disc. Faraday Soc.*, 24 (1957) 171

APPENDIX 12.1

Ionization Constants of Weak Electrolytes and their Temperature Variation

Table 1

Aqueous Solutions

$$pK_a = -\log K_a = A_1/T - A_2 + A_3T$$

Aqueous Solution	pK_a *at 25°*	A_1	A_2	A_3	*Reference*
Acetic acid	4·756	1170·48	3·1649	0·013399	1
Acetyl-α-alanine	3·715	908·48	2·8416	0·011771	2
Acetyl-β-alanine	4·445	1279·32	3·9494	0·013763	2
*Acetyl-α-amino-*n-*butyric acid*	3·716	906·43	2·9315	0·012096	2
Acetylglycine	3·670	1248·54	4·8146	0·014411	2
α-Alanine K_1	2·348	1383·06	6·3639	0·013662	3
α-Alanine K_2	9·866	2941·55	1·8171	0·006095	3
β-Alanine K_1	3·552	1487·31	5·6516	0·014138	4
β-Alanine K_2	10·237	2799·04	0·3062	0·003877	4
Allothreonine K_1	2·108	1111·7	4·7982	0·010657	5
Allothreonine K_2	9·096	2764·3	1·8531	0·005631	5
4-Aminobenzophenone	2·166	1917·9	7·312	0·01022	6
*α-Amino-*n-*butyric acid* K_1	2·286	1174·74	5·3735	0·012487	3
*α-Amino-*n-*butyric acid* K_2	9·830	2879·31	1·6446	0·006095	3
*α-Amino-*iso-*butyric acid* K_1	2·357	1344·95	6·3053	0·013924	3
*α-Amino-*iso-*butyric acid* K_2	10·205	3010·95	1·5404	0·005520	3
γ-Aminobutyric acid K_1	4·031	1209·07	3·7820	0·012605	7
γ-Aminobutyric acid K_2	10·556	2804·84	−0·5879	0·001880	7
ε-Aminocaproic acid K_1	4·373	1803·5	7·6874	0·020166	8
ε-Aminocaproic acid K_2	10·804	2708·6	−2·5445	−0·002757	8
2-Aminoethanol-1-phosphoric acid K_2	5·838	1228·34	2·7328	0·01493	9
*α-Amino-*n-*valeric acid* K_1	2·318	1222·02	5·5238	0·012553	3
*α-Amino-*n-*valeric acid* K_2	9·808	2618·57	−0·1669	0·002869	3
Ammonium ion	9·245	2835·76	0·6322	0·001225	10
Aspartic acid K_1	1·990	1109·6	4·1563	0·008138	8
Aspartic acid K_2	3·900	1706·3	6·7436	0·016506	8
Aspartic acid K_3	10·002	2880·3	2·6890	0·010173	8

APPENDIX 12.1

Table 1—(Contd.)

Aqueous Solution	pK_a *at 25°*	A_1	A_2	A_3	*Reference*
Benzoic acid	4·203	819·63	1·287	0·00919	11
Boric acid	9·234	2237·94	3·305	0·016883	12
Bromoacetic acid	2·901	939·55	4·2803	0·013515	13
n-*Butyric acid*	4·820	1033·39	2·6215	0·013334	14
n-*Butylammonium ion*	10·640	2942·44	−1·078	−0·001032	30
iso-*Butyric acid*	4·848	950·27	2·1032	0·012625	15, 16
N-Carbamoylalanine	3·892	1088·10	3·5768	0·012810	17
N-Carbamoyl-β-alanine	4·487	1152·77	3·1036	0·012490	17
*N-Carbamoyl-α-amino-*n*-butyric acid*	3·886	1018·65	3·3079	0·012670	17
*N-Carbamoyl-α-amino-*iso*-butyric acid*	4·463	1125·63	2·9285	0·012129	17
N-Carbamoyl-γ-aminobutyric acid	4·683	1074·06	2·6066	0·012364	17
N-Carbamoylglycine (hydantoic acid)	3·876	1364·94	5·0675	0·014640	17
Carbonic acid K_1	6·352	3404·71	14·8435	0·032786	18
Carbonic acid K_2	10·329	2902·39	6·4980	0·02379	19
Chloroacetic acid	2·865	1229·13	6·1715	0·016486	13, 20
Citric acid K_1	3·128	1255·6	4·5635	0·011673	21
Citric acid K_2	4·761	1585·2	5·4460	0·016399	21
Citric acid K_3	6·396	1814·9	6·3664	0·022389	21
Cyanoacetic acid	2·469	1029·79	5·0481	0·013626	22
Diethylacetic acid	4·734	492·16	0·0453	0·010493	15, 16
5:5 *Diethylbarbituric acid*	7·980	2324·47	3·3491	0·011856	23
Dimethylammonium ion	10·774	1932·6	−6·495	−0·007389	24
(*Dimethylethylene—* K_1	5·987	86·4	−9·048	−0·01124	25
(*diaminodiacetic acid* K_2	9·977	−458·8	−18·809	−0·02446	25
N-Dimethylglycine K_2	9·940	1209·07	−7·4407	−0·005217	26
2·2′*Dipyridinium ion*	4·352	−64·4	−7·264	−0·00902	27
Ephedrinium ion	9·544	1834·51	−5·1480	−0·005894	28
γ-Ephedrinium ion	9·706	1832·21	−5·5189	0·006562	28
(*Ethylenediamine—* K_3	6·273	1396·6	0·506	0·00704	29
(*tetraacetic acid* K_4	10·948	1143·4	−7·241	−0·00043	29

APPENDIX 12.1

Table 1—(Contd.)

	pK_a *at 25°*	A_1	A_2	A_3	*Reference*
Ethylenediammonium ion K_1	6·838	1925·89	−1·964	−0·005308	30
Ethylenediammonium ion K_2	9·960	2492·90	−2·010	−0·001371	30
Fluoroacetic acid	2·584	877·22	4·2999	0·013223	13
Formic acid	3·752	1342·85	5·2743	0·015168	31
Germanic acid	8·775	−19841·0	−150·561	−0·25235	32
*Glucose-*1-*phosphoric acid* K_2	6·503	1432·16	3·4213	0·017177	33
Glutamic acid K_1	2·162	1399·4	7·258	0·01587	34
Glutamic acid K_2	4·288	1257·2	3·931	0·01341	34
Glutamic acid K_3	9·387	437·2	−13·352	−0·01823	34
*Glycerol-*1-*phosphoric acid* K_2	6·657	1411·37	3·3605	0·017718	35
*Glycerol-*2-*phosphoric acid* K_1	1·329	1891·91	13·4799	0·02841	36
*Glycerol-*2-*phosphoric acid* K_2	6·650	1667·40	4·8394	0·01978	36
Glycine K_1	2·350	1332·17	5·8870	0·012643	37
Glycine K_2	9·780	2686·95	0·5103	0·004286	37
Glycolic acid	3·831	1303·26	4·7845	0·014236	38
Glycylalanine	3·153	691·79	1·8996	0·0091655	39
*Glycyl-α-amino-*n-*butyric acid*	3·155	727·94	2·2214	0·0098400	39
Glycylasparagine	2·942	1176·54	4·8520	0·012907	39
Glycylglycine K_1	3·140	1003·35	3·5670	0·011207	39
Glycylglycine K_2	8·252	2902·3	3·4932	0·006749	8
Glycylglycylglycine K_1	3·225	—	—	—	40
Glycylglycylglycine K_2	8·090	—	—	—	40
Glycylleucine	3·180	718·40	2·1910	0·0099294	39
Glycylserine	2·981	1090·92	4·1993	0·011087	39
(*Hexamethylene—* K_1	9·840	2543·96	−2·776	−0·004935	30
(*diammonium ion* K_2	10·931	3416·24	1·882	0·004527	30
Hexanoic acid	4·857	966·12	2·1256	0·012550	15, 16
iso-*Hexanoic acid*	4·845	935·01	1·9378	0·012230	15
Hydroxyproline K_1	1·818	1156·7	5·2753	0·010777	5
Hydroxyproline K_2	9·662	2442·9	−0·1266	0·004500	5

Table 1—(Contd.)

Aqueous Solution	pK_a *at 25°*	A_1	A_2	A_3	*Reference*
Iodoacetic acid	3·174	716·15	2·6357	0·011429	13
Lactic acid	3·860	1286·49	4·8607	0·014776	41
Leucine K_1	2·328	1283·60	6·0027	0·013505	3
Leucine K_2	9·744	2819·38	1·2396	0·005127	3
iso-*Leucine* K_1	2·318	1298·09	6·1967	0·013959	3
iso-*Leucine* K_2	9·758	2933·52	2·0479	0·006578	3
Malonic acid K_1	2·847	2019·38	10·6810	0·022657	42
Malonic acid K_2	5·696	1565·82	5·6639	0·020485	42, 43
Metanilic acid	3·738	1327·59	1·5533	0·002813	44
Methylaminediacetic acid K_1	2·148	0·9	−2·023	0·00041	25
Methylaminediacetic acid K_2	10·006	1648·8	−4·068	0·00137	25
Methoxyacetic acid	3·570	974·26	3·6704	0·013325	45
3-*Methoxyalanine* K_1	2·038	1290·20	6·0134	0·012490	45
3-*Methoxyalanine* K_2	9·175	2788·60	2·0924	0·006423	45
Monoethanolammonium ion	9·499	2309·66	−2·9209	−0·003915	46
Monomethylammonium ion	10·624	2568·3	−2·990	−0·003285	24
Nitrilotriacetic acid K_1	1·651	2422·5	14·936	0·02838	47
Nitrilotriacetic acid K_2	2·948	1404·8	6·829	0·01699	47
Nitrilotriacetic acid K_3	10·280	2148·6	1·047	0·01382	47
Norleucine K_1	2·335	1193·30	5·2850	0·012130	3
Norleucine K_2	9·834	2851·89	1·2891	0·005218	3
Orthanilic acid	2·459	1106·68	3·2314	0·006634	48
Oxalic acid K_1	1·271	—	—	—	49
Oxalic acid K_2	4·266	1423·8	6·5007	0·020095	50
Phenolsulphonic acid K_2	9·053	1961·2	1·1436	0·012139	51
Phosphoric acid K_1	2·148	799·31	4·5535	0·013486	52
Phosphoric acid K_2	7·198	1979·5	5·3541	0·019840	53
o-*Phthalic acid* K_1	2·950	561·57	1·2843	0·007883	54
o-*Phthalic acid* K_2	5·408	2175·83	9·5508	0·025694	55
Piperidinium ion	11·123	2105·6	−6·3535	−0·007687	56
Proline K_1	1·952	1512·6	7·9217	0·016094	5

Table 1—(Contd.)

Aqueous Solution	pK_a *at 25°*	A_1	A_2	A_3	*Reference*
Proline K_2	10·640	2230·4	−3·1592	0·000010	5
Propionic acid	4·874	1213·26	3·3860	0·014055	57
Propionylglycine	3·718	1101·03	3·7708	0·012730	2
n-*Propylammonium ion*	10.568	2742·67	−2·188	−0·002746	30
Sacrosine K_2	10·200	2213·06	−2·4514	−0·001094	26
Serine K_1	2·186	1311·2	5·6397	0·011496	5
Serine K_2	9·208	2594·1	0·6031	0·003725	5
Succinic acid K_1	4·207	1206·25	3·3266	0·011697	58
Succinic acid K_2	5·638	1679·13	5·7043	0·019153	58
Sulphamic acid	0·988	3792·8	24·122	0·041544	59
Sulphanilic acid	3·227	1143·71	1·2979	0·002314	60
Tartaric acid K_1	3·033	1525·59	6·6558	0·015336	61
Tartaric acid K_2	4·366	1765·35	7·3015	0·019376	61
Taurine K_2	9·061	2458·49	0·0997	0·003069	62
Telluric acid	7·637	−1870·0	−27·276	−0·044833	63
Tetramethylene diammonium ion K_1	9·216	2610·48	−1·325	−0·002899	30
Tetramethylene diammonium ion K_2	10·753	2772·77	−2·122	−0·002245	30
Threonine K_1	2·088	1716·1	8·5867	0·016504	5
Threonine K_2	9·100	2631·1	1·2866	0·005237	5
Triethanolammonium ion	7·762	1341·16	−4·6252	−0·004567	64
Triethylammonium ion	10·715	806·43	−13·050	−0·01690	30
Trifluorobutyric acid	4·156	—	—	—	65
Trifluorovaleric acid	4·495	—	—	—	66
Trimethylacetic acid (*Pivalic acid*)	5·032	1044·58	2·4939	0·013490	15, 16
Trimethylammonium ion	9·800	541·4	−12·611	−0·015525	24
Valeric acid	4·842	921·38	1·8574	0·012105	15, 16
iso-*Valeric acid*	4·780	768·87	1·2582	0·011603	15, 16
Valine K_1	2·286	1245·31	6·0251	0·013868	3
Valine K_2	9·719	2776·46	1·1033	0·005056	3

All values are with reference to the molality scale. Data for bases are expressed as acidic ionization constants, e.g., for ammonia we quote pK at 25° = 9·245 for the ammonium ion:

$$NH_4^+ + H_2O \rightarrow NH_3 + H_3O^+$$

The basic ionization constant of the reaction:

$$NH_3 + H_2O \rightarrow NH_4^+ + OH^-$$

is got from the relation:

$$pK_a \text{ (acidic)} + pK_b \text{ (basic)} = pK_w \text{ (water)}$$

pK_w (water) being 13·997 at 25°.

REFERENCES

[1] HARNED, H. S. and EHLERS, R. W., *J. Amer. chem. Soc.*, 54 (1932) 1350; 55 (1933) 652; MACINNES, D. A. and SHEDLOVSKY, T., *ibid.*, 54 (1932) 1429 found pK = 4·755 at 25° from conductance data

[2] KING, E. J. and KING, G. W., *ibid.*, 78 (1956) 1089

[3] SMITH, P. K., TAYLOR, A. C. and SMITH, E. R. B., *J. biol. Chem.*, 122 (1937) 109

[4] MAY, M. and FELSING, W. A., *J. Amer. chem. Soc.*, 73 (1951) 406; data recalculated by Dr. E. J. KING

[5] SMITH, P. K., GORHAM, A. T. and SMITH, E. R. B., *J. biol. Chem.*, 144 (1942) 735

[6] SAGER, E. E. and SIEWERS, I. J., *J. Res. nat. Bur. Stand.*, 45 (1950) 489

[7] KING, E. J., *J. Amer. chem. Soc.*, 76 (1954) 1006

[8] SMITH, E. R. B. and SMITH, P. K., *J. biol. Chem.*, 146 (1942) 187

[9] CLARKE, H. B., DATTA, S. P. and RABIN, B. R., *Biochem. J.*, 59 (1955) 209

[10] BATES, R. G. and PINCHING, G. D., *J. Res. nat. Bur. Stand.*, 42 (1949) 419; *J Amer. chem. Soc.*, 72 (1950) 1393; see also EVERETT, D. H. and LANDSMAN, D. A., *Trans. Faraday Soc.*, 50 (1954) 1221 who, from cells with liquid junction, got results in good agreement with those of BATES and PINCHING over the range 15° to 45°

[11] JONES, A. V. and PARTON, H. N., *Trans. Faraday Soc.*, 48 (1952) 8; SMOLYAKOV, B. S. and PRIMANCHUK, M. P., *Russ. J. phys. Chem.*, 40 (1966) 331. Many studies have been made on benzoic acid at 25°: from conductance measurements IVES, D. J. G., LINSTEAD, R. P. and RILEY, H. L., *J. chem. Soc.*, (1933) 561, found pK = 4·190, BROCKMAN, F. G. and KILPATRICK, M., *J. Amer. chem. Soc.*, 56 (1934) 1483 pK = 4·199, SAXTON, B. and MEIER, H. F., *ibid.*, 56 (1934) 1918, pK = 4·200, DIPPY, J. F. J. and WILLIAMS, F. R., *J. chem. Soc.*, (1934) 1888 pK = 4·203, JEFFERY, G. H. and VOGEL, A. I., *Phil. Mag.*, 18 (1934) 901 pK = 4·196. From e.m.f. measurements BRISCOE, H. T. and PEAKE, J. S., *J. phys. Chem.*, 42 (1938) 637 give pK = 4·218 and BRIEGLEB, G. and BIEBER, A. (see Table 4) give pK = 4·212 whilst from spectrophotometric methods, HALBAN, H. VON and BRÜLL, J., *Helv. chim. Acta*, 27 (1944) 1719 give pK = 4·216, KILPATRICK, M. and ARENBERG, C. A., *J. Amer. chem. Soc.*, 75 (1953) 3812 pK = 4·208 and ROBINSON, R. A. and BIGGS, A. I., *Aust. J. Chem.*, 10 (1957) 128 pK = 4·203; see also ELLIS, A. J., *J. chem. Soc.*, (1963) 2299

[12] MANOV, G. G., DELOLLIS, N. J. and ACREE, S. F., *J. Res. nat. Bur. Stand.*, 33 (1944) 287

[13] IVES, D. J. G. and PRYOR, J. H., *J. chem. Soc.*, (1955) 2104

[14] HARNED, H. S. and SUTHERLAND, R. O., *J. Amer. chem. Soc.*, 56 (1934) 2039; the conductance data of BELCHER, D., *ibid.*, 60 (1938) 2744 and SAXTON, B. and DARKEN, L. S., *ibid.*, 62 (1940) 846 give pK = 4·823 and 4·818 respectively at 25°

[15] EVERETT, D. H., LANDSMAN, D. A. and PINSENT, B. R. W., *Proc. roy. Soc.*, 215A (1952) 403

[16] DIPPY, J. F. J., *J. chem. Soc.*, (1938) 1222, obtained the following pK values from conductance data at 25°: *iso*butyric acid, 4·860, valeric acid, 4·860; *iso*valeric acid, 4·777; trimethylacetic acid, 5·050; hexanoic acid, 4·879; diethylacetic acid, 4·751

[17] KING, E. J., *J. Amer. chem. Soc.*, 78 (1956) 6020

[18] HARNED, H. S. and DAVIS, R., *ibid.*, 65 (1943) 2030; SHEDLOVSKY, T. and MACINNES, D. A., *ibid.*, 57 (1935) 1705 found pK = 6·583, 6·429, 6·366 and 7·317 at 0°, 15° 25° and 38° respectively from conductance data whilst Näsänen, R., *Acta chem. scand.*, 1 (1947) 204 got pK values within 0·002 of those of HARNED and DAVIS. These pK values are for the reaction:

$$CO_2 + H_2O \rightarrow H^+ + HCO_3^-$$

The true ionization constant for the reaction:

$$H_2CO_3 \rightarrow H^+ + HCO_3^-$$

has been found to be $2{\cdot}4 \times 10^{-4}$ at 15° and $1{\cdot}32 \times 10^{-4}$ at 25° by ROUGHTON, F. J. W., *J. Amer. chem. Soc.*, 63 (1941) 2930 and by BERG, D. and PATTERSON, A., *ibid.*, 75 (1953) 5197 respectively. Thus at 25° only about 0·3 per cent of the dissolved carbon dioxide is in the form of H_2CO_3 molecules

[19] HARNED, H. S. and SCHOLES, S. R., *ibid.*, 63 (1941) 1706

[20] WRIGHT, D. D., *ibid.*, 56 (1934) 314, using the quinhydrone electrode obtained pK values lower by about 0·011 over the range 0° to 40°; the conductance measurements of SAXTON, B. and LANGER, T. W., *ibid.*, 55 (1933) 3638 give pK = 2·854 at 25°

[21] BATES, R. G. and PINCHING, G. D., *ibid.*, 71 (1949) 1274

[22] FEATES, F. S. and IVES, D. J. G., *J. chem. Soc.*, (1956) 2798. The authors propose a more complicated equation to represent their results

[23] MANOV, G. G., SCHUETTE, K. E. and KIRK, F. S., *J. Res. nat. Bur. Stand.*, 48 (1952) 84

[24] EVERETT, D. H. and WYNNE-JONES, W. F. K., *Proc. roy. Soc.*, 177A (1941) 499. The cell had a liquid junction but the potential at this junction should be small

[25] OCKERBLOOM, N. E. and MARTELL, A. E., *J. Amer. chem. Soc.*, 78 (1956) 267

[26] DATTA, S. P. and GRZYBOWSKI, A. K., *Trans. Faraday Soc.*, 54 (1958) 1179, 1188

[27] NÄSÄNEN, R., *Suomen Kem.*, 28B (1955) 161

[28] EVERETT, D. H. and HYNE, J. B., *J. chem. Soc.*, (1958) 1636

[29] CARINI, F. F. and MARTELL, A. E., *J. Amer. chem. Soc.*, 75 (1953) 4810

[30] EVERETT, D. H. and PINSENT, B. R. W., *Proc. roy. Soc.*, 215A (1952) 416; COX, M. C., EVERETT, D. H., LANDSMAN, D. A. and MUNN, R. J., *J. chem. Soc.*, B (1968)1373

[31] HARNED, H. S. and EMBREE, N. D., *J. Amer. chem. Soc.*, 56 (1934) 1042; SAXTON, B. and DARKEN, L. S., *ibid.*, 62 (1940) 846 found pK = 3·738 by conductance measurements at 25° and PRUE, J. E. and READ, A. J., *Trans. Faraday Soc.*, 62 (1966) 1271 found pK = 3·739 by e.m.f. measurements

[32] ANTIKAINEN, P. J., *Suomen Kem.*, 30B (1957) 123; the interpretation of the results is complicated by the formation of pentagermanic acid, $H_2Ge_5O_{11}$ in alkaline solution

[33] ASHBY, J. H., CLARKE, H. B., CROOK, E. M. and DATTA, S. P., *Biochem. J.*, 59 (1955) 203

[34] LLOPIS, J. and ORDONEZ, D., *J. electroanal. Chem.*, 5 (1963) 129

[35] DATTA, S. P. and GRZYBOWSKI, A. K., *Biochem. J.*, 69 (1958) 218

[36] ASHBY, J. H., CROOK, E. M. and DATTA, S. P., *ibid.*, 56 (1954) 198

[37] King, E. J., *J. Amer. chem. Soc.*, 73 (1951) 155; these results are in good agreement with the earlier results of Owen, B. B., *ibid.*, 56 (1934) 24
[38] Nims, L. F., *ibid.*, 58 (1936) 987
[39] King, E. J., *ibid.*, 79 (1957) 6151
[40] Evans, W. P. and Monk, C. B., *Trans. Faraday Soc.*, 51 (1955) 1244
[41] Nims, L. F. and Smith, P. K., *J. biol. Chem.*, 113 (1936) 145 (e.m.f.); Martin, A. W. and Tartar, H. V., *J. Amer. chem. Soc.*, 59 (1937) 2672 (conductance); the A_1, A_2 and A_3 parameters have been calculated from both sets of results
[42] Das, S. N. and Ives, D. J. G., *Proc. chem. Soc.*, (1961) 373
[43] Hamer, W. J., Burton, J. O. and Acree, S. F., *J. Res. nat. Bur. Stand.*, 24 (1940) 269
[44] McCoy, R. D. and Swinehart, D. F., *J. Amer. chem. Soc.*, 76 (1954) 4708
[45] King, E. J., *ibid.*, 82 (1960) 3575
[46] Datta, S. P. and Grzybowski, A. K., *J. chem. Soc.*, (1962) 3068; Bates, R. G. and Pinching, G. D., *J. Res. nat. Bur. Stand.*, 46 (1951) 349; Sivertz, V., Reitmeier, R. E. and Tartar, H. V., *J. Amer. chem. Soc.*, 62 (1940) 1379 give pK = 9·501 from conductance data at 25°
[47] Hughes, V. L. and Martell, A. E., *ibid.*, 78 (1956) 1319
[48] Diebel, R. N. and Swinehart, D. F., *J. phys. Chem.*, 61 (1957) 333
[49] Darken, L. S., *J. Amer. chem. Soc.*, 63 (1941) 1007; Parton, H. N. and Nicholson, A. J. C., *Trans. Faraday Soc.*, 35 (1939) 546, give 1·336, 1·258 and 1·271 at 25°, 30° and 35° respectively
[50] Pinching, G. D. and Bates, R. G., *J. Res. nat. Bur. Stand.*, 40 (1948) 405; they show that the data of Harned, H. S. and Fallon, L. D., *J. Amer. chem. Soc.*, 61 (1939) 3111 and of Parton, H. N. and Gibbons, R. C., *Trans. Faraday Soc.*, 35 (1939) 542 are in substantial agreement with theirs
[51] Bates, R. G., Siegel, G. L. and Acree, S. F., *J. Res. nat. Bur. Stand.*, 31 (1943) 205
[52] Bates, R. G., *ibid.*, 47 (1951) 127; the pK values of Nims, L. F., *J. Amer. chem. Soc.*, 56 (1934) 1110 are about 0·02 lower whilst Lugg, J. W. H., *ibid.*, 53 (1931) 1 calculated from conductance data pK = 2·09 at 18° (interpolated from Bates' data pK = 2·118)
[53] Bates, R. G. and Acree, S. F., *J. Res. nat. Bur. Stand.*, 30 (1943) 129; Grzybowski, A. K., *J. phys. Chem.*, 62 (1958) 550, 555; Ender, F., Teltschik, W., and Schäfer, K., *Z. Elektrochem.*, 61 (1957) 775
[54] Hamer, W. J., Pinching, G. D. and Acree S. F., *J. Res. nat. Bur. Stand.*, 35 (1945) 539
[55] Hamer, W. J. and Acree, S. F., *ibid.*, 35 (1945) 381
[56] Bates, R. G. and Bower, V. E., *ibid.*, 57 (1956) 153
[57] Harned, H. S. and Ehlers, R. W., *J. Amer. chem. Soc.*, 55 (1933) 2379; Belcher, D., *ibid.*, 60 (1938) 2744 found pK = 4·872 at 25°
[58] Pinching, G. D. and Bates, R. G., *J. Res. nat. Bur. Stand.*, 45 (1950) 322, 444
[59] King, E. J. and King, G. W., *J. Amer. chem. Soc.*, 74 (1952) 1212
[60] MacLaren, R. O. and Swinehart, D. F., *ibid.*, 73 (1951) 1822
[61] Bates, R. G. and Canham, R. G., *J. Res. nat. Bur. Stand.*, 47 (1951) 343
[62] King, E. J., *J. Amer. chem. Soc.*, 75 (1953) 2204
[63] Antikainen, P. J., *Suomen Kem.*, 28B (1955) 135; 30B (1957) 201
[64] Bates, R. G. and Allen, G. F., *J. Res. nat. Bur. Stand.*, 64A (1960) 343
[65] Henne, A. L. and Fox, C. J., *J. Amer. chem. Soc.*, 73 (1951) 2323; they found *p*K = 4·167 and 3·068 at 35° for trifluorobutyric acid and trifluoropropionic acid respectively; trifluoroacetic acid and heptafluorobutyric acid are much stronger acids with ionization constants about 0·5
[66] Henne, A. L. and Fox, C. J., *ibid.*, 75 (1953) 5750

APPENDIX 12.1

Table 2

Acidic Ionization Constants of Some Amines in Aqueous Solution at 25°

Amine	pK_a	*Reference*	*Amine*	pK_a	*Reference*
tris (*Hydroxymethyl*) *aminomethane*	8·076	(1)	*Nonylammonium ion*	10·64	(4)
Ethylammonium ion	10·631	(2)	*Decylammonium ion*	10·64	(4)
Diethylammonium ion	10·933	(2)	*Undecylammonium ion*	10·63	(4)
Triethylammonium ion	10·867	(3)	*Dodecylammonium ion*	10·63	(4)
n-*Propylammonium ion*	10·530	(2)	*Didodecylammonium ion*	10·99	(4)
n-*Butylammonium ion*	10·597	(2)	*Tridecylammonium ion*	10·63	(4)
iso-*Butylammonium ion*	10·43	(4)	*Ditridecylammonium ion*	10·99	(4)
n-*Amylammonium ion*	10·63	(4)	*Tetradecylammonium ion*	10·62	(4)
iso-*Amylammonium ion*	10·60	(4)	*Pentadecylammonium ion*	10·61	(4)
Hexylammonium ion	10·64	(4)	*Dipentadecylammonium ion*	11·00	(4)
Dihexylammonium ion	11·01	(4)	*Hexadecylammonium ion*	10·61	(4)
Heptylammonium ion	10·66	(4)	*Heptadecylammonium ion*	10·60	(4)
Octylammonium ion	10·65	(4)	*Octadecylammonium ion*	10·60	(4)
Dioctylammonium ion	11·01	(4)	*Dioctadecylammonium ion*	10·99	(4)
			Docosylammonium ion	10·60	(4)

REFERENCES

[1] BATES, R. G. and PINCHING, G. D., *J. Res. nat. Bur. Stand.*, 43 (1949) 519; they also found pK = 8·221 and 7·937 at 20° and 30° respectively

[2] EVANS, A. G. and HAMANN, S. D., *Trans. Faraday Soc.*, 47 (1951) 34. Although the cell had a liquid junction, the potential there should be small

[3] ABLARD, J. E., MCKINNEY, D. S. and WARNER, J. C., *J. Amer. chem. Soc.*, 62 (1940) 2181; at 40° and 50°, pK = 10·455 and 10·203 respectively

[4] HOERR, C. W., MCCORKLE, M. R. and RALSTON, A. W., *ibid.*, 65 (1943) 328

APPENDIX 12.1

Table 3

Acidic Ionization Constants of Some Polyamines in Aqueous Solution at 20°

	pK_1	pK_2	pK_3	pK_4	pK_5
Hydrazine	−0·88	8·11	—	—	— (1)
1:2-*Ethanediamine*	7·00	10·09	—	—	— (2)
1:3-*Propanediamine*	8·64	10·62	—	—	— (2)
1:4-*Butanediamine*	9·35	10·80	—	—	— (2)
1:5-*Pentanediamine*	9·74	11·05	—	—	— (3)
1:8-*Octanediamine*	10·10	11·00	—	—	— (3)
cis-1:2-*Cyclohexanediamine*	6·13	9·93	—	—	— (2)
trans-1:2-*Cyclohexanediamine*	6·47	9·94	—	—	— (2)
1:3 *Diamino-2-propanol*	7·93	9·69	—	—	— (2)
2:2′:2″ *Triaminotriethylamine*	7·98	9·26	10·15	—	— (2)
Tetraethylenepentamine (25°)	2·65	4·25	7·87	9·08	9·92 (4)

The pK values quoted are those of the conjugate acid.

REFERENCES

1 SCHWARZENBACH, G., *Helv. chim. Acta.*, 19 (1936) 178
2 BERTSCH, C. R., FERNELIUS, W. C. and BLOCK, B. P., *J. phys. Chem.*, 62 (1958) 444; measurements were also made at 10°, 30° and 40°
3 SCHWARZENBACH, G., *Helv. chim. Acta.*, 16 (1933) 522; his values for the ethane, propane and butanediamines agree with those recorded within about 0·05 pK units
4 JONASSEN, H. B., FREY, F. W. and SCHAAFSMA, A., *J. phys. Chem.*, 61 (1957) 504

Table 4

Ionization Constants of some Substituted Benzoic Acids (pK_a *values*)

	15°	20°	25°	30°	35°	40°	45°
Benzoic acid	4·214	4·213	4·212	4·215	4·221	4·232	4·241
m-*Hydroxybenzoic acid*	—	—	4·079	—	—	—	—
m-*Chlorobenzoic acid*	3·839	3·831	3·824	3·826	3·828	3·830	3·833
m-*Bromobenzoic acid*	3·819	3·813	3·809	3·810	3·810	3·813	3·818
m-*Iodobenzoic acid*	—	—	3·857	—	—	—	—
m-*Cyanobenzoic acid*	3·609	3·599	3·598	3·596	3·599	3·604	3·613
m-*Nitrobenzoic acid*	—	—	3·450	—	—	—	—
m-*Toluic acid*	—	—	4·243	—	—	—	—
p-*Hydroxybenzoic acid*	4·597	4·585	4·582	4·576	4·577	4·580	4·583
p-*Chlorobenzoic acid*	4·000	3·991	3·986	3·981	3·981	3·981	3·985
p-*Bromobenzoic acid*	4·012	4·005	4·002	4·002	4·001	4·005	4·007
p-*Cyanobenzoic acid*	3·558	3·551	3·551	3·553	3·554	3·561	3·567
p-*Nitrobenzoic acid*	3·449	3·444	3·442	3·440	3·444	3·445	3·450
p-*Toluic acid*	—	—	4·344	—	—	—	—

BRIEGLEB, G. and BIEBER, A., *Z. Elektrochem.*, 55 (1951) 250. Measurements were made of the potentials of cells without liquid junction, using the quinhydrone electrode.

APPENDIX 12.1

Table 5

Ionization Constants of Organic Acids in Aqueous Solution at 25° (pK_a *values*)

Cyano (cyclohexyl) acetic acid	2·366	(1)
Nitratoacetic acid	2·26	(2)
Sulphoacetic acid	4·07	(3)
β-Cyanopropionic acid	3·991	(1)
α-Nitratopropionic acid	2·39	(2)
β-Nitratopropionic acid	3·97	(2)
β-Sulphopropionic acid	4·52	(3)
Malonamic acid	3·641	(4)
Pyruvic acid	2·490	(5)
Acrylic acid	4·257	(6) (7)
γ-Cyanobutyric acid	4·436	(1)
Cyanoisobutyric acid	2·420	(1)
α-Nitratobutyric acid	2·39	(2)
Succinamic acid	4·539	(4)
Vinylacetic acid	4·342	(6) (8)
α-Crotonic acid (trans)	4·698	(7) (9)
2-*Butynoic acid* (*Tetrolic acid*)	2·652	(7)
Glutaramic acid	4·600	(4)
β-β-Dimethylacrylic acid	5·120	(7) (8)
Penta-2-*en*-1-*oic acid*	4·695	(8)
Penta-3-*en*-1-*oic acid*	4·507	(8)
Penta-4-*en*-1-*oic acid*	4·678	(8)
2-*Ethylbutan*-1-*oic acid*	4·751	(10)
Adipamic acid	4·628	(4)
Hexa-2-*en*-1-*oic acid*	4·703	(8)
Hexa-3-*en*-1-*oic acid*	4·516	(8)
Hexa-4-*en*-1-*oic acid*	4·719	(8)
4-*Methylpenta*-2-*en*-1-*oic acid*	4·701	(8)
4-*Methylpenta*-3-*en*-1-*oic acid*	4·600	(8)
cis-3-*Methylpenta*-2-*en*-1-*oic acid*	5·149	(8)
trans-3-*Methylpenta*-2-*en*-1-*oic acid*	5·131	(8)
Heptanoic acid	4·893	(10)
5-*Methylhexa*-4-*en*-1-*oic acid*	4·799	(8)
Octanoic acid	4·894	(10)
Nonanoic acid	4·955	(10)
Cyclohexanecarboxylic acid	4·900	(11)
1-*Methylcyclohexanecarboxylic acid*	5·131	(11)
cis-2-*Methylcyclohexanecarboxylic acid*	5·036	(11)
trans-2-*Methylcyclohexanecarboxylic acid*	4·735	(11)
cis-3-*Methylcyclohexanecarboxylic acid*	4·883	(11)
trans-3-*Methylcyclohexanecarboxylic acid*	5·02	(11)
cis-4-*Methylcyclohexanecarboxylic acid*	5·036	(11)
trans-4-*Methylcyclohexanecarboxylic acid*	4·886	(11)
trans-1-*Cyanocyclohexane*-2-*carboxylic acid*	3·865	(1)
Tartronic acid K_1	2·366	(12)
Tartronic acid K_2	4·735	(12)
Oxaloacetic acid K_1	2·555	(5)
Oxaloacetic acid K_2	4·370	(5)
Dihydroxytartaric acid K_1	1·947	(12)
Dihydroxytartaric acid K_2	4·004	(12)
Methylmalonic acid K_1	3·072	(13) (14)
Methylmalonic acid K_2 (20°)	5·87	(14)

Table 5—(Contd.)

Maleic acid K_1	1·921	(15)
Maleic acid K_2	6·225	(15)
Fumaric acid K_1	3·019	(15)
Fumaric acid K_2	4·384	(15)
Glutaric acid K_1	4·343	(16) (17) (18)
Glutaric acid K_2	5·272	(17) (18)
Ethylmalonic acid K_1	2·961	(13) (14)
Ethymalonic acid K_2 (20°)	5·90	(14)
Dimethylmalonic acid K_1	3·151	(13) (14)
Dimethylmalonic acid K_2 (20°)	6·20	(14)
Adipic acid K_1	4·430	(16) (17) (18)
Adipic acid K_2	5·277	(17) (18)
n-*Propylmalonic acid* K_1	2·989	(13) (14)
n-*Propylmalonic acid* K_2 (20°)	5·89	(14)
β-Methylglutaric acid K_1	4·235	(19)
Methylethylmalonic acid K_1	2·812	(13)
trans-Glutaconic acid K_1 (*Propene*-1:3-*dicarboxylic acid*)	3·767	(7)
trans-Glutaconic acid K_2	5·077	(7)
Pimelic acid K_1	4·509	(16) (17)
Pimelic acid K_2	5·312	(17)
Diethylmalonic acid K_1	2·150	(13) (14)
Diethylmalonic acid K_2 (20°)	7·47	(14)
β-Ethylglutaric acid K_1	4·285	(19)
β-β-Dimethylglutaric acid K_1	3·718	(19)
Suberic acid K_1	4·524	(16) (17)
Suberic acid K_2	5·327	(17) (18)
β-n-Propylglutaric acid K_1	4·309	(19)
Ethyl-n-*propylmalonic acid* K_1	2·106	(13)
β:*β-Methylethylglutaric acid* K_1	3·632	(19)
Azelaic acid K_1	4·550	(17)
Azelaic acid K_2	5·333	(17)
Di-n-*propylmalonic acid* K_1 (20°)	2·19	(14)
Di-n-*propylmalonic acid* K_2 (20°)	7·69	(14)
β:*β-Diethylglutaric acid* K_1	3·483	(19)
β:*β-Methyl*-n-*propylglutaric acid* K_1	3·626	(19)
β:*β-Ethyl*-n-*propylglutaric acid* K_1	3·510	(19)
Cyclopropane-1 : 1-*dicarboxylic acid* K_1	1·824	(20)
Cyclopropane-1 : 1-*dicarboxylic acid* K_2	7·431	(20)
Cyclobutane-1 : 1-*dicarboxylic acid* K_1	3·127	(20)
Cyclobutane-1 : 1-*dicarboxylic acid* K_2	5·879	(20)
Cyclopentane-1 : 1-*dicarboxylic acid* K_1	3·230	(20)
Cyclopentane-1 : 1-*dicarboxylic acid* K_2	6·081	(20)
Cyclohexane-1 : 1-*dicarboxylic acid* K_1	3·451	(20)
Cyclohexane-1 : 1-*dicarboxylic acid* K_2	6·108	(20)

APPENDIX 12.1

Table 5—(Contd.)

		Ortho	*Meta*	*Para*
Fluorobenzoic acid		3·267[21]	3·865[21]	4·141[22]
Chlorobenzoic acid		2·943[22]	3·830[22]	3·977[22]
Bromobenzoic acid		2·854[22]	3·812[21]	3·971[22]
Iodobenzoic acid		2·863[21]	3·851[21]	3·93[22a]
Nitrobenzoic acid		2·173[23]	3·493[21]	3·425[21]
Hydroxybenzoic acid		2·996[24]	4·082[24]	4·530[24, 25]
Aminobenzoic acid K_1		2·108[26]	3·124[26]	2·413[26, 27]
Aminobenzoic acid K_2		4·946[26]	4·744[26]	4·853[26, 27]
Toluic acid		3·908[23, 28]	4·272[21]	4·373[21]
Ethylbenzoic acid		3·793[29]	—	4·353[30]
iso-*Propylbenzoic acid*		3·635[29]	—	4·354[30]
tert-*Butylbenzoic acid*		3·535[29]	—	4·400[30]
Anisic acid		4·094[23]	4·088[21]	4·471[31]
Acetylbenzoic acid		4·126[32]	3·825[32]	3·700[32]
Phenylbenzoic acid		3·460[23]	—	—
Phenoxybenzoic acid		3·527[23]	3·951[23]	4·523[23]
Benzoylbenzoic acid		3·536[32]	—	—
p-*Toluoylbenzoic acid*		3·644[32]	—	—
Phthalic acid K_1		2·950[33]	3·70[34, 35]	3·54[35]
Phthalic acid K_2		5·408[33]	4·60[34, 35]	4·46[35]
Salicylaldoxime		$pK_1 = 1·37$[36]	$pK_2 = 9·180$[36]	$pK_3 = 12·11$[36]

APPENDIX 12.1

Table 5—(Contd.)

		2 : 3	2 : 4	2 : 5	2 : 6	3 : 4	3 : 5
Dinitrobenzoic acid		1·851[37]	1·425[37]	1·622[37]	1·140[37]	2·818[37]	2·824[37]
Dichlorobenzoic acid		—	2·76[38]	—	1·82[38]	3·64[38]	—
Dimethylbenzoic acid		3·738[29]	4·182[29]	3·977[29]	3·246[29]	4·408[29]	4·301[29]
Dimethoxybenzoic acid		—	—	—	3·44[38]	—	—
Chloronitrobenzoic acid		2·022[37]	1·963[37]	2·167[37]	1·342[37]	—	—
Bromonitrobenzoic acid		—	—	—	1·373[37]	—	—
Hydroxynitrobenzoic acid		1·873[24]	2·231[24]	2·121[24]	2·236[24]	—	—
Hydroxychlorobenzoic acid		—	—	2·629[24]	2·627[24]	—	—
Hydroxybromobenzoic acid		—	—	2·613[24]	—	—	—
Hydroxymethylbenzoic acid		—	—	—	3·321[24]	—	—
2:4:6-*Trinitrobenzoic acid*	0·654[37]						
2:4:6-*Trimethylbenzoic acid*	3·437[29]						
2:4:6-*Trimethoxybenzoic acid*	3·58[39]						
4-*Methyl*-3:5-*dinitrobenzoic acid*	2·971[37]						
2-*Hydroxy*-3:5-*dinitrobenzoic acid*	0·697[24]						
4-tert-*Butyl*-2:6-*dimethylbenzoic acid*	3·442[29]						
1-*Naphthoic acid*	3·695[29]						
2-*Napthoic acid*	4·161[29]						
3-*Hydroxy*-2-*naphthoic acid*	2·708[24]						
3-*Methoxy*-2-*naphthoic acid*	3·824[24]						
Phenylacetic acid	4·3074[40, 41]	*Ortho*		*Meta*		*Para*	
(*Ethylphenyl*) *acetic acid*		—		—		4·373[30]	
(iso*Propylphenyl*) *acetic acid*		—		—		4·391[30]	
(tert*Butylphenyl*) *acetic acid*		—		—		4·417[30]	
(*Fluorophenyl*) *acetic acid*		—		—		4·246[22]	
(*Chlorophenyl*) *acetic acid*		4·068[31]		4·140[31]		4·190[40]	

Table 5—(Contd.)

		Ortho	Meta	Para
(Bromophenyl) acetic acid		4·053[31]	—	4·188[40]
(Iodophenyl) acetic acid		4·038[21]	4·159[21]	4·178[40]
(Nitrophenyl) acetic acid		4·004[23]	3·967[31]	3·851[40]
Tolylacetic acid		—	—	4·370[6]
Anisylacetic acid		—	—	4·360[31]
Phenoxyacetic acid	3·171[31]	—	—	—
(Fluorophenoxy) acetic acid		3·085[42]	3·082[42]	3·130[42]
(Chlorophenoxy) acetic acid		3·051[42]	3·070[42]	3·103[42]
(Bromophenoxy) acetic acid		3·123[42]	3·095[42]	3·132[42]
(Iodophenoxy) acetic acid		3·173[42]	3·128[42]	3·159[42]
(Cyanophenoxy) acetic acid		2·975[42]	3·034[42]	2·932[42]
(Nitrophenoxy) acetic acid		2·896[42]	2·951[42]	2·893[42]
(Methylphenoxy) acetic acid		3·227[42]	3·203[42]	3·215[42]
(Methoxyphenoxy) acetic acid		3·231[42]	3·141[42]	3·212[42]
α-Naphthylacetic acid	4·236[11]			
β-Naphthylacetic acid	4·256[11]			
Cinnamylideneacetic acid	4·426[6]			
(2:4-Dinitrophenyl) acetic acid	3·502[31]			
(3:4-Dimethoxyphenyl) acetic acid	4·333[31]			
(2:6-Dimethylphenoxy) acetic acid	3·356[42]			
(3-Nitro-4-Chlorophenoxy) acetic acid	2·959[42]			
Diphenylacetic acid	3·939[40]			
β-Phenylpropionic acid	4·6644[41]			
Mandelic acid	3·411[43]			
cis-Cinnamic acid	3·879[6]			
trans-Cinnamic acid	4·438[6]			
Chlorocinnamic acid		4·234[23]	4·293[44]	4·413[23]
Nitrocinnamic acid		4·151[44]	4·120[44]	4·046[44]
Hydroxycinnamic acid		4·613[44]	4·397[44]	—
Methylcinnamic acid		4·500[44]	4·442[44]	4·564[44]

APPENDIX 12.1

Table 5—(Contd.)

		Ortho		*Meta*		*Para*
Methoxycinnamic acid		4·462[44]		4·376[44]		4·539[44]
β-Phenylpropionic acid	4·660[6]					
β-(Chlorophenyl) propionic acid		4·577[44]		4·585[44]		4·607[44]
β-(Nitrophenyl) propionic acid		4·504[44]		—		4·473[44]
β-(Methylphenyl) propionic acid		4·663[44]		4·677[44]		4·684[44]
β-Anisylpropionic acid		4·804[44]		4·654[44]		4·689[44]
γ-Phenylbutyric acid	4·757[6]					
Anilinium ion	4·596[45, 46]	—		—		—
Fluoroanilinium ion		—		3·391[46]		4·532[46]
Chloroanilinium ion		2·636[45, 46]		3·337[46]		3·99[47, 48]
Nitroanilinium ion		—0·260[46]		2·463[46, 47, 49]		0·991[46]
Toluidinium ion		4·394[46, 50]		4·683[46, 50]		5·091[46, 48, 50]
m-*Nitro*-p-*toluidinium ion*	2·959[46]					
N-Methylanilinium ion	4·848[51]					
N-N-Dimethylanilinium ion	5·150[51]					
N-N-Dimethylanilinium ion (20°)	5·178[51a]	—		—		—
Chloro-N-N-dimethylanilinium ion (20°)				3·829[51a]		4·389[51a]
Bromo-N-N-dimethylanilinium ion (20°)				—		4·282[51a]
Nitro-N-N-dimethylanilinium ion (20°)				2·658[51a]		0·670[51a]
Methyl-N-N-dimethylanilinium ion (20°)						5·611[51a]
Methoxy-N-N-dimethylanilinium ion (20°)						5·893[51a]
$H^+(CH_3)_2NC_6H_4N(CH_3)_2H^+$ (20°) K_1						6·051[51a]
K_2						2·500[51a]
N-N-Diethyltoluidinium ion						7·24[47]
Benzylammonium	9·35[52]					
2:3-*Dimethoxybenzylammonium ion*	9·41[52]					
3:4-*Dimethoxybenzylammonium ion*	9·39[52]					
3:4-*Methylenedioxybenzylammonium ion*	9·37[52]					
		pK_1				pK_1
2-*Hydroxy-3-methoxybenzylammonium ion*		8·70[52]				11·06[52]
2-*Methoxy-3-hydroxybenzylammonium ion*		8·89[52]				10·54[52]
3-*Methoxy-4-hydroxybenzylammonium ion*		8·96[52]				10·42[52]
		2 : 3	2 : 4	2 : 5	2 : 6	3 : 4
Dimethylanilinium ion		4·70[50]	4·89[50]	4·53[50]	3·95[50]	5·17[50]
2:4:6-*Trimethylanilinium ion*		4·38[50]				

Table 5—(Contd.)

		Ortho	*Meta*	*Para*
Phenol	9·998[53]			
Fluorophenol		—	—	9·922[54]
Chlorophenol		8·477[55]	9·023[55]	9·378[55]
Bromophenol		8·425[55]	—	—
Nitrophenol		7·234[55, 56]	8·399[53]	7·149[47, 55, 57]
Cresol		10·287[53]	10·091[53]	10·262[53]
Phenolsulphonic acid, K *salt*		—	—	9·03[25]
Methoxyphenol		9·984[53]	9·649[53]	10·209[53]
Hydroxybenzaldehyde		8·374[52]	9·016[52]	7·615[52]
Vanillin		7·912[58]	8·889[58]	7·396[58]
cis-*Hydroxycinnamic acid* K_2	10·66[59]			
trans-*Hydroxycinnamic acid* K_2	9·63[59]			
5-*Hydroxy*-4-*methylcoumarin*	8·26[60]			
6-*Hydroxy*-4-*methylcoumarin*	9·14[60]			
7-*Hydroxy*-4-*methylcoumarin*	7·80[60]			
2 : 4-*Dichlorophenol*	7·850[55]			
Dinitrophenol		2 : 4 4·11[47, 61]	2 : 5 5·216[55] 2 : 6 3·706[37]	3 : 4 5·424[55]
2 : 4 : 6-*Trinitrophenol*	0·708[37]			
4-*Chloro*-2 : 6-*dinitrophenol*	2·97[47]			
Catechol K_1	9·449[62]			
Catecholdisulphonic acid K_3	8·32[62]			
Catecholdisulphonic acid K_4	13·07[62]			
Xylenol		2 : 3 10·54[63, 64] 2 : 4 10·60[63, 64]	2 : 5 10·41[63, 64] 2 : 6 10·63[63, 64]	3 : 4 10·36[63, 64] 3 : 5 10·19[63, 64]
2 : 3 : 5-*Trimethylphenol* (20°)		10·69[64]		
2 : 4 : 6-*Trimethylphenol* (20°)		10·99[64]		
Nitromesitol		8·984[55]		
1-*Nitroso*-2-*naphthol*	7·77[64a]			
2-*Nitroso*-1-*naphthol*	7·38[64a]			

Table 5—(Contd.)

Furoic acid	3·169[7]		
Barbituric acid	4·035[65]		
5-*Allyl*-5-*isobutyl barbituric acid*	7·79[65]		
5-*Allyl*-5-*isopropyl barbituric acid*	7·99[65]		
5-*Butyl*-5-*ethylbarbituric acid*	7·98[65]		
5·5-*Diallylbarbituric acid*	7·77[65]		
5·5-*Diethylbarbituric acid*	7·971[65]		
1:3-*Dimethylbarbituric acid*	4·678[65]		
5-*Ethyl*-5-*isopentylbarbituric acid*	7·96[65]		
5-*Ethyl*-5-*phenylbarbituric acid*	7·45[65]		
5-*cycloHexenyl*-1:5-*dimethylbarbituric acid*	8·37[65]		
1-*Methylbarbituric acid*	4·348[65]		
5-*Methyl*-5-*phenylbarbituric acid*	7·73[65]		
5-iso*Propylbarbituric acid*	4·940[65]		
		pK_1	pK_2
Picolinic acid		1·01[66, 67]	5·32[66, 67]
Picolinic acid, methyl ester	2·21[66]		
Nicotinic acid		2·07[66, 67]	4·81[66, 67]
Nicotinic acid amide (20°)	3·328[68]		
Nicotinic acid, methyl ester	3·13[66]		
Trigonelline (*nicotinic acid, methyl betaine*)	2·04[66]		
iso*Nicotinic acid*		1·84[66, 67]	4·86[66, 67]
iso*Nicotinic acid, methyl ether*	3·26[66]		
8-*Hydroxyquinoline*		4·910[69, 70]	9·813[69, 70]
8-*Hydroxy*-5-*nitrosoquinoline* (14°)		2·40[70]	7·75[70]
5:7-*Dichloro*-8-*hydroxyquinoline*		2·887[71]	7·617[71]
8-*Hydroxyquinoline*-5-*sulphonic acid*		4·108[72]	8·753[72]
7-*Nitro*-8-*hydroxyquinoline*-5-*sulphonic acid*		1·950[73]	5·750[73]

Table 5—(Contd.)

			pK$_1$	pK$_2$
7-*Iodo*-8-*hydroxyquinoline*-5-*sulphonic acid*			2·514[74]	7·417[74]
7-*Phenylazo*-8-*hydroxyquinoline*-5-*sulphonic acid*			3·41[74]	7·845[74]
7-(4-*Nitrophenylazo*)-8-*hydroxyquinoline*-5-*sulphonic acid*			3·14[74]	7·494[74]
8-*Hydroxyquinazoline* (14°)			3·36[70]	8·54[70]
8-*Hydroxy*-2:4-*dimethylquinazoline* (14°)			3·79[70]	9·41[70]
8-*Hydroxyquinoxaline* (14°)			0·8[70]	8·75[70]
1:10-*Phenanthroline*	4·857[75]			

The ionization constants of weak organic acids have been discussed by Dippy, J. F. J., *Chem. Rev.*, 25 (1939) 151, by Ingold, C. K., 'Structure and Mechanism in Organic Chemistry', Chap. XIII, G. Bell and Sons, Ltd., London (1953) and by Brown, H. C., McDaniel, D. H. and Häfliger, O., 'Determination of Organic Structures by Physical Methods', Chap. 14, edited by Braude, E. A. and Nachod, F. C., Academic Press Inc., New York (1955). It is difficult to assess the probable error of these results but, when independent determinations of the ionization constant have been made, the agreement is often within 0·005 in pK$_a$.

APPENDIX 12.1

REFERENCES TO TABLE 5

[1] Ives, D. J. G. and Sames, K., *J. chem. Soc.*, (1943) 513
[2] McCallum, K. S. and Emmons, W. D., *J. org. Chem.*, 21 (1956) 367
[3] Banks, C. V. and Zimmerman, J., *J. org. Chem.*, 21 (1956) 1439
[4] Jeffery, G. H. and Vogel, A. I., *J. chem. Soc.*, (1934) 1101
[5] Pedersen, K. J., *Acta chem. scand.*, 6 (1952) 243; at 37° he found pK = 2·420 for pyruvic acid, pK_1 = 2.450, pK_2 = 4·359 for oxaloacetic acid
[6] Dippy, J. F. J. and Lewis, R. H., *J. chem. Soc.*, (1937) 1008
[7] German, W. L., Jeffery, G. H. and Vogel, A. I., *ibid.*, (1937) 1604
[8] Ives, D. J. G., Linstead, R. P. and Riley, H. L., *ibid.*, (1933) 561
[9] Saxton, B. and Waters, G. W., *J. Amer. chem. Soc.*, 59 (1937) 1048
[10] Dippy, J. F. J., *J. chem. Soc.*, (1938) 1222
[11] Dippy, J. F. J., Hughes, S. R. C. and Laxton, J. W., *ibid.*, (1954) 4102
[12] Pedersen, K. J., *Acta chem. scand.*, 9 (1955) 1634; at 37° he found pK_1 = 2·380 and pK_2 = 4·758 for tartronic acid
[13] Jeffery, G. H. and Vogel, A. I., *J. chem. Soc.*, (1936) 1756
[14] Schwarzenbach, G., *Helv. chim. Acta.*, 16 (1933) 529; he found for the first ionization constants at 20° of methyl-, ethyl-, *n*-propyl-, dimethyl- and diethylmalonic acid, pK = 3·12, 2·94, 3·05, 3·20 and 2·29 respectively; measurements were also made in water-methanol mixtures
[15] German, W. L., Jeffery, G. H. and Vogel, A. I., *Phil. Mag.*, 22 (1936) 790
[16] Jeffery, G. H. and Vogel, A. I., *J. chem. Soc.*, (1935) 21
[17] Gane, R. and Ingold, C. K., *ibid.*, (1928) 1594; for the first ionization constants of glutaric, adipic, pimelic and suberic acid they give pK = 4·337, 4·409, 4·478 and 4·513 respectively
[18] Speakman, J. C., *ibid.*, (1940) 855 found pK_1 = 4·39 and 4·43, pK_2 = 5·50 and 5·42 for glutaric and adipic acid respectively at 20°; Schwarzenbach, G., *Helv. chim. Acta*, 16 (1933) 522 found pK_1 = 4·35 and 4·54, pK_2 = 5·49 and 5·58 for glutaric and suberic acid respectively at 20°
[19] Jeffery, G. H. and Vogel, A. I., *J. chem. Soc.*, (1939) 446
[20] German, W. L., Jeffery, G. H. and Vogel, A. I., *ibid.*, (1935) 1624
[21] Dippy, J. F. J. and Lewis, R. H., *ibid.*, (1936) 644
[22] Dippy, J. F. J., Williams, F. R. and Lewis, R. H., *ibid.*, (1935) 343; Saxton, B. and Meier, H. F., *J. Amer. chem. Soc.*, 56 (1934) 1918 found pK = 2·921, 3·821 and 3·982 for *o*-, *m*- and *p*-chlorobenzoic acid respectively
[22a] Robinson, R. A. and Ang, K. P., *J. chem. Soc.*, (1959) 2314
[23] Dippy, J. F. J. and Lewis, R. H., *J. chem. Soc.*, (1937) 1426
[24] Bray, L. G., Dippy, J. F. J., Hughes, S. R. C. and Laxton, L. W., *ibid.*, (1957) 2405
[25] Sager, E. E., Schooley, M. R., Carr, A. S. and Acree, S. F., *J. Res. nat. Bur. Stand.*, 35 (1945) 521 found pK_1 = 4·57 and pK_2 = 9·46 for *p*-hydroxybenzoic acid and pK = 8·47, 8·50, 8·47 and 8·41 for the methyl, ethyl, butyl and benzyl esters respectively
[26] Lumme, P. O., *Suomen Kem.*, 30B (1957) 176
[27] Robinson, R. A. and Biggs, A. I., *Aust. J. Chem.*, 10 (1957) 128 give pK_1 = 2·45, pK_2 = 4·85 for *p*-aminobenzoic acid and 2·472, 2·508, 2·465 and 2·487 for the butyl, ethyl, methyl and propyl esters respectively
[28] Halban, H. von and Brüll, J., *Helv. chim. Acta*, 27 (1944) 1719
[29] Dippy, J. F. J., Hughes, S. R. C. and Laxton, J. W., *J. chem. Soc.*, (1954) 1470
[30] Baker, J. W., Dippy, J. F. J. and Page, J. F., *ibid.*, (1937) 1774
[31] Dippy, J. F. J. and Williams, F. R., *ibid.*, (1934) 1888
[32] Bray, L. G., Dippy, J. F. J. and Hughes, S. R. C., *ibid.*, (1957) 265
[33] Hamer, W. J., Pinching, G. D. and Acree, S. F., *J. Res. nat. Bur. Stand.*, 35 (1945) 539; Hamer, W. J. and Acree, S. F., *ibid.*, 35 (1945) 381
[34] Ang, K. P., *J. phys. Chem.*, 62 (1958) 1109
[35] Thamer, B. J. and Voigt, A. F., *ibid.*, 56 (1952) 225; for *m*-phthalic acid they give pK_1 = 3·62 and pK_2 = 4·60
[36] Lumme, P. O., *Suomen Kem.*, 30B (1957) 194

[37] Dippy, J. F. J., Hughes, S. R. C. and Laxton, J. W., *J. chem. Soc.*, (1956) 2995
[38] Davis, M. M. and Hetzer, H. B., *J. phys. Chem.*, 61 (1957) 123, 125
[39] Schubert, W. M., Zahler, R. E. and Robins, J., *J. Amer. chem. Soc.*, 77 (1955) 2293
[40] Dippy, J. F. J. and Williams, F. R., *J. chem. Soc.*, (1934) 161
[41] King, E. J. and Prue, J. E., *ibid.*, (1961) 275
[42] Hayes, N. V. and Branch, G. E. K., *J. Amer. chem. Soc.*, 65 (1943) 1555
[43] Banks, W. H. and Davies, C. W., *J. chem. Soc.*, (1938) 73
[44] Dippy, J. F. J. and Page, J. F., *ibid.*, (1938) 357
[45] Pedersen, K. J., *K. danske vidensk. Selsk.*, 14 (1937) No. 9; 15 (1937) No. 3; for aniline he gives pK = 4·780 at 14·8° and 4·428 at 34·9°, for *o*-chloroaniline pK = 2·788 at 14·8° and 2·490 at 34·9°
[46] Kilpatrick, M. and Arenberg, C. A., *J. Amer. chem. Soc.*, 75 (1953) 3812. Their value for *p*-chloroaniline is low compared with the recorded mean from the next two references
[47] Bates, R. G. and Schwarzenbach, G., *Helv. chim. Acta*, 37 (1954) 1069
[48] James, J. C. and Knox, J. G., *Trans. Faraday Soc.*, 46 (1950) 254; They also found pK = 5·11 for *p*-toluidine
[49] Bryson, A., *ibid.*, 45 (1949) 257 gives pK = 2·65 for *m*-nitroaniline at 21°
[50] Beale, R. N., *J. chem. Soc.*, (1954) 4494: he gives pK = 4·42, 4·73 and 5·08 for *o*-, *m*- and *p*-toluidine respectively
[51] Bacarella, A. L., Grunwald, E., Marshall, H. P. and Purlee, E. L., *J. org. Chem.*, 20 (1955) 747
[51a] Willi, A. V., *Helv. chim. Acta*, 40 (1957) 2019
[52] Robinson, R. A. and Kiang, A. K., *Trans. Faraday Soc.*, 52 (1956) 327
[53] Biggs, A. I., *ibid.*, 52 (1956) 35
[54] Robinson, R. A., unpublished results
[55] Judson, C. M. and Kilpatrick, M., *J. Amer. chem. Soc.*, 71 (1949) 3110
[56] Dippy, J. F. J., *et al.*, ref. 37 found pK = 7·229 and Biggs, A. I., ref. 53, pK = 7·210
[57] Robinson, R. A. and Biggs, A. I., *Trans. Faraday Soc.*, 51 (1955) 901
[58] Robinson, R. A. and Kiang, A. K., *ibid.*, 51 (1955) 1398
[59] Mattoo, B. N., *ibid.*, 53 (1957) 760
[60] Mattoo, B. N., *ibid.*, 54 (1958) 19
[61] Bale, W. D. and Monk, C. B., *ibid.*, 53 (1957) 450 give pK = 4·078
[62] Näsänen, R. and Markkanen, R., *Suomen Kem.*, 29B (1956) 119; Näsänen, R., *ibid.*, 30B (1957) 61
[63] Herington, E. F. G. and Kynaston, W., *Trans. Faraday Soc.*, 53 (1957) 138
[64] Riccardi, R. and Franzosini, P., *Ann. chim. (Rome)*, 47 (1957) 977; they find that the pK values for xylenols are about 0·04 higher at 20° than those reported at 25°
[64a] Dyrssen, D. and Johansson, E., *Acta chem. scand.*, 9 (1955) 763
[65] Biggs, A. I., *J. chem. Soc.*, (1956) 2485
[66] Green, R. W. and Tong, H. K., *J. Amer. chem. Soc.*, 78 (1956) 4896
[67] Lumme, P. O., *Suomen Kem.*, 30B (1957) 168 gives pK_1 = 1·03, pK_2 = 5·397 for picolinic acid, pK_1 = 1·982, pK_2 = 4·817 for nicotinic acid and pK_1 = 1·676, pK_2 = 4·913 for *iso*nicotinic acid
[68] Willi, A. V., *Helv. chim. Acta*, 37 (1954) 602
[69] Näsänen, R., Lumme, P. O. and Mukula, A. L., *Acta chem. scand.*, 5 (1951) 1199; pK_1 = 5·017 at 20°
[70] Irving, H., Rossotti, H. S. and Harris, G., *Analyst*, 80 (1955) 83 give pK_1 = 5·00, pK_2 = 9·85 for 8-hydroxyquinoline at 14°
[71] Näsänen, R., *Suomen Kem.*, 26B (1953) 69
[72] Näsänen, R. and Uusitalo, E., *Acta chem. scand.*, 8 (1954) 112
[73] Näsänen, R. and Uusitalo, E., *Suomen Kem.*, 28B (1955) 17
[74] Uusitalo, E., *Ann. Acad. Scient. Fennicae AII* (1957) 87
[75] Näsänen, R. and Uusitalo, E., *Suomen Kem.*, 29B (1956) 11; they found pK = 5·079 at 0° and 4·641 at 50°

APPENDIX 12.1

Table 6

Ionization Constants of Weak Acids in Mixed Solvents

$$pK_a = -\log K_a = A_1/T - A_2 + A_3T$$

	pK *at 25°*	A_1	A_2	A_3	*Reference*
Acetic acid in 20% *dioxan*	5·292	1423·45	4·2934	0·016136	(1)
,, ,, 45% ,,	6·307	1568·31	4·5387	0·018736	(1)
,, ,, 70% ,,	8·321	1549·12	2·5194	0·018933	(1)
,, ,, 82% ,,	10·509	4168·33	19·238	0·052857	(2) (3)
,, ,, 10% *methanol*	4·904	1417·19	4·5806	0·015874	(4)
,, ,, 20% ,,	5·078	1572·21	5·3447	0·017279	(4)
,, ,, 50% ,,	5·660	1755·02	6·1509	0·019870	(4a)
,, ,, 50% *glycerol*	5·271	1321·43	3·4148	0·014268	(5)
β-Alanine (K_1) in 5% *isopropanol*	3·599	1594·69	6·1904	0·014887	(6)
,, ,, 10% ,,	3·642	1753·58	7·1209	0·016375	(6)
,, ,, 20% ,,	3·723	2351·54	10·9051	0·022608	(6)
4-Aminopyridine in 50% *methanol*	8·520	3179·2	4·9361	0·009362	(6a)
Ammonia in 60% *methanol*	8·591	2326·0	−2·2600	−0·004928	(7)
Benzoic acid in 10% *methanol*	4·387	1600·1	6·2358	0·01763	(8)
,, ,, 20% ,,	4·721	1503·0	5·3008	0·01671	(8)
n-*Butyric acid in* 5% *isopropanol*	4·946	1048·52	2·6719	0·013753	(9)
,, ,, 10% ,,	5·052	1217·52	3·6903	0·015625	(9)
,, ,, 20% ,,	5·341	1459·37	5·0187	0·018325	(9)
Formic acid in 20% *dioxan*	4·180	1339·04	5·0628	0·015938	(10)
,, ,, 45% ,,	5·292	1333·79	4·6393	0·017634	(10)
,, ,, 70% ,,	7·016	1181·65	1·9920	0·016922	(10)
,, ,, 82% ,,	9·141	3360·38	15·952	0·046343	(2) (10)
Glycine (K_1) *in* 20% *dioxan*	2·629	1368·94	5·6875	0·012493	(11)

APPENDIX 12.1

Table 6—(Contd.)

	pK *at 25°*	A_1	A_2	A_3	*Reference*
Glycine (K_1) *in* 45% *dioxan*	3·105	1273·49	4·7113	0·011887	(11)
,, ,, 70% ,,	3·965	1187·30	3·3894	0·011322	(11)
Glycine (K_2) *in* 20% *dioxan*	9·907	3076·89	2·9308	0·008441	(11)
,, ,, 45% ,,	10·237	3276·18	4·5232	0·012645	(11)
,, ,, 70% ,,	11·280	3482·20	5·557	0·017298	(11)
Methylamine in 60% *methanol*	9·712	3129·4	1·8590	0·003607	(7)
Phosphoric acid (K_2) *in* 10% *methanol*	7·365	2336·12	7·4808	0·023514	(12)
,, ,, ,, 20% ,,	7·600	2192·68	6·2536	0·021800	(12)
,, ,, ,, 50% ,,	8·443	2092·05	5·2254	0·022316	(4a)
Propionic acid in 20% *dioxan*	5·466	1356·57	3·6704	0·015384	(13)
,, ,, 45% ,,	6·553	1480·12	3·5287	0·017163	(13)
,, ,, 70% ,,	8·612	1508·10	1·6539	0·017466	(13)
,, ,, 82% ,,	10·752	3748·69	15·935	0·047314	(2) (14)
,, , 10% *methanol*	5·042	1753·20	7·039	0·02080	(14)
,, ,, 20% ,,	5·238	1113·72	2·853	0·01462	(14)
,, ,, 10% *ethanol*	5·046	421·94	−2·325	0.00438	(14)
,, ,, 20% ,,	5·107	1932·86	7·674	0·02111	(14)
,, ,, 5% *isopropanol*	4·978	1191·09	3·1909	0.013995	(15)
,, ,, 10% ,,	5·086	1403·86	4·5143	0.016406	(15)
,, ,, 20% ,,	5·332	1771·13	6·7087	0.020457	(15)
Tris (hydroxymethyl) aminomethane in 50% *methanol*	7·818	3689·57	8·5466	0·013381	(15a)

APPENDIX 12.1

REFERENCES TO TABLE 6

[1] HARNED, H. S. and KAZANJIAN, G. L., *J. Amer. chem. Soc.*, 58 (1936) 1912

[2] DANYLUK, S. S., TANIGUCHI, H. and JANZ, G. J., *J. phys. Chem.*, 61 (1957) 1679

[3] HARNED, H. S. and FALLON, L. D., *J. Amer. chem. Soc.*, 61 (1939) 2377; HARNED, H. S., *J. phys. Chem.*, 43 (1939) 275

[4] HARNED, H. S. and EMBREE, N. D., *J. Amer. chem. Soc.*, 57 (1935) 1669

[4a] PAABO, M., ROBINSON, R. A. and BATES, R. G., *ibid.*, 87 (1965) 415

[5] HARNED, H. S. and NESTLER, F. H. M., *ibid.*, 68 (1946) 966

[6] MAY, M. and FELSING, W. A., *ibid.*, 73 (1951) 406

[6a] PAABO, M., ROBINSON, R. A. and BATES, R. G., *Anal. chem.*, 38 (1966) 1573

[7] EVERETT, D. H. and WYNNE-JONES, W. F. K.. *Trans.Faraday Soc.*, 48 (1952) 531

[8] PARTON, H. N. and ROGERS, J., *ibid.*, 38 (1942) 238

[9] FELSING, W. A. and MAY, M., *J. Amer. chem. Soc.*, 70 (1948) 2904

[10] HARNED, H. S. and DONE, R. S., *ibid.*, 63 (1941) 2579

[11] HARNED, H. S. and BIRDSALL, C. M., *ibid.*, 65 (1943) 54, 1117; the second ionization constants are given in the second of these papers in the form of basic ionization constants. They are given as acidic ionization constants, K_2, in this appendix, the conversion being made with the aid of the ionization constants of water in dioxan-water mixtures, determined by HARNED, H. S. and FALLON, L. D. *ibid.*, 61 (1939) 2374

[12] ENDER, F., TELTSCHIK, W. and SHÄFER, K., *Z. Elektrochem.*, 61 (1957) 775

[13] HARNED, H. S. and DEDELL, T. R., *J. Amer. chem. Soc.*, 63 (1941) 3308

[14] PATTERSON, A. and FELSING, W. A., *ibid.*, 64 (1942) 1480

[15] MOORE, R. L. and FELSING, W. A., *ibid.*, 69 (1947) 2420

[15a] WOODHEAD, M., PAABO, M., ROBINSON, R. A. and BATES, R. G., *J. Res. nat. Bur. Stand.*, 69A (1965) 263

APPENDIX 12.1

Table 7

pK_a of some acids in mixed solvents at 25°: ref. (1)

Wt % methanol	16·47	34·47	54·20	75·94	93·74
Vol % methanal	20	40	60	80	95
Formic acid	3·906	4·133	4·553	5·214	6·448
Acetic acid	4·998	5·308	5·764	6·432	7·764
Propionic acid	5·137	5·546	6·009	6·745	—
Butyric acid	5·110	5·571	6·037	6·748	—
Benzoic acid	4·501	4·941	5·492	6·218	7·379
Anilinium ion	4·450	4·296	4·124	4·000	4·519
N-Methylanilinium ion	4·685	4·450	4·133	3·832	4·055
N-N-Dimethylanilinium ion	4·951	4·700	4·262	3·753	3·859

% Methanol	10	20	40	60	80	90	95	100	*Reference*
Acetic acid	4·907	5·068	5·450	5·899	6·633	7·306	7·992	9·524	(2)
Oxalic Acid K_1	1·514	1·647	—	—	—	—	—	—	(3)
,, ,, K_2	4·538	4·808	—	—	—	—	—	—	(3)

% Ethanol	10	20	40	60	80	94·64	99·67	*Ref.*
Acetic acid	4·915	5·159	5·538	6·052	6·909	8·180	10·035	(3a)
Chloroacetic acid	—	3·188	—	—	—	—	—	(4)
2:4-Dinitrophenol	—	4·083	—	—	—	—	—	(4)
Propionic acid	—	5·291	—	—	—	—	—	(4)

APPENDIX 12.1

Table 7—(Contd.)

% Dioxan	20	30	40	45	50	60	70	82	*Reference*
Anilinium ion	4·45	—	—	4·02	—	—	3·60	3·43	(5)
p-*Chloroanilinium ion*	3·66	—	—	3·09	—	—	2·70	2·57	(5)
m-*Nitroanilinium ion*	2·09	1·84	1·61	1·45	1·40	1·15	1·05	1·04	(5)
Benzoic acid	4·869	5·282	5·794	—	6·38	—	—	—	(6)
$C_6H_5{\cdot}N(CH_3)_2H^+$	4·853	—	—	4·194	—	—	—	—	(7)
m-$Cl{\cdot}C_6H_4{\cdot}N(CH_3)_2H^+$	3·333	—	—	2·487	—	—	—	—	(7)
p-$Cl{\cdot}C_6H_4{\cdot}N(CH_3)_2H^+$	3·925	—	—	3·140	—	—	—	—	(7)
p-$BrC_6H_4N(CH_3)_2H^+$	3·760	—	—	—	—	—	—	—	(7)
m-$NO_2C_6H_4N(CH_3)_2H^+$	2·057	—	—	1·220	—	—	—	—	(7)
p-$CH_3C_6H_4N(CH_3)_2H^+$	—	—	—	4·697	—	—	—	—	(7)
p-$CH_3OC_6H_4N(CH_3)_2H^+$	5·567	—	—	5·177	—	—	—	—	(7)
p-$H^+(CH_3)_2NC_6H_4N(CH_3)_2H^+$	5·965	—	—	—	—	—	—	—	(7)
p-*Toluidinium ion*	4·93	—	—	4·55	—	—	4·18	4·07	(5)

100% Formamide: Acetic acid 6·82, Formic acid $5{\cdot}49_5$ (Ref. 8)

The results recorded under ref. (7) were obtained at 20°.

All solvent compositions are, unless otherwise mentioned, expressed as weight percentages; for bases, it is the acidic ionization constant which is recorded; the constants refer to the molality scale and can be converted to the molarity scale by: $pK_c = pK_m - \log d_o$.

APPENDIX 12.1

REFERENCES TO TABLE 7

[1] Bacarella, A. L., Grunwald, E., Marshall, H. P. and Purlee, E. L., *J. org. Chem.*, 20 (1955) 747

[2] Shedlovsky, T. and Kay, R. L., *J. phys. Chem.*, 60 (1956) 151

[3] Parton, H. N. and Nicholson, A. J. C., *Trans. Faraday Soc.*, 35 (1939) 546

[3a] Spivey, H. O. and Shedlovsky, T., *J. phys. Chem.*, 71 (1967) 2171; data are also given at 0° and 35°; the original data at 25° have been converted to the molal scale

Bale, W. D. and Monk, C. B., *Trans. Faraday Soc.*, 53 (1957) 450; their value for propionic acid is higher than that recorded by Patterson, A. and Felsing W. A., *J. Amer. chem., Soc.* 64 (1942) 1480

[5] James, J. C. and Knox, J. G., *Trans. Faraday Soc.*, 46 (1950) 254

[6] Dunsmore, H. S. and Speakman, J. C., *ibid.*, 50 (1954) 236

[7] Willi, A. V., *Helv. chim. Acta*, 40 (1957) 2019

[8] Mandel, M. and Decroly, P., *Trans. Faraday Soc.*, 56 (1960) 29; data are given for temperatures from 15° to 45°

APPENDIX 12.2

Table 1

Ionization Constants of Water

Temp.	$-\log K_w$	*Temp.*	$-\log K_w$
0°	$14{\cdot}943_5$	35°	$13{\cdot}680_1$
5°	$14{\cdot}733_8$	40°	$13{\cdot}534_8$
10°	$14{\cdot}534_9$	45°	$13{\cdot}396_0$
15°	$14{\cdot}346_3$	50°	$13{\cdot}261_7$
20°	$14{\cdot}166_8$	55°	$13{\cdot}136_9$
25°	$13{\cdot}996_5$	60°	$13{\cdot}017_1$
30°	$13{\cdot}833_0$		

From Harned, H. S. and Robinson, R. A., *Trans. Faraday Soc.*, 36 (1940) 973

The ionization constant of deuterium oxide at 25° is given as $-\log K_w = 14{\cdot}955$ (molal scale). The data from 0° to 50° can be represented by the equation: $-\log K_w = 4913{\cdot}14/T - 7{\cdot}5117 + 0{\cdot}0200854T$
Covington, A. K., Robinson, R. A., and Bates, R. G., *J. Phys. Chem.*, 70 (1966) 3820

The ionization constant of tritium oxide at 25° is given as $K_w = 6{\cdot}1 \times 10^{-16}$ with concentrations expressed as aquamolalities (moles per 55·51 moles of solvent); this corresponds to pK_w (molal scale) = 15·39. Goldblatt, M. and Jones, W. M., *J. chem. Phys.*, 51 (1969) 1881

APPENDIX 12.3

Table 1

pa_{H^+} of Some Standard Solutions

°C	0·05 M *Potassium tetraoxalate*	*Potassium hydrogen tartrate (saturated at 25°)*	0·05 M *Potassium hydrogen phthalate*	0·025 M Na_2HPO_4 + 0·025 M KH_2PO_4	0·03043 M Na_2HPO_4 + 0·008695 M KH_2PO_4	0·01 M *Borax*	$Ca(OH)_2$ *(saturated at 25°)*
0	1·666	—	4·003	6·984	7·534	9·464	13·423
5	1·668	—	3·999	6·951	7·500	9·395	13·207
10	1·670	—	3·998	6·923	7·472	9·332	13·003
15	1·672	—	3·999	6·900	7·448	9·276	12·810
20	1·675	—	4·002	6·881	7·429	9·225	12·627
25	1·679	3.557	4·008	6·865	7·413	9·180	12·454
30	1·683	3·552	4·015	6·853	7·400	9·139	12·289
35	1·688	3·549	4·024	6·844	7·389	9·102	12·133
38	1·691	3·548	4·030	6·840	7·384	9·081	12·043
40	1·694	3·547	4·035	6·838	7·380	9·068	11·984
45	1·700	3·547	4·047	6·834	7·373	9·038	11·841
50	1·707	3·549	4·060	6·833	7·367	9·011	11·705
55	1·715	3·554	4·075	6·834	—	8·985	11·574
60	1·723	3·560	4·091	6·836	—	8·962	11·449
70	1·743	3·580	4·126	6·845	—	8·921	—
80	1·766	3·609	4·164	6·859	—	8·885	—
90	1·792	3·650	4·205	6·877	—	8·850	—
95	1·806	3·674	4·227	6·886	—	8·833	—
100		3·68	4·25	6·91		8·84	
125		3·79	4·37	7·02		8·77	
150		3·90	4·54	7·14		8·68	

These pa_{H^+} values are sometimes referred to as pH(*S*) values.

REFERENCES

BATES, R. G., *J. Res. natn. Bur. Stand.*, 66A (1962) 179

PERKOVETS, V. D. and KRYUKOV, P. A., *Izv. Sib. Otd. Akad. Nauk., SSSR., Ser. Khim. Nauk.* (1968) No. 6, 22; PAVLYUK, L. A., SMOLYAKOV, B. S. and KRYUKOV, P. A., *ibid.* (1969) No. 3, 13

APPENDIX 12.3

Table 2

pH *Values of Some Solutions at 25°*

Solution	pH	*Reference*
0·1 M — HCl	1·092	1
0·01 M — HCl + 0·09 M — KCl	2·102	1
0·01 M — *Sulphamic acid*	2·083	1
0·05 M *Citric acid*	2·238	1
0·01 M *Citric acid*	2·624	1
0·1 M — KH_2Cit	3·717	1
0·02 M — KH_2Cit	3·836	1
0·01 M — $H\bar{F}$ + 0·01247 M $K\bar{F}$ + 0·01079 M — KCl	3·800	2
0·02 M — H_2Suc + 0·01 M — NaHSuc + 0·02 M — NaCl	3·823	3
0·01 M — HAc + 0·01 M — NaAc	4·718	1
0·01 M — H_2Mal + 0·01 M — NaHMal + 0·01 M — NaCl	5·444	4
0·01 M — NaHSuc + 0·01 M — Na_2Suc	5·474	1
0·01 M — HB + 0·009554 M — NaB + 0·01592 M — NaCl	7·903	5
0·01 M — $NaHCO_3$ + 0·01 M — Na_2CO_3	10·112	1
0·01 M — Na_2CO_3	11·006	1
0·01 M — Na_3PO_4	11·719	1
0·01 M — NaOH	11·939	1
0·05 M — NaOH	12·619	1

Cit = citrate $\bar{F}$ = formate Suc = succinate Ac = acetate
Mal = malonate B = diethylbarbiturate

REFERENCES

1 BATES, R. G., PINCHING, G. D. and SMITH, E. R., *J. Res. nat. Bur. Stand.*, 45 (1950) 418; data are also given for 0°, 10° and 38°

2 Calculated from the data of HARNED, H. S. and EMBREE, N. D., *J. Amer. chem. Soc.*, 56 (1934) 1042

3 Calculated from the data of PINCHING, G. D. and BATES, R. G., *J. Res. nat. Bur. Stand.*, 45 (1950) 322, 444

4 HAMER, W. J., BURTON, J. O. and ACREE, S. F., *ibid.*, 24 (1940) 269

5 Calculated from the data of MANOV, G. G., SCHUETTE, K. E. and KIRK, F. S., *ibid.*, 48 (1952) 84

APPENDIX 12.3

Table 3

Solutions giving round values of $y = -\log \gamma_{H^+} m_{H^+}$ *at* 25°

A		*B*		*C*		*D*		*E*	
y	*x*	*y*	*x*	*y*	*x*	*y*	*x*	*y*	*x*
1·00	67·0	2·20	49·5	4·10	1·3	5·80	3·6	7·00	46·6
1·10	52·8	2·30	45·8	4·20	3·0	5·90	4·6	7·10	45·7
1·20	42·5	2·40	42·2	4·30	4·7	6·00	5·6	7·20	44·7
1·30	33·6	2·50	38·8	4·40	6·6	6·10	6·8	7·30	43·4
1·40	26·6	2·60	35·4	4·50	8·7	6·20	8·1	7·40	42·0
1·50	20·7	2·70	32·1	4·60	11·1	6·30	9·7	7·50	40·3
1·60	16·2	2·80	28·9	4·70	13·6	6·40	11·6	7·60	38·5
1·70	13·0	2·90	25·7	4·80	16·5	6·50	13·9	7·70	36·6
1·80	10·2	3·00	22·3	4·90	19·4	6·60	16·4	7·80	34·5
1·90	8·1	3·10	18·8	5·00	22·6	6·70	19·3	7·90	32·0
2·00	6·5	3·20	15·7	5·10	25·5	6·80	22·4	8·00	29·2
2·10	5·1	3·30	12·9	5·20	28·8	6·90	25·9	8·10	26·2
2·20	3·9	3·40	10·4	5·30	31·6	7·00	29·1	8·20	22·9
		3·50	8·2	5·40	34·1	7·10	32·1	8·30	19·9
		3·60	6·3	5·50	36·6	7·20	34·7	8·40	17·2
		3·70	4·5	5·60	38·8	7·30	37·0	8·50	14·7
		3·80	2·9	5·70	40·6	7·40	39·1	8·60	12·4
		3·90	1·4	5·80	42·3	7·50	40·9	8·70	10·3
		4·00	0·1	5·90	43·7	7·60	42·4	8·80	8·5
						7·70	43·5	8·90	7·0
						7·80	44·5	9·00	5·7
						7·90	45·3		
						8·00	46·1		

F		*G*		*H*		*I*		*J*	
y	*x*	*y*	*x*	*y*	*x*	*y*	*x*	*y*	*x*
8·00	20·5	9·20	0·9	9·60	5·0	10·90	3·3	12·00	6·0
8.10	19·7	9·30	3·6	9·70	6·2	11·00	4·1	12·10	8·0
8·20	18·8	9·40	6·2	9·80	7·6	11·10	5·1	12·20	10·2
8.30	17·7	9·50	8·8	9·90	9·1	11·20	6·3	12·30	12·8
8.40	16·6	9·60	11·1	10·00	10·7	11·30	7·6	12·40	16·2
8·50	15·2	9·70	13·1	10·10	12·2	11·40	9·1	12·50	20·4
8·60	13·5	9·80	15·0	10·20	13·8	11·50	11·1	12·60	25·6
8·70	11·6	9·90	16·7	10·30	15·2	11·60	13·5	12·70	32·2
8·80	9·4	10·00	18·3	10·40	16·5	11·70	16·2	12·80	41·2
8·90	7·1	10·10	19·5	10·50	17·8	11·80	19·4	12·90	53·0
9·00	4·6	10·20	20·5	10·60	19·1	11·90	23·0	13·00	66·0
9·10	2·0	10·30	21·3	10·70	20·2	12·00	26·9		
		10·40	22·1	10·80	21·2				
		10·50	22·7	10·90	22·0				
		10·60	23·3	11·00	22·7				
		10·70	23·8						
		10·80	24·25						

APPENDIX 12.3

The buffer solutions are made by mixing two solutions and diluting to 100 ml. The solutions to be mixed are as follows:

A 25 ml 0·2 M – KCl + *x* ml 0·2 M – HCl
B 50 ml 0·1 M – Potassium hydrogen phthalate + *x* ml 0·1 M – HCl
C 50 ml 0·1 M – Potassium hydrogen phthalate + *x* ml 0·1 M – NaOH
D 50 ml 0·1 M – KH_2PO_4 + *x* ml 0·1 M – NaOH
E 50 ml 0·1 M – *tris* (Hydroxymethyl)aminomethane + *x* ml 0·1 M – HCl
F 50 ml 0·025 M – Borax + *x* ml 0·1 M – HCl
G 50 ml 0·025 M – Borax + *x* ml 0·1 M – NaOH
H 50 ml 0·05 M – $NaHCO_3$ + *x* ml 0·1 M – NaOH
I 50 ml 0·05 M – Na_2HPO_4 + *x* ml 0·1 M – NaOH
J 25 ml 0·2 M – KCl + *x* ml 0·2 M – NaOH

The relevant references are BOWER, V. E. and BATES, R. G., *J. Res. Nat. Bur. Stand.*, 55 (1955) 197 for the first four and BATES, R. G. and BOWER, V. E., *Anal. Chem.*, 28 (1956) 1322 for the remainder.

APPENDIX 14.1

Values of the definite integral $Q(b)$ *in Equation* (14.2)

$$Q(b) = \int_2^b x^{-4} e^x \mathrm{d}x$$

b	$Q(b)$	b	$Q(b)$	b	$\log Q(b)$
2·0	0	5	0·771	15	1·97
2·1	0·0440	6	1·041	17	2·59
2·2	0·0843	7	1·42	20	3·59
2·4	0·156	8	2·00	25	5·35
2·6	0·218	9	2·95	30	7·19
2·8	0·274	10	4·63	40	11·01
3·0	0·326	12	13·41	50	14·96
3·5	0·442	14	47·0	60	18·98
4·0	0·550	15	93·0	70	23·05

APPENDIX 14.2

Limiting Equivalent Conductivities and Dissociation Constants of some Salts in Organic Solvents at 25°

In molarity units

Salt	*Nitrobenzene*		*Acetone*		*Pyridine*		*Ethylene Dichloride*		*Ethylidene Chloride*	
	Λ^0	$K \times 10^4$	Λ^0	$K \times 10^4$	Λ^0	$K \times 10^4$	Λ^0	$K \times 10^4$	Λ^0	$K \times 10^4$
LiPi	—	0·0006	158·1	10·3	58·6	0·83	—	—	—	—
NaPi	32·30	0·28	163·7	13·5	60·5	0·43	—	—	—	—
KPi	33·81	6·86	165·9	34·3	65·7	1·0	—	—	—	—
NH_4Pi	34·4	1·46	180·2	11·1	80·5	2·8	—	—	—	—
NaI	—	—	—	—	75·2	3·7	—	—	—	—
KI	—	—	192·8	80·2	80·4	2·1	—	—	—	—
NH_4I	—	—	—	—	95·2	2·4	—	—	—	—
$AgNO_3$	—	—	—	—	86·9	9·3	—	—	—	—
$AgClO_4$	—	—	—	—	81·9	19·1	—	—	—	—
AgPi	—	—	—	—	68·0	30·6	—	—	—	—
KCNS	—	—	201·6	38·3	—	—	—	—	—	—
$(CH_3)_4NPi$	33·3	400	183·1	112	76·7	6·7	73·8	0·32	—	—
$(C_2H_5)_4NPi$	32·4	1400	176·5	175	—	—	69·4	1·59	116·6	0·348
$(C_2H_5)_4NCl$	$38·5_5$	125	—	—	—	—	77·4	0·510	—	—
$(C_2H_5)_4NBr$	—	—	—	—	—	—	72·1	0·697	—	—
$(n\text{-}C_3H_7)_4NPi$	29·5	—	—	—	—	—	62·7	1·94	103·7	0·397
$(n\text{-}C_3H_7)_4NI$	—	—	190·6	49·8	—	—	—	—	—	—
$(n\text{-}C_4H_9)_4NPi$	27·9	—	152·4	223	57·7	12·8	57·4	2·28	96·9	0·454
$(n\text{-}C_4H_9)_4N{\cdot}ClO_4$	—	—	182·4	95·8	—	—	66·2	1·53	—	—
$(n\text{-}C_4H_9)_4N{\cdot}NO_3$	34·5	250	187·2	54·6	76·6	3·7	66·3	1·18	—	—
$(n\text{-}C_4H_9)_4NCl$	—	—	172·3	22·8	—	—	—	—	—	—
$(n\text{-}C_4H_9)_4N{\cdot}Br$	33·5	162	183·0	32·9	75·3	2·5	—	—	—	—
$(n\text{-}C_4H_9)_4NI$	—	—	179·4	64·8	73·1	4·1	—	—	—	—
$(n\text{-}C_4H_9)_4N{\cdot}Ac$	35·5	67	—	—	76	1·7	53·5	1·34	—	—
$(n\text{-}C_5H_{11})_4NPi$	26·8	—	—	—	—	—	54·5	2·38	90·4	0·494
$(n\text{-}C_5H_{11})_4NBr$	—	—	174·4	42·5	—	—	—	—	—	—
$(n\text{-}C_4H_9)_4N$ $(C_6H_5)_3BF$	23·4	—	134·2	197	48·0	13·2	52·4	2·03	—	—

Pi = picrate   Ac = Acetate

REFERENCES

KRAUS, C. A. *et al.*, *J. Amer. chem. Soc.*, 69 (1947) 451, 454, 814, 1016, 1731, 2472, 2481; 70 (1948) 706, 1709; 73 (1951) 2459, 3293; HEALEY, F. H. and MARTELL, A. E., *ibid.*, 73 (1951) 3296

APPENDIX 14.3

Fuoss' Theory of Ion Pair Formation

The derivation of Fuoss' equation can be outlined as follows:

The cations of an electrolyte are represented by charged spheres of radius a and the anions by charged point masses. Only those anions which are on the surface of or within the sphere of volume $v = \frac{4}{3}\pi a^3$ of a cation are counted as ion-pairs. From equation (4.13) its potential energy is:

$$u = -\frac{|z_1 z_2| e^2}{\varepsilon a(1 + \kappa a)}$$

It is useful at this stage to note that:

$$\exp\left(-\frac{u}{kT}\right) = f_{\pm}^2 e^b$$

where $b = \dfrac{|z_1 z_2| e^2}{\varepsilon k T a}$, and $f_{\pm}$ is the rational activity coefficient of free ions.

Let the solution contain n_B cations per unit volume of which n_{B+} are free ions and n_{B0} are ion pairs. If the electrolyte is symmetrical, there must be a similar distribution of $n_{B-}(= n_{B+})$ free ions and n_{B0} ion pairs among the n_B anions per unit volume. On adding δn_B ions of each kind, the probability that an anion will remain free is proportional to $(1 - n_{B+}v)$ and the probability that it forms an ion pair is proportional not to $n_{B+}v$ but to $n_{B+}v \exp\left(-\frac{u}{kT}\right)$. But an equal number of the added cations must form ion pairs to preserve electrical neutrality. Hence:

$$\frac{\delta n_{B0}}{\delta n_{B-}} = \frac{2 n_{B+} v \exp\left(-\frac{u}{kT}\right)}{1 - n_{B}^{+} v}$$

Integration gives, noting that for dilute solutions

$$(1 - n_{B+}v) \approx 1 \text{ and } f_{\pm} \approx y_{\pm}$$

$$n_{B0} = n^2{}_{B+} v y_{\pm}{}^2 e^b$$

and introducing the usual molarity scale of concentration

$$\frac{1}{K} = \frac{1 - \alpha}{\alpha^2 y_{\pm}{}^2 c} = \frac{4\pi N a^3 e^b}{3000}$$

Fuoss has found that this equation represents satisfactorily the data for tetra*iso*amylammonium nitrate in dioxan-water mixtures. The results for lanthanum ferricyanide mentioned in Chapter 14 can also be represented within $\pm 0{\cdot}04$ in pK by this equation with $a = 7{\cdot}75$ Å.

The equation however implies a curious type of dependence of the ionization-constant K on the ion-size parameter a: differentiation with respect to a, at constant εT (i.e. for a given medium and temperature) reveals that K (molar-scale) has a *maximum* given by

$$\log K_{\max} = -3 \log L - 21{\cdot}274 \text{ at an ion-size } a = L/3$$

where $L = |z_1 z_2| e^2/(\varepsilon k T)$

(The characteristic length L is twice Bjerrum's critical distance.)

For a 1:1 electrolyte in water at 25° this maximum is $K_{\max} = 1{\cdot}46$ mole l^{-1}, and occurs at $a = 2{\cdot}38$ Å. Increasing the ion-size beyond this value leads to smaller values of K, in contradiction of the experimental result that ion-pairing decreases rapidly as ion-size increases. In media of low dielectric constant, or with polyvalent salts, the maximum occurs at much larger a values (e.g. 22 Å for a 3:3 salt in water) and in these cases the a values of any imaginable salt will lie on the inside of the maximum, so that the necessary increase of K with a will occur. It will also be noted that, provided a can be treated as constant for a given salt, the results for different media and temperatures reduce on Fuoss' equation to the simple form

$$\ln K = A - B/(\varepsilon T)$$

where $A = \ln \dfrac{3000}{4\pi N a^3}$ and $B = \dfrac{|z_1 z_2| e^2}{k a}$

Ref.: Fuoss, R. M., *J. Amer. Chem. Soc.* 80 (1958) 5059

ADDENDUM

CHAPTER 1

Important developments in the study of electrolytes have come from the relaxation techniques of EIGEN and his collaborators[1]. In these methods the solution is subjected to a sudden perturbation in one of the external parameters affecting a chemical equilibrium in the solution, and the subsequent return to equilibrium is followed by suitably rapid means such as spectrophotometry, polarimetry, or conductivity. The perturbation may be brought about by several methods: a temperature-jump may be produced by the sudden discharge of a capacitor through the solution; a pressure-jump by the sudden rupture of a metal membrane to release a hydrostatic pressure; or a high-intensity pulse of electric field may be applied. By the last method, for example, the rates of dissociation and recombination of the proton and hydroxyl ion in water and ice have been determined, though the recombination rate in water has the extremely high value of $1{\cdot}3 \times 10^{11}$ l mole^{-1} sec^{-1}. For a full account of these important techniques see reference (1), where the original papers are cited.

Recent work on the structure of water and aqueous solutions is fully discussed in a book by KAVANAU[2].

CHAPTER 2

The various frames of reference which may be used in describing diffusion, and the relations between them, are discussed by KIRKWOOD *et al.*[3].

CHAPTER 3

Much more comprehensive tables of energies and entropies of hydration of ions are given by NOYES[4]. STOKES[5] considers that the conventional Pauling crystal radii of ions (Appendix 3.1) are too small for isolated ions *in vacuo*, and proposes larger values deduced from the Van der Waals radii of the noble-gas atoms. Using these for the gaseous ions, but the Pauling radii for the aqueous ions, the free energies of solvation of all the cations of the noble-gas electronic structure are accounted for satisfactorily. GOURARY and ADRIAN[6] find Pauling's radii inconsistent with the electron-densities in the sodium chloride crystal determined by WITTE and WÖLFEL[7], and propose an alternative division of the cation–anion internuclear distance, leading to a different set of ionic radii.

CHAPTER 4

Distribution functions are considered from a more fundamental point of view by FRIEDMAN[8].

CHAPTER 5

STOKES[9,10] has shown that significant errors may arise from thermal diffusion when conductance-cells of the design shown in *Figures 5.3(a)* and *(b)* are used at temperatures far above or below room temperature, and advises the provision of an auxiliary mixing-chamber to overcome this effect. A cell suitable for measurements on solutions which attack glass significantly during a conductance measurement is described by MARSH and STOKES[11]. PRUE[12] has studied the 'shaking effect', which causes differences in conductance between stagnant and rapidly-stirred or freshly-shaken dilute solutions, and suggests that it arises from ion-exchange at the glass surface. It can be minimized by waxing the glass surfaces or by keeping the electrodes and current-carrying region well away from the cell walls.

CHAPTER 6

For hydrochloric acid $\Lambda^0 = 426{\cdot}50$ cm^2 Int.Ω^{-1} equiv^{-1} ($\lambda^0_{H^+} = 350{\cdot}15$) at 25° and 580·9 at 50°[10,13]. Values of 165·93, 199·18, 284·35 and 368·82 cm^2 Int. Ω^{-1} equiv^{-1} have been found[11] for the hydroxyl ion at 15°, 25°, 50° and 75°, respectively.

The data illustrated in *Figure 6.2* can be supplemented by[14] $t^0_{Cl^-(KCl)} = 0{\cdot}5046$ at 0°. The transport number of chloride ions in 0·1 N potassium chloride solution has been found[15] to be 0·4834, 0·4818, 0·4808 and 0·4783 at 70°, 86°, 100° and 115°, respectively.

CHAPTER 7

Further elaborations of the theory of conductance in dilute solutions of symmetrical valence-type electrolytes are being made by FUOSS and ONSAGER[15a].

The compatibility of the Fuoss–Onsager and of Pitts equation has been considered by FERNÁNDEZ-PRINI and PRUE[15b]. For a 1:1 electrolyte, Pitts equation can be converted into the form

$$\Lambda = \Lambda^0 - S\sqrt{c} + Ec \ln c + J_1 c - J_2 c^{3/2}$$

The second term on the right is common to both equations, S being the $(B_1\Lambda^0 + B_2)$ term in the nomenclature of Chapter 7. The third term is also common to both equations and can be written

$$\frac{\kappa \mathbf{e}^2}{16\varepsilon kT}\left\{\frac{2\kappa \mathbf{e}^2\Lambda^0}{3\varepsilon kT} - \beta\sqrt{c}\right\}\ln c$$

β being the B_2 coefficient of Chapter 7. For an aqueous solution at 25° ($\varepsilon = 78{\cdot}30$),

$$Ec \ln c = (0{\cdot}5326\ \Lambda^0 - 20{\cdot}55)\ c \log c$$

The last two terms appear in each equation insofar as there is agreement that there are terms in c and $c^{3/2}$ and that J_1 and J_2 are functions of Λ^0 and a, but J_1 and J_2 are defined somewhat differently. The term in $c^{3/2}$ is sometimes omitted from the Fuoss–Onsager treatment.

The contributions of these four terms can be seen by considering a 1:1 electrolyte in aqueous solution at 25° with $a = 4$ Å, $\Lambda^0 = 150$ cm² Int. Ω^{-1} equiv.$^{-1}$ and $c = 0{\cdot}01$ mole litre^{-1}.

	$S\sqrt{c}$	$Ec \ln c$	J_1c	$J_2c^{3/2}$	Λ
Fuoss–Onsager	9·52	−1·19	2·55	0·13	141·71
Pitts	9·52	−1·19	2·94	0·54	141·69

In other instances the predictions of the two equations differ somewhat; for example, hydrochloric acid in aqueous solution and potassium iodide in dimethylformamide. It is clear that results of very high accuracy are needed. The present position seems to favour Pitts equation for aqueous and dimethylformamide solutions, whereas the Fuoss–Onsager treatment leads to more realistic values of the a parameter when methanol is the solvent.

CHAPTER 8

Several determinations of standard cell potentials have been made covering a range of about 50°. The following values of E^0 refer to 25°:

H_2|HBr in water|AgBr, Ag	0·07106 V[16]
H_2|HBr in water|Hg_2Br_2, Hg	0·13923 V[17]
H_2|HI in water|AgI, Ag	−0·15244 V[18]
H_2|HCl in 50 per cent methanol|AgCl, Ag	0·19058 V[19]
D_2|DCl in deuterium oxide|AgCl, Ag	0·21266 V[20]

The standard potentials of the cells:

$H_2|HCl$ in water$|AgCl, Ag$
$D_2|DCl$ in deuterium oxide$|AgCl, Ag$
$H_2|HBr$ in water$|AgBr, Ag$

have been measured up to 275°, 225° and 200°, respectively[21].

HARNED[22] has suggested a method of calculating the activity coefficients of salts in water–methanol solvents. It seems that the ratio $\gamma_{HCl}/\gamma_{salt}$ in water and in water–methanol, all activity coefficients being taken at the same concentration, depends only on the nature of the salt and not on the solvent composition. Thus if the activity coefficient of hydrochloric acid is known in a number of water–methanol mixtures, then the activity coefficient of a salt in these mixtures can be calculated from its activity coefficient in water.

The activity coefficients of a number of compounds in aqueous solution have been measured: ammonium perchlorate[23]; indium chloride and sulphate[24]; methanesulphonic acid and ethanesulphonic acid and their lithium, sodium, potassium, ammonium, tetramethylammonium, tetraethylammonium and tetrabutylammonium salts[25]; rubidium and caesium fluoride[26]; tetra-alkylammonium salts[25]; thallous sulphate[28]. Isopiestic vapour pressure measurements have been made on a number of salts at 140·3°[29]. Direct vapour pressure measurements have been made of sodium chloride solutions at high temperatures which give the following osmotic coefficients[30]:

°C	$m = 0{\cdot}5$	$m = 1$	$m = 2$	$m = 3$
60	0·921	0·943	0·999	1·057
80	0·920	0·942	0·996	1·053
100	0·917	0·938	0·989	1·042
125	0·912	0·929	0·972	1·016
150	0·904	0·917	0·953	0·989
175	0·893	0·902	0·929	0·957
200	0·877	0·882	0·901	0·920
225	0·854	0·854	0·865	0·877
250	0·824	0·818	0·823	0·831

CHAPTER 12

The Born equation for the change in free energy when an ion is transferred from one solvent to another has been modified[31]. Dielectric saturation is assumed around the ion (up to 1·5 Å for water as solvent), the macroscopic dielectric constant is used at distances greater than 4 Å from the centre of the ion and, in the intermediate region, the dielectric constant is assumed to vary linearly with the distance.

The ionization constants of a number of acids, many of them of considerable biological interest, have been measured. Some of these are recorded here, the figures in parentheses being the value of pK_a at 25°: ascorbic acid (4·25)[32], *iso*citric acid (3·287, 4·714, 6·396)[33], methionine (2·125, 9·28)[34]. Values for some other acids are given at the end of this section [p. 561].

The heat of ionization of the reaction:

$$H_2O \rightarrow H^+ + OH^-$$

is given (p. 363) as 13·52 kcal $mole^{-1}$ from e.m.f. measurements. Recent direct calorimetric determinations[35] give 13·34 kcal $mole^{-1}$; the difference is worthy of further study.

Direct calorimetric measurements are being made[36,37,38] of enthalpy changes on the dissociation of weak acids; until recently, most of our information about such enthalpy changes has come from the temperature coefficient of the dissociation constant (p. 357) and calorimetry provides a welcome alternative method. For example, calorimetry gives 5·65 kcal $mole^{-1}$ for the enthalpy change on the dissociation of phenol[36], whereas 5·62 kcal $mole^{-1}$ has been found[57] from the temperature dependence of the dissociation constant.

CHAPTER 14

The relation between the ion-pair concept and solutions of the non-linear form of the Poisson–Boltzmann equation is discussed by GUGGENHEIM[50] and by SKINNER and FUOSS[39].

CHAPTER 15

In addition to studies of the excess free energy of mixing of electrolyte solutions, considerable interest is to be found in the enthalpy change on mixing[41] and also in the corresponding volume changes[42]. The method of obtaining activity coefficients by ultracentrifugation (p. 211) has been applied to aqueous mixtures of hydrochloric acid and barium chloride[43].

There is another method of evaluating the activity coefficients of both solutes in a mixed solution which is analogous to the McKay–Perring method (p. 443) but is particularly useful if at least one of the solutes is a non-electrolyte. Starting with the cross-differentiation relation, equation (15.1), let us suppose that for two non-electrolytes:

$$\left(\frac{\partial \ln \gamma_B}{\partial m_C}\right)_{m_B} = \left(\frac{\partial \ln \gamma_C}{\partial m_B}\right)_{m_C} = A m_B^p m_C^q$$

Then

$$\ln \gamma_B = \ln \gamma_B^0 + \frac{A}{q+1} m_B^p\, m_C^{q+1}$$

Here m_B and m_C are the molalities of B and C in the mixture and γ_B^0 is the activity coefficient of B in its own solution at molality $m_B(m_C = 0)$.
Similarly,

$$\ln \gamma_C = \ln \gamma_C^0 + \frac{A}{p+1} m_B^{p+1}\, m_C^q$$

We now use the Gibbs–Duhem relation:

$$\begin{aligned} -\,55{\cdot}51 \,\mathrm{d} \ln a_w &= m_B \,\mathrm{d} \ln m_B\gamma_B + m_C \,\mathrm{d} \ln m_C\gamma_C \\ &= m_B \,\mathrm{d} \ln m_B\gamma_B^0 + m_C \,\mathrm{d} \ln m_C\gamma_C^0 \\ &\quad + \frac{A m_B}{q+1} \,\mathrm{d}\,(m_B^p\, m_C^{q+1}) + \frac{A}{p+1} \,\mathrm{d}\,(m_B^{p+1} m_C^q) \end{aligned}$$

The first two terms are equal to $\mathrm{d}(m_B\phi_B^0) + \mathrm{d}(m_C\phi_C^0)$ where ϕ_B^0 and ϕ_C^0 are the osmotic coefficients of solutions of B only and of C only, respectively. Let a new function be defined:

$$\Delta = -\,55{\cdot}51 \ln a_w - m_B\phi_B^0 - m_C\phi_C^0$$

Then,

$$\begin{aligned} \Delta = \frac{pA}{q+1} \int m_B^p m_C^{q+1} \mathrm{d}m_B + A \int m_B^{p+1}\, m_C^q \,\mathrm{d}m_C \\ + A \int m_B^p\, m_C^{q+1} \mathrm{d}m_B + \frac{qA}{p+1} \int m_B^{p+1}\, m_C^q \,\mathrm{d}m_C \end{aligned}$$

or

$$\frac{\Delta}{m_B m_C} = A \frac{p+q+1}{(p+1)\,(q+1)} m_B^p m_C^q$$

Δ can be determined by isopiestic measurements. If it can be expressed as a function of m_B, m_C with no cross-products ($p = 0$ or $q = 0$), then:

$$\frac{\Delta}{m_B m_C} = \left(\frac{\partial \ln \gamma_B}{\partial m_C}\right)_{m_B} = \left(\frac{\partial \ln \gamma_C}{\partial m_B}\right)_{m_C}$$

For example, if:

$$\Delta/(m_B m_C) = A + Bm_B + Cm_C + Dm_B^2$$

then $\qquad \ln \gamma_B = \ln \gamma_B^0 + m_C(A + Bm_B + \frac{1}{2}Cm_C + Dm_B^2)$

and $\qquad \ln \gamma_C = \ln \gamma_C^0 + m_B(A + \frac{1}{2}Bm_B + Cm_C + \frac{1}{3}Dm_B^2)$

If there are cross-products the situation can still be handled. For example, if:

$$\Delta/(m_B m_C) = A + Bm_B + Cm_B m_C,$$

then,

$$\left(\frac{\partial \ln \gamma_B}{\partial m_C}\right)_{m_B} = \left(\frac{\partial \ln \gamma_C}{\partial m_B}\right)_{m_C} = A + Bm_B + 4/3\, m_B m_C$$

$$\ln \gamma_B = \ln \gamma_B^0 + m_C(A + Bm_B + \tfrac{2}{3}\, Cm_B m_C)$$

$$\ln \gamma_C = \ln \gamma_C^0 + m_B(A + \tfrac{1}{2}Bm_B + \tfrac{2}{3}Cm_B m_C)$$

The following aqueous mixtures have been studied by this method: sucrose–mannitol[44], mannitol–sodium chloride[45], mannitol–potassium chloride[46], urea–sodium chloride[47], glycine–potassium chloride[48].

The trace-activity coefficient of one electrolyte in the presence of another, i.e., the activity coefficient $\gamma_{(0)C}$ of C when present in vanishing concentration in a solution of B, can be determined by the following method, which has been used for the hydrochloric acid–calcium perchlorate system[49]. The e.m.f. of the cell:

$$H_2 \left| \begin{matrix} \text{HCl} & (m_B) \\ \text{Ca(ClO}_4)_2 & (m_C) \end{matrix} \right| \text{AgCl, Ag}$$

is measured at constant m_B and increasing m_C; for example, $m_B = 0{\cdot}7702$, $m_C = 0{\cdot}0216$, $0{\cdot}0763$, $0{\cdot}1948$. This is accomplished by adding a solution of calcium perchlorate containing hydrochloric acid at molality m_B to the solution in the cell which initially contained only hydrochloric acid at molality m_B. Since

$$E = E^0 - 2k \log \gamma_B m_B$$

$$-\left(\frac{\partial E}{\partial m_C}\right)_{m_B} = 2k\left(\frac{\partial \log \gamma_B}{\partial m_C}\right)_{m_B} = 3k\left(\frac{\partial \log \gamma_C}{\partial m_B}\right)_{m_C}$$

then in the limit when $m_C \to 0$,

$$3k\frac{\mathrm{d} \log \gamma_{(0)C}}{\mathrm{d}m_B} = -\lim_{m_C \to 0}\left(\frac{\partial E}{\partial m_C}\right)_{m_B} = -Y$$

Y can be evaluated graphically from a few e.m.f. measurements.

The experiment is repeated at different values of m_B. Then

$$3k \log \gamma_{(0)C} = -2 \int_0^{m_B} Y\sqrt{m_B} \, . \, \mathrm{d}\sqrt{m_B}$$

this integrand being convenient because it remains finite as

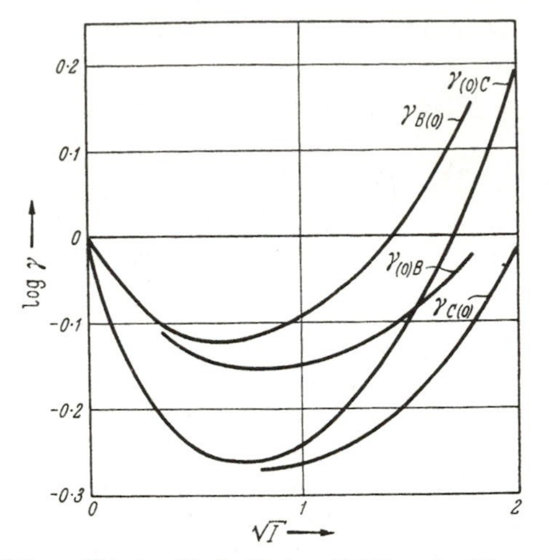

Figure A.1. Activity coefficients of hydrochloric acid (B) and calcium perchlorate (C) in their own solutions, and trace-activity coefficients in the presence of one another.

$m_B \rightarrow 0$. This limiting value follows from the Debye–Hückel equation:

$$\log \gamma_C = -2A(m_B + 3m_C)^{1/2}$$

$$\frac{\mathrm{d} \log \gamma_{(0)C}}{\mathrm{d}m_B} = -2A\sqrt{m_B}$$

$$\lim_{m_B \rightarrow 0} (Y\sqrt{m_B}) = 3kA$$

With this limiting value, $\log \gamma_{(0)C}$ can be calculated by tabular integration to finite values of m_B.

Figure A.1 compares these trace-activity coefficients of calcium perchlorate with those of the salt in its own solution and similar values for hydrochloric acid obtained in an analogous way. This can be compared with *Figure 15.1* although the situation is now more complicated because the (limiting) Debye–Hückel contribution is twice as great for calcium perchlorate as it is for hydrochloric acid. Nevertheless, the same general pattern can be seen; the high activity coefficient of hydrochloric acid is lowered on addition of calcium perchlorate and the comparatively low activity coefficient of this salt is increased on addition of hydrochloric acid.

ADDENDUM

Measurements of pK_a over a Temperature Range

$$pK_a = A_1/T - A_2 + A_3T$$

Aqueous Solution	pK_a at 25°	A_1	A_2	A_3	*Reference*
Acetic acid (CD_3COOH in H_2O)	4·772	1079·37	2·5200	0·012313	50
Acetic acid (CH_3COOD in D_2O)	5·313	1316·56	3·3181	0·014135	51
Acetic acid (CD_3COOD in D_2O)	5·325	1278·92	3·0490	0·013702	50
2-Aminoethyl phosphate	10·638	2998·17	1·3841	0·006596	52
2-Aminoethyl sulphate	9·182	2530·12	−0·2782	−0·001402	52
2-Amino-2-methyl-1,3-propanediol	8·801	2952·00	2·2652	0·003909	53
4-Aminopyridine	9·114	2575·8	−0·0828	0·001309	54
Arginine K_1	1·822	1087·6	4·7526	0·009819	55
Arginine K_2	8·994	2643·6	0·8783	0·003363	55
t-Butylammonium ion	10·685	3021·63	−0·9376	−0·001303	56
o-Chlorophenol	8·527	2443·32	4·800	0·017199	57
Creatine	2·6308	1175·02	4·9883	0·012334	58
Creatinine	4·829	198·23	−7·0705	−0·009753	59
o-Cresol	10·330	2180·87	0·0933	0·010449	57
m-Cresol	10·098	2127·24	0·1280	0·010367	57
p-Cresol	10·276	2127·40	−0·0404	0·010409	57
Diethanolammonium ion	8·883	1830·15	−4·0302	−0·004326	60
N,N-Di(2-hydroxyethyl)-glycine	8·333	1329·60	−4·0159	−0·000476	61
2,4-Dinitrophenol	4·084	982·9	0·8585	0·005528	62
2,5-Dinitrophenol	5·231	1271·6	1·1061	0·006937	62
2,6-Dinitrophenol	3·725	1031·4	2·2933	0·008582	62
Di*iso*propylcyanoacetic acid	2·5557	222·64	1·4224	0·010839	63
Hydrocyanic acid	9·21	3807·5	8·889	0·01792	64
Imidazole	6·994	1906·88	−0·6225	−0·000085	65
Malic acid K_1	3·459	1358·85	5·1382	0·013550	66
Malic acid K_2	5·097	1658·53	6·2364	0·019353	66
Morpholinium ion	8·492	1663·29	−4·1724	−0·004224	67
4-Nitro-3-methylphenol	7·409	2075·02	3·1531	0·012082	68
o-Nitrophenol	7·230	2223·12	4·3092	0·013709	68
m-Nitrophenol	8·355	1723·10	−0·5593	0·006741	68
p-Nitrophenol	7·156	2150·69	3·8133	0·01260	69
Phenol	10·020	2119·82	0·0468	0·009918	57
Phenylacetic acid	4·305	628·08	0·547	0·009208	70
Phosphoric acid K_2 (in D_2O)	7·780	2202·11	5·9823	0·021388	71
Phosphoric acid K_3	12·375	—	—	—	72
Piperazinium ion K_1	5·333	952·11	−4·3919	−0·007555	73
Piperazinium ion K_2	9·731	1656·59	−6·1316	−0·006556	73
Pyrrolidinium ion	11·305	2318·85	−5·2942	−0·005923	74
Quinolinium ion	4·882	3038·4	11·2126	0·019798	75
Tris(hydroxymethyl)aminomethane	8·069	3037·61	3·9321	0·006085	76

REFERENCES

[1] Eigen, M. and De Maeyer, L., 'Technique of Organic Chemistry', (edited by Friess, S. L., Lewis, E. S. and Weissberger, A.), Vol. 8, Part 2, Chap. 18, p. 895, John Wiley and Sons, Inc., New York (1963); 'Structure of electrolytic solutions', (edited by Hamer, W. J.), Chap.5, p. 64, John Wiley and Sons, Inc., New York (1959)

[2] Kavanau, J. L., 'Water and solute–water interactions', Holden–Day, San Francisco (1964)

[3] Kirkwood, J. G., Baldwin, R. L., Dunlop, P. J., Gosting, L. J. and Kegeles, G., *J. chem. Phys.*, 33 (1960) 1505

[4] Noyes, R. M., *J. Amer. chem. Soc.*, 84 (1962) 513; 86 (1964) 971

[5] Stokes, R. H., *ibid.*, 86 (1964) 979

[6] Gourary, B. S. and Adrian, F. J., *Solid St. Phys.*, 10 (1960) 127

[7] Witte, H. and Wölfel, E., *Z. phys. Chem. Frankf. Ausg.*, 3 (1955) 296

[8] Friedman, H. L., 'Ionic solution theory', Interscience, New York (1962)

[9] Stokes, R. H., *J. phys. Chem.*, 65 (1961) 1277

[10] — *ibid.*, 65 (1961) 1242

[11] Marsh, K. N. and Stokes, R. H., *Aust. J. Chem.*, 17 (1964) 740

[12] Prue, J. E., *J. phys. Chem.*, 67 (1963) 1152

[13] Cook, B. M. and Stokes, R. H., *ibid.*, 67 (1963) 511

[14] Steel, B. J., *ibid.*, 69 (1965) 3208

[15] Smith, J. E. and Dismukes, E. B., *J. phys. Chem.*, 67 (1963) 1160; 68 (1964) 1603

[15a] Fuoss, R. M. and Accascina, F., 'Electrolytic Conductance', Interscience, New York (1959)

[15b] Fernández-Prini, R. and Prue, J. E., *Z. phys. Chem.*, 228 (1965) 373

[16] Hetzer, H. B., Robinson, R. A. and Bates, R. G., *J. phys. Chem.*, 66 (1962) 1423

[17] Gupta, S. R., Hills, G. J. and Ives, D. J. G., *Trans. Faraday Soc.*, 59 (1963) 1886

[18] Hetzer, H. B., Robinson, R. A. and Bates, R. G., *J. phys. Chem.*, 68 (1964) 1929

[19] Paabo, M., Robinson, R. A. and Bates, R. G., *J. chem. Engng Data*, 9 (1964) 374

[20] Gary, R., Bates, R. G. and Robinson, R. A., *J. phys. Chem.*, 68 (1964) 1186

[21] Greeley, R. S., Smith, W. T., Stoughton, R. W. and Lietzke, M. H., *ibid.*, 64 (1960) 652; Lietzke, M. H. and Stoughton, R. W., *ibid.*, 68 (1964) 3043; Towns, M. B., Greeley, R. S. and Lietzke, M. H., *ibid.*, 64 (1960) 1861

[22] Harned, H. S., *ibid.*, 66 (1962) 589

[23] Esval, O. E. and Tyree, S. Y., *ibid.*, 66 (1962) 940

[24] Covington, A. K., Hakeem, M. A. and Wynne-Jones, W. F. K., *J. chem. Soc.*, (1963) 4394

[25] Gregor, H. P., Rothenberg, M. and Fine, N., *J. phys. Chem.*, 67 (1963) 1110

[26] Tien, H. T., *ibid.*, 67 (1963) 532

[27] Ebert, L. and Lange, J., *Z. phys. Chem.*, 139A (1928) 584; Lindenbaum, S. and Boyd, G. E., *J. phys. Chem.*, 68 (1964) 911; Bower, V. E. and Robinson, R. A., *Trans. Faraday Soc.*, 59 (1963) 1717

[28] CREETH, J. M., *J. phys. Chem.*, 64 (1960) 920
[29] SOLDANO, B. A. and MEEK, M., *J. chem. Soc.*, (1963) 4424
[30] GARDNER, E. R., JONES, P. J. and DE NORDWALL, H. J., *Trans. Faraday Soc.*, 59 (1963) 1994; GARDNER, E. R., *ibid.*, 65 (1969) 91; see also FABUSS, B. M. and KOROSI, A., *Desalination*, 1 (1966) 139; LIU, CHIA-TSUN and LINDSAY, W. T., *J. phys. Chem.*, 74 (1970) 341
[31] HEPLER, L. G., *Aust. J. Chem.*, 17 (1964) 587; see also WOODHEAD, M., PAABO, M., ROBINSON, R. A. and BATES, R. G., *J. Res. nat. Bur. Stand.*, 69A (1965) 263
[32] BELL, R. P. and ROBINSON, R. R., *Trans. Faraday Soc.*, 57 (1961) 965
[33] HITCHCOCK, D. I., *J. phys. Chem.*, 62 (1958) 1337
[34] PELLETIER, S. and QUINTIN, M., *C.R. Acad. Sci., Paris*, 244 (1957) 894
[35] HALE, J. D., IZATT, R. M. and CHRISTENSEN, J. J., *J. phys. Chem.*, 67 (1963) 2605; VANDERZEE, C. E. and SWANSON, J. A., *ibid.*, 67 (1963) 2608
[36] FERNANDEZ, L. P. and HEPLER, L. G., *J. Amer. chem. Soc.*, 81 (1959) 1783
[37] FERNANDEZ, L. P. and HEPLER, L. G., *J. phys. Chem.*, 63 (1959) 110; O'HARA, W. F. and HEPLER, L. G., *ibid.*, 65 (1961) 2107; O'HARA, W. F., HU, T. and HEPLER, L. G., *ibid.*, 67 (1963) 1933; KO, H. C., O'HARA, W. F., HU, T. and HEPLER, L. G., *J. Amer. chem. Soc.*, 86 (1964) 1003; MILLERO, F. J., AHLUWALIA, J. C. and HEPLER, L. G., *J. chem. Engng Data*, 9 (1964) 192, 319; 10 (1965) 199; *J. phys. Chem.*, 68 (1964) 3435
[38] ERNST, Z. L., IRVING, R. J. and MENASHI, J., *Trans. Faraday Soc.*, 60 (1964) 56
[39] GUGGENHEIM, E. A., *Discuss. Faraday Soc.*, 24 (1957) 53
[40] SKINNER, J. F. and FUOSS, R. M., *J. Amer. chem. Soc.*, 86 (1964) 3423
[41] STERN, J. H., ANDERSON, C. W. and PASSCHIER, A. A., *J. phys. Chem.*, 69 (1965) 207; WU, Y. C., SMITH, M. B. and YOUNG, T. F., *ibid.*, 69 (1965) 1868, 1873
[42] WIRTH, H. E., LINDSTROM, R. E. and JOHNSON, J. N., *ibid.*, 67 (1963) 2339
[43] RUSH, R. M. and JOHNSON, J. S., *ibid.*, 68 (1964) 2321
[44] ROBINSON, R. A. and STOKES, R. H., *ibid.*, 65 (1961) 1954
[45] KELLY, F. J., ROBINSON, R. A. and STOKES, R. H., *ibid.*, 65 (1961) 1958
[46] ROBINSON, R. A. and STOKES, R. H., *ibid.*, 66 (1962) 506
[47] BOWER, V. E. and ROBINSON, R. A., *ibid.*, 67 (1963) 1524
[48] — — *J. Res. nat. Bur. Stand.*, 69A (1965) 131
[49] STOKES, J. M. and STOKES, R. H., *J. phys. Chem.*, 67 (1963) 2442
[50] PAABO, M., BATES, R. G. and ROBINSON, R. A., *ibid.*, 70 (1966) 540, 2073
[51] GARY, R., BATES, R. G. and ROBINSON, R. A., *ibid.*, 69 (1965) 2750
[52] DATTA, S. P. and GRZYBOWSKI, A. K., *J. chem. Soc.* (1962) 3068
[53] HETZER, H. B. and BATES, R. G., *J. phys. Chem.*, 66 (1962) 308
[54] BATES, R. G. and HETZER, H. B., *J. Res. nat. Bur. Stand.*, 64A (1960) 427
[55] DATTA, S. P. and GRZYBOWSKI, A. K., *Biochem. J.*, 78 (1961) 289
[56] HETZER, H. B., ROBINSON, R. A. and BATES, R. G., *J. phys. Chem.*, 66 (1962) 2696
[57] CHEN, D. T. Y. and LAIDLER, K. J., *Trans. Faraday Soc.*, 58 (1962) 480; they also give data for the xylenols
[58] DATTA, S. P. and GRZYBOWSKI, A. K., *J. chem. Soc.* (1963) 6004
[59] GRZYBOWSKI, A. K. and DATTA, S. P., *ibid.* (1964) 187
[60] BOWER, V. E., ROBINSON, R. A. and BATES, R. G., *J. Res. nat. Bur. Stand.*, 66A (1962) 71

[61] DATTA, S. P., GRZYBOWSKI, A. K. and BATES, R. G., *J. phys. Chem.*, 68 (1964) 275
[62] RICCARDI, R. and BRESESTI, M., *Ann. Chim.* (*Rome*), 50 (1960) 1305; see also SMOLYAKOV, B. S. and PRIMANCHUK, M. P., *Russ. J. phys. Chem.*, 40 (1966) 989
[63] IVES, D. J. G. and MARSDEN, P. D., *J. chem. Soc.* (1965) 649
[64] IZATT, R. M., CHRISTENSEN, J. J., PACK, R. T. and BENCH, R., *Inorg. Chem.*, 1 (1962) 828; see also ANG, K. P., *J. chem. Soc.* (1959) 3822
[65] DATTA, S. P. and GRZYBOWSKI, A. K., *J. chem. Soc.*, B, (1966) 136
[66] EDEN, M. and BATES, R. G., *J. Res. nat. Bur. Stand.*, 62 (1959) 161
[67] HETZER, H. B., BATES, R. G. and ROBINSON, R. A., *J. phys. Chem.*, 70 (1966) 2869
[68] ROBINSON, R. A. and PEIPERL, A., *ibid.*, 67 (1963) 1723, 2860
[69] ALLEN, G. F., ROBINSON, R. A. and BOWER, V. E., *ibid.*, 66 (1962) 171
[70] SMOLYAKOV, B. S. and PRIMANCHUK, M. P., *Russ J. phys. Chem.*, 40 (1966) 463
[71] GARY, R , BATES, R. G. and ROBINSON, R. A., *J. phys. Chem.*, 68 (1964) 3806
[72] VANDERZEE, C. E. and QUIST, A. S., *ibid.*, 65 (1961) 118
[73] HETZER, H. B., ROBINSON, R. A. and BATES, R. G., *J. phys. Chem.*, 72 (1968) 2081
[74] HETZER, H. B., BATES, R. G. and ROBINSON, R. A., *J. phys. Chem.*, 67 (1963) 1124
[75] RICCARDI, R. and BRESESTI, M., *Ann. Chim.* (*Rome*), 48 (1958) 826; 49 (1959) 1891; data are also given for the methylquinolines
[76] DATTA, S. P., GRZYBOWSKI, A. K. and WESTON, B. A., *J. chem. Soc.*, (1963) 792; BATES, R. G. and HETZER, H. B., *J. phys. Chem.*, 65 (1961) 667

INDEX

Note: Since references to authors quoted are given in full at the end of each chapter, authors' names are indexed only when regularly associated with some law or effect, e.g. "Walden's rule". Substances appearing only in the tables in the appendices, and not separately mentioned in the text, are not indexed, but substances discussed in the text appear in the index. Page numbers given in black type indicate that tables relevant to the topic appear there.

A CATALOG OF SELECTED
DOVER BOOKS
IN SCIENCE AND MATHEMATICS

A CATALOG OF SELECTED

DOVER BOOKS

IN SCIENCE AND MATHEMATICS

Astronomy

BURNHAM'S CELESTIAL HANDBOOK, Robert Burnham, Jr. Thorough guide to the stars beyond our solar system. Exhaustive treatment. Alphabetical by constellation: Andromeda to Cetus in Vol. 1; Chamaeleon to Orion in Vol. 2; and Pavo to Vulpecula in Vol. 3. Hundreds of illustrations. Index in Vol. 3. 2,000pp. 6⅛ x 9¼.
23567-X, 23568-8, 23673-0 Three-vol. set

THE EXTRATERRESTRIAL LIFE DEBATE, 1750–1900, Michael J. Crowe. First detailed, scholarly study in English of the many ideas that developed from 1750 to 1900 regarding the existence of intelligent extraterrestrial life. Examines ideas of Kant, Herschel, Voltaire, Percival Lowell, many other scientists and thinkers. 16 illustrations. 704pp. 5⅜ x 8½. 40675-X

A HISTORY OF ASTRONOMY, A. Pannekoek. Well-balanced, carefully reasoned study covers such topics as Ptolemaic theory, work of Copernicus, Kepler, Newton, Eddington's work on stars, much more. Illustrated. References. 521pp. 5⅜ x 8½.
65994-1

AMATEUR ASTRONOMER'S HANDBOOK, J. B. Sidgwick. Timeless, comprehensive coverage of telescopes, mirrors, lenses, mountings, telescope drives, micrometers, spectroscopes, more. 189 illustrations. 576pp. 5⅝ x 8¼. (Available in U.S. only.)
24034-7

STARS AND RELATIVITY, Ya. B. Zel'dovich and I. D. Novikov. Vol. 1 of *Relativistic Astrophysics* by famed Russian scientists. General relativity, properties of matter under astrophysical conditions, stars, and stellar systems. Deep physical insights, clear presentation. 1971 edition. References. 544pp. 5⅝ x 8¼. 69424-0

Chemistry

CHEMICAL MAGIC, Leonard A. Ford. Second Edition, Revised by E. Winston Grundmeier. Over 100 unusual stunts demonstrating cold fire, dust explosions, much more. Text explains scientific principles and stresses safety precautions. 128pp. 5⅜ x 8½. 67628-5

THE DEVELOPMENT OF MODERN CHEMISTRY, Aaron J. Ihde. Authoritative history of chemistry from ancient Greek theory to 20th-century innovation. Covers major chemists and their discoveries. 209 illustrations. 14 tables. Bibliographies. Indices. Appendices. 851pp. 5⅜ x 8½. 64235-6

CATALYSIS IN CHEMISTRY AND ENZYMOLOGY, William P. Jencks. Exceptionally clear coverage of mechanisms for catalysis, forces in aqueous solution, carbonyl- and acyl-group reactions, practical kinetics, more. 864pp. 5⅜ x 8½.
65460-5

THE HISTORICAL BACKGROUND OF CHEMISTRY, Henry M. Leicester. Evolution of ideas, not individual biography. Concentrates on formulation of a coherent set of chemical laws. 260pp. 5⅜ x 8½. 61053-5

A SHORT HISTORY OF CHEMISTRY, J. R. Partington. Classic exposition explores origins of chemistry, alchemy, early medical chemistry, nature of atmosphere, theory of valency, laws and structure of atomic theory, much more. 428pp. 5⅜ x 8½. (Available in U.S. only.) 65977-1

GENERAL CHEMISTRY, Linus Pauling. Revised 3rd edition of classic first-year text by Nobel laureate. Atomic and molecular structure, quantum mechanics, statistical mechanics, thermodynamics correlated with descriptive chemistry. Problems. 992pp. 5⅜ x 8½. 65622-5

Engineering

DE RE METALLICA, Georgius Agricola. The famous Hoover translation of greatest treatise on technological chemistry, engineering, geology, mining of early modern times (1556). All 289 original woodcuts. 638pp. 6¾ x 11. 60006-8

FUNDAMENTALS OF ASTRODYNAMICS, Roger Bate et al. Modern approach developed by U.S. Air Force Academy. Designed as a first course. Problems, exercises. Numerous illustrations. 455pp. 5⅜ x 8½. 60061-0

DYNAMICS OF FLUIDS IN POROUS MEDIA, Jacob Bear. For advanced students of ground water hydrology, soil mechanics and physics, drainage and irrigation engineering and more. 335 illustrations. Exercises, with answers. 784pp. 6⅛ x 9¼. 65675-6

ANALYTICAL MECHANICS OF GEARS, Earle Buckingham. Indispensable reference for modern gear manufacture covers conjugate gear-tooth action, gear-tooth profiles of various gears, many other topics. 263 figures. 102 tables. 546pp. 5⅜ x 8½. 65712-4

MECHANICS, J. P. Den Hartog. A classic introductory text or refresher. Hundreds of applications and design problems illuminate fundamentals of trusses, loaded beams and cables, etc. 334 answered problems. 462pp. 5⅜ x 8½. 60754-2

MECHANICAL VIBRATIONS, J. P. Den Hartog. Classic textbook offers lucid explanations and illustrative models, applying theories of vibrations to a variety of practical industrial engineering problems. Numerous figures. 233 problems, solutions. Appendix. Index. Preface. 436pp. 5⅜ x 8½. 64785-4

STRENGTH OF MATERIALS, J. P. Den Hartog. Full, clear treatment of basic material (tension, torsion, bending, etc.) plus advanced material on engineering methods, applications. 350 answered problems. 323pp. 5⅜ x 8½. 60755-0

A HISTORY OF MECHANICS, René Dugas. Monumental study of mechanical principles from antiquity to quantum mechanics. Contributions of ancient Greeks, Galileo, Leonardo, Kepler, Lagrange, many others. 671pp. 5⅜ x 8½. 65632-2

METAL FATIGUE, N. E. Frost, K. J. Marsh, and L. P. Pook. Definitive, clearly written, and well-illustrated volume addresses all aspects of the subject, from the historical development of understanding metal fatigue to vital concepts of the cyclic stress that causes a crack to grow. Includes 7 appendixes. 544pp. 5⅜ x 8½. 40927-9

STATISTICAL MECHANICS: Principles and Applications, Terrell L. Hill. Standard text covers fundamentals of statistical mechanics, applications to fluctuation theory, imperfect gases, distribution functions, more. 448pp. 5⅜ x 8½. 65390-0

THE VARIATIONAL PRINCIPLES OF MECHANICS, Cornelius Lanczos. Graduate level coverage of calculus of variations, equations of motion, relativistic mechanics, more. First inexpensive paperbound edition of classic treatise. Index. Bibliography. 418pp. 5⅜ x 8½. 65067-7

THE VARIOUS AND INGENIOUS MACHINES OF AGOSTINO RAMELLI: A Classic Sixteenth-Century Illustrated Treatise on Technology, Agostino Ramelli. One of the most widely known and copied works on machinery in the 16th century. 194 detailed plates of water pumps, grain mills, cranes, more. 608pp. 9 x 12. 28180-9

ORDINARY DIFFERENTIAL EQUATIONS AND STABILITY THEORY: An Introduction, David A. Sánchez. Brief, modern treatment. Linear equation, stability theory for autonomous and nonautonomous systems, etc. 164pp. 5⅜ x 8¼. 63828-6

ROTARY WING AERODYNAMICS, W. Z. Stepniewski. Clear, concise text covers aerodynamic phenomena of the rotor and offers guidelines for helicopter performance evaluation. Originally prepared for NASA. 537 figures. 640pp. 6⅛ x 9¼. 64647-5

INTRODUCTION TO SPACE DYNAMICS, William Tyrrell Thomson. Comprehensive, classic introduction to space-flight engineering for advanced undergraduate and graduate students. Includes vector algebra, kinematics, transformation of coordinates. Bibliography. Index. 352pp. 5⅜ x 8½. 65113-4

HISTORY OF STRENGTH OF MATERIALS, Stephen P. Timoshenko. Excellent historical survey of the strength of materials with many references to the theories of elasticity and structure. 245 figures. 452pp. 5⅜ x 8½. 61187-6

ANALYTICAL FRACTURE MECHANICS, David J. Unger. Self-contained text supplements standard fracture mechanics texts by focusing on analytical methods for determining crack-tip stress and strain fields. 336pp. 6⅛ x 9¼. 41737-9

Mathematics

HANDBOOK OF MATHEMATICAL FUNCTIONS WITH FORMULAS, GRAPHS, AND MATHEMATICAL TABLES, edited by Milton Abramowitz and Irene A. Stegun. Vast compendium: 29 sets of tables, some to as high as 20 places. 1,046pp. 8 x 10½. 61272-4

FUNCTIONAL ANALYSIS (Second Corrected Edition), George Bachman and Lawrence Narici. Excellent treatment of subject geared toward students with background in linear algebra, advanced calculus, physics and engineering. Text covers introduction to inner-product spaces, normed, metric spaces, and topological spaces; complete orthonormal sets, the Hahn-Banach Theorem and its consequences, and many other related subjects. 1966 ed. 544pp. 6⅛ x 9¼. 40251-7

ASYMPTOTIC EXPANSIONS OF INTEGRALS, Norman Bleistein & Richard A. Handelsman. Best introduction to important field with applications in a variety of scientific disciplines. New preface. Problems. Diagrams. Tables. Bibliography. Index. 448pp. 5⅜ x 8½. 65082-0

FAMOUS PROBLEMS OF GEOMETRY AND HOW TO SOLVE THEM, Benjamin Bold. Squaring the circle, trisecting the angle, duplicating the cube: learn their history, why they are impossible to solve, then solve them yourself. 128pp. 5⅜ x 8½. 24297-8

VECTOR AND TENSOR ANALYSIS WITH APPLICATIONS, A. I. Borisenko and I. E. Tarapov. Concise introduction. Worked-out problems, solutions, exercises. 257pp. 5⅝ x 8¼. 63833-2

THE ABSOLUTE DIFFERENTIAL CALCULUS (CALCULUS OF TENSORS), Tullio Levi-Civita. Great 20th-century mathematician's classic work on material necessary for mathematical grasp of theory of relativity. 452pp. 5⅝ x 8¼. 63401-9

AN INTRODUCTION TO ORDINARY DIFFERENTIAL EQUATIONS, Earl A. Coddington. A thorough and systematic first course in elementary differential equations for undergraduates in mathematics and science, with many exercises and problems (with answers). Index. 304pp. 5⅜ x 8½. 65942-9

FOURIER SERIES AND ORTHOGONAL FUNCTIONS, Harry F. Davis. An incisive text combining theory and practical example to introduce Fourier series, orthogonal functions and applications of the Fourier method to boundary-value problems. 570 exercises. Answers and notes. 416pp. 5⅜ x 8½. 65973-9

COMPUTABILITY AND UNSOLVABILITY, Martin Davis. Classic graduate-level introduction to theory of computability, usually referred to as theory of recurrent functions. New preface and appendix. 288pp. 5⅜ x 8½. 61471-9

ASYMPTOTIC METHODS IN ANALYSIS, N. G. de Bruijn. An inexpensive, comprehensive guide to asymptotic methods–the pioneering work that teaches by explaining worked examples in detail. Index. 224pp. 5⅜ x 8½ 64221-6

ESSAYS ON THE THEORY OF NUMBERS, Richard Dedekind. Two classic essays by great German mathematician: on the theory of irrational numbers; and on transfinite numbers and properties of natural numbers. 115pp. 5⅜ x 8½. 21010-3

THE FUNCTIONS OF MATHEMATICAL PHYSICS, Harry Hochstadt. Comprehensive treatment of orthogonal polynomials, hypergeometric functions, Hill's equation, much more. Bibliography. Index. 322pp. 5⅜ x 8½. 65214-9

THREE PEARLS OF NUMBER THEORY, A. Y. Khinchin. Three compelling puzzles require proof of a basic law governing the world of numbers. Challenges concern van der Waerden's theorem, the Landau-Schnirelmann hypothesis and Mann's theorem, and a solution to Waring's problem. Solutions included. 64pp. 5⅜ x 8½. 40026-3

CALCULUS REFRESHER FOR TECHNICAL PEOPLE, A. Albert Klaf. Covers important aspects of integral and differential calculus via 756 questions. 566 problems, most answered. 431pp. 5⅜ x 8½. 20370-0

THE PHILOSOPHY OF MATHEMATICS: An Introductory Essay, Stephan Körner. Surveys the views of Plato, Aristotle, Leibniz & Kant concerning propositions and theories of applied and pure mathematics. Introduction. Two appendices. Index. 198pp. 5⅜ x 8½. 25048-2

INTRODUCTORY REAL ANALYSIS, A.N. Kolmogorov, S. V. Fomin. Translated by Richard A. Silverman. Self-contained, evenly paced introduction to real and functional analysis. Some 350 problems. 403pp. 5⅜ x 8½. 61226-0

APPLIED ANALYSIS, Cornelius Lanczos. Classic work on analysis and design of finite processes for approximating solution of analytical problems. Algebraic equations, matrices, harmonic analysis, quadrature methods, much more. 559pp. 5⅜ x 8½. 65656-X

AN INTRODUCTION TO ALGEBRAIC STRUCTURES, Joseph Landin. Superb self-contained text covers "abstract algebra": sets and numbers, theory of groups, theory of rings, much more. Numerous well-chosen examples, exercises. 247pp. 5⅜ x 8½. 65940-2

SPECIAL FUNCTIONS, N. N. Lebedev. Translated by Richard Silverman. Famous Russian work treating more important special functions, with applications to specific problems of physics and engineering. 38 figures. 308pp. 5⅜ x 8½. 60624-4

QUALITATIVE THEORY OF DIFFERENTIAL EQUATIONS, V. V. Nemytskii and V.V. Stepanov. Classic graduate-level text by two prominent Soviet mathematicians covers classical differential equations as well as topological dynamics and ergodic theory. Bibliographies. 523pp. 5⅜ x 8½. 65954-2

NUMBER THEORY AND ITS HISTORY, Oystein Ore. Unusually clear, accessible introduction covers counting, properties of numbers, prime numbers, much more. Bibliography. 380pp. 5⅜ x 8½. 65620-9

THEORY OF MATRICES, Sam Perlis. Outstanding text covering rank, nonsingularity and inverses in connection with the development of canonical matrices under the relation of equivalence, and without the intervention of determinants. Includes exercises. 237pp. 5⅜ x 8½. 66810-X

POPULAR LECTURES ON MATHEMATICAL LOGIC, Hao Wang. Noted logician's lucid treatment of historical developments, set theory, model theory, recursion theory and constructivism, proof theory, more. 3 appendixes. Bibliography. 1981 edition. ix + 283pp. 5⅜ x 8½. 67632-3

CALCULUS OF VARIATIONS, Robert Weinstock. Basic introduction covering isoperimetric problems, theory of elasticity, quantum mechanics, electrostatics, etc. Exercises throughout. 326pp. 5⅜ x 8½. 63069-2

THE CONTINUUM: A Critical Examination of the Foundation of Analysis, Hermann Weyl. Classic of 20th-century foundational research deals with the conceptual problem posed by the continuum. 156pp. 5⅜ x 8½. 67982-9

CHALLENGING MATHEMATICAL PROBLEMS WITH ELEMENTARY SOLUTIONS, A. M. Yaglom and I. M. Yaglom. Over 170 challenging problems on probability theory, combinatorial analysis, points and lines, topology, convex polygons, many other topics. Solutions. Total of 445pp. 5⅜ x 8½. Two-vol. set.
Vol. I: 65536-9 Vol. II: 65537-7

A SURVEY OF NUMERICAL MATHEMATICS, David M. Young and Robert Todd Gregory. Broad self-contained coverage of computer-oriented numerical algorithms for solving various types of mathematical problems in linear algebra, ordinary and partial, differential equations, much more. Exercises. Total of 1,248pp. 5⅜ x 8½. Two volumes. Vol. I: 65691-8 Vol. II: 65692-6

INTRODUCTION TO PARTIAL DIFFERENTIAL EQUATIONS WITH APPLICATIONS, E. C. Zachmanoglou and Dale W. Thoe. Essentials of partial differential equations applied to common problems in engineering and the physical sciences. Problems and answers. 416pp. 5⅜ x 8½. 65251-3

THE THEORY OF GROUPS, Hans J. Zassenhaus. Well-written graduate-level text acquaints reader with group-theoretic methods and demonstrates their usefulness in mathematics. Axioms, the calculus of complexes, homomorphic mapping, p-group theory, more. Many proofs shorter and more transparent than older ones. 276pp. 5⅜ x 8½. 40922-8

DISTRIBUTION THEORY AND TRANSFORM ANALYSIS: An Introduction to Generalized Functions, with Applications, A. H. Zemanian. Provides basics of distribution theory, describes generalized Fourier and Laplace transformations. Numerous problems. 384pp. 5⅜ x 8½. 65479-6

Math–Decision Theory, Statistics, Probability

ELEMENTARY DECISION THEORY, Herman Chernoff and Lincoln E. Moses. Clear introduction to statistics and statistical theory covers data processing, probability and random variables, testing hypotheses, much more. Exercises. 364pp. 5⅜ x 8½. 65218-1

Math–Geometry and Topology

ELEMENTARY CONCEPTS OF TOPOLOGY, Paul Alexandroff. Elegant, intuitive approach to topology from set-theoretic topology to Betti groups; how concepts of topology are useful in math and physics. 25 figures. 57pp. $5\frac{3}{8}$ x $8\frac{1}{2}$. 60747-X

COMBINATORIAL TOPOLOGY, P. S. Alexandrov. Clearly written, well-organized, three-part text begins by dealing with certain classic problems without using the formal techniques of homology theory and advances to the central concept, the Betti groups. Numerous detailed examples. 654pp. $5\frac{3}{8}$ x $8\frac{1}{2}$. 40179-0

EXPERIMENTS IN TOPOLOGY, Stephen Barr. Classic, lively explanation of one of the byways of mathematics. Klein bottles, Moebius strips, projective planes, map coloring, problem of the Koenigsberg bridges, much more, described with clarity and wit. 43 figures. 210pp. $5\frac{3}{8}$ x $8\frac{1}{2}$. 25933-1

CONFORMAL MAPPING ON RIEMANN SURFACES, Harvey Cohn. Lucid, insightful book presents ideal coverage of subject. 334 exercises make book perfect for self-study. 55 figures. 352pp. $5\frac{5}{8}$ x $8\frac{1}{4}$. 64025-6

THE GEOMETRY OF RENÉ DESCARTES, René Descartes. The great work founded analytical geometry. Original French text, Descartes's own diagrams, together with definitive Smith-Latham translation. 244pp. $5\frac{3}{8}$ x $8\frac{1}{2}$. 60068-8

THE THIRTEEN BOOKS OF EUCLID'S ELEMENTS, translated with introduction and commentary by Sir Thomas L. Heath. Definitive edition. Textual and linguistic notes, mathematical analysis. 2,500 years of critical commentary. Unabridged. 1,414pp. $5\frac{3}{8}$ x $8\frac{1}{2}$. Three-vol. set.
Vol. I: 60088-2 Vol. II: 60089-0 Vol. III: 60090-4

GEOMETRY OF COMPLEX NUMBERS, Hans Schwerdtfeger. Illuminating, widely praised book on analytic geometry of circles, the Moebius transformation, and two-dimensional non-Euclidean geometries. 200pp. $5\frac{5}{8}$ x $8\frac{1}{4}$. 63830-8

DIFFERENTIAL GEOMETRY, Heinrich W. Guggenheimer. Local differential geometry as an application of advanced calculus and linear algebra. Curvature, transformation groups, surfaces, more. Exercises. 62 figures. 378pp. $5\frac{3}{8}$ x $8\frac{1}{2}$. 63433-7

CURVATURE AND HOMOLOGY: Enlarged Edition, Samuel I. Goldberg. Revised edition examines topology of differentiable manifolds; curvature, homology of Riemannian manifolds; compact Lie groups; complex manifolds; curvature, homology of Kaehler manifolds. New Preface. Four new appendixes. 416pp. $5\frac{3}{8}$ x $8\frac{1}{2}$.
40207-X

TOPOLOGY, John G. Hocking and Gail S. Young. Superb one-year course in classical topology. Topological spaces and functions, point-set topology, much more. Examples and problems. Bibliography. Index. 384pp. $5\frac{5}{8}$ x $8\frac{1}{4}$. 65676-4

LECTURES ON CLASSICAL DIFFERENTIAL GEOMETRY, Second Edition, Dirk J. Struik. Excellent brief introduction covers curves, theory of surfaces, fundamental equations, geometry on a surface, conformal mapping, other topics. Problems. 240pp. 5⅜ x 8½. 65609-8

Math–History of

A SHORT ACCOUNT OF THE HISTORY OF MATHEMATICS, W. W. Rouse Ball. One of clearest, most authoritative surveys from the Egyptians and Phoenicians through 19th-century figures such as Grassman, Galois, Riemann. Fourth edition. 522pp. 5⅜ x 8½. 20630-0

THE HISTORY OF THE CALCULUS AND ITS CONCEPTUAL DEVELOPMENT, Carl B. Boyer. Origins in antiquity, medieval contributions, work of Newton, Leibniz, rigorous formulation. Treatment is verbal. 346pp. 5⅜ x 8½. 60509-4

THE HISTORICAL ROOTS OF ELEMENTARY MATHEMATICS, Lucas N. H. Bunt, Phillip S. Jones, and Jack D. Bedient. Fundamental underpinnings of modern arithmetic, algebra, geometry and number systems derived from ancient civilizations. 320pp. 5⅜ x 8½. 25563-8

A HISTORY OF MATHEMATICAL NOTATIONS, Florian Cajori. This classic study notes the first appearance of a mathematical symbol and its origin, the competition it encountered, its spread among writers in different countries, its rise to popularity, its eventual decline or ultimate survival. Original 1929 two-volume edition presented here in one volume. xxviii+820pp. 5⅜ x 8½. 67766-4

GAMES, GODS & GAMBLING: A History of Probability and Statistical Ideas, F. N. David. Episodes from the lives of Galileo, Fermat, Pascal, and others illustrate this fascinating account of the roots of mathematics. Features thought-provoking references to classics, archaeology, biography, poetry. 1962 edition. 304pp. 5⅜ x 8½. (Available in U.S. only.) 40023-9

OF MEN AND NUMBERS: The Story of the Great Mathematicians, Jane Muir. Fascinating accounts of the lives and accomplishments of history's greatest mathematical minds–Pythagoras, Descartes, Euler, Pascal, Cantor, many more. Anecdotal, illuminating. 30 diagrams. Bibliography. 256pp. 5⅜ x 8½. 28973-7

HISTORY OF MATHEMATICS, David E. Smith. Nontechnical survey from ancient Greece and Orient to late 19th century; evolution of arithmetic, geometry, trigonometry, calculating devices, algebra, the calculus. 362 illustrations. 1,355pp. 5⅜ x 8½. Two-vol. set. Vol. I: 20429-4 Vol. II: 20430-8

A CONCISE HISTORY OF MATHEMATICS, Dirk J. Struik. The best brief history of mathematics. Stresses origins and covers every major figure from ancient Near East to 19th century. 41 illustrations. 195pp. 5⅜ x 8½. 60255-9

Physics

OPTICAL RESONANCE AND TWO-LEVEL ATOMS, L. Allen and J. H. Eberly. Clear, comprehensive introduction to basic principles behind all quantum optical resonance phenomena. 53 illustrations. Preface. Index. 256pp. $5\frac{3}{8}$ x $8\frac{1}{2}$. 65533-4

ULTRASONIC ABSORPTION: An Introduction to the Theory of Sound Absorption and Dispersion in Gases, Liquids and Solids, A. B. Bhatia. Standard reference in the field provides a clear, systematically organized introductory review of fundamental concepts for advanced graduate students, research workers. Numerous diagrams. Bibliography. 440pp. $5\frac{3}{8}$ x $8\frac{1}{2}$. 64917-2

QUANTUM THEORY, David Bohm. This advanced undergraduate-level text presents the quantum theory in terms of qualitative and imaginative concepts, followed by specific applications worked out in mathematical detail. Preface. Index. 655pp. $5\frac{3}{8}$ x $8\frac{1}{2}$. 65969-0

ATOMIC PHYSICS (8th edition), Max Born. Nobel laureate's lucid treatment of kinetic theory of gases, elementary particles, nuclear atom, wave-corpuscles, atomic structure and spectral lines, much more. Over 40 appendices, bibliography. 495pp. $5\frac{3}{8}$ x $8\frac{1}{2}$. 65984-4

AN INTRODUCTION TO HAMILTONIAN OPTICS, H. A. Buchdahl. Detailed account of the Hamiltonian treatment of aberration theory in geometrical optics. Many classes of optical systems defined in terms of the symmetries they possess. Problems with detailed solutions. 1970 edition. xv + 360pp. $5\frac{3}{8}$ x $8\frac{1}{2}$. 67597-1

THIRTY YEARS THAT SHOOK PHYSICS: The Story of Quantum Theory, George Gamow. Lucid, accessible introduction to influential theory of energy and matter. Careful explanations of Dirac's anti-particles, Bohr's model of the atom, much more. 12 plates. Numerous drawings. 240pp. $5\frac{3}{8}$ x $8\frac{1}{2}$. 24895-X

ELECTRONIC STRUCTURE AND THE PROPERTIES OF SOLIDS: The Physics of the Chemical Bond, Walter A. Harrison. Innovative text offers basic understanding of the electronic structure of covalent and ionic solids, simple metals, transition metals and their compounds. Problems. 1980 edition. 582pp. $6\frac{1}{8}$ x $9\frac{1}{4}$. 66021-4

HYDRODYNAMIC AND HYDROMAGNETIC STABILITY, S. Chandrasekhar. Lucid examination of the Rayleigh-Benard problem; clear coverage of the theory of instabilities causing convection. 704pp. $5\frac{3}{8}$ x $8\frac{1}{4}$. 64071-X

INVESTIGATIONS ON THE THEORY OF THE BROWNIAN MOVEMENT, Albert Einstein. Five papers (1905–8) investigating dynamics of Brownian motion and evolving elementary theory. Notes by R. Fürth. 122pp. $5\frac{3}{8}$ x $8\frac{1}{2}$. 60304-0

THE PHYSICS OF WAVES, William C. Elmore and Mark A. Heald. Unique overview of classical wave theory. Acoustics, optics, electromagnetic radiation, more. Ideal as classroom text or for self-study. Problems. 477pp. $5\frac{3}{8}$ x $8\frac{1}{2}$. 64926-1

CATALOG OF DOVER BOOKS